glencoe
teen**health**

Mc
Graw
Hill
Education

Bothell, WA • Chicago, IL • Columbus, OH • New York, NY

Meet the Authors

Mary H. Bronson, Ph.D. recently retired after teaching for 30 years in Texas public schools. Dr. Bronson taught health education in grades K–12 as well as health education methods classes at the graduate and undergraduate levels. As Health Education Specialist for the Dallas School District, Dr. Bronson developed and implemented a district-wide health eduation program. She has been honored as Texas Health Educator of the Year by the Texas Association for Health, Physical Education, Recreation, and Dance and selected Teacher of the Year twice by her colleagues. Dr. Bronson has assisted school districts throughout the country in developing local health education programs. She is also the coauthor of Glencoe Health.

Michael J. Cleary, Ed.D., C.H.E.S. is a professor at Slippery Rock University, where he teaches methods courses and supervises field experiences. Dr. Cleary taught health education at Evanston Township High School in Illinois and later served as the Lead Teacher Specialist at the McMillen Center for Health Education in Fort Wayne, Indiana. Dr. Cleary has published widely on curriculum development and assessment in K–12 and college health education. Dr. Cleary is also coauthor of Glencoe Health.

Betty M. Hubbard, Ed.D., C.H.E.S. has taught science and health education in grades 6–12 as well as undergraduate- and graduate-level courses. She is a professor at the University of Central Arkansas, where in addition to teaching she conducts in-service training for health education teachers in school districts throughout Arkansas. In 1991, Dr. Hubbard received the university's teaching excellence award. Her publications, grants, and presentations focus on research-based, comprehensive health instruction. Dr. Hubbard is a fellow of the American Association for Health Education and serves as the contributing editor of the Teaching Ideas feature of the American Journal of Health Education.

Contributing Author

Dinah Zike, M.Ed. is an international curriculum consultant and inventor who has designed and developed educational products and three-dimensional, interactive graphic organizers for more than 35 years. As president and founder of Dinah-Might Adventures, L.P., Dinah is author of more than 100 award-winning educational publications. Dinah has a B.S. and an M.S. in educational curriculum and instruction from Texas A&M University. Dinah Zike's Foldables® are an exclusive feature of McGraw-Hill.

MHEonline.com

Send all inquiries to:
McGraw-Hill Education
STEM Learning Solutions Center
8787 Orion Place
Columbus, OH 43240

ISBN: 978-0-02-138540-9
MHID: 0-02-138540-8

Printed in the United States of America.

12 13 14 15 QVS 21 20 19 18

STEM McGraw-Hill is committed to providing instructional materials in Science, Technology, Engineering, and Mathematics (STEM) that give all students a solid foundation, one that prepares them for college and careers in the 21st century.

Reviewers

Professional Reviewers

Amy Eyler, Ph.D., CHES
Washington University in St. Louis
St. Louis, Missouri

Shonali Saha, M.D.
Johns Hopkins School of Medicine
Baltimore, Maryland

Roberta Duyff
Duyff & Associates
St. Louis, MO

Teacher Reviewers

Lou Ann Donlan
Altoona Area School District
Altoona, PA

Steve Federman
Loveland Intermediate School
Loveland, Ohio

Rick R. Gough
Ashland Middle School
Ashland, Ohio

Jacob Graham
Oblock Junior High
Plum, Pennsylvania

William T. Gunther
Clarkston Community Schools
Clarkston, MI

Ellie Hancock
Somerset Area School District
Somerset, PA

Diane Hursky
Independence Middle School
Bethel Park, PA

Veronique Javier
Thomas Cardoza Middle School
Jackson, Mississippi

Patricia A. Landon
Patrick F. Healy Middle School
East Orange, NJ

Elizabeth Potash
Council Rock High School South
Holland, PA

The Path to Good Health

Your health book has many features that will aid you in your learning. Some of these features are listed below. You can use the map at the right to help you find these and other special features in the book.

* The **Big Idea** can be found at the start of each lesson.

* Your **Foldables**® help you organize your notes.

* The **Quick Write** at the start of each lesson will help you think about the topic and give you an opportunity to write about it in your journal.

* The **Bilingual Glossary** contains vocabulary terms and definitions in Spanish and English.

* **Health Skills Activities** help you learn more about each of the 10 health skills.

* **Infographs** provide a colorful, visual way to learn about current health news and trends.

* The **Fitness Zone** provides an online fitness resource that includes podcasts, videos, activity cards, and more!

* **Hands-On Health Activities** give you the opportunity to complete hands-on projects.

* **Videos** encourage you to explore real life health topics.

* **Audio** directs you to online audio chapter summaries.

* **Web Quest** activities challenge you to relate lesson concepts to current health news and research.

* **Review** your understanding of health concepts with lesson reviews and quizzes.

What's the word on the street? The **glossary** lists vocabulary terms in English and Spanish.

Quick! Write about your good health habits using a **Quick Write** activity.

Think big! Start your journey with a **Big Idea** and increase your pace with **Foldables**®.

Sharpen your skills with **Health Skills Activities**.

Got a nose for news? Check out each chapter's **infographs** for health news and trends.

Get into the zone –the **Fitness Zone!** Listen to podcasts, watch videos, and more.

Show what you know by completing a **Hands-On Health Activity**.

Stop! Look and Listen! Watch a Health eSpotlight **video** and explore real life health topics. Listen to the **audio** summaries to review the chapter.

Go on a quest. Take a **Web Quest** to learn more about health news and research.

Finish strong! **Review** your understanding of health concepts with lesson reviews and quizzes.

Contents

Unit 1

Building Healthy Relationships

Contents

Unit 2

Building Character and Preventing Bullying

chapter 4 | Bullying and Cyberbullying

Contents

Unit 3
Mental and Emotional Health

6 Mental and Emotional Disorders

Contents

Unit 4

Conflict Resolution and Violence Prevention

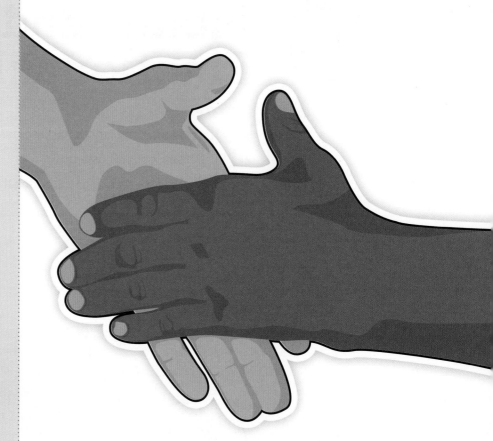

chapter 7 Conflict Resolution

chapter 8

Violence Prevention

Contents

Unit 5
Nutrition and Physical Activity

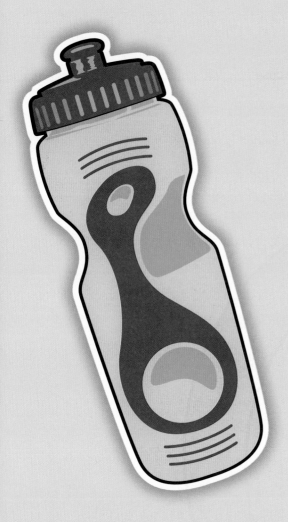

Contents

Unit 6

Health During the Life Cycle

chapter 12 Personal Health Care

Contents

Unit 7
Your Body Systems

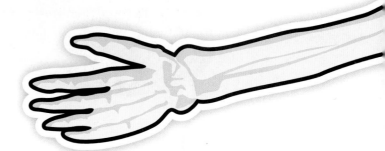

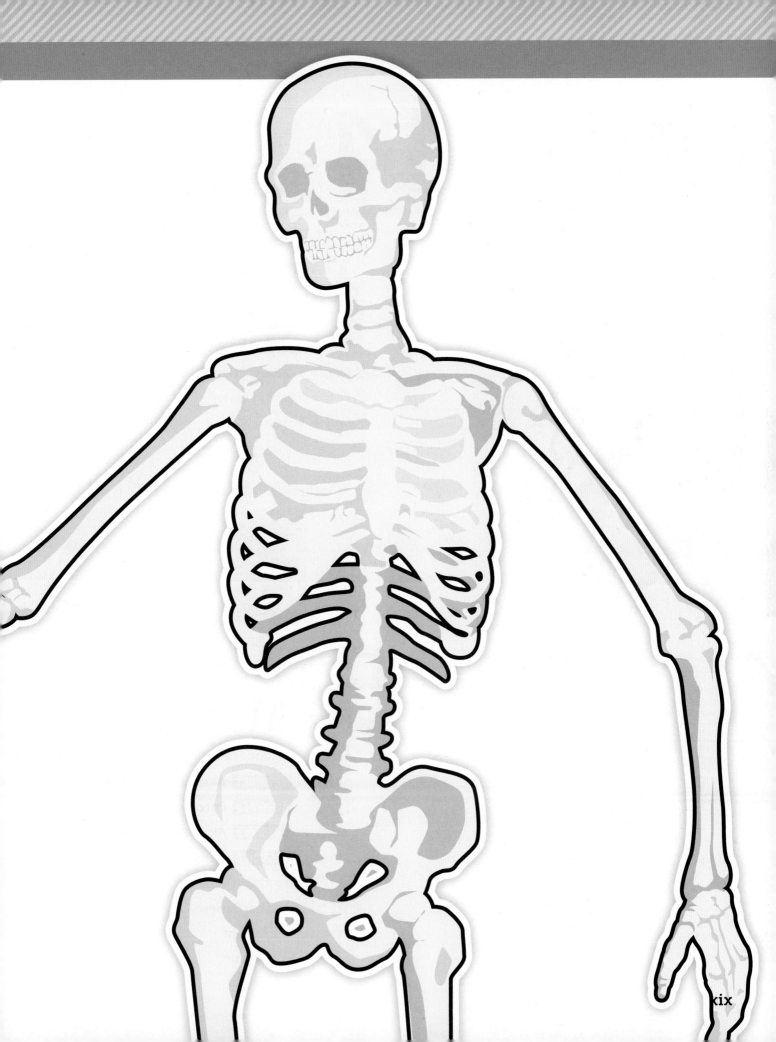

Contents

Unit 8

Tobacco, Alcohol, and Other Drugs

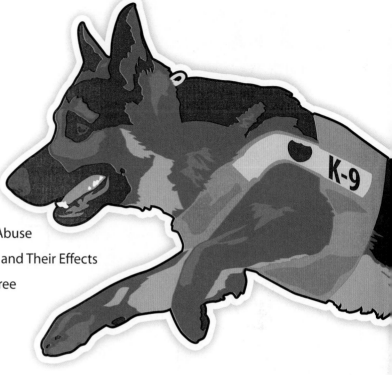

Contents

Unit 9

Preventing Disease

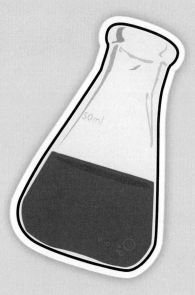

19

Noncommunicable Diseases

Contents

Unit 10
Safety and A Healthy Environment

chapter **20** **Safety**

21 Green Schools + Environmental Health

Your Total Health

WHAT IS HEALTH?

Do you know someone you would describe as "healthy"? What kinds of traits do they have? Maybe they are involved in sports. Maybe they just "look" healthy. Looking fit and feeling well are important, but there is more to having good health. Good health also includes getting along well with others and feeling good about yourself.

Your **physical**, **emotional**, and **social** *health* are all **related** and make up your *total* **health.**

Health, the *combination of physical, mental/emotional, and social well-being,* may look like the sides of a triangle. You need all three sides to make the triangle. Each side supports the other two sides. Your physical health, mental/emotional health, and social health are all related and make up your total health.

Physical Health

Physical health is one side of the health triangle. Engaging in physical activity every day will help to build and maintain your physical health. Some of the ways you can improve your physical health include the following:

EATING HEALTHY FOODS Choose nutritious meals and snacks.

VISITING THE DOCTOR REGULARLY Get regular checkups from a doctor and a dentist.

CARING FOR PERSONAL HYGIENE Shower or bathe each day. Brush and floss your teeth at least twice every day.

WEARING PROTECTIVE GEAR When playing sports, using protective gear and following safety rules will help you avoid injuries.

GET ENOUGH SLEEP Most teens need about nine hours of sleep every night.

You can also have good physical health by avoiding harmful behaviors, such as using alcohol, tobacco, and other drugs. The use of tobacco has been linked to many diseases, such as heart disease and cancer.

Mental/Emotional Health

Another side of the health triangle is your mental/emotional health. How do you handle your feelings, thoughts, and emotions each day? You can improve your mental/emotional health by talking and thinking about yourself in a healthful way. Share your thoughts and feelings with your family, a trusted adult, or with a friend.

If you are mentally and emotionally healthy, you can face challenges in a positive way. Be patient with yourself when you try to learn new subjects or new skills. Remember that everybody makes mistakes—including you! Next time you can do better.

Taking action to reach your goals is another way to develop good mental/emotional health. This can help you focus your energy and give you a sense of accomplishment. Make healthful choices, keep your promises, and take responsibility for what you do, and you will feel good about yourself and your life.

Social Health

A third side of the health triangle is your social health. Social health means how you relate to people at home, at school, and everywhere in your world. Strong friendships and family relationships are signs of good social health.

Do you get along well with your friends, classmates, and teachers? Do you spend time with your family? You can develop skills for having good relationships. Good social health includes supporting the people you care about. It also includes communicating with, respecting, and valuing people. Sometimes you may disagree with others. You can disagree and express your thoughts, but be thoughtful and choose your words carefully.

Your total health is made up of three parts, like a triangle.

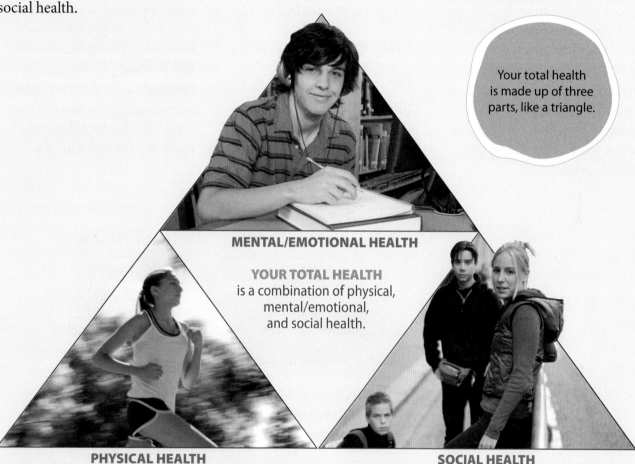

MENTAL/EMOTIONAL HEALTH

YOUR TOTAL HEALTH is a combination of physical, mental/emotional, and social health.

PHYSICAL HEALTH

SOCIAL HEALTH

ACHIEVING WELLNESS

What is the difference between health and wellness? Wellness is *a state of well-being or balanced health over a long period of time.* Your health changes from day to day. One day you may feel tired if you did not get enough sleep. Maybe you worked very hard at sports practice. The next day, you might feel well rested and full of energy because you rested. Your emotions also change. You might feel sad one day but happy the next day.

Your overall health is like a snapshot of your physical, mental/emotional, and social health. Your wellness takes a longer view. Being healthy means balancing the three sides of your health triangle over weeks or months. Wellness is sometimes represented by a continuum, or scale, that gives a picture of your health at a certain time. It may also tell you how well you are taking care of yourself.

(t)Brand X Pictures/Punchstock, (bl)Photodisc/Getty Images, (br)©PhotoAlto

The Mind-Body Connection

Your emotions have a lot to do with your physical health. Think about an event in your own life that made you feel sad. How did you deal with this emotion? Sometimes people have a difficult time dealing with their emotions. This can have a negative effect on their physical health. For example, they might get headaches, backaches, upset stomachs, colds, the flu, or even more serious diseases. Why do you think this happens?

Your mind and body connect through your nervous system. This system includes thousands of miles of nerves. The nerves link your brain to your body. Upsetting thoughts and feelings sometimes affect the signals from your brain to other parts of your body.

Your emotions have *a lot* to do with *your* physical health.

The mind-body connection describes *how your emotions affect your physical and overall health and how your overall health affects your emotions.* This connection shows again how important it is to keep the three sides of the health triangle balanced. If you become very sad or angry, or if you have other strong emotions, talk to someone. Sometimes talking to a good friend helps. Sometimes you may need the services of a counselor or a medical professional.

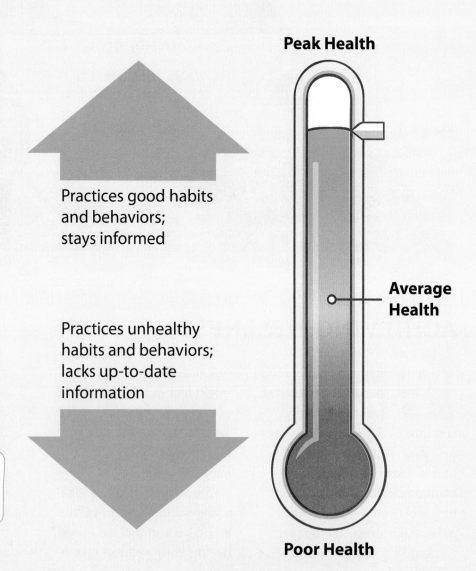

Practices good habits and behaviors; stays informed

Practices unhealthy habits and behaviors; lacks up-to-date information

Peak Health

Average Health

Poor Health

The Wellness Scale identifies how healthy you are at a given point in time.

Health Influences *and* Risk Factors

WHAT INFLUENCES YOUR HEALTH?

What are your favorite foods or activities? Your answers reflect your personal tastes, or likes and dislikes. Your health is influenced by your personal tastes and by many other factors such as:

- heredity
- environment
- family and friends
- culture
- media
- attitudes
- behavior

Heredity

You can control some of these factors, but not all of them. For example, you cannot control the natural color of your hair or eyes. Heredity (huh•RED•i•tee) is *the passing of traits from parents to their biological children.* Heredity determines the color of your eyes and hair, and other physical traits, or parts of your appearance. Genes are the basic units of heredity. They are made from chemicals called DNA, and they create the pattern for your physical traits. You inherited, or received, half of your DNA from your mother and half from your father.

Environment

Think about where you live. Do you live in a city, a suburb, a small town, or in a rural area? Where you live is the physical part of your environment (en•VY•ruhn•mehnt), or *all the living and nonliving things around you.*

Environment is another factor that affects your personal health. Your physical environment includes the home you live in, the school you attend, and the air and water around you.

Your *social environment* includes the people in your life. They can be friends, classmates, and neighbors. Your friends and peers, or *people close to you in age who are a lot like you,* may influence your choices.

You may feel pressure to think and act like them. Peer pressure can also influence health choices. The influence can be positive or negative. Helping a friend with homework, volunteering with a friend, or simply listening to a friend are examples of positive peer influence. A friend who wants you to drink alcohol, for example, is a negative influence. Recreation is also a part of your social environment. Playing games and enjoying physical activities with others can have a positive effect on your health.

Traits such as eye and hair color are inherited from parents.

Culture

Your family is one of the biggest influences on your life. It shapes your **cultural background,** or *the beliefs, customs, and traditions of a specific group of people.* You learned that your family influences your health. In addition to your family, your **culture,** or *the collected beliefs, customs, and behaviors of a group,* also affects your health. Your family and their culture may influence the foods you eat as well as the activities and special events you celebrate with special foods. Some families fast (do not eat food) during religious events. Ahmed's family observes the holiday of Ramadan.

During this holiday, members of his family fast until sundown. Your family might also celebrate traditions that include dances, foods, ceremonies, songs, and games. Your culture can also affect your health. Knowing how your lifestyle and family history relate to health problems can help you stay well.

Media

What do television, radio, movies, magazines, newspapers, books, billboards, and the Internet have in common? They are all forms of **media,** or *various methods for communicating information.* The media is another factor that affects your personal health.

The media provide powerful sources of information and influence.

You may learn helpful new facts about health on the Internet or television. You might also see a commercial for the latest video game or athletic shoes. The goal of commercials on television or the Internet, as well as advertisements in print, is to make you want to buy a product. The product may be good or bad for your health. You can make wise health choices by learning to **evaluate,** or *determine the quality* of everything you see, hear, or read.

The celebration of Kwanzaa is a tradition in many African American families.

YOUR BEHAVIOR AND YOUR HEALTH

Do you protect your skin from the sun? Do you get enough sleep so that you are not tired during the day? Do you eat healthful foods? Do you listen to a friend who needs to talk about a problem? Your answers to these questions reflect your personal lifestyle factors, or *the behaviors and habits that help determine a person's level of health.* Positive lifestyle factors promote good health. Negative lifestyle factors promote poor health.

Positive lifestyle factors promote **good** health.

Your attitude, or your *feelings and beliefs,* toward your personal lifestyle factors plays an important role in your health. You will also have greater success in managing your health if you keep a positive attitude. Teens who have a positive attitude about their health are more likely to practice good health habits and take responsibility for their health.

Risk Behaviors

"Dangerous intersection. Proceed with caution." "Don't walk." "No lifeguard on duty." You have probably seen these signs or similar signs. They are posted to warn you about possible risks or dangers and to keep you safe.

 Eating well-balanced meals, starting with a good breakfast.

 Getting at least 60 minutes of physical activity daily.

 Sleeping at least eight hours every night.

 Doing your best in school and other activities.

 Avoiding tobacco, alcohol, and other drugs.

 Following safety rules and wearing protective gear.

 Relating well to family, friends, and classmates.

Lifestyle factors affect your personal health.

Risk, or *the chance that something harmful may happen to your health and wellness,* is part of everyday life. Some risks are easy to identify. Everyday tasks such as preparing food with a knife or crossing a busy street both carry some risk. Other risks are more hidden. Some foods you like might be high in fat.

You cannot avoid every kind of risk. However, the risks you can avoid often involve risk behavior. A risk behavior is an action or behavior that might cause injury or harm to you or others. Playing a sport can be risky, but if you wear protective gear, you may avoid injury. Wear a helmet when you ride a bike to avoid the risk of a head injury if you fall. Smoking cigarettes is another risk behavior that you can avoid. Riding in a car without a safety belt is a risk behavior you can avoid by buckling up. Another risk behavior is having a lifestyle with little physical activity, such as sitting in front of the TV or a computer instead of being active. You can avoid many kinds of risk by taking responsibility for your personal health behaviors and avoiding risk.

RISKS AND CONSEQUENCES

All risk behaviors have consequences. Some consequences are minor or short-term. You might eat a sweet snack just before dinner so that you lose your appetite for a healthy meal. Other risk behaviors may have serious or life-threatening consequences. These are long-term consequences.

Experimenting with alcohol, tobacco, or other drugs has long-term consequences that can seriously damage your health. They can affect all three sides of your health triangle. They can lead to dangerous addictions, which are physical and mental dependencies.

These substances can confuse the user's judgment and can increase the risks he or she takes. Using these substances may also lead to problems with family and friends, and problems at school.

Risks that affect your health are more complicated when they are **cumulative risks** (KYOO•myuh•luh•tiv), which occur *when one risk factor adds to another to increase danger.* For example, making unhealthy food choices is one risk. Not getting regular physical activity is another risk. Add these two risks together over time, and you raise your risk of developing diseases such as heart disease and cancer.

Many choices you make affect your health. Knowing the consequences of your choices and behaviors can help you take responsibility for your health.

Reducing Risks

Practicing **prevention**, *taking steps to avoid something,* is the best way to deal with risks. For example, wear a helmet when you ride a bike to help prevent head injury. Slow down when walking or running on wet or icy pavement to help prevent a fall. Prevention also means watching out for possible dangers. When you know dangers are ahead, you can avoid them and prevent accidents.

Physical injury can be a consequence of risk behaviors.

ERproductions Ltd/Blend Images LLC

STAYING INFORMED You can take responsibility for your health by staying informed. Learn about developments in health to maintain your own health. Getting a physical exam at least once a year by a doctor is another way to stay informed about your health.

CHOOSING ABSTINENCE

If you practice abstinence from risk behaviors, you care for your own health and others' health by preventing illness and injury. Abstinence is *the conscious, active choice not to participate in high-risk behaviors.* By choosing not to use tobacco, you may avoid getting lung cancer. By staying away from alcohol, illegal drugs, and sexual activity, you avoid the negative consequences of these risk behaviors.

Abstinence is good for all sides of your health triangle. It promotes your physical health by helping you avoid injury and illness. It protects your mental/emotional health by giving you peace of mind. It also benefits your relationships with family members, peers, and friends. Practicing abstinence shows you are taking responsibility for your personal health behaviors and that you respect yourself and others. You can feel good about making positive health choices, which will strengthen your mental/emotional health as well as your social health.

- ✓ Plan ahead.
- ✓ Think about consequences.
- ✓ Resist negative pressure from others.
- ✓ Stay away from risk takers.
- ✓ Pay attention to what you are doing.
- ✓ Know your limits.
- ✓ Be aware of dangers.

Reducing risk behaviors will help maintain your overall health.

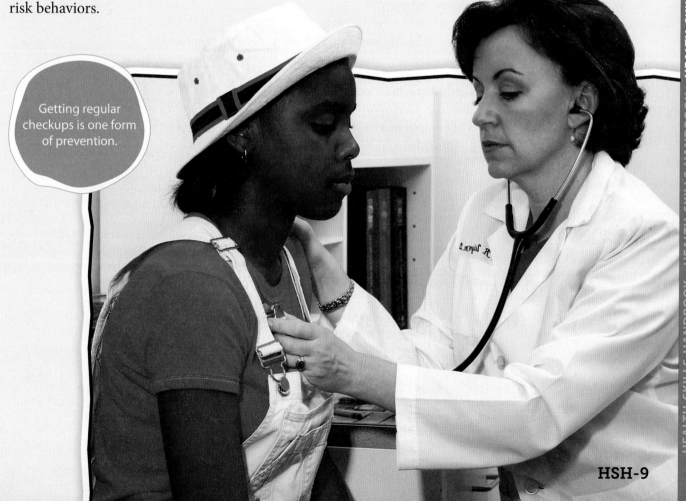

Getting regular checkups is one form of prevention.

Getty Images

Building Health Skills

SKILLS FOR A HEALTHY LIFE

Health skills are *skills that help you become and stay healthy.* Health skills can help you improve your physical, mental/emotional, and social health. Just as you learn math, reading, sports, and other kinds of skills, you can learn skills for taking care of your health now and for your entire life.

> These ten skills affect your physical, mental/emotional, and social health and can benefit you throughout your life.

Health Skills	What It Means to You
Accessing Information	You know how to find valid and reliable health information and health-promoting products and services.
Practicing Healthful Behaviors	You take action to reduce risks and protect yourself against illness and injury.
Stress Management	You find healthy ways to reduce and manage stress in your life.
Analyzing Influences	You recognize the many factors that influence your health, including culture, media, and technology.
Communication Skills	You express your ideas and feelings and listen when others express theirs.
Refusal Skills	You can say no to risky behaviors.
Conflict-Resolution Skills	You can work out problems with others in healthful ways.
Decision Making	You think through problems and find healthy solutions.
Goal Setting	You plan for the future and work to make your plans come true.
Advocacy	You take a stand for the common good and make a difference in your home, school, and community.

SELF-MANAGEMENT SKILLS

When you were younger, your parents and other adults decided what was best for your health. Now that you are older, you make many of these decisions for yourself. You take care of your personal health. You are developing your self-management skills. Two key self-management skills are practicing healthful behaviors and managing stress. When you eat healthy foods and get enough sleep, you are taking actions that promote good health. Stress management is learning to cope with challenges that put a strain on you mentally or emotionally.

Practicing Healthful Behaviors

Your behaviors affect your physical, mental/emotional, and social health. You will see benefits quickly when you practice healthful behaviors. If you exercise regularly, your heart and muscles grow stronger. When you eat healthful foods and drink plenty of water, your body works well.

Getting a good night's sleep will help you wake up with more energy. Respecting and caring for others will help you develop healthy relationships. Managing your feelings in positive ways will help you avoid actions you may regret later.

Staying *positive* is a **good health** *habit.*

Practicing healthful behaviors can help prevent injury, illness, and other health problems. When you practice healthful actions, you can help your total health. Your total health means your physical, mental/emotional, and social health. This means you take care of yourself and do not take risks. It means you learn health-promoting habits. When you eat well-balanced meals and

healthful snacks and get regular physical checkups you are practicing good health habits. Staying positive is another good health habit.

Managing Stress

Learning ways to deal with stress, *the body's response to real or imagined dangers or other life events,* is an important self-management skill. Stress management can help you learn ways to deal with stress. Stress management means identifying sources of stress. It also means you learn how to handle stress in ways that support good health. Relaxation is a good way to deal with stress. Exercise is another way to positively deal with stress.

Studying for a test can cause stress.

Making Decisions *and* Setting Goals

The path to good health begins with making good decisions. You may make more of your own decisions now. Some of those decisions might be deciding which clothes to buy or which classes to take.

As you grow older, you gain more freedom, but with it comes more responsibility. You will need to understand the short-term and long-term consequences of decisions.

Another responsibility is goal setting. You also need to plan how to reach those goals.

When you learn how to set realistic goals, you take a step toward health and well-being. Learning to make decisions and to set goals will help give you purpose and direction in your life.

ACCESSING INFORMATION

Knowing how to get reliable, or *trust-worthy and dependable,* health information is an important skill. Where can you find all this information? A main source is from adults you can trust. Community resources give you other ways to get information. These include the library and government health agencies. Organizations such as the American Red Cross can also provide good information.

Reliable Sources

You can find facts about health and health-enhancing products or services through media sources such as television, radio, and the Internet. TV and radio interviews with health professionals can give you information about current scientific studies related to health.

Web sites that end in .gov and .edu are often the most reliable sites. These sites are maintained by government organizations and educational institutions.

Getting health information is important, but so is analyzing whether that health information is valid, or reliable. Carefully review web sites ending in .org.

Many of these sites are maintained by organizations, such as the American Cancer Society or American Diabetes Association. However, some sites ending in .org may not be legitimate.

The Internet can be a good source of health information.

SW Productions/Getty Images

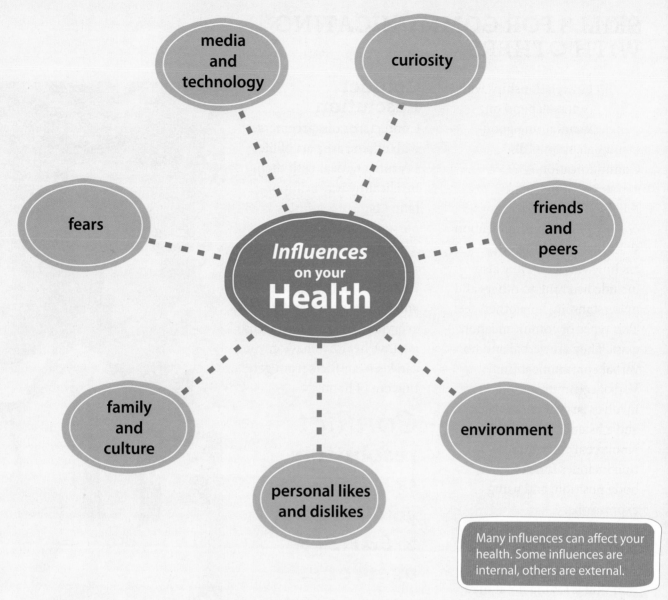

media and technology

curiosity

fears

Influences on your Health

friends and peers

family and culture

environment

personal likes and dislikes

Many influences can affect your health. Some influences are internal, others are external.

Analyzing Influences

Learning how to analyze health information, products, and services will help you act in ways that protect your health. The first step in analyzing an influence is to identify its source. A TV commercial may tell you a certain food has health benefits.

Your **decisions** have to do with your *own* **values** and **beliefs.**

Ask yourself who is the source of the information. Next, think about the motive, or reason, for the influence. Does the advertiser really take your well-being into consideration? Does the ad make you curious about the product?

Does it try to scare you into buying the product? Analyzing influences involves recognizing factors that affect or influence your health.

Your decisions also have to do with your own values and beliefs. The opinions of your friends and family members affect your decisions. Your culture and messages from the media also affect your decisions.

SKILLS FOR COMMUNICATING WITH OTHERS

Your relationships with others depend on maintaining good communication skills. Communication is *the exchange of information through the use of words or actions.* Good communication skills include telling others how you feel. They also include listening to others and understanding how others feel. Two types of communication exist. They are verbal and nonverbal communication. Verbal communication involves a speaker or writer, and a listener or reader. Nonverbal communication includes tone of voice, body position, and using expressions.

Refusal Skills

An important communication skill is saying no. It may be something that is wrong. It may be something that you are not comfortable doing. You may worry what will happen if you don't go along with the group. Will your friends still like you? Will you still be a part of the group? It is at these times that refusal skills, or *strategies that help you say no effectively,* can help. Using refusal skills can sometimes be challenging, but they can help you stay true to yourself and to your beliefs. Also, other people will have respect for you for being honest.

Conflict Resolution

Conflicts, or disagreements with others, are part of life. Learning to deal with them in a healthy way is important. Conflict resolution is *a life skill that involves solving a disagreement in a way that satisfies both sides.* Conflict-resolution skills can help you find a way to satisfy everyone. Also, by using this positive health behavior, you can keep conflicts from getting out of hand.

Conflict **resolution** skills can help you find a way to **satisfy** everyone.

Advocacy

People with advocacy skills *take action in support of a cause.* They work to bring about a change by speaking out for something like health and wellness. When you speak out for health, you encourage other people to live healthy lives. Advocacy also means keeping others informed.

Using refusal skills effectively can help you avoid potentially dangerous situation.

Image Source/Getty Images

Making Decisions *and* Setting Goals

DECISIONS AND YOUR HEALTH

As you grow up, you usually gain more privileges. Along with privileges comes responsibility. You will make more of your own decisions. The choices and decisions you make can affect each part of your health triangle.

As you get older, you will learn to make more important decisions. You will need to understand the short-term and long-term consequences of the decisions you make.

You can learn the skill of making good decisions.

Some decisions may help you avoid harmful behaviors. These questions can help you understand some of the consequences of health-related decisions.

- How will this decision affect my health?
- Will it affect the health of others? If so, how?
- Is the behavior I might choose harmful or illegal?
- How will my family feel about my decision?
- Does this decision fit with my values?
- How will this decision affect my goals?

THE DECISION-MAKING PROCESS

You make decisions every day. Some decisions are easy to make. Other decisions are more difficult. Understanding the process of **decision making,** or *the process of making a choice or solving a problem,* will help you make the best possible decisions. The decision-making process can be broken down into six steps. You can apply these six steps to any decision you need to make.

Step 1: State the Situation
Identify the situation as you understand it. When you understand the situation and your choices you can make a sound decision. Ask yourself: What choice do you need to make? What are the facts? Who else is involved?

Step 2: List the Options
When you feel like you understand your situation, think of your options. List all of the possibilities you can think of. Be sure to include only those options that are safe. It is also important to ask an adult you trust for advice when making an important decision.

Step 3: Weigh the Possible Outcomes

After listing your options, you need to evaluate the consequences of each option. The word H.E.L.P. can be used to work through this step of the decision-making process.

- **H** (Healthful) What health risks will this option present to me and to others?
- **E** (Ethical) Does this choice reflect what my family and I believe to be ethical, or right? Does this choice show respect for me and others?
- **L** (Legal) Will I be breaking the law? Is this legal for someone my age?
- **P** (Parent Approval) Would my parents approve of this choice?

Step 4: Consider Your Values

Always consider your values or the beliefs that guide the way you live. Your values reflect what is important to you and what you have learned is right and wrong. Honesty, respect, consideration, and good health are values.

Step 5: Make a Decision and Act

You've weighed your options. You've considered the risks and consequences. Now you're ready for action. Choose the option that seems best for you. Remember that this step is not complete until you take action.

Step 6: Evaluate the Decision

Evaluating the results can help you make better decisions in the future. To evaluate the results, ask yourself: Was the outcome positive or negative? Were there any unexpected outcomes? Was there anything you could have done differently? How did your decision affect others? Do you think you made the right decision? What have you learned from the experience? If the outcome was not what you expected, try again.

> Understanding the decision-making process will help you make sound decisions.

Step 1
State the situation.

Step 2
List the options.

Step 3
Weigh the possible outcomes.

Step 4
Consider your values.

Step 5
Make a decision and act.

Step 6
Evaluate the decision.

SETTING REALISTIC GOALS

When you think about your future, what do you see? Do you see someone who has graduated from college and has a good job? Are there things you want to achieve? Answering these questions can give you an idea of your goals in life. A goal is something you want to accomplish.

Goal setting is *the process of working toward something you want to accomplish.* When you have learned to set realistic goals, they can help you focus on what you want to accomplish in life. Realistic goals are goals you can reach.

Setting goals can benefit your health. Many goals can help to improve your overall health. Think about all you want to accomplish in life. Do you need to set some health-related goals to be able to accomplish those things?

Goals can become milestones and can tell you how far you have come. Reaching goals can be a powerful boost to your self-confidence. Improving your self-confidence can help to strengthen your mental/emotional health.

Types of Goals

There are two basic types of goals—short-term goals, *goals that you can achieve in a short length of time,* and long-term goals, *goals that you plan to reach over an extended period of time.* As the names imply, short-term goals can be accomplished more quickly than long-term goals.

Reaching *goals* can be a **powerful** *boost* to your **self confidence.**

Getting your homework turned in on time might be a short-term goal. Long-term goals are generally accomplished over months or years. Getting a college education might be a long-term goal. Often long-term goals are made up of short-term goals.

Reaching Your Goals

To accomplish your short-term and long-term goals, you need a plan. A goal-setting plan that has a series of steps for you to take can be very effective in helping you accomplish your goals. Following a plan can help you make the best use of your time, energy, and other resources. Here are the steps of a goal-setting plan:

- Step 1: Identify a specific goal and write it down. Write down exactly what your goal is. Be sure the goal is realistic.
- Step 2: List the steps to reach your goal. Breaking big goals into smaller goals can make them easier to accomplish.
- Step 3: Get help and support from others. There are many people in your life who can help you reach your goals. They may be parents, teachers, coaches, or other trusted adults.
- Step 4: Evaluate your progress. Check periodically to see if you are actually progressing toward your goal. You may have to identify and consider how to overcome an obstacle before moving toward your goal.
- Step 5: Celebrate when you reach your goal. Give yourself a reward.

Jamie has set a goal to be chosen for the all-star team

Choosing Health Services

WHAT IS HEALTH CARE?

You will probably at some point need to seek health care services. Health care provides services that promote, maintain, or restore health to individuals or communities. The health care system is all the medical care available to a nation's people, the way they receive the care, and the way the care is paid for. It is through the health care system that people receive medical services.

←Emergency

©Royalty-Free/Corbis

HEALTH CARE PROVIDERS

Many different professionals can help you with your health care. You may be most familiar with your own doctor who is your primary care provider: a health care professional who provides checkups and general care. Nurse practitioners and physician's assistants can also provide primary care.

In addition to doctors, nurse practitioners, and physician's assistants, many other health care professionals provide care. Nurses, pharmacists, health educators, counselors, mental health specialists, dentists, and nutritionists are all health care providers.

Preventive Care

Getting regular checkups is one way to prevent health problems and maintain wellness. During a checkup, your health care provider will check you carefully. She or he will check your heart and lungs and vision and hearing. You may also receive any immunizations you need. During your visit, your doctor may talk to you about healthful eating, exercise, avoiding alcohol, tobacco, and drugs, and other types of preventive care, or steps taken to keep disease or injury from happening or getting worse.

Specialists

Sometimes your primary care provider is not able to help you. In that case, he or she will refer you to a specialist, or health care professional trained to treat a special category of patients or specific health problems. Some specialists treat specific types of people. Other specialists treat specific conditions or body systems.

Different specialists treat different conditions.

Specialist	Specialty
Allergist	Asthma, hay fever, other allergies
Cardiologist	Heart problems
Dermatologist	Skin conditions and diseases
Oncologist	Cancer
Ophthalmologist	Eye diseases
Orthodontist	Tooth and jaw irregularities
Orthopedist	Broken bones and similar problems
Otolaryngologist	Ear, nose, and throat
Pediatrician	Infants, children, and teens

HEALTH CARE SETTINGS

Years ago, people were very limited as to where they could go for health care. In more recent years, new types of health care delivery settings have been developed. People now can go to their doctors' offices, hospitals, surgery centers, hospices, and assisted living communities.

Doctors' Offices

Doctors' offices are probably the most common setting for receiving health care. Your doctor, nurse practitioner, or physician's assistant has medical equipment to help them diagnose illnesses and to give checkups. Most of your medical needs can be met at your doctor's office.

Hospitals

If your medical needs cannot be met at your doctor's office, you may need to go to the hospital. Hospitals have much more medical equipment for diagnosing and treating illnesses. They have rooms for doing surgery and for emergency medicine. They have rooms for patients to stay overnight, if necessary. Hospitals have staff on duty around the clock every day of the year.

Surgery Centers

Your doctor may recommend that you go to a surgery center rather than a hospital. Surgery centers are facilities that offer outpatient surgical care. This means that the patients do not stay overnight. They go home the same day they have the surgery. Serious surgeries cannot be done in a surgery center. They would be done in a hospital where the patient can stay and recover. For general outpatient care, many people go to clinics.

Clinics

Clinics are similar to doctors' offices and often have primary care physicians and specialists on staff. If you go to a clinic, you might not see the same doctor each time you go. You might see whoever is on duty that day. This might make it more difficult for the doctor to get to know you and your health issues. However, for people who do not need to go to the doctor often, a clinic might be a good fit.

Hospice Care

Hospice care provides a place where terminally ill patients can live out the remainder of their lives. Terminally ill patients will not recover from their illness. Hospice workers are specially trained and are experts in pain management. They are also trained and skilled at giving emotional support to the family and the patient. Many terminally ill patients receive hospice care in their own homes. Nurses visit the patient in their own home and provide medications for pain. They also spend time with family members, helping them learn to cope during the emotionally difficult time.

Pixtal/AGE Fotostock

Assisted Living Communities

As people get older, they may not be able to take care of themselves as well as they used to. Assisted living communities offer older people an alternative to nursing homes. In nursing homes, all of the resident's needs are taken care of. In assisted living communities, the residents can choose which services they need. They may be unable to drive and need transportation. They may need reminders to take medications. They may need to have food prepared for them. In an assisted living community, the residents are able to live in their own apartments as long as they are able. Medical staff is available when the residents need help.

PAYING FOR HEALTH CARE

Health care costs can be expensive. Many people buy health insurance to help pay for medical costs. Health insurance is a plan in which a person pays a set fee to an insurance company in return for the company's agreement to pay some or all medical expenses when needed. They pay a monthly premium, or fee, to the health insurance company for the policy. There are several different options when choosing health insurance.

Private Health Care Plans

One health insurance option is managed care. Health insurance plans emphasize preventative medicine and work to control the cost and maintain the quality of health care. Using managed care, patients save money when they visit doctors who participate in the managed care plan. There are several different managed care plans such as a health maintenance organization (HMO), a preferred provider organization (PPO), and a point-of-service (POS) plan.

Government Public Health Care Plans

The government currently offers two types of health insurance—Medicaid and Medicare. Medicaid is for people with limited income. Medicare is for people over the age of 65 and for people of any age with certain disabilities.

Following a logical plan can help you achieve many goals.

Unit 1

building healthy relationships

Building Healthy Relationships

LESSONS

PREMIUM ONLINE RESOURCES >
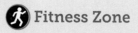 Audio
 Videos
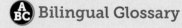 Bilingual Glossary
 Fitness Zone
 Web Quest
 Review

2

HAVE A GREAT DAY AT SCHOOL!!!

- Mom's Party SAT.9/3 @ 6PM
- "BIG GAME" THIS WEEKEND
- Need: Juice, Apples + Bread

AT PRACTICE BE HOME AROUND 7PM!

Many teens enjoy spending time together. *What activities do you enjoy with friends?*

Practicing Communication Skills

BIG IDEA Healthy relationships depend on good communication.

Before You Read

QUICK WRITE List Make a list of all the people with whom you've communicated today. Have all your communications been with words?

 ▶ Video

As You Read

FOLDABLES Study Organizer

Make the Foldable® found in the FL pages in the back of the book to record the information presented in Lesson 1.

Vocabulary

› communication
› body language
› mixed message
› "I" messages
› active listening
› assertive
› aggressive
› passive

 🔊 Audio

🔠 Bilingual Glossary

WHAT IS COMMUNICATION?

MAIN IDEA Communication involves your words, postures, gestures, and facial expressions.

You communicate with different people every day. Communication is *the exchange of information through the use of words or actions.* Communication involves sending and receiving messages, but it is more than just talking face-to-face. You communicate on the phone, in writing, and with your actions.

Communication is *more than* just **talking** *face-to-face*.

You use the phone to communicate with people you may not see every day. You communicate in writing with your teachers and online. E-mails and text messages let you share quick information, such as dates or directions. When you communicate by phone, in writing, and online, you use only your words. When you communicate in person, you also use your facial expressions and body language.

Verbal *and* Nonverbal Communication

When you talk to someone in person, you use your words. This is verbal communication. When you talk in person, others can see you, so your body language is also part of your communication. Body language is *postures, gestures, and facial expressions.* Your posture is how you hold your body. Gestures are motions you make with your hands. Body language is nonverbal communication—messages you send with your expressions and gestures.

When you talk to another person, do you look right at the other person? Do you show you care about the conversation? Your facial expressions show that you are truly interested in talking with the other person. Take the time to watch people as they talk. What does their body tell you about them as they talk?

> **Reading Check**
> **DEFINE** *What is body language?*

©Creatas/PunchStock

Mixed Messages

If your words and your body language do not match, you may send mixed messages. A mixed message is *a situation in which your words say one thing but your body language says another.* If you feel shy or nervous, you might talk very quietly, or you might fidget with your hands.

You might not look directly at the person you are speaking to. An angry person might talk with his or her arms folded. If you are feeling embarrassed, you might look down instead of at the person you are talking to.

If your body language does not match the tone of your voice, you may send a mixed message.

For example, saying "I'm fine" in an angry tone might make others think something is wrong. A good communicator will look directly at someone and speak in a caring voice. Sometimes it can be difficult to understand a person who says one thing but who displays body language that sends a different message.

GOOD COMMUNICATION SKILLS

MAIN IDEA Good communication includes listening and showing that you understand what the other person is saying.

Good communication requires sending and receiving information that is understood by everyone involved. You send messages and receive messages by speaking, listening, and writing. It takes skills to be a good communicator. You need to be a good speaker, listener and writer to send and receive messages.

It takes *skills* to be a **good communicator**.

Speaking and listening in person is the most direct way to communicate. In person, the other person hears your words, and also sees your face and body. Your facial expressions and body language can tell a lot. If you look directly at the other person, you show attention and respect. If you smile, it looks as if you care.

If you look around or past the other person as you speak, you may look like you don't care about the conversation. If you look bored, you look like you don't care. If you hold your arms folded and stiff, you may appear too firm or disrespectful.

>>> **Reading Check**

DESCRIBE *Name three examples of body language.*

Body language is part of communication. *What message does this teen send with body language?*

Steve Smith/Getty Images

Speaking and Listening Skills

Outbound ("Sending")	Inbound ("Receiving")
■ **Think, then speak.** Avoid saying the first thing that comes to mind. Plan what you're going to say. Think it through.	■ **Listen actively.** Recognize the difference between hearing and listening. Hearing is just being aware of sound. Listening is paying attention to it. Use your mind as well as your ears.
■ **Use "I" messages.** Express your concerns in terms of yourself. You'll be less likely to make others angry or feel defensive.	■ **Ask questions.** This is another way to show you are listening. It also helps clear up anything you don't understand. It prevents misunderstandings, which are a roadblock to successful communication.
■ **Make clear, simple statements.** Be specific and accurate. Stick to the subject. Give the other person a chance to do the same.	■ **Mirror thoughts and feelings.** Pay attention to what is being said. Repeat what someone says to show that you understand.
■ **Be honest with thoughts and feelings.** Say what you really think and feel, but be polite. Respect the feelings of your listener.	■ **Use appropriate body language.** Even if you disagree, listen to what the other person has to say. Make eye contact, and don't turn away.
■ **Use appropriate body language.** Make eye contact. Show that you are involved as a speaker. Avoid mixed messages. Beware of gestures, especially when speaking with people of different cultural backgrounds. Some gestures, such as pointing, are considered rude in certain cultures.	■ **Wait your turn. Avoid interrupting.** Let the person finish speaking. You'll expect the same courtesy when it's your turn.

Different skills are involved in sending and receiving messages. *Explain how these skills are related to one another.*

> If you want to **get your** *message* **across,** you need **good** *speaking* **skills.**

Speaking Skills

If you want to get your message across, you will need to develop good speaking skills. You can practice these guidelines to help you become a better speaker:

- **Think before speaking.** When you speak without thinking, you risk being misunderstood. Think about what you will say. You do not want to start talking without thinking about your words first.
- **Make clear, simple statements.** Be specific when you speak. Stay focused on the topic and on your message. Use examples when necessary to make your point more clear.

- **Use "I" messages.** "I" mes-sages *speak from your point of view to send a message.* For example, rather than saying, "You're not making sense," instead try saying, "I'm not sure what you mean."
- **Be honest.** Tell the truth to describe your thoughts and feelings. Be polite and kind.
- **Use appropriate body language.** Make eye contact. Think about your expressions and gestures. Show that you are paying attention.

Listening Skills

Listening skills are as important as speaking skills. Be an active listener. Active listening means *hearing, thinking about, and responding to another person's message.* You can develop these skills to become a good listener:

- **Pay attention.** Listen to what the speaker is saying. Think about that person's message as you are listening.
- **Use body language.** Face the speaker. Look at the speaker. Focus on the words you hear. Use your posture, gestures, and facial expressions to show that you are listening.
- **Wait your turn.** Before you ask questions or respond to what someone is saying, let that person finish speaking. Avoid interrupting someone.

- **Ask questions.** If it is appropriate, ask questions. Use "I" messages when you speak.
- **Mirror thoughts and feelings.** After the other person finishes speaking, repeat back in your own words what you believe that person said. This will show that you are listening. It will also help you better understand the message.

Writing Skills

When you write an e-mail, a text, or a note, remember that the person you are writing to cannot see you or hear your tone of voice. This may lead to misunderstandings. You can also practice basic skills for written communication:

- **Write clear, simple statements.** Be sure to state your thoughts and feelings clearly.
- **Reread your words before you send a message.** Again, the other person cannot see you. Make sure your thoughts can be clearly understood.

Make sure your written words say what you mean. *How can you be sure your message is clear?*

Russell Glenister/Corbis

Communicating *with* Parents *or* Guardians

Tyler has been dreaming of playing football for a while now. Tryouts for the local team are in two weeks. Tyler needs to talk to his parents first, but he is worried they will not understand how he feels about playing the sport. They want him to have enough time to do homework and study, and practices will take up a lot of time. Tyler is unsure about how to talk to his parents about playing football.

* State clear reasons for your request.

* Use "I" messages.

* Use a respectful tone and stay calm.

* Use appropriate listening skills.

* Be willing to compromise.

HAVE A GREAT DAY AT SCHOOL!!!

- Mom's Party SAT. 9/3 @ 6PM
- "BIG GAME" THIS WEEKEND
- Need: Juice, Apples + Bread

AT PRACTICE BE HOME AROUND 7PM!

With A Group

Write a script showing how Tyler talks to his parents about playing football. Use the techniques above to effectively communicate ideas, thoughts, needs, and feelings. Role-play your conversation for the class.

YOUR COMMUNICATION STYLE

MAIN IDEA Communication styles include assertive, aggressive, and passive.

Once you know *how* to communicate clearly, you can choose your communication *style*. A communicator can be *assertive*, *aggressive*, or *passive*. Each of these three communication styles has its own traits which can make communication either more or less effective.

A communicator can be **assertive**, **aggressive**, or **passive**.

An assertive communicator is friendly but firm. An assertive communicator *states his or her position in a firm but positive way.* For example, you tell a friend, "I need to be home on time because I promised my parents." An assertive communicator shows respect for himself or herself and others.

An aggressive communicator is someone who tends to be *overly forceful, pushy, hostile, or otherwise attacking in approach.*

Aggressive communicators may think too much about themselves and not show respect to others. For example, you tell a friend, "You'd better get me home on time." An aggressive communicator can hurt feelings or make others angry.

A passive communicator *has a tendency to give up, give in, or back down without standing up for his or her rights and needs.* For example, a passive communicator who has a firm curfew might tell a friend, "It doesn't matter when I get home." A passive communicator may care too much about what others think of him or her. Sometimes a passive communicator may not feel confidence or self-respect. You can change from being a passive communicator to an assertive communicator by reminding yourself that your thoughts and opinions have value.

>>> **Reading Check**

IDENTIFY *Which communication style is the most effective? Why?* ■

Cultural Perspectives

Differences in Body Language Body language is used in different ways in different cultures. For example, in many cultures of Southeast Asia, it is not considered appropriate to show facial expressions of sadness or anger. People from these cultures might smile to mask negative feelings or a negative statement. In other cultures, it is considered impolite to make eye contact when communicating with a member of the other gender. Certain hand and arm gestures often have different meanings in different cultures.

LESSON 1

REVIEW

>>> **After You Read**

1. **DEFINE** What does *communication* mean?
2. **EXPLAIN** What is a mixed message?
3. **EXPLAIN** Which of the three communication styles do you think is most effective? Explain your answer.

>>> **Thinking Critically**

4. **HYPOTHESIZE** Imagine you have a friend who wants to copy your homework. Use "I" messages to respond.
5. **SYNTHESIZE** What are some nonverbal ways to show consideration for others?

>>> **Applying Health Skills**

6. **ANALYZING INFLUENCES** Pay attention to various conversations you see in school or on TV. Watch for an example of mixed messages in conversation. Describe the example.

🄬 Review

🔊 Audio

Family Relationships

BIG IDEA Your relationships with family members are some of the most important in your life.

Before You Read

QUICK WRITE Name the people you can talk to if you have a problem.

▶ Video

As You Read

STUDY ORGANIZER Make the study organizer found in the FL pages in the back of the book to record the information presented in Lesson 2.

Vocabulary

› family
› nurture
› role
› abuse
› physical abuse
› sexual abuse
› neglect

🔊 Audio

🔤 Bilingual Glossary

WHAT MAKES A FAMILY?

MAIN IDEA Family members support one another.

The way you relate with your family prepares you for how you relate to others for the rest of your life. A **family** is *the basic unit of society and includes two or more people joined by blood, marriage, adoption, or a desire to support each other.* Families come in many shapes, sizes, and types:

- A **couple** is two people who do not have children.
- A **nuclear family** is two parents and one or more children.
- An **extended family** is a nuclear family plus other relatives.
- A **blended family** has two adults and one or more children from a previous marriage.
- A **foster family** has adults caring for one or more children born to different parents.
- An **adoptive family** is a couple plus adopted children.
- A **joint-custody family** has two parents living apart and sharing custody of children.
- A **single-custody family** has parents living apart and children living with one parent.

Families Meet Needs

The main role of a family is to meet the needs of its members. Families provide food, clothing and shelter. They should also provide support and comfort. Healthy families nurture each member of the family. To **nurture** means *to fulfill physical needs, mental/emotional needs, and social needs.* A healthy family helps to nurture all sides of its members' health triangles:

- **Physical** Families care for their members by providing food, clothing and shelter.
- **Mental/Emotional** Family members offer love, acceptance, and support. They also pass along traditions, values, and beliefs. Your greatest influences in these areas often come from your family.
- **Social** Families teach their members how to get along with each other and with people outside the family..

Reading Check

DEFINE *What is a family?*

Roles *and* Responsibilities *in the* Family

Every family member has a special role in the family. A **role** is *a part you play when you interact with another person.* Each role has certain responsibilities. Parents and other adults in a family have the responsibility of meeting the basic needs of the family. Parents also teach and model healthful behaviors and good communication skills.

Your role as a family member is to follow rules at home and to accept certain responsibilities.

If you have younger brothers or sisters, you may be a role model. Your role includes showing respect, caring, and appreciation. You can show that you respect your family by accepting your responsibilities. You can show appreciation by saying "thank you" to family members. You can also show caring by giving your time and attention to members of your family.

Family roles can change. As you get older, you will take on more responsibilities. You may be asked to help with younger or older members of your family.

You may take on more responsibilities in the home. Each time you accept a new responsibility, you show respect, love, and support for your family.

Every family **member** has a *special role* in the family.

>>> **Reading Check**

RECALL *What are a parent's main family responsibilities?*

BUILDING STRONG FAMILIES

MAIN IDEA A strong family is built on good relationships.

People with strong family relationships feel connected. They feel safe and secure. This list offers guidelines for making and keeping strong family relationships:

- **Support one another.** Does your family believe in you? Knowing that you have support adds meaning when you succeed. Support can also help when you face a challenge.

- **Show appreciation for one another.** Families grow stronger when each member shows appreciation to the other members. For example, a child may say "Thank you for dinner," to parents or guardians. Another way to show appreciation for your family members is to help with tasks such as doing the dishes or folding the laundry.

- **Follow family rules.** Some rules are related to responsibilities, such as when to do your homework or when to take out the trash. Others rules can include what time to be home or when the TV can be on. Following the rules at home helps build trust and respect in a family.

- **Spend quality time together.** Take time for activities that include the whole family. Some families always have dinner together so they can talk about events of the day. Others plan evenings at home or weekend outings together.

Family responsibilities can change. *How is this teen showing a willingness to take responsibility as part of a family?*

- **Use good communication skills.** Talking openly helps solve problems and disagreements. Communication helps to develop trust and respect.
- **Show responsibility.** Do your tasks and chores without being asked. When you accept your family responsibilities, you show and model respect.
- **Show respect.** Speak to family members in a respectful tone of voice. Show respect for differences in family members.

Show respect for privacy and personal belongings.

>>> **Reading Check**

RECALL *Name three ways to build strong family relationships.*

CHANGES IN THE FAMILY

MAIN IDEA > Families deal with change.

Changes and challenges affect every family. Change may occur with the birth of a baby, separation, divorce, or remarriage. You may take on new responsibilities, go to college, or start a job. Change may also occur with the loss of a job, illness or injury, military service, or moving to a new home. Strong relationships make it easier to cope with changes.

Communicating *openly* to family members can help *reduce stress*.

Sometimes family changes can cause you to worry or feel stress or sadness. When a family experiences change, it becomes more important for family members to communicate with one another.

Communicating openly with family members can help reduce stress. Talk about ways your role in the family might change. Sometimes outside help is needed to deal with serious change in the family structure. Family members may seek help from experts such as counselors, health care workers, religious leaders, legal experts, or law enforcement personnel.

Changes that happen in a family affect all family members. *What are some serious changes that can affect families?*

CHANGE	POSITIVE WAYS TO COPE
Moving to a new home	Before the move, look at a map of your new neighborhood. Find your new house, your school, and nearby parks. When in a new neighborhood, try to meet other teens.
Separation, divorce, or remarriage of parents	Tell both parents you love them. Talk to them or to another trusted adult about how you feel. A separation or divorce is not your fault.
Job change or job loss	If the family needs to limit spending for a while, ask how you can help.
Birth or adoption of a new sibling	Spend time with your new sibling. Ask your parents how you can help. Imagine what your relationship might be like in the future.
Illness or injury	Show that you care about a sick or injured family member by spending time with him or her and asking how you can help.
Death, loss, and grief	Accept the ways family members express grief. Don't expect their ways of coping to be the same as yours. Pay extra attention to younger members of the family.

Juan Silva/Stockbyte/Getty Images

SERIOUS FAMILY PROBLEMS

MAIN IDEA Serious family problems may require help from counselors or others.

Family problems can sometimes be serious and require outside help. Drug or alcohol addiction is a serious problem. While it may be that only one family member is addicted to drugs or alcohol, the whole family suffers. A family needs outside help to deal with drug or alcohol addiction.

A *parent* or *guardian* is **responsible** for caring for a *child's* **physical, mental/ emotional** and **social needs**.

A serious family situation that may require the help of police or other authorities is abuse. Abuse is *a pattern of mistreatment of another person.* Abuse can affect children or adults. Physical abuse involves *the use of physical force, such as hitting or pushing.* A person who is physically abused may shows bruises, scratches, burns, or broken bones. Emotional abuse can also be serious. Emotional abuse occurs when someone always yells or puts down another person. Emotional abuse can damage a person's self-esteem.

Sexual abuse is *any mistreatment of a child or adult involving sexual activity.* Sexual abuse is unwanted use of forced sexual activity including touching private body parts or being forced to touch body parts. Sexual abuse also includes showing sexual materials to a child. Abuse often includes secrets, and threats to keep secrets.

A parent or guardian is responsible for caring for a child's physical, mental/emotional and social needs. When a parent fails to provide proper care, he or she may be charged with neglect. Neglect is *failure to provide for the basic physical and emotional needs of a dependent.* Physical neglect can include not providing food, shelter, clothing, or medical care. Emotional neglect involves failing to give love and respect.

All forms of abuse and neglect are against the law. These kinds of serious family problems may also require outside help. Any person who feels abused or neglected must find someone who can help. The process of getting help can start by talking to another trusted adult, a teacher, a school counselor, or a medical professional.

>>> **Reading Check**

DEFINE *What is abuse and what forms can it take?* ■

LESSON 2
REVIEW

>>> **After You Read**

1. **VOCABULARY** Define the word *neglect.*
2. **IDENTIFY** What are three ways to build and keep strong family relationships?
3. **EXPLAIN** Identify some family problems that would require outside help.

>>> **Thinking Critically**

4. **SYNTHESIZE** Give some examples of ways you can use good communication skills with your family.
5. **COMPARE AND CONTRAST** Explain the difference between nurture and neglect.

>>> **Applying Health Skills**

6. **ADVOCACY** Imagine you have a friend who you think may have a problem at home. Your friend is behaving in an unusual way. He or she seems sad, quiet and withdrawn. How can you communicate your concern to your friend?

🔄 Review
🔊 Audio

Peer Relationships

BIG IDEA Strong relationships will have a positive effect on your physical, mental/emotional, and social health.

BananaStock/JupiterImages

>>> **Before You Read**

QUICK WRITE List the people you consider your friends.

 Video

>>> **As You Read**

STUDY ORGANIZER Make the study organizer found in the FL pages in the back of the book to record the information presented in Lesson 3.

>>> **Vocabulary**

> peers
> acquaintance
> friendship
> reliable
> loyal
> sympathetic
> peer pressure
> assertive response

 Audio

 Bilingual Glossary

WHO ARE YOUR PEERS?

MAIN IDEA Your peer group is made up of people who are close in age and have things in common with you.

As a teen, you spend a lot of time among your peers. Your **peers** are *people close to you in age who are a lot like you.* The students at your school and other teens you know from your outside activities are your peers. Peers will be an important part of your life throughout your lifetime. Think about all the peers you encounter in your daily life. Your peers may be your friends, classmates, teammates, or neighbors.

> *Peers* will be an **important** part of your *life* throughout your **lifetime**.

Your peers also include your acquaintances. An **acquaintance** is *someone you see occasionally or know casually.* Sometimes your acquaintances become friends. A peer can be someone you have never met but with whom you have something in common.

For example, if you volunteer for an agency that works for a clean environment, you are peers with other teens who volunteer for that agency. They may live in other cities or other countries, but they are still your peers.

Friendships During *the* Teen Years

During your teen years, you develop friendships with some of your peers and acquaintances. A **friendship** is *a relationship with someone you know, trust, and regard with affection.* Friendships usually begin with a common interest, such as a sport, a class in school, or conversations on the bus to or from school.

Your peers and acquaintances are the people around you. Your friends are the relationships you choose. Your friends may have a shared interest or the same values as you do. Having friends is an important part of your social health and growth. In strong friendships, you appreciate the values of loyalty, honesty, trust, and mutual respect.

What Makes a Good Friend?

Friendships share positive qualities and usually grow stronger with time. The traits of strong friendships apply equally to you and your friends. Strong, healthy friendships have a number of qualities in common:

- **Shared Values** Friendships can begin with shared values. If you are an honest, responsible person, you will appreciate honesty and responsibility in your friends. For example, if you work hard at school and are eager to learn, you share interests with other good students. If you take some responsibility for any of your younger siblings at home, your friends are likely to also have strong family relationships.

- **Reliability** A good friend is reliable. Reliable means *trustworthy and dependable.* Reliable friends do what they say they will do. They do not talk negatively to other people about their friends.

- **Loyalty** A good friend is loyal, or *faithful,* to his or her friends. In friendships that are strong, friends are loyal to each other. They agree to stick together not just in good times but also through any disagreements they may have. Good friends will respect and honor each other's personal interests, values, beliefs, and differences.

Friendships grow **stronger** with **time**.

- **Sympathy** Good friends share sympathy for one another. A sympathetic friend is *aware of how you may be feeling at a given moment.* For example, if you studied hard for a test but you did not do as well as you had hoped, you may feel sad or disappointed. A good friend will understand how you feel. The other person will listen if you want to talk about it and offer support.

- **Caring** A good friend cares about you. A friend can show caring by being interested in your feelings, your values, and your beliefs. A caring friend is a good listener and pays attention to you and your interests.

- **Trust** Good friends trust each other. They learn through their friendship that trust is important. Trust is proven through reliability and loyalty.

- **Respect** A good friend has self-respect. A good friend also has respect for his or her family, school, and friends. Good friends show respect by giving their time and attention to each other. You can show respect by displaying all the traits of friendship: reliability, loyalty, sympathy, caring and trust. Good friends also show their respect for each other's values and differences.

> **》》 Reading Check**
>
> **RECALL** *Name three qualities of a good friend.*

> Your peers are close to you in age and have a lot in common with you. *Describe some of the things this group of teens has in common with one another.*

Making New Friends

Teens typically find most of their friends at school, in their neighborhood, or in shared activities. Sometimes it can be hard to find new friends, such as when you change grades, schools, or move to a new neighborhood. However, you can develop skills to help make new friends.

- **Be yourself.** Identify your values, beliefs, and special interests. What would make you a good friend? You want to make friends who value you for who you are.
- **Break the ice.** Start a conversation with a compliment or a question. Show your interest. If the other person shares the same interest, you may be able to begin a new friendship.
- **Seek out teens who share your interests.** Join a club, sports team, or community group. There you will find peers who share some of the same interests you do.
- **Join a group that works for a cause you support.** You can show your citizenship and giving qualities to people who share your values. You will also help your community.

Sometimes it can be **hard to find** new friends.

Strengthening Friendships

Once you make a new friend, you will want to strengthen that friendship and help make it long-lasting. You can use a number of key strategies that can not only help you make new friendships but also help you make existing friendship stronger.

- **Spend time together.** The more time you spend with someone, the better you get to know each other. Do your homework together, share a special interest, practice a sport, or work on a school or community project together.
- **Communicate openly and honestly.** Open and honest communication will build trust and respect—qualities you want in your friendships.

An important part of keeping friendships strong is identifying problems and working to solve them. *These teens appear to be having a disagreement. What are they doing to resolve their dispute in a positive way?*

- **Help each other through hard times.** Good friendships aren't only about the fun times. Good friends also share their time and sympathy when a friend has a problem and may need support.
- **Respect each other's differences.** No two people are exactly alike. Show respect for the ways your friends may be different from you.
- **Encourage each other to reach goals.** An important part of friendship is sharing the interests and goals of others. Be giving of your time and attention, and be supportive of your friend's goals.
- **Identify problems and work to solve them.** A part of communication includes discussing problems and expressing your interest in solving a problem. It could be a problem your friend shares with you or a problem between you and a friend.

In order to communicate your interest in a friend, remember to think before speaking. Also practice being a good listener. Be honest and truthful about your feelings and opinions.

Talking about a problem with a friend can help both you and the other person understand your differences. It can also help you both better understand the issue and help find a solution to whatever problem the two of you may have.

PEER PRESSURE

Teens spend much of their time among peers at school and in other activities. Peers influence some of the decisions that you make. *The influence that your peer group has on you* is peer pressure. Teens want to fit in and be accepted. Sometimes, without even knowing it, they are influenced by their peers. For example, if you notice that almost everyone at school wears zippered sweatshirts, you may want a zippered sweatshirt too.

You are influenced by what you see your peers do. This is called indirect peer pressure. No one is making you get a zippered sweatshirt, but you want one because you see everyone around you wears one.

You **always** have a **choice** to say no.

At other times, you may feel direct peer pressure. A peer might tell you what you should do to fit in or be accepted.

If you choose not to do what that person suggests, you may worry whether you will fit in and be part of the group. Remember that while your peers have a big influence on your life, you always have a choice to say no if you believe that a behavior or action will be harmful.

>>> **Reading Check**

DEFINE *Explain what indirect peer pressure is and give an example.*

Positive Peer Pressure

Like all influences, peer pressure can have a positive or negative effect. Positive peer pressure helps you make healthful choices. For example, if someone says "You are such a good dancer, we wish you would join the dance team," that is a positive suggestion. Dance may strengthen your physical health. You may also make new friends who have a shared interest.

Imagine that many of your peers volunteer at a food bank. They enjoy the sense of citizenship and caring. You may choose to volunteer too, based on your peers' positive experiences. Volunteering may make you feel good about yourself because you are helping others. You may make new friends. You are using an example you see in others to make a positive choice for yourself. This is positive peer pressure, or positive influence.

Positive examples set by others can have a major influence on your choices and decisions. *What are some other ways in which peer pressure can have a positive influence on you?*

Negative Peer Pressure

Negative peer pressure may make decisions difficult. Your peers may urge you to do something you do not agree with or do not want to do. When you face negative peer pressure, you have a choice to make. It helps to think about your values. If you feel you have to choose between making a healthful choice or fitting in with a group, think about the consequences. True friends will respect your decision.

Encouraging a person to act in a way that is harmful or illegal is one form of negative peer pressure. Other forms may include dares, threats, teasing, or name-calling. You can recognize negative peer pressure by using the **H.E.L.P.** guidelines. **H.E.L.P.** stands for **H**ealthful, **E**thical, **L**egal, and **P**arent-approved. Would your choice affect your well-being or that of others? Would it show respect? Would your choice break the law? Would your parents approve?

If what your friends tell you to do does not meet the **H.E.L.P.** guidelines, you can refuse. All of your actions are your own choices, but you can learn ways to resist negative peer pressure.

- **Avoid the situation.** If you can tell that a situation might be unsafe, harmful, or against rules, do not participate.
- **Use assertive responses.** If your peers suggest a dangerous behavior or situation, say no. Use an assertive response, which *states your position strongly and confidently.*
- **Focus on the issue**. State your reasons for your choice. Avoid responding if your peers tease you. Avoid trading insults.
- **Walk away.** It is best to try to talk things out with peers who try to pressure you. If anyone gets angry, though, walk away.

>>> **Reading Check**

EXPLAIN *Tell why negative peer pressure can be harmful or hurtful.* ■

©McGraw-Hill Education/Ken Karp

Resisting peer pressure is a skill you can learn. *What are some effective ways you have found to handle negative peer pressure?*

>>> **After You Read**

1. **VOCABULARY** Define *friendship.*
2. **IDENTIFY** What are three ways to strengthen friendships?
3. **EXPLAIN** Define negative peer pressure and give an example.

>>> **Thinking Critically**

4. **COMPARE AND CONTRAST** Explain the consequences that both positive and negative influences from peers can have on a teen's life.
5. **SYNTHESIZE** What does H.E.L.P. stand for and why is it important?

>>> **Applying Health Skills**

6. **ANALYZING INFLUENCES** Do you think adults experience as much peer pressure as teens? Write a brief paragraph explaining your opinion.

⟳ Review

🔊 Audio

Hands-On HEALTH ACTIVITY

WHAT YOU WILL NEED

* Pencil or pen
* Paper

WHAT YOU WILL DO

1 Working in pairs, imagine a situation in which "you" messages might occur. Think of your own "you" message. Write the situation on the top of the paper. Write the "you" message below on the left. Change that into an "I" message, writing the "I" version on the right.

2 Here are three sample situations:
• Your older brother was late picking you up. He had no excuse.
• A classmate told a lie about you.
• Your sister borrowed something and returned it in poor condition.

3 Read each "you" message to the rest of the class. Then read the corresponding "I" message.

WRAPPING IT UP

Was the "you" message or "I" message most effective? Explain why. Think of a disagreement you have had with a family member or friend. How could using "I" messages have helped resolve the conflict? How does practicing positive behaviors, such as using "I" messages when you communicate, benefit your overall health?

"I" Messages

Communicating effectively is especially important when there is a disagreement.

When you use "I" messages, you express your feelings. "I" messages are unlike "you" messages, which place blame on the other person. To help see the difference, compare these two statements:

• **"You" message:** You always get your way! You're selfish!

• **"I" message:** Sometimes I would like to have a say in what we do.

This activity will allow you to practice sending "I" messages. If you practice this skill, you will become a better communicator.

© McGraw-Hill Education/Ken Karp

READING REVIEW

FOLDABLES and Other Study Aids

Take out the Foldable® that you created and any study organizers that you created. Find a partner and quiz each other using these study aids.

LESSON 1 Practicing Communication Skills

BIG IDEA Healthy relationships depend on good communication.

* Communication is the exchange of information through the use of words or actions.
* Communication involves your words, postures, gestures, and facial expressions.
* Body language is postures, gestures, and facial expressions.
* A mixed message is a situation in which your words say one thing but your body language says another.
* Good communication includes listening and showing that you understand what the other person is saying.
* "I" messages speak from your point of view to send a message.
* Active listening means hearing, thinking about, and responding to another person's message.
* Communications styles include assertive, aggressive, and passive.

LESSON 2 Family Relationships

BIG IDEA Your relationships with family members are some of the most important in your life.

* A family is the basic unit of society and includes two or more people joined by blood, marriage, adoption, or a desire to support each other.
* Family members support one another.
* To nurture means to fulfill physical needs, mental/emotional needs, and social needs.
* A strong family is built on good relationships.
* Families learn to deal with change.
* Serious family problems, such as abuse, addiction, or neglect, may require help from counselors or others.
* Neglect is the failure of parents to provide their children with basic physical and emotional care and protection.

LESSON 3 Peer Relationships

BIG IDEA Strong relationships will have a positive effect on your physical, mental/emotional, and social health.

* Your peer group is made up of people who are close in age and have things in common with you.
* Qualities of friendship include shared values, reliability, loyalty, sympathy, caring, trust, and respect.
* Peer pressure can affect you in positive or negative ways.
* Using H.E.L.P. guidelines can help you make healthful choices when you face negative peer pressure.

 Review

 Web Quest

ASSESSMENT

Reviewing Vocabulary *and* Main Ideas

> family   > verbal   > nurture   > neglect

> body language   communication   > role

≫ On a sheet of paper, write the numbers 1–6. After each number, write the term from the list that best completes each statement.

LESSON 1 Practicing Communication Skills

1. Expressing your feelings, thoughts, and experiences in words, through speaking or writing, is _____.

2. _____ includes posture, gestures, and facial expressions that send messages.

3. The part you play when you interact with another person is known as a _____.

LESSON 2 Family Relationships

4. The _____ is the basic unit of society and includes two or more people brought together by blood, marriage, adoption, or a desire for mutual support.

5. To _____ is to fulfill physical, mental/emotional, and social needs.

6. _____ is failure to provide for the basic physical and emotional needs of a dependent.

≫ On a sheet of paper, write the numbers 7–12. Write *True* or *False* for each statement below. If the statement is false, change the underlined word or phrase to make it true.

LESSON 3 Peer Relationships

7. The influence that people your own age have on you is called <u>nurturing</u>.

8. You can use the <u>H.E.L.P.</u> criteria to resist negative peer pressure.

9. <u>Trust</u> is a trait of good friendships.

10. Friends urging you to come to a party where there might be alcohol is an example of <u>positive</u> peer pressure.

11. An <u>acquaintance</u> is someone you see occasionally or know casually.

12. An <u>aggressive response</u> states your position strongly and confidently.

 eAssessment

 Using complete sentences, answer the following questions on a sheet of paper.

Thinking Critically

13. ANALYZE How do positive relationships affect your physical and mental/emotional health? Explain your answer.

14. ASSESS What are some qualities you look for in a friend? How would those qualities affect your mental/emotional and social health?

15. INFER How can learning how to resist negative peer pressure improve all sides of your health triangle?

Write About It

16. DESCRIPTIVE WRITING Design a greeting card for someone who has had a positive influence on your life. Write a message for the inside of the card that includes specific examples of how that person has positively influenced your life.

17. NARRATIVE WRITING Write a skit that shows a situation in which a teen uses each step of the H.E.L.P. guidelines to resist negative peer pressure and make a healthful choice.

STANDARDIZED TEST PRACTICE

Writing
Read the directions in the left column below and refer to the example in the right column.

Directions:
Write a letter to persuade your principal to organize more after-school activities. Begin your letter by stating your viewpoint. Then provide concrete examples of activities you think would be valuable to students. In addition, list benefits you feel these activities would have, and explain them in your letter. Use a respectful but firm tone to persuade the principal to see your point of view. Use the example in the next column to give you ideas about the way you should structure your letter.

Example:
Did you know that strong friendships and positive peer pressure add to a student's school success and overall good health? It's true! One of the best ways to forge strong friendships is through after-school activities. This is why I think our school should provide more after-school activities such as clubs and organizations. I think this is a good idea for several reasons: First, after-school activities would strengthen relationships between students. Second, students would have a safe place to go to after school. Third, teens would have fun and make new friends.

BUILDING HEALTHY RELATIONSHIPS

Have you ever considered all the time and energy you put into maintaining positive, healthy relationships with your friends, parents, siblings, teachers—even your girlfriend or boyfriend? Healthy relationships require attention.

FAMILY

Your family is the single greatest influence on developing your values.

TYPES OF FAMILIES

FAMILY TYPE	MAKEUP
Couple	Two people who do not have children
Nuclear	Two parents and one or more children
Extended	A nuclear family plus other relatives such as grandparents
Single-parent	One parent and one or more children
Blended	Two people, one or both with children from previous marriages
Foster	Adults caring for one or more children born to different parents
Adoptive	A couple plus one or more adopted children
Joint-custody	Two parents living apart, sharing custody of their children
Single-custody	Two parents living apart and one or more children living with only one parent

WHAT MAKES A SOLID FAMILY?

TOGETHERNESS

APPRECIATION

SUPPORT

RESPECT

FOLLOWING FAMILY RULES

FRIENDS

Your friends are those in your peer group with whom you have the most in common and most enjoy spending time with. Friendships, unlike family relationships, are ones that you freely choose.

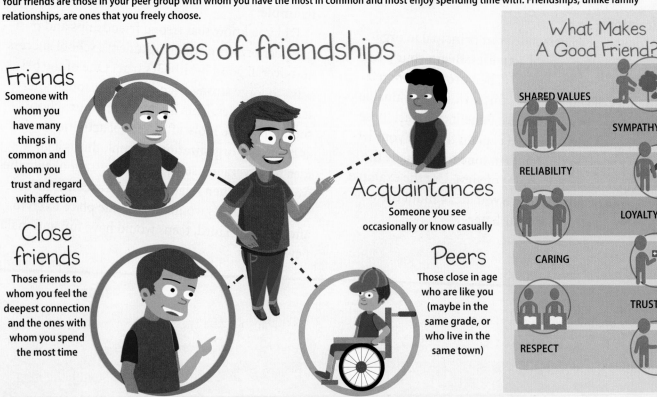

Types of friendships

Friends
Someone with whom you have many things in common and whom you trust and regard with affection

Close friends
Those friends to whom you feel the deepest connection and the ones with whom you spend the most time

Acquaintances
Someone you see occasionally or know casually

Peers
Those close in age who are like you (maybe in the same grade, or who live in the same town)

What Makes A Good Friend?

SHARED VALUES

SYMPATHY

RELIABILITY

LOYALTY

CARING

TRUST

RESPECT

FORMING CLOSER RELATIONSHIPS

During your teen years, you will become interested in forming closer friendships and new kinds of relationships.

ADOLESCENCE BRINGS CHANGE

Relationships become deeper and more important during the teen years.

Your friends may also begin to include more members of the opposite gender.

Consider your future

The places you go and the people you spend time with can have consequences that affect your safety, health, and plans for your future.

List the goals you want to achieve to get a better idea of what limits you need to set to help you reach your goals.

STARTING TO DATE
Group dating

Spending time with a group of your peers is a way to get to know others in a low-stress environment.

Going out with a group can help you discover new interests and learn more about activities you enjoy.

Healthy dating relationships

* Mutual respect
* Caring
* Honesty
* Loyalty
* Commitment

SAYING NO TO RISK BEHAVIORS

Using the S.T.O.P. strategy can help you deal with negative peer pressure. You can use one or all of the following steps:

SAY NO IN A FIRM VOICE.
State your feelings firmly but politely. Make your "no" sound like you mean it. Use body language to support your words.

TELL WHY NOT.
Explain why you feel the way you do. Do not apologize. Just say, "No, thanks. I care about my health."

OFFER OTHER IDEAS.
Suggest an alternative activity that is safe and fun.

PROMPTLY LEAVE.
If all else fails, let your actions match your words. Simply walk away from the situation.

Dating Relationships and Abstinence

PREMIUM ONLINE RESOURCES

 Audio

 Videos

 Bilingual Glossary

 Fitness Zone

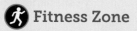 Web Quest

 Review

Beginning *to* Date

BIG IDEA Your friendships become more important during adolescence.

Before You Read

QUICK WRITE Write a short paragraph that describes how your friendships with boys and girls have changed over the last few years.

 Video

As You Read

FOLDABLES Study Organizer

Make the Foldable® found in the FL pages in the back of the book to record the information presented in Lesson 1 about beginning to date.

Vocabulary
> commitment
> affection

 Audio

🅰️🅱️🅲️ Bilingual Glossary

Developing Good Character

Respect Every relationship benefits from respect. One way to show respect is through simple courtesies—for example, holding the door for the person behind you. Using good manners at someone else's home is another sign of respect. What ways can you think of to show respect for someone you are dating?

CHANGING FRIENDSHIPS

MAIN IDEA Changes during the teen years include making new friends and forming new types of friendships.

You have learned that a friendship is a relationship between two people that is based on trust, caring, and consideration. People often choose friends with similar interests and values. Good friends can benefit your health in many ways. They can have a positive influence and help you resist harmful behaviors.

The teen years are a time of growth. This time, or adolescence, brings physical, mental/emotional, and social changes. During puberty, your body starts to develop the physical traits of an adult. Your brain also changes, and you start to see the world in more complex ways.

You may begin to try to understand yourself and how you fit into society. Teens will also discover new interests, including making new friends. You may also start to develop different types of friendships.

Thinking *about* Dating

Relationships become more important during the teen years than they did when you were younger. As you grow and mature, your friendships may change. You still share good times and have fun with your friends and acquaintances, but you also begin seeking deeper qualities in the people you choose as friends. These qualities may include loyalty and trust. Your group of friends may also begin to include more members of the opposite gender.

How have your *friendships* **changed** as you have **grown?**

During the teen years, you may also develop feelings of attraction. For example, it may be that the girl across the street was just another neighbor. Now you might pay more attention to her when she is around.

Brand X Pictures/PunchStock

When you were younger, maybe your brother's best friend used to annoy you. Now you might find yourself worrying about how you look when he visits.

These kinds of new feelings cause some teens to begin to think about dating. Dating is a way to get to know people better. There is no specific time when you are supposed to start dating. Some people feel ready to date while in their teens. Others do not feel ready to start dating until much later.

Group Dating

Choosing to spend time with a group of your peers is one way to date and get to know others. For example, a mixed group of teens may get together to watch a movie or play a game. Other fun group activities might include hiking, dancing, skating, or playing sports. These kinds of activities also have the benefit of keeping you physically active. Some other advantages to going out as part of a group include:

- **Conversation** It is generally easier to keep conversation going when several teens are out together. Everyone can talk less and listen more.
- **Less Pressure** You will likely feel less pressure to engage in sexual activity and other risk behaviors when you go out as part of a group.
- **Less Expense** A group date can be less expensive for each individual because everyone in the group can help share the cost of an activity.

Group dating offers chances to grow and mature. You can learn how to communicate with different types of people. Going out with a group can help you learn more about activities you enjoy. It may help you discover new interests as you learn about other people. Group dating may even lead you to meet someone you would like to get know to better as an individual.

Spending time in a group activity is one alternative to individual dating. *Tell why you think these teens enjoy one another's company.*

Individual Dating

As you continue to grow and become more mature, you may want to go out on a date with just one person. Individual dating is a big step that should be taken for the right reasons. Peer pressure is not a good reason to begin individual dating. You should wait until you are ready to date. Your parents may also let you know when they feel you are ready for individual dating.

How will you know whether you would like to date someone? You may find that you enjoy being around one person in particular. You may also find that you share common interests and values. At some point, you may agree that you would like to spend time together by going on a date. You can think of dating as a special form of friendship. As is the case with any other friendship, a healthy dating relationship is based on caring, honesty, and respect.

Dating is a way to **get to know people** better.

Individual dating is not always stress-free. You may feel nervous going out for the first time. You may worry about what your date will think of you. These thoughts and feelings may be new to you, but many people think and feel the same way before a date. Dating as a teen involves developing a different type of friendship than those you had when you were younger. A date does not have to mean the start of a lifelong commitment, which is *a pledge or a promise.* Going on an individual date may turn out to be just the first step in making a special new friend.

>> **Reading Check**

EXPLAIN *Why is spending time in a group a good alternative to individual dating?*

Andersen Ross/Brand X Pictures/Getty Images

Healthful Ways *to* Show Affection

Another change that happens during the teen years is the development of feelings of affection, or *feelings of love for another person.* Most people want to find someone special to care about deeply. It is one of the great gifts and joys of life. Showing affection can take many different forms. One form of showing affection is sexual intimacy. However, it is more healthful to postpone sexual activity until adulthood and marriage.

Showing affection in a healthy way lets someone know that you care. *Explain why you think it is important to show someone that you care.*

Show that you are a **good friend** by *listening* and **being sympathetic.**

However, teens can show affection in healthful ways. Holding hands and hugging are physical ways to show affection, but you can also do something thoughtful for another person. You might give a friend a card or a small gift as a way to show affection. For example, Bethany has a big soccer game coming up. Her friend Justin likes to create healthful meals and snacks, so he makes Bethany a tray of fruits and vegetables to share with her teammates. Bethany knows Justin has an important test later this week, so she sends him an encouraging note.

You can show that you are a good friend by listening and being sympathetic to the other person's thoughts and ideas. These kinds of actions deepen the bonds of affection. They also display good character, which is a sign that you are maturing. ■

>>> **After You Read**

1. **DEFINE** What does the term *relationships* mean?
2. **IDENTIFY** Name three advantages of dating in a group.
3. **LIST** Name two healthful ways for teens to show affection.

>>> **Thinking Critically**

4. **APPLY** If you were thinking of dating someone, what characteristics might you look for in the other person?
5. **EXPLAIN** Discuss some reasons that dating may not be right for everyone.

>>> **Applying Health Skills**

6. **EVALUATE** Adam and Emily have been dating for a while. Adam is expressing that he thinks they should engage in more physical affection. What would you advise Emily to tell him?

ⓒ Review

⑩ Audio

Healthy Dating Relationships

BIG IDEA Engaging in unhealthful behaviors, such as sexual activity, carry consequences that can seriously affect your future.

>>> **Before You Read**

QUICK WRITE Identify some risks involved in sexual activity. Write a couple of sentences explaining them.

▶ Video

>>> **As You Read**

STUDY ORGANIZER Make the study organizer found in the FL pages in the back of the book to record the information presented in Lesson 2.

>>> **Vocabulary**

› sympathetic
› consequences
› limits
› dating violence
› abstinence

🔊 Audio

🔤 Bilingual Glossary

Myth vs. Fact

Myth: People with good character and values never get STDs.

Fact: Anyone can get an STD. If you do not know your partner's history of sexual activity, you cannot know whether that person has an STD.

HEALTHY DATING RELATIONSHIPS

MAIN IDEA Healthy dating relationships involve healthful boundaries and healthful ways of showing affection.

Remember that a dating relationship is a special kind of friendship. Qualities of good friendships include reliability and loyalty. Good friends support you and keep their promises. Good friends are also trustworthy and **sympathetic,** or *aware of how you may be feeling at a given moment.* They allow you to share your thoughts and emotions. When you decide to date individually, it is important that you and your dating partner establish healthful boundaries. The qualities of healthful dating relationships have much in common with those of good friendships.

Thinking *about* Your Future

As you grow and mature, you will become more independent. Adults will not always be present to set limits for you and make sure that you stay within them. Setting your own limits will become very important. What you do can have serious **consequences,** or *the results of actions.*

The places you go and the people you spend time with can affect your safety, health, and plans for the future. You will need to evaluate situations and avoid negative influences. Remember to use refusal skills when you are pressured to do something you are not ready to do.

A *dating relationship* is a special kind of **friendship.**

One way to establish limits in your own life and avoid these consequences is to write down the goals you want to achieve. Once you understand what you want to achieve, you will have a better idea of what limits, or boundaries, you need to set to help you reach your goals.

>>> **Reading Check**

EXPLAIN *What is one way to help establish limits in your own life? How would setting limits help you?*

Setting Limits

Imagine playing a game or sport that had no rules. The activity would seem confusing, and it could also be dangerous. Rules bring order and purpose to games, and they serve a similar purpose in daily life. Rules can take the form of limits, or *invisible boundaries that protect you.* Among the many common examples of limits are the laws that society sets and lives by.

You probably have limits that your parents or guardians set at home. Those limits might include what TV shows you can watch, what websites you can visit, and how late you can stay up at night. Like laws, these limits are intended to keep you safe and to protect you. Rules for using the Internet, for example, may keep you away from dangerous or inappropriate sites.

Good communication skills are one characteristic of healthy dating relationships. *Describe how these teens are demonstrating good communication skills.*

In addition, having a set bedtime helps to ensure that you will get the sleep you need as a growing teen. These are all examples of ways your parents may try to promote good health and prevent illness. However, while your parents set certain limits, you must also learn to identify healthful limits for yourself.

Respecting Yourself *and* Your Date

Healthy friendships tend to bring out the best in each person. Healthy dating relationships should do the same thing. Qualities of healthy friendships and healthy dating relationships include mutual respect, caring, honesty, and commitment.

You can also use communication, cooperation, and compromise to help build a healthy dating relationship. For example, you may want to go for a hike, while your date may want to play disc golf. A compromise might be a trip to a park that has a disc golf course. Sometimes compromise is not the best choice to resolve a disagreement.

You should not compromise if the result would be harmful or unlawful. You should also avoid compromising on things that really matter to you, such as your values and beliefs.

Healthy dating *relationships* should **bring out the best** in each *person*.

Practicing abstinence also reflects the respect you have for yourself and for others. Remember, if you want to show someone that you care for that person, you can do so in ways that do not involve sexual activity. You can practice being kind and considerate. You can also offer support by talking and listening to the other person.

>>> **Reading Check**

EXPLAIN *What is one way to help establish limits in your own life? How would setting limits help you?*

Respect Both you and your date deserve to be treated with consideration and respect.

Communication When you are with someone you are dating, you should be yourself and communicate your thoughts and feelings honestly.

No Pressure You should never feel pressured to do anything that goes against your values or your family's guidelines.

©McGraw-Hill Education/Ken Karp

Dating Violence

Healthy dating relationships are built on respect. Violence of any type is a sign of an unhealthy relationship and also shows a lack of respect. For example, *when a person in a dating relationship tries to control his or her partner,* he or she is committing dating violence. Dating violence can include physical, emotional, or psychological abuse.

Teens who are just starting to date are not always sure how to have a healthy dating relationship. If a dating relationship feels uncomfortable or becomes violent, it is unhealthy. If your dating partner becomes violent or abusive, find help to get away from the person. Ask a parent or other trusted adult for help if you are concerned about breaking off the dating relationship. Another form of dating violence is date rape, *one person in a dating relationship forces the other person to take part in sexual activity.* A dating partner might use force, teasing, or intimidation to convince another person to become sexually active. In other cases, a dating partner might add date rape drugs to food or a soft drink while the victim is not looking or has left the area for a short time. After eating the food or drinking the drink, the victim may become unconscious.

Alcohol might also be used as a date rape drug. Using alcohol makes it harder to think clearly, to make good choices, to say *no,* or to resist an assault. Dating violence and date rape are crimes. To protect yourself, follow some simple rules:

- Stay with your drink at all times. If your drink has been left unattended, throw the drink out and get a new one.
- Avoid drinking from open containers or punch bowls. Drugs may have been added to the bowl.
- If your drink smells or tastes odd, throw it out.
- If you feel drugged or drunk after drinking a soft drink, get help immediately.

Sometimes, a victim who was drugged may not be sure that a crime has been committed. If you suspect that you were drugged and raped, tell someone. To learn more about how to prevent date rape, go to www.womenshealth.gov and search for "date rape" or "date rape facts."

When a Dating Relationship Ends

Most dating relationships formed during the teen years do not last. One or both dating partners may simply change or outgrow the relationship. Whatever the reason for a breakup, the loss of a special relationship can be difficult to cope with. Breaking up can result in stress and depression. When a dating relationship ends, it is natural and normal to feel lonely and hurt. However, these feelings fade with time.

If it is the other person who decides to break up, it can be even more painful. However, the healthiest thing you can do is to respect the other person's wishes. Accept that person's decision and find a way to move on. It may not be healthful to start another dating relationship right away. Eventually, though, you will find another person with similar interests, values, and goals who you would like to get to know better.

>>> **Reading Check**

IDENTIFY *What are two examples of limits for teens?*

Vicky Kasala/Getty Images

CONSEQUENCES OF EARLY SEXUAL ACTIVITY

MAIN IDEA Choosing abstinence is a way to avoid the physical consequences of sexual activity.

As teens begin individual dating, they may face new pressures. Among these is the pressure to engage in sexual activity. The Internet, movies, TV, popular music, and magazines may suggest that sexual activity among young people is common. The media and other influences may make it look like normal behavior.

The choice to be *sexually abstinent* promotes good health.

In truth, most teens avoid sexual activity. The Centers for Disease Control and Prevention (CDC) conduct a Youth Risk Behavior Surveillance Survey (YRBSS) every two years. In 2011, the survey showed that just six percent of teens under the age of 13 have engaged in sexual activity.

You may read and hear about sexual activity among teens in magazines, music, and online. Maybe you feel pressure from your friends. However, you can choose not to engage in sexual activity. You can practice sexual **abstinence,** or *the conscious, active choice not to participate in high-risk behaviors.*

The choice to be sexually abstinent as a teen promotes good health. It helps teens avoid the risks that accompany sexual activity. Sexual abstinence shows that you are focusing on your current goals and your plans for the future. It also shows respect for the physical and emotional well-being of others. It is important to understand that being sexually active can have serious consequences. Sexual activity can affect all three sides of your health triangle—physical, mental/emotional, and social.

The choices you make have consequences that will affect your future. *Identify some healthful limits you can set for yourself to help you achieve your goals.*

Physical Consequences

Teens who become sexually active expose themselves to a number of risks. Becoming sexually active brings the risk of being infected with a sexually transmitted disease (STD). STDs can damage the reproductive system and prevent a person from ever having children. Some STDs remain in the body for life—even after they are diagnosed and treated. Other sexually transmitted diseases, especially HIV/AIDS, can result in death. Any type of sexual activity can result in an STD.

Another risk of early sexual activity is unplanned pregnancy. Most teens do not have the emotional maturity to become parents. Teens usually do not have the financial resources to take care of a baby. The teen years are a time for thinking about what you want to do with the rest of your life. Teens who become parents usually must put their own education and career plans on hold. When people wait until adulthood to become parents, they are better able to achieve their long-term goals.

>>> **Reading Check**

IDENTIFY *What are some specific physical consequences of early sexual activity?*

Chase Jarvis/Getty Images

Mental/Emotional Consequences

Teens who become sexually active may also experience other consequences. Some of these consequences may affect a teen's mental/emotional health. The consequences can include:

- **Emotional distress** because one or both partners are not committed to each other.

When a **dating relationship** ends, it is *natural* and *normal* to feel **lonely and hurt.**

- **Loss of self-respect** because sexual activity may go against their personal values and those of their families.
- **Guilt** over concealing their sexual activity from their parents and others.
- **Regret and anxiety** if sexual activity results in an unplanned pregnancy, an STD, or the breakup of the relationship with the partner.

Teen parenthood can be difficult both emotionally and financially. *List some other challenges that this teen might face.*

PhotoDisc/Getty Images

Social Consequences

Sexual activity can also affect a teen's social health. Becoming sexually active can limit a teen's interest in forming new friendships. The teen years are a time to meet new people and explore new interests. A teen who is involved in an exclusive relationship with one other person may not be open to meeting new people. In addition, an unplanned pregnancy can limit a teen's plans for the future. For teens who become parents, caring for a baby and raising a child must become their first priority.

>>> **Reading Check**

EXPLAIN *Why might teens experience regret and anxiety from a sexual relationship?* ■

LESSON 2

REVIEW

>>> **After You Read**

1. **DEFINE** What does *consequences* mean? Use the term in an original sentence.
2. **LIST** Name the types of consequences that can result from engaging in sexual activity.
3. **IDENTIFY** What are three possible mental/emotional consequences of teens engaging in sexual activity?

>>> **Thinking Critically**

4. **ANALYZE** What are some qualities that good friendships and healthy dating relationships have in common?
5. **EXPLAIN** What are three benefits of abstaining from sexual activity?

>>> **Applying Health Skills**

6. **EVALUATE** How can sexually transmitted diseases (STDs) affect a teen both now and in the future?

🄲 Review

🔊 Audio

Abstinence *and* Saying No

BIG IDEA Practicing refusal skills will help you deal with peer pressure.

Before You Read

QUICK WRITE List three ways of saying no when someone pressures you to do something dangerous or unhealthy.

▶ Video

As You Read

STUDY ORGANIZER Make the study organizer found in the FL pages in the back of the book to record the information presented in Lesson 3.

Vocabulary

› risk behaviors
› refusal skills

🔊 Audio

🅰🅱🅲 Bilingual Glossary

🏃 Fitness Zone

Physical Fitness Plan One way a friend can give positive peer pressure is by promoting healthful activities such as exercise. You might set up a regular time to play sports or go for a run or hike with your friends. Sharing a fitness routine can help build stronger relationships and improve everyone's physical and emotional well-being.

CHOOSING ABSTINENCE

MAIN IDEA Choosing abstinence involves communication with your dating partner, self-control, avoiding risky situations, and using refusal skills.

One of the most important limits you can set for yourself is choosing abstinence from **risk behaviors,** or *actions that might cause injury or harm to you or others.* This includes avoiding sexual activity. When you begin dating, choosing abstinence is the healthful choice. Many teens are making this choice. The CDC's Youth Risk Behavior Survey shows that percentage of teens choosing abstinence is steadily increasing.

> One of the most **important limits** you can set is *choosing abstinence.*

Committing *to* Abstinence

Abstinence is a choice you will have to recommit to each time you face pressure to engage in sexual activity. Even if you have been sexually active in the past, you can still choose abstinence.

It is important to talk about your decision with the person you date. These tips may help the conversation go more smoothly:

- Choose a relaxed and comfortable time and place.
- Begin on a positive note, perhaps by talking about your affection for the other person.
- Be clear about your reasons for choosing abstinence.
- Be firm in setting limits in your physical relationship.

To stay firm in your decision, continue to remind yourself of your limits and the reasons you are choosing abstinence in the first place.

Dealing *with* Sexual Feelings

Practicing abstinence requires planning and self-control. Sexual feelings are normal and healthy. You cannot prevent them, but you do have control over how you deal with them. Teens can learn ways to manage these feelings. The tips on the next page can help you to maintain self-control and practice abstinence.

©Radius Images/Corbis

- **Set limits on expressing affection.** Think about your priorities and set limits for your behavior before you are in a situation where sexual feelings may develop.
- **Communicate with your partner.** Make sure your dating partner understands and respects your limits.
- **Talk with a trusted adult.** Consider asking a parent or other trusted adult for suggestions on ways to manage your feelings and emotions.
- **Avoid risky situations.** Choose safe, low-pressure activities, such as a group date.
- **Date someone who respects and shares your values.** A dating partner who respects you and has similar values will understand your commitment to abstinence.

>>> **Reading Check**

ANALYZE *Why is it important to discuss your commitment to abstinence with your dating partner?*

Avoiding Risky Situations

Where you go and what you do can have a big impact on your health and safety. Here are some basic precautions:

- **Before you go on a date, know where you are going and what you will be doing.** Find out who else will be there, and discuss with your parents or guardians what time they expect you home.
- **Avoid places where alcohol and other drugs are present.** These substances can impair a person's judgment. People under the influence of alcohol or other drugs are more likely to take part in risk behaviors, including sexual activity.
- **Avoid being alone on a date.** You may find it more difficult to maintain self-control when you and your date are alone together. One-on-one situations also increase the risk of being forced or pressured into sexual activity. Group dating can be a healthful alternative.

Communication and respect are signs of a good relationship. *Explain how setting limits with your dating partner can help strengthen your relationship.*

SW Productions/Getty Images

Saying No to Risk Behaviors

It's Saturday night, and Zoe is at a party at a friend's house. It seems like fun at first, but Zoe soon realizes that her friend's parents are not home and that some of the people at the party are older. Zoe notices a lot of people smoking and drinking alcohol. Some couples are finding places in the house to be alone. Zoe decides that this party is not for her and calls her parents to pick her up. While she's waiting, someone offers her a cigarette and a beer and asks her to go upstairs.

With A Group

Write a paragraph explaining how Zoe could use the S.T.O.P strategy to refuse the cigarette and beer and the offer to go off alone with someone else. The strategy below is detailed on the next page:

- Say no in a firm voice.
- Tell why not.
- Offer another idea.
- Promptly leave.

USING REFUSAL SKILLS

MAIN IDEA You can use and practice refusal skills to help you keep your commitment to abstinence.

Think about some of the reasons you decided to commit to abstinence. You will need to remember them if you are ever pressured. Abstinence is not a decision you can make once and never think about again. You can use refusal skills, or *strategies that help you say no effectively,* to help maintain your decision.

Peer pressure can be **negative** or **positive.**

Taking basic precautions to avoid high-risk situations can help you enjoy dating in your teens. *Explain why it is important to know where you are going and what you will be doing on a date.*

Refusal skills take practice. The S.T.O.P. strategy can be an effective way to say no to risk behaviors. The letters in S.T.O.P. represent the four steps of the strategy. You can use one or all of the following steps:

- **Say no in a firm voice.** State your feelings firmly but politely. Say, "No, I don't want to." Make your no sound like you really mean it. Use body language to support your words. Make eye contact with the person.
- **Tell why not.** If the other person keeps up the pressure, explain why you feel the way you do. You do not need to use phony excuses or make up reasons. Just say, "No, thanks, I care about my health." Do not apologize. You have done nothing wrong.

- **Offer other ideas.** If the person pressuring you to do something is a friend, you might suggest alternatives. Instead of something that has risks, suggest an activity that is safe and fun instead.
- **Promptly leave.** Sometimes, firmly saying no, explaining why not, and suggesting alternatives may not work. If all else fails, just walk away and leave the situation. Let your actions match your words. If you need a ride home, phone a parent or other trusted adult to come and pick you up.

⟩⟩ Reading Check

IDENTIFY *What is a refusal strategy you might use when being pressured to do something you do not want to do?*

Know where you are going, what you will be doing, and when you will be home.

Avoid places where alcohol and other drugs are present.

Avoid being alone on a date at home or in an isolated place.

Purestock/Getty Images

DEALING WITH PRESSURE

MAIN IDEA Knowing how to respond to people who want you to do something you do not want to do will help you resist that pressure.

Most of your friends are probably your peers—the people close to you in age who have a lot in common with you. Sometimes teens worry about what their friends think. Your friends' opinions can affect how you act. The influence your peer group has on you is called peer pressure. Peer pressure can be negative or positive.

Negative Peer Pressure

Friends should not pressure you to do something that is unsafe or unhealthful. They should not pressure you to do something that goes against your values or your family's values. True friends will respect your choices.

By using refusal skills, you stand up for your values and build self-respect. You also show others you have strength and character. *Identify the S.T.O.P. strategy step this teen appears to be taking.*

Encouraging a person to act in a way that is harmful or illegal is one form of negative peer pressure. It can also come in the form of dares, threats, teasing, name-calling, or bullying. Use refusal skills and have responses ready for those who try to get you to do something negative.

Positive Peer Pressure

Positive peer pressure is when true friends suggest that you do the right thing. They may encourage you to study more, work on a project together, or welcome new people into the group. Friends can also help you say no to risk behaviors. Positive peer pressure can be good for you. It can improve your health and safety and also help you feel better about yourself.

>>> **Reading Check**

DEFINE *What is* peer pressure? ∎

>>> **After You Read**

1. **DEFINE** Explain what the term *risk behavior* means. Use it in an original sentence.
2. **LIST** List four ways to help keep a commitment to abstinence.
3. **IDENTIFY** Give one example of positive peer pressure and one of negative peer pressure.

>>> **Thinking Critically**

4. **EVALUATE** Jay has told Ron twice that he does not want to sneak into the school basketball game. Jay has even offered to pay for Ron's ticket, but Ron insists on sneaking in. What should Jay do next?

>>> **Applying Health Skills**

5. **APPLY** Write a skit about a situation in which a teen uses each step of the S.T.O.P. strategy to refuse to participate in a risk behavior.

⟳ Review

◀) Audio

A Taste *of* Parenthood

WHAT YOU WILL NEED

* Help of a parent, guardian, or sibling
* Clock or watch with an alarm feature
* Paper and pencil or pen

Many teens do not understand how demanding parenthood can be. This brief experiment will give you a glimpse of the responsibilities.

WHAT YOU WILL DO

1 Select a time when you have a light workload. Arrange with a family member to help you.

2 Go about your normal activities, such as watching TV or speaking to a friend on the phone.

3 Without warning, your partner is to set off the alarm. Stop what you are doing and turn off the alarm. The interruption represents the attention a baby requires.

4 Note the time you heard the alarm and what you were doing when it went off.

5 Return to your normal activities. Your partner is to set the alarm off a minimum of five times.

WRAPPING IT UP

Write about your experiment. What would it be like to respond to such an alarm every day? How might it be similar to caring for a child? Share your report with classmates and compare your experiences.

Ingram Publishing/SuperStock

READING REVIEW

FOLDABLES and Other Study Aids

Take out the Foldable® that you created and any study organizers that you created. Find a partner and quiz each other using these study aids.

LESSON 1 Beginning to Date

BIG IDEA Your friendships become more important during adolescence.

* Changes during the teen years include making new friends and forming new types of friendships.
* Dating is a way to get to know other people better.
* Some people start dating while in their teens, but others do not feel ready to date until much later.

LESSON 2 Healthy Dating Relationships

BIG IDEA Healthy dating relationships involve healthful boundaries and healthful ways of showing affection.

* Qualities of healthy relationships include mutual respect, caring, honesty, and commitment.
* Growing up involves establishing limits in order to avoid risks and consequences so that you can achieve your goals in life.
* Choosing abstinence is a way to avoid the physical consequences of sexual activity.

LESSON 3 Abstinence and Saying No

BIG IDEA Practicing refusal skills will help you deal with peer pressure.

* Choosing abstinence involves communication with your dating partner, self-control, avoiding risky situations, and using refusal skills.
* Abstinence is a choice you will have to recommit to each time you face pressure to engage in sexual activity..
* It is normal and healthy to have sexual feelings, but you can control them instead of letting sexual feelings control you.
* You can use and practice refusal skills to help you keep your commitment to abstinence.
* Knowing how to respond to people who want you to do something you do not want to do will help you resist that pressure.
* The S.T.O.P strategy (Say no in a firm voice; Tell why not; Offer other ideas; Promptly leave) can be an effective way to say no to risk behaviors.

 Review

 Web Quest

ASSESSMENT

Reviewing **Vocabulary** *and* **Main Ideas**

- › abstinence
- › sympathetic
- › affection
- › commitment
- › consequences
- › limits

» On a sheet of paper, write the numbers 1–6. After each number, write the term from the list that best completes each statement.

LESSON 1 **Beginning to Date**

1. A _____ is a pledge or a promise to someone.

2. Feelings of love for another person are also known as _____.

LESSON 2 **Healthy Dating Relationships**

3. Good friends are _____, meaning they understand how you may be feeling at a given moment.

4. The invisible boundaries that protect you, or _____, include the laws that society uses.

5. The places you go and the people you spend time with can have _____ that affect your personal safety, health, and plans for your future.

6. _____ is the conscious, active choice not to participate in high-risk behaviors.

» On a sheet of paper, write the numbers 7–12. Write True or False for each statement below. If the statement is false, change the underlined word or phrase to make it true.

LESSON 3 **Abstinence and Saying No**

7. <u>Consequences</u> are actions that might cause injury or harm to you or others.

8. Teens should practice <u>abstinence</u> when it comes to risk behaviors.

9. Encouraging a person to act in a way that is harmful or illegal is one form of <u>sympathy.</u>

10. Part of developing good refusal skills is learning how to apply the <u>S.T.O.P. formula.</u>

11. A teen who practices abstinence from risk behaviors <u>will</u> experience negative legal consequences.

12. Abstinence from <u>risk behaviors</u> includes avoiding tobacco, alcohol, drug use, and sexual activity.

✔ eAssessment

>> Using complete sentences, answer the following questions on a sheet of paper.

☁ *Thinking* **Critically**

13. ANALYZE Della was invited to a get-together at the home of a friend's friend. When she arrived, Della found out that the girl's parents were not home and saw others drinking and smoking. What would be an assertive way for Della to let her friend know she is not interested in staying at the party?

14. INFER What effect does abstaining from risk behaviors have on your relationships?

✏ *Write* **About It**

15. NARRATIVE WRITING Write a skit that shows a situation in which a teen uses each step of the S.T.O.P. formula to refuse to participate in a risk behavior.

16. DESCRIPTIVE WRITING Write a blog post describing ways in which choosing abstinence from risk behaviors can benefit all sides of your health triangle.

Ⓐ Ⓑ Ⓒ Ⓓ STANDARDIZED TEST PRACTICE

Writing
Read the prompts below. On a separate sheet of paper, write an essay that addresses each prompt. Use information from the chapter to support your writing.

1. Imagine that a friend is considering engaging in a sexual activity. Write a dialogue in which you describe setting personal limits and the benefits of remaining abstinent before marriage.

2. Write an essay discussing why it is important to use assertive refusal skills rather than passive responses when dealing with peer pressure.

Unit 2

building character + preventing bullying

Building Character

LESSONS

 *PREMIUM* ONLINE RESOURCES

 Audio

 Videos

 Bilingual Glossary

 Fitness Zone

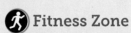 Web Quest

 Review

Working together to help others is one way to build character. *Explain how you think volunteering can help build character.*

What *is* Character?

BIG IDEA Character is the way a person thinks, feels, and acts.

Before You Read

QUICK WRITE Write a short description of a person who has been a positive role model for you.

 Video

As You Read

FOLDABLES | Study Organizer

Make the Foldable® found in the FL pages in the back of the book to record the information presented in Lesson 1.

Vocabulary

› character
› integrity
› tolerance
› prejudice
› accountability
› advocacy
› role model

🔊 Audio

🆎 Bilingual Glossary

UNDERSTANDING CHARACTER

MAIN IDEA A person's character demonstrates his or her values and beliefs.

Do you tell the truth? Do you respect others? These are some signs of good character, or *the way a person thinks, feels, and acts.* Good character is an important part of a healthy identity.

Six Traits *of* Good Character

You are a member of many groups, like a family, a sports team, or friends. In order for people to get along, they need to have good character traits, including: trustworthiness, respect, responsibility, fairness, caring, and citizenship.

> *Good* character is an **important part** of a healthy **identity**.

TRUSTWORTHINESS If you are honest, loyal, and reliable, you earn people's trust. This is trustworthiness. You show the courage to do the right thing.

For example, a friend may ask you to do something that goes against your values, such as using tobacco. If you say no to remain true to your values, you show integrity. Integrity is *being true to your ethical values* and doing what you know is right.

RESPECT You show respect by being considerate of others and accepting their differences. It does not mean that you must agree with everything that another person says or does. With respect you show tolerance (TAHL·er·ence), *the ability to accept other people as they are.* You avoid *negative and unjustly formed opinions,* or prejudice (PREH·juh·dis), that includes fears formed without having facts or firsthand knowledge.

RESPONSIBILITY You hold yourself accountable for your choices, decisions, and actions. Accountability is *a willingness to answer for your actions and decisions.* You think about consequences before you act.

Design Pics/SW Productions

FAIRNESS You play by the rules, take turns, and share. You listen to others. You do not take advantage of other people or blame them when things don't go the way you expect.

CARING You show that you are kind and considerate of others.

Caring includes showing gratitude and helping others.

CITIZENSHIP You show respect for authority and interest in the world around you, including the health and safety at school and in your community. You can show this respect through **advocacy,** *taking action in support of a cause,* and taking a stand to make a difference.

>>> **Reading Check**

IDENTIFY *What are the six traits of good character?*

WHAT SHAPES YOUR CHARACTER?

MAIN IDEA Your life experiences and role models shape your character.

Many influences shape your character. From an early age, you learn values from your family members. As you mature, you make your own choices about what kind of character you will have.

Do you value honesty in others? You may choose to always be honest. Do you see fairness around you? You may choose to be fair. You may learn that you appreciate kindness in others, and choose to be kind. You choose and practice to have good character.

What influences character? *Many factors can influence a person's character. List other facts that are not shown here.*

Life Experiences

Character is shaped by your family and the people around you. If you learned to share and respect your siblings, you learned fairness and respect. As you mature, your teachers, other adults, and friends can shape your character.

Do you *value* honesty in others?

When you find someone you trust, you learn about being trustworthy. At school, you learn to be responsible for your schoolwork, to follow the rules, and to care for others. As you learn about your community, you learn about citizenship.

Role Models

One way to learn about character is to watch and listen to others. You learn to model your actions by examples. A **role model** is *a person who inspires you to think or act a certain way* and whose behavior serves as a good example. A teacher may be a model of responsibility and fairness. A coach might model trustworthiness and respect. Family members may model caring.

>>> **Reading Check**

EXPLAIN *Who are the first teachers of character?*

Parents or Guardians	Stories	Life Experiences	Examples Set by Others
The earliest influence on your character was likely a parent or guardian. Parents and guardians are our first teachers of character.	Have you heard the expression, that's the moral of the story? Many stories contain a moral. The moral teaches about values or character traits.	You learn from things you do in your life. You may have done something that turned out to be a mistake. Did you learn from that mistake?	Role models inspire us to act or think in a certain way. They set good examples. Who do you look up to for inspiration?

DEVELOPING GOOD CHARACTER

MAIN IDEA Good character means making good choices.

Mia and her older sister, Megan, are in the grocery store shopping for groceries. As they walk down an aisle, Megan looks down on the ground and notices a $20 dollar bill. Megan quickly takes the money to a store manager. The manager thanks Megan and Mia for turning in the money. Mia is proud of her sister for demonstrating responsibility and trustworthiness.

As you watch, listen to, and learn from others, you choose the traits you want to demonstrate your character. You develop character that becomes a way of life and a part of who you are. For example, if you do your family chores, you demonstrate responsibility. If you listen to different points of view, you demonstrate respect. When you help a person in need, you demonstrate caring and kindness.

Good character is something anyone can **choose**.

Demonstrating good character is a choice. If you choose to be honest, you will tell the truth and you will not cheat. If you obey the rules and respect authority, you are choosing citizenship. You practice making choices to demonstrate good character. Good character is something anyone can choose. The more you practice the traits you value, the more they become a part of your character.

>>> **Reading Check**

ANALYZE *What are two ways to develop good character?* ■

>>> **After You Read**

1. **VOCABULARY** Define the term *character*. Use it in a sentence.
2. **EXPLAIN** Tell how good character contributes to physical, mental/emotional, and social health.
3. **IDENTIFY** Name two traits that show good character.

>>> **Thinking Critically**

4. **HYPOTHESIZE** Think of an act of citizenship you know about. It can be an act of someone you know, or someone you have read about or seen on TV. Tell how the act demonstrates citizenship. Does the example include any other traits of good character? If so, tell how.

>>> **Applying Health Skills**

5. **ANALYZING INFLUENCES** Name two people you know or know about who could be good role models. Explain how each person demonstrates good character and what you might learn from each person.

⟳ Review

🔊 Audio

Working together can build good character. *Explain what traits of good character can show when working together.*

McGraw-Hill Education/Digital Light Source, Richard Hutchings

Trustworthiness *and* Respect

BIG IDEA Good character is built on trustworthiness and respect.

>>> **Before You Read**

QUICK WRITE Write your own definition of trust. Can you name people in your life who are trustworthy?

▶ Video

>>> **As You Read**

STUDY ORGANIZER Make the study organizer found in the FL pages in the back of the book to record the information presented in Lesson 2.

>>> **Vocabulary**

› loyal

🔊 Audio

Ⓐ Bilingual Glossary

JUPITERIMAGES/Brand X/Alamy

TRUSTWORTHINESS

MAIN IDEA Trustworthiness is a trait of good character.

Gabriel and Tuan are taking a quiz. Gabriel studied for the quiz and feels prepared. Tuan tries to look at Gabriel's quiz, but Gabriel covers his answers. Gabriel shows trustworthiness and integrity by not allowing Tuan to look at his answers.

If you are honest, other people will trust you. You are trustworthy. If you work on a project with other students, and every day you show up on time with your part of the work complete, your team members will trust you. When you borrow a dollar for lunch and return it the next day, you show trustworthiness. If you tell the truth, do not cheat or steal, you are trustworthy.

Trustworthy people are reliable. They can be relied on to do what is expected and what they say they will do. They show up on time, always tell the truth, and demonstrate that they can be trusted.

If you always bring in your completed homework, your teacher will rely on you to come to class prepared. You demonstrate reliability.

Trustworthy people have integrity. Integrity is the quality of doing what you know is right. Imagine you see a student leave her wallet on the lunch table at school. If you return it immediately, you show integrity. You are doing the right thing.

Integrity is the *quality* of **doing** what you know is *right*.

Being loyal, or *faithful*, makes you trustworthy. Good friends are loyal, or faithful. A loyal friend will not say, or allow others to say, untrue or unkind things about you.

>>> **Reading Check**

IDENTIFY *Name two characteristics of trustworthiness.*

Developing Good Character

Respect When you are faced with a difficult choice, you want to earn the respect of others, but more importantly, you want to respect yourself. Making healthy decisions shows that you respect yourself and your health. *What are some other ways of showing respect for yourself?*

RESPECT

MAIN IDEA Good character is demonstrated through respect.

Dipali and Lorinda are good friends. Lorinda asks Dipali to come to her house for a party on Friday. When Dipali arrives, she sees that everyone, including Lorinda, is drinking alcohol. Dipali knows she shouldn't be drinking alcohol and decides she needs to leave. Dipali shows that she respects herself and doesn't want to do anything that might risk her health.

Respect means having consideration for the feelings of others. Think about how you want to be treated, and treat others the same way. You show respect with good manners. You show respect by listening to the points of view of others, even when they are different from yours. An important element of respect is tolerance, the ability to accept other people as they are. Our world is made up of people with different cultures and backgrounds.

Learning about people who have different backgrounds can enrich our lives.

Tolerance also prevents prejudice. Prejudice is an opinion or fear formed without facts or full knowledge.

Respect starts with your self. You show self-respect by leading a healthy life. You stay away from high-risk behaviors, such as sexual activity, or tobacco use, alcohol use, and other drug use. You show respect for your body by eating healthful foods, and getting plenty of physical activity and rest.

>>> Reading Check

RECALL *Name two ways to show respect for others.* ■

Demonstrations of trustworthiness and respect contribute to good character. *Which of these traits can describe you?*

Trustworthiness	Respect
☐ I always tell the truth.	☐ I am polite to others.
☐ I do what I say I will do.	☐ I listen to other people's opinions.
☐ I do what I know is the right thing.	☐ I accept people who are different from me.
☐ I am loyal to my friends.	☐ I take care of my health.
☐ I never cheat or steal.	☐ I am tolerant of differences in people.

>>> After You Read

1. **VOCABULARY** Define the term *integrity*. Use it in a sentence.
2. **IDENTIFY** Name three traits of trustworthiness.
3. **EXPLAIN** Describe how you can demonstrate respect for yourself.

>>> Thinking Critically

4. **APPLY** Think about a time you were faced with a decision whether to tell the truth or stay quiet. What did you do?

>>> Applying Health Skills

5. **SETTING GUIDELINES** Work with a small group to make a poster that outlines rules for respectful behavior in your school community. Use what you learned in this lesson. When your poster is complete, share it with your class.

Ⓒ Review

🔊 Audio

Responsibility *and* Fairness

BIG IDEA Acts of responsibility and fairness demonstrate good character.

>>> Before You Read

QUICK WRITE List responsibilities that you believe lie ahead. What responsibilities are you looking forward to?

▶ **Video**

>>> As You Read

STUDY ORGANIZER Make the study organizer found in the FL pages in the back of the book to record the information presented in Lesson 3.

>>> Vocabulary

› abstinence

🔊 **Audio**

🔤 **Bilingual Glossary**

RESPONSIBILITY

MAIN IDEA Your character shows in your responsibilities.

Yuri has a new puppy. He promises to help take care of the dog. Yuri's friend, Jacob, asks him to go to see a movie after school. Yuri wants to go, but it is his turn to walk the dog. Yuri calls Jacob and asks if they can go to a movie on another day. He accepts responsibility for taking care of his pet.

Showing **responsibility** means doing what you say you will do.

What are some of your responsibilities right now? Do you take responsibility for completing your schoolwork on time? At home, do you make your bed, help with the dishes or the trash? As you get older, you take on more responsibilities. Accepting responsibility includes accountability. When you are accountable for your actions, you do not blame others for your mistakes. You accept the consequences for your actions.

To accept responsibility means to be willing to take on duties and tasks. Showing responsibility means doing what you say you will do. You take credit for things done well and not done so well. When you have a responsibility, you follow through without being asked or reminded.

Taking Responsibility *for* Your Health

As you mature, you take responsibility for your health by making good decisions. You learn that making healthful food choices improves your physical and mental health. You learn to make time for physical activity and to avoid risk behaviors. You learn to practice **abstinence** (AB·stuh·nuhns), which is your *conscious, active choice not to participate in high-risk behaviors,* such as sexual activity, tobacco use, alcohol use, and drug use.

> ### >>> Reading Check
>
> **EXPLAIN** *What does it means to be responsible?*

Responsibility		Fairness	
____	Does chores and tasks without reminders.	____	Treats people equally.
____	Keeps promises.	____	Considers new ideas.
____	Thinks before acting.	____	Shares.
____	Makes good decisions about physical and mental health.	____	Takes turns.
____	Does school work on time.	____	Shows good sportsmanship, win or lose.

These traits describe responsibility and fairness.
List the ways in which you demonstrate responsibility and fairness.

FAIRNESS

MAIN IDEA A person who is fair treats everyone equally and honestly.

Abby and her friend Taylor are judging a school art contest. Abby wants to award the prize to Paul because his painting is very good. Taylor wants to award the prize to her friend Katy. "I know Katy is your friend, but Paul really deserves to win," says Abby. "It's only fair to give it to the person who deserves to win." Abby understands the importance of fairness.

Fairness means treating people **equally** and **honestly**.

Good character is also demonstrated with fairness. Fairness means treating people equally and honestly. A person with fairness is open-minded and patient.

A fair person does not take advantage of another person. Many things that you do, such as taking turns and sharing, demonstrate fairness. Fairness includes being a good sport, win or lose.

Many of your actions show responsibility and fairness. For example, you can show fairness while playing sports. Pretend you are playing football with some friends. While you are running with the ball, you accidentally step out of bounds near the goal line. A person on the other team sees you step out of bounds. He or she says that you did not score. What would you do? If you admit that you step out of bounds, you are displaying fairness.

>>> Reading Check

IDENTIFY *Name two traits of fairness.* ■

LESSON 3

REVIEW

>>> **After You Read**

1. **VOCABULARY** Define the term *responsibility*. Use it in an original sentence.
2. **EXPLAIN** Describe ways a teen can be responsible for his or her physical and mental health.
3. **IDENTIFY** Name two traits of fairness.

>>> **Thinking Critically**

4. **APPLY** Describe a time you lost a game or competition. What did you do? Describe whether your actions showed fairness.

>>> **Applying Health Skills**

5. **PRACTICING HEALTHFUL BEHAVIORS** Tell about a time you took responsibility for your physical or mental health by making a good decision.

 Review

Audio

Being *a* Good Citizen

BIG IDEA Good character includes caring and citizenship.

Before You Read

QUICK WRITE Write a short paragraph to explain how caring for others can benefit your health.

▶ Video

As You Read

STUDY ORGANIZER Make the study organizer found in the FL pages in the back of the book to record the information presented in Lesson 4.

Vocabulary

› empathy

 Audio

 Bilingual Glossary

CARING FOR OTHERS

MAIN IDEA Good character shows in acts of caring.

Caring means treating other people with kindness and understanding. You show you care when you help and support others. Caring people show gratitude to others who help them. They show forgiveness to those who have hurt them. We show that we care by listening, offering to help others, and looking at others' points of views.

Sometimes, just listening to a person talk about what is bothering him or her is an important act of caring. You may not be able to help the friend with his or her problem. Taking the time to listen shows that you care.

The Spirit *of* Giving

One quality found in caring people is the spirit of giving. This doesn't mean giving of "things," it means giving of yourself: your time, your attention, your help and support. When you show caring and understanding, you are giving your time and attention. You can also give your time and attention to friends and others in your community.

You can also share your skills and experience. If you like to cook, you can share food with classmates or a local shelter. If you are good at a sport, you can play with younger students to help them improve their skills. If you speak Spanish, you can give your time by tutoring other students at school.

Taking the **time** to **listen** shows that you *care*.

Showing Sympathy *and* Empathy

When you care about others, you are kind to them. You show consideration for their feelings. You can show **empathy,** *the ability to identify with and share another person's feelings.* If a friend is feeling sad or disappointed, you can show empathy by sharing your friend's feelings.

>>> **Reading Check**

DEFINE *What is empathy?*

CITIZENSHIP

MAIN IDEA Good citizens help make a community a better place.

Every individual is a citizen of a community. A community can be your neighborhood, your school, your city, or your country. The way you conduct yourself as a member of a community is citizenship. Citizenship is a part of good character. Teens who obey the rules and follow the laws are good citizens.

Citizenship means doing what you can to help your community. Good citizens work to protect the environment. They keep the environment clean, and recycle paper, glass, and aluminum.

They take a stand to prevent violence and bullying. Good citizens also volunteer in helping the community and environment. They advocate, or take action to support a cause and to help the community.

Caring and citizenship work together. You are a good citizen because you care about other people, the community, and the environment.

> ### Reading Check
>
> **IDENTIFY** *List two ways to demonstrate citizenship.* ■

These teens help to keep their community clean. *How do these teens demonstrate citizenship?*

REVIEW

After You Read

1. **VOCABULARY** Define the term *citizenship*. Use it in an original sentence.
2. **EXPLAIN** Explain what it means to advocate.
3. **IDENTIFY** Name two ways to show good citizenship.

Thinking Critically

4. **HYPOTHESIZE** Discuss some ways you can give to your community.

Applying Health Skills

5. **ACCESSING INFORMATION** Search the Internet for "teen volunteer organizations." Some sites will give you the opportunity to type in your city, your ZIP code, or a special interest. Find one site that interests you and share what you learned on the site with your classmates.

⟳ Review

◀) Audio

Caia Image/Image Source

Making a Difference

BIG IDEA Your words and actions show your character.

Before You Read

QUICK WRITE List three ways you show that you care.

 Video

As You Read

STUDY ORGANIZER Make the study organizer found in the FL pages in the back of the book to record the information presented in Lesson 5.

Vocabulary

> constructive criticism
> I-messages
> cliques

 Audio

 Bilingual Glossary

© McGraw-Hill Education/Ken Karp

Developing Good Character

Respect Older adults have a lot of wisdom and experience to share. You can show respect for older adults by listening and speaking in a polite manner. *What are some other ways you can show respect to older adults?*

CHARACTER IN ACTION

MAIN IDEA Good character helps you develop and maintain healthy relationships.

Your good character shows in your actions and words. Good character improves your own life, and affects everyone around you. Think about all of your relationships: with your family, your friends and classmates, and your community. When your actions and words show good character, you have many good and strong relationships.

Making a Difference at Home

Family relationships are built on caring, respect, fairness, trust, and responsibility. Your actions and words make a difference at home. You can show your character by treating your family members with care and respect. Show appreciation for the things your family does for you. Help out with chores without being asked. Take responsibility for your own chores or tasks. Be patient and kind to your siblings.

You can keep your **good character** even if you have a *problem*

You can keep your good character even if you have a problem at home. Discuss your problem with respect for others. You may want to offer constructive criticism. Constructive criticism is *using a positive message to make a suggestion.* For example, if your younger brother is having trouble tying his shoe, you might offer to demonstrate. You can also make suggestions using I-messages, which are *messages in which you offer a suggestion from your own point of view.* For example, "I can't understand you when you talk with your mouth full."

Reading Check

LIST *Name two ways to build good character at home.*

Making a Difference at School

Good character can make a difference at school. At school, you have relationships with friends, classmates, teachers, and other adults. Good character means showing positive values.

It also means setting good examples of fairness, trustworthiness, caring, and responsibility. You can build your character by accepting others and showing tolerance toward differences.

People tend to feel comfortable around others like themselves.

This sometimes leads to the formation of **cliques,** *groups of friends who hang out together and act in similar ways* and who have same interests and values. Being a part of a clique can provide a person with a sense of belonging. However, cliques can be negative if they exclude others or show prejudice toward those whose interests are different.

You show your character at school by taking responsibility for your schoolwork, obeying school rules, and treating teachers and students with respect.

Teens show good character to their classmates. *How can you show good character to your classmates?*

>>> **Reading Check**

DESCRIBE *How can cliques be harmful? How can they be helpful?*

Health SKILLS ACTIVITY

Advocacy

Relating to Your *Community*

Building a healthy relationship with the community includes the following skills:

* Respect. Show your respect for your community by taking pride in it. Help keep it clean. Avoid littering or defacing property.

* Tolerance. Demonstrate tolerance by getting to know people in your community who are different from you. When you demonstrate tolerance, you encourage others around you to do the same.

With A Group

Create several posters encouraging teens to show respect for their community. Include posters that discuss intolerance and how it affects the community. Your posters should also encourage teens to demonstrate tolerance to others.

©Creatas/PunchStock

Making *a* Difference *in* Your Community

You read about how character shows in citizenship. You can make a difference in your community with your acts of citizenship. You protect your environment when you throw away trash and recycle glass, aluminum and papers. You can also help your community by volunteering your time or skills to others.

Most communities have volunteer programs to help others. Volunteers make and serve food to shelters, collect clothes and books to donate to others, and work to help others in different ways.

Volunteers collect and box goods to send to service men and women. Volunteers tutor and coach children. Volunteers take part in activities, such as bake sales or car washes, to collect money for people in need. When you volunteer your time and skills to help others, you make a difference in your community.

As a volunteer, your acts of citizenship show that you advocate fairness, equality, caring, and giving. You are advocating for a better community and better world. ■

Teens give their time to help the community. *Explain how you think it might feel to provide food to hungry people.*

⟩⟩⟩ After You Read

1. **VOCABULARY** Define the term *constructive criticism*. Use it in an original sentence.
2. **EXPLAIN** Describe how a clique can be harmful.
3. **IDENTIFY** List three places to develop your good character.

⟩⟩⟩ Thinking Critically

4. **APPLY** Think of a public figure who works to help world situations. Use the Internet to learn about how that person works to advocate for change.

⟩⟩⟩ Applying Health Skills

5. **ADVOCACY** Imagine that a clique of students is bullying another student at lunch and in the school halls. What steps can you take to advocate against bullying? What traits of good character does it take to advocate against bullying?

⟳ Review

◉ Audio

Hands-On HEALTH ACTIVITY

Developing Good Character

WHAT YOU WILL NEED

* poster board
* markers or crayons

WHAT YOU WILL DO

1 Your teacher will divide the class into six small groups and assign each group one of the six traits of character: trustworthiness, respect, responsibility, fairness, caring, or citizenship.

2 Brainstorm and list examples of how teens can develop the assigned character trait. For example, if your group was assigned trustworthiness, you might list telling the truth and keeping promises.

3 Create a poster featuring the examples you listed in Step 2. As a group, explain to the class how your examples can help a teen develop good character.

Character is formed every day by your thoughts and actions. Developing good character is important to your health. It will help you develop positive relationships and behaviors. A person of good character is trustworthy; treats people with respect; is responsible, fair, and caring; and is a good citizen. In this activity, you will create a poster with examples of how to develop one of the six traits of character.

WRAPPING IT UP

After all the groups have presented their posters, discuss these questions as a class: How can teens help other teens develop good character? How can good character affect all sides of the health triangle? Display your posters where your classmates can see them.

Christopher Futcher/Getty Images

READING REVIEW

FOLDABLES and Other Study Aids

Take out the Foldable® that you created and any study organizers that you created. Find a partner and quiz each other using these study aids.

LESSON 1 What is Character?

BIG IDEA Character is the way a person thinks, feels, and acts.

* Good character is part of a healthy identity.
* A person's character demonstrates his or her values and beliefs.
* Your life experiences and role models shape your character.
* Good character means making good choices.

LESSON 2 Trustworthiness and Respect

BIG IDEA Good character is built on trustworthiness and respect.

* Trustworthiness is a trait of good character.
* Trustworthy people can be relied upon to do what they say they will do.
* Integrity is the quality of doing what you know is right.
* Good character is demonstrated through respect.

LESSON 3 Responsibility and Fairness

BIG IDEA Acts of responsibility and fairness demonstrate good character.

* Your character shows in your responsibilities.
* Accepting responsibility includes accountability.
* Taking responsibility for your health includes making good decisions.
* A fair person treats everyone equally and honestly.

LESSON 4 Being a Good Citizen

BIG IDEA Good character includes caring and citizenship.

* Good character shows in acts of caring.
* Caring involves having the spirit of giving and showing sympathy and empathy.
* Good citizens help to make their community a better place.
* Caring and citizenship go together in helping to build character.

LESSON 5 Making a Difference

BIG IDEA Your words and actions show your character.

* Good character helps you develop and maintain healthy relationships.
* Traits of good character include communication skills, respect, tolerance, and citizenship.
* You can demonstrate good character by working to make a difference at home, at school, and in your community.

 Review

 Web Quest

ASSESSMENT

Reviewing Vocabulary *and* Main Ideas

- › role model
- › tolerance
- › character
- › loyal
- › fairness
- › respect
- › abstinence
- › integrity

» On a sheet of paper, write the numbers 1–8. After each number, write the term from the list that best completes each statement

LESSON 1 What is Character?

1. The way a person thinks, feels, and acts is known as _____.

2. The quality of always doing what you know is right is _____.

3. _____ is the ability to accept other people as they are.

4. A _____ is a person who inspires you to think or act a certain way.

LESSON 2 Trustworthiness and Respect

5. To be _____ means you are faithful.

6. Having consideration for the feelings of others is _____.

LESSON 3 Responsibility and Fairness

7. The conscious, active choice not to participate in high-risk behaviors is _____.

8. _____ means treating people equally and honestly.

» On sheet of paper, write the numbers 9-14. Write *True* or *False* for each statement below. If the statement is false, change the underlined word or phrase to make it true.

LESSON 4 Being a Good Citizen

9. If a friend feels sad or depressed, you can show <u>empathy</u> by sharing your friend's feelings.

10. <u>Abstinence</u> means you support a cause.

11. <u>Citizenship</u> means doing what you can to support your community.

LESSON 5 Making a Difference

12. <u>Destructive</u> criticism is using a positive message to make a suggestion.

13. Messages in which you offer a suggestion from your own point of view are called <u>I-messages</u>.

14. Groups of friends who hang out together and act in similar ways are called <u>cliques</u>.

✓ eAssessment

>> Using complete sentences, answer the following questions on a sheet of paper.

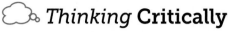 *Thinking* **Critically**

15. ANALYZE List the six traits of good character. Next to each trait, give an example of how someone might demonstrate the trait.

Write **About It**

16. OPINION Imagine you are writing an article on advocating for a school recycling program. In your article, explain why a school recycling program would benefit the students, teachers, and the community.

STANDARDIZED TEST PRACTICE

Reading
Read the passage below and then answer the questions that follow.

Did you know that one in seven people in the world goes to bed hungry every night? Or that people suffer from diseases that were stamped out in this country long ago? We are citizens of many different communities, such as cities or countries. However, we are all citizens of the world. Unfortunately, not all members of the world community share its resources equally.

One way to help suffering communities is by acting as a health advocate. Tell others about the suffering of other nations. Become and stay informed about world events. The World Health Organization is currently taking action to help people in need. Visit their Web site for further information.

1. Which statement best sums up the main point of the passage?
   A. We all live in many different communities.
   B. The health skill of advocacy can be used to help nations in need.
   C. Other countries are suffering from diseases that we have already overcome.

2. The passage notes that "not all members of this global community share its resources equally." Of the following quotes, which is *not* a detail that supports that comment?
   A. "Did you know that one in seven people in the world goes to bed hungry every night?"
   B. "Or that people suffer from diseases that were stamped out in this country long ago?"
   C. "Become and stay informed about world events."

BUILDING CHARACTER

Character is how a person thinks, feels, and acts. Someone with good character has the following traits:

- Trustworthiness
- Responsibility
- Respect
- Fairness
- Citizenship
- Caring

WHAT DO THEY MEAN?

TRUSTWORTHINESS
You are honest, loyal and do what you say you are going to do.

RESPECT
You are considerate of others and accept their differences.

RESPONSIBILITY
You hold yourself accountable for your choices, decisions, and actions.

FAIRNESS
You play by the rules, take turns, and share.

CARING
You show that you are kind and considerate of others.

CITIZENSHIP
You show respect for authority and interest in the world around you.

HOW IS CHARACTER DEVELOPED?

Life experiences

Your family, friends, teachers, and others in your community can shape your character.

Role models

A UCLA School of Public Health study asked 750 teens if they had a role model. Fifty-six percent said they did. Girls most often named a parent or relative as a role model, while boys named a sports star or other public figure.

PUTTING IT INTO ACTION

Your good character shows in your actions and words.

AT HOME

Treat parents and family members with care and respect.

Show appreciation.

Help with chores.

Be patient and kind to siblings.

AT SCHOOL

Take responsibility for your schoolwork.

Obey school rules.

Respect teachers and students.

Take a stand against violence and bullying.

IN YOUR COMMUNITY

Pick up trash and throw away into garbage cans.

Recycle bottles, cans, and paper.

Volunteer your time and talents.

THE BATTLE AGAINST BULLYING

Bullying and cyberbullying are both dangerous and damaging forms of teasing, taunting, intimidation, or harassment.

WHY DOES BULLYING HAPPEN?

BULLIES

- To fit in with a group
- To feel superior
- To avoid being bullied themselves
- Lack of tolerance for those who are different
- Lack of parental supervision

BULLIED

- Being overweight/underweight, being uninterested in sports, wearing different clothing, being new to a school
- Weak or unable to defend themselves
- Depressed, anxious, or insecure with low self-esteem
- Less popular than others and having few friends
- Unable to get along well with others

Cyberbullying

Cyberbullying uses electronic means, like text messages and the Internet.

INSTANT MESSAGES   **SOCIAL MEDIA SITES**

WEBSITES   **ONLINE GAMING SITES**

E-MAILS   **TEXT MESSAGES**

HOW TO PREVENT AND STOP BULLYING
On-the-spot strategies

No one deserves to be bullied. If you are a target, here are some ways to prevent bullying or stop it when it occurs:

TELL THE BULLY TO STOP.   STOP!

TRY HUMOR.

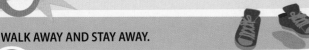

WALK AWAY AND STAY AWAY.

AVOID PHYSICAL VIOLENCE.

FIND AN ADULT.

CYBERBULLYING

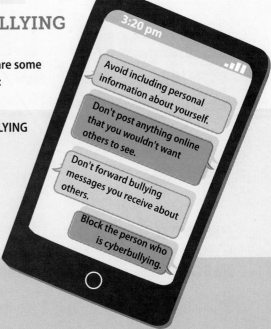

3:20 pm

Avoid including personal information about yourself.

Don't post anything online that you wouldn't want others to see.

Don't forward bullying messages you receive about others.

Block the person who is cyberbullying.

Bullying + Cyberbullying

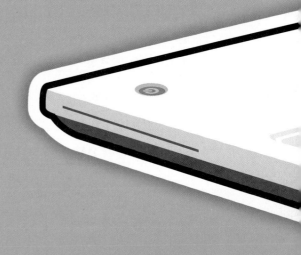

 **PREMIUM ONLINE RESOURCES** 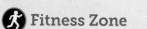 Audio  Videos Bilingual Glossary Fitness Zone Web Quest Review

People who are bullied can be hurt physically and emotionally. *List two forms of bullying.*

Bullying *and* Harassment

BIG IDEA Anyone can experience bullying and harassment, but there are effective ways to stop both.

Before You Read

QUICK WRITE Write about three factors you think are responsible for bullying. Share one factor with the class.

 Video

As You Read

FOLDABLES | Study Organizer

Make the Foldable® found in the FL pages in the back of the book to record the information presented in Lesson 1.

Vocabulary

> bullying
> labeling
> intimidation
> harassment
> sexual harassment

 Audio

🔤 Bilingual Glossary

WHAT IS BULLYING?

MAIN IDEA Most students have been bullied at one time or another.

Josh was waiting in the school lunch line. Suddenly, he felt someone shove him. Josh then heard Rick and his friends laughing "Move out of the way, loser," as the four cut in front of Josh. Josh felt an urge to push Rick back, but he noticed a teacher was walking toward them. He was glad that the teacher intervened. What would you do if you saw this happening at your school?

This disrespectful behavior is an example of bullying. Bullying is *a type of violence in which one person uses threats, taunts, or violence to intimidate another again and again.* Bullies may tease their victims or try to keep them out of a group. They may attack them physically. Three out of four students have been bullied at one time or another.

What *are* Bullying Behaviors?

Bullies often taunt people who are shy or stand out in some way. Male bullies often use threats of physical violence.

Female bullies often use verbal put-downs that hurt other people's feelings. Different kinds of bullying include:

- **physical bullying**—hitting, kicking, pinching, spitting, tripping/pushing, taking personal belongings, or making mean or rude hand gestures.

Three out of **four** students have been *bullied* at one time or another.

- **verbal bullying**—teasing, labeling or *name-calling,* taunting, or making threats to physically harm.
- **psychological bullying**—intimidation or *purposely frightening another person through threatening words, looks, or body language,* spreading rumors, isolating a person, or threatening to use force, embarrassing in public.

Where Bullying Occurs

Bullying can happen almost anywhere during or after school hours. It can also occur on the school playground or bus, while going to or from school, in your neighborhood, or on the Internet.

What *is* Harassment?

Any behavior that is directed toward another person because of race, nationality, skin color, gender, age, disability, and/or religion becomes harassment or the *ongoing conduct that offends another person by criticizing his or her race, color, religion, physical disability, or gender.* When harassment becomes violent, it might be a hate crime. *A hate crime is committed against another person because she or he is a member of a certain social group.*

Harassment, though, is defined by Federal civil rights law. This law refers to harassment that is severe, persistent, or creates a hostile environment. It protects against harassment based on a person's:

- race
- nationality
- skin color
- gender
- age
- religion

Harassment that involves behavior or remarks of a sexual nature is called sexual harassment. It is *uninvited and unwelcome sexual conduct directed at another person.* Obscene or inappropriate e-mails, text messages, or voice mails with a sexual meaning can also be sexual harassment. This behavior is illegal. Conduct related to harassment is gender discrimination. *It occurs by singling out or excluding a person based on gender.*

How Can I Respond *to* Harassment?

Our differences are what make us interesting. Making a joke about someone else because of that person's race, gender, ethnic identity, religion, or a

The facts about bullying are sometimes misunderstood. *Predict who is bullied more often—boys or girls.*

physical disability is harassment. Harassment is a crime. If you are a target of harassment, you can:

- Tell the person to stop. Say that if it continues, you will report the harassment.
- Be assertive. Speak firmly, looking the person in the eye.
- Tell your parents or other guardians. Ask for advice on handling the harassment.
- Report it to an adult. Harassment is illegal and charges can be filed against the person who is harassing you.

>>> **Reading Check**

IDENTIFY *Federal law protects against harassment based on what protected classes?*

Why Do Teens Get Bullied?

The reasons teens get bullied have little to do with the teen who is being bullied. A bully will target another teen because he or she is trying to fill an unmet need. Some teens may bully others to feel good about him- or herself. Being different in any way is one reason why a teen is bullied. It is common during the teen years to want to fit in with a group. However, as you become an adult, you will begin to value your individuality more.

Our differences are what make us interesting. Some people, though, are uncomfortable with differences. They may choose to become bullies and target teens who are different. Other reasons that teens may be targeted by bullies include those considered to be:

- different: being overweight or underweight, being uninterested in sports or athletics, wearing glasses or different clothing, being new to a school, or not having popular items.
- weak or unable to defend themselves.
- annoying or provoking.
- depressed, anxious, or insecure with low self-esteem.
- less popular than others and having few friends.
- unable to get along well with others.

Why Do Teens Become Bullies?

Teens who bully use their "power" to hurt people. The bully's power does not always mean he or she is bigger or stronger. The bully might be popular or smart—or the bully may know a secret about the person being bullied. Bullying is not a healthy behavior. Teens may become bullies because they feel that bullying behavior will help them

- fit in with a group.
- feel superior.
- avoid being bullied themselves.

Other times, a teen may become a bully because of behaviors that he or she learned from parents or others. In this situation, a teen may not believe that he or she is being a bully. He or she may feel that bullying behavior is okay because of a

- lack of tolerance of others who are different.
- lack of parental supervision.

Bullied teens may not be aware that bullies actually have low self-esteem. They pick on others to feel better or more important. Bullying others may make the teen feel superior. Also, a bully almost always needs an audience that supports his or her actions.

Many bullies have been bullied by other teens. So, they might bully as a way to be part of a group or to keep from being bullied themselves. However, this means that the negative cycle of bullying repeats itself unless the cycle is broken.

> **》》 Reading Check**
>
> **EXPLAIN** *What is one main reason that bullies hurt others?*

Bullies are not always confident and may have low self-esteem. *Describe the negative cycle of bullying.*

WHAT ARE THE EFFECTS OF BULLYING?

MAIN IDEA Anyone involved with bullying is affected in negative ways.

Everyone involved with bullying is affected—the bullies, those being bullied, and the people who watch the bullying. Bullying can contribute to many problems, including negative mental health, substance use, and suicide.

Teens who are bullied can experience negative physical and mental/emotional health. Teens who are bullied need to find someone to talk to about the bullying. Teens who are bullied are more likely to:

- feel fear, helplessness, depression, and loneliness.
- have low self-esteem.
- miss school or skip school.
- drop out of school.
- have various health problems.
- have trouble sleeping.
- inflict self-harm.

Bullies also experience effects and consequences for their bullying behavior. Many times, the bully may be seeking acceptance by bullying others.

Even though the bully is the aggressor, he or she may also experience mental/emotional and social health problems.

Bullies are more likely to:

- have low self-esteem.
- drop out of school.
- have problems with violence.
- have problems with substance abuse.
- have problems with criminal behavior.

It may seem that the teens who witness bullying are not affected by it. Unfortunately, even a teen who witnesses bullying, but does not participate will be affected by bullying. Teens who witness bullying when it happens can experience:

- increased use of tobacco, alcohol, or other drugs.
- increased mental health problems, including depression and anxiety.
- missing or skipping school.

> ### >>> Reading Check
>
> **EXPLAIN** *What are some negative health effects of bullying?* ◼

>>> After You Read

1. **VOCABULARY** Define the term *bullying*. Use it in an original sentence.
2. **IDENTIFY** Name the forms of bullying.
3. **RECALL** What are some forms of harassment?

>>> Thinking Critically

4. **APPLY** Your cousin writes to tell you about a "really funny kid" who just came to his school. He explains that this new person gets a laugh by knocking other students' books out of their hands. How would you explain to your cousin that this action is inappropriate behavior?

>>> Applying Health Skills

5. **ACCESSING INFORMATION** Harassment is considered a hate crime in 46 of the 50 states. Find out what the laws are in your community regarding harassment. Make a poster explaining the penalties for this behavior.

🄲 Review

🔊 Audio

Bullying affects everyone. *Describe the effects of bullying experienced by teens who witness the bullying.*

L. Mouton/PhotoAlto

Cyberbullying

BIG IDEA ❯ Cyberbullying is a growing problem that causes harm and humiliation.

 Before You Read

QUICK WRITE Write a poem or short story about a cyberbully. Give your poem or story a positive ending.

▶ Video

 As You Read

STUDY ORGANIZER Make the study organizer found in the FL pages in the back of the book to record the information presented in Lesson 2.

 Vocabulary

› cyberbullying

🔊 Audio

🔤 Bilingual Glossary

HOW IS TECHNOLOGY USED TO BULLY?

MAIN IDEA ❯ Cyberbullying is more difficult to avoid than face-to-face bullying.

Cyberbullying is *the electronic posting of mean-spirited messages about a person, often done anonymously.* Technology includes devices and equipment such as cell phones, computers, and tablets. Cyberbullies use technology to harass people, threaten them, or spread rumors. Examples of the communication tools that cyberbullies use to send mean messages or spread embarrassing rumors, photos, videos, or fake profiles include:

- social media sites.
- text messages.
- e-mails.
- instant messages.
- chat rooms.
- websites.
- online gaming sites.

This type of bullying allows one person to bully another without ever seeing him or her in person. The person being bullied may not even be able to identify the cyberbully.

Because of this lack of face-to-face contact, the bully might not realize how much his or her actions hurt the other person.

Cyberbullies use **technology** to **harass** people, **threaten** them, or **spread rumors**.

Cyberbullying *vs.* Bullying

Both types of bullying target one person with the intention to harm or humiliate him or her. However, differences exist between cyberbullying and bullying:

- Cyberbullying can be anonymous and can be difficult to trace.
- Cyberbullying can reach a wider audience very quickly.
- Cyberbullying can happen at any time.

©PhotoAlto/PunchStock

- Inappropriate or harassing messages, texts, and pictures are extremely difficult to delete after they have been sent or posted online.

These differences can make cyberbullying more difficult to avoid than face-to-face bullying. Teens who are cyberbullied have a harder time getting away from the behavior. In addition, teens who are cyberbullied may often be bullied in person as well.

Effects *of* Cyberbullying

Cell phones and computers themselves are not the cause of cyberbullying.

You can use social media sites for positive activities like connecting with friends and family, getting help with school work, and for entertainment. Unfortunately, these communication tools are also used to hurt other people.

Cell phones and *computers* themselves are **not the cause** of **cyberbullying.**

With more adolescents and teens using technology, the opportunities for cyberbullying have increased.

Whether bullying is done in person or through technology, it has negative effects on a person's physical, mental/emotional, and social health. Teens who are cyberbullied are more likely to:
- use alcohol and drugs.
- skip school.
- experience in-person bullying.
- be unwilling to attend school.
- receive poor grades.
- have lower self-esteem.
- have more health problems.

>>> **Reading Check**

LIST *What are the effects on a teen who is being cyberbullied?*

WHAT ARE SOME TYPES OF CYBERBULLYING?

MAIN IDEA Cyberbullies use several types of technology to attack another person.

Today, many people have access to technology whether it's at school, home, or through community resources, such as your local library. Many teens also have a cell phone that they use to keep in touch with family and friends. Having access to these tools has many benefits. Using computers and cell phones, however, can also make you vulnerable to cyberbullies.

Teens spend much of their social lives online, and so cyberbullying is more of a threat. *List other technologies that might be used for cyberbullying.*

The use of technology in everyday life has provided bullies with a number of new ways to attack another person. A cyberbully can use a computer to post false information on social media sites, in blogs, e-mails, and instant messages. Text messages can also be used to cyberbully.

As the use of technology increases, the ways that cyberbullies attack others will increase, too. Rather than avoiding technology use, think about what you can do to prevent cyberbullies from targeting you.

How Can I Prevent Cyberbullying?

The best way to avoid becoming the victim of a cyberbully is to do what you can to prevent it from occurring. Avoid including personal information about yourself in text messages, e-mail, or social media sites.

Another important rule to remember is that any photo that you post online will remain online forever. Even if you delete a photo, a person with good computer skills can retrieve the image. This includes photos that are sent via e-mail, posted to a social media site, or sent via text.

In some cases, cyberbullying begins as a result of a photo or message that one teen sends to another.

> You can avoid being a target of cyberbullying and also help others to protect themselves. *Evaluate whether you should respond immediately to a hateful text message from a cyberbully.*

Sexting, or the sending of explicit e-mails and photos, is always risky. It's also against the law if the person in the photo is under 18 years old. Any message or picture you send to another person is no longer in your control. The person receiving your message or picture can post it online or share it with others without your consent.

How Can I Stop Cyberbullying?

When cyberbullying happens, keep all evidence of it. Write down the dates, times, and descriptions of incidents. Save and print screenshots, e-mails, text messages, etc.

- Do not respond to or forward cyberbullying messages. Block the person who is cyberbullying. Visit the web site's help page to learn how to block users.

Stop Cyberbullying

Do Keep a record and evidence of attacks.	**Don't** Respond to the cyberbully.
Do Tell an adult if you receive harassing messages.	**Don't** Forward messages or images sent by a cyberbully.
Do Block messages from the cyberbully, if possible.	**Don't** Share or post personal information with strangers or people you do not know well.
Do Report the incident to your social media site, Internet provider, cell phone service, school, and/or local law enforcement agency.	**Don't** Visit websites that are unsafe.
Do Keep your passwords safe and do not share them with anyone except your parent(s).	**Don't** Post any text or images online that could hurt or embarrass you or those you know.

>>> After You Read

1. **VOCABULARY** Define the term *cyberbullying*.
2. **LIST** Name the communication tools used by cyberbullies.
3. **EXPLAIN** What are three negative effects of cyberbullying on the victim?

>>> Thinking Critically

4. **APPLY** Find Internet ads or watch television commercials designed to stop bullying. How are the ads effective? How might the ads be more effective?

>>> Applying Health Skills

5. **COMMUNICATION** A friend of yours has received humiliating text messages. What advice would you give her on handling this cyberbullying? Write a "To Do" list of strategies for your friend.
6. **DECISION-MAKING** A friend of yours has been receiving hurtful text messages. You know who is sending the messages. You want to be a good friend. You think about confronting the bully but worry that the bully will target you. Use the steps in the decision-making process to decide what to do.

 Review

 Audio

Strategies *to* Stop Bullying

BIG IDEA Every person can take steps to stop bullying behavior.

Buccina Studios/Getty Images

>>> **Before You Read**

QUICK WRITE What steps do you take to keep yourself safe from bullying? Write a short paragraph about these strategies.

 Video

>>> **As You Read**

STUDY ORGANIZER Make the study organizer found in the FL pages in the back of the book to record the information presented in Lesson 3.

>>> **Vocabulary**

› bullying behavior

 Audio

 Bilingual Glossary

Myth vs. Fact

Myth: With more people using computers, cell phones, and wireless Internet, cyberbullying is on the rise.
Fact: Traditional forms of bullying are still more common than cyberbullying. Bullied people being hit, shoved, kicked, gossiped about, intimidated, or excluded.

HOW SHOULD I STOP A BULLY?

MAIN IDEA You can stop bullying now and in the future by using strategies.

Have you ever heard someone say, "What's the matter—can't you take a joke? You're just too sensitive." Comments like these might be directed at people who are targets of bullying. These remarks can make the bullied person feel as though he or she has no right to be sensitive. In some cases, the person who is bullied may feel that he or she deserves to be bullied.

No one deserves to be **bullied**.

On-the-Spot Strategies

No one deserves to be bullied. If you are a target, here are some ways to stop the bullying when it is happening:

- **Tell the bully to stop.** Look at the person and speak in a firm, positive voice with your head up. Say that if the behavior continues, you will report the bullying.
- **Try humor.** This works best if joking is easy for you. Respond to the bully by agreeing with him or her in a humorous way. It could catch the bully off guard.
- **Walk away and stay away.** Do this if speaking up seems too difficult or unsafe.
- **Avoid physical violence.** Try to walk away and get help if you feel physically threatened. If violence does occur, protect yourself but do not escalate the violence.
- **Find an adult.** If the bullying is taking place at school, tell a teacher or a school official immediately.

>>> **Reading Check**

LIST *What are three ways to handle bullying on the spot?*

Do's

- Do keep control of yourself.
- Do stay calm and speak softly.
- Do walk away if necessary.
- Do apologize if necessary.
- Do try to turn the other person's attention somewhere else.
- Do use your sense of humor.
- Do give the other person a way out.
- Do try to understand how the other person thinks or feels.
- Do tell an adult.

Don'ts

- Don't let your emotions get the better of you.
- Don't let the other person force you into a fight.
- Don't try to get even.
- Don't tease.
- Don't be hostile, rude, or sarcastic.
- Don't threaten or insult the person.

Putting these tips to use can reduce your chances of being bullied. *Explain how you think these tips can help you avoid being bullied.*

Strategies *for* the Future

There are also several ways to avoid being bullied in the future:

- **Talk to an adult you trust.** A family member, teacher, or other adult can help. Telling someone can help you feel less alone. They can also help you make a plan to stop the bullying.
- **Avoid places where you know bullies target other students.** Stairwells, hallways, courtyards without supervision, and playground areas can be risky locations.
- **Stay with a group or find a safe place to go.** Bullies are less likely to target a student who is with his or her friends.

Stop Bullying Behavior

At times, a teen may not realize that his or her actions or words may actually hurt feelings. If you are someone who likes to joke with your friends, watch how the person responds to your jokes. If he or she seems hurt, stop the behavior.

Sometimes, a bully may not intend to hurt the feelings of another person. Some teens may need help recognizing bullying behavior, or *actions or words that are designed to hurt another person.*

If you have been called a bully or you think you might have bullied another person, follow these steps:

- Stop and think before you say or do something that could hurt someone.
- Talk to an adult you trust. Describe what upsets you about the other person.
- Remember that everyone is different, and that our differences make us interesting and unique.
- If you think you have bullied someone in the past, apologize to that person.

>>> **Reading Check**

IDENTIFY *What are some ways to avoid bullying?* ■

REVIEW

>>> **After You Read**

1. **DEFINE** Define the term *bullying behavior.*
2. **LIST** Name two ways to stop bullying behavior.
3. **IDENTIFY** How might a bully try to correct his or her behavior?

>>> **Thinking Critically**

4. **ANALYZE** Shayna is being teased repeatedly by Dejon. His remarks bother her. She doesn't know what to do. What advice do you have for Shayna?

>>> **Applying Health Skills**

5. **COMMUNICATION SKILLS** Your new classmate, Seth, is having trouble with a student who is bullying and teasing him. Seth feels uncomfortable facing the bully. What strategies would you offer Seth to help him deal with this problem? Explain why.

 Review

Audio

Image Source/Getty Images

Promoting Safe Schools

BIG IDEA Students, teachers, and parents can promote schools that are safe from bullying.

>>> **Before You Read**

QUICK WRITE Write a short paragraph about ways to stay safe from bullying in school.

▶ Video

>>> **As You Read**

STUDY ORGANIZER Make the study organizer found in the FL pages in the back of the book to record the information presented in Lesson 4.

>>> **Vocabulary**

› zero tolerance policy

🔊 Audio

🔤 Bilingual Glossary

Character Check

Caring for Others If someone you know has been bullied, you can demonstrate caring by showing concern and empathy for that person. Listen if the person wants to talk. Help him or her know when to seek help from a parent or other trusted adult. *Describe some other ways you could show the person you care.*

HOW CAN TEENS PROMOTE SAFE SCHOOLS?

MAIN IDEA By recognizing the signs of bullying, students can take a stand against it.

Keeping schools safe takes effort by students, as well as parents, teachers, and school officials. You can help by being aware of bullying and recognizing the signs of bullying. Then, if you are confronted by a bully or witness bullying, you can take a stand against it.

What Are *the* Warning Signs?

Bullying and harassment are not always obvious. Many teens who are bullies or are being bullied do not ask for help. Those being bullied may try to hide the problem from friends. They may be afraid or embarrassed to talk about it. However, many warning signs can indicate someone is being affected by bullying—either being bullied or bullying others. Recognizing these warning signs is an important first step for taking action against bullying.

It is important to talk with students who show the signs of being bullied. It is also important to talk to those who show signs of bullying others. These warning signs can also point to other issues or problems, such as depression or substance abuse. Talking to the person can help identify the cause of the problem.

Many teens who are **bullies** or are **bullied** do not ask for help.

Look for changes in behavior. Not all teens who are bullied show warning signs. However, certain signs may mean someone is experiencing a bullying problem. If you know someone in distress or danger, talk to a trusted adult right away. Likewise, students may be bullying others if they demonstrate several warning signs that you can learn to recognize.

Take a Stand Against Bullying

If you see someone else being bullied you may not know what to do to stop it. Using the following strategies can help the person who is being bullied.

- Tell the bully to leave the person alone.
- Offer an escape to the person being bullied by saying a teacher needs to see him or her.
- Avoid using violence and insults.
- Tell an adult.
- Tell the person who is bullied, "I'm here for you."
- Ask the person being bullied what you can do to help.
- Spend time with the person being bullied.

Warning signs can help identify teens who are being bullied or who may be bullies themselves. *List three signs of a bullying problem.*

Warning Signs of Bullying Behavior

Signs of Being Bullied	Signs of a Bully
Unexplained injuries	Getting into physical or verbal fights
Lost or destroyed clothing, books, electronics, or jewelry	Having friends who bully others
Frequent headaches or stomach aches, feeling sick or faking illness	Increased aggressive behavior
Changes in eating habits, skipping meals or binge eating. Coming home from school hungry because the student did not eat lunch.	Sent to the principal's office or to detention frequently
Difficulty sleeping or having frequent nightmares	Unexplained extra money or new belongings
Declining grades, loss of interest in schoolwork, or not wanting to go to school	Blaming others for his or her problems
Sudden loss of friends or avoiding social situations	Not accepting responsibility for his or her actions
Feelings of helplessness or decreased self esteem	Competitiveness and worry about his or her reputation or popularity
Self-destructive behaviors, such as running away from home, harming him- or herself, or talking about suicide	

Health SKILLS ACTIVITY

Conflict Resolution

Stand Up to *Bullying*

While walking to class, Molly sees a large group of students gathered in the hallway. Molly sees that two students, Tracey and Sarah are involved in an argument. Tracey is teasing Sarah about her outfit and calling her names. Sarah tries to walk away, but Tracey starts to push her against her locker. Sarah is very upset and needs help.

What Would You Do?

Write a dialogue that shows how Molly can help Sarah escape this situation.

WHEN SHOULD I REPORT BULLYING?

MAIN IDEA At times, the best option for dealing with bullying is reporting the behavior to an adult.

Many schools are developing a zero tolerance policy, *a policy that makes no exceptions for anybody for any reason,* to bullying. In some cases, the policy may extend to bullying and cyberbullying that takes place off of school property. This type of policy provides students who feel helpless and isolated a way to stop the bullying.

Any teen who is bullied, or who has a friend who is being bullied, should tell a parent or teacher about the bullying. If necessary, tell more than one adult about the bullying. Stopping the bullying quickly can prevent it from becoming worse. If the bullying becomes violent, it should be reported immediately to a parent, teacher, or the police.

You should **tell a parent** or **teacher** if you or a friend is *being bullied.*

All bullying is wrong. If you see another teen being bullied, support that teen. Avoid teens who tend to bully others. Talk to your school administrators about starting an anti-bullying campaign at your school. Help educate other students about cyberbullying by talking about treating others with respect on social sites or the Internet.

>>> **Reading Check**

DEFINE *What does zero tolerance policy mean?* ■

>>> **After You Read**

1. **DEFINE** Define the term *zero tolerance policy.*
2. **IDENTIFY** Who can help prevent bullying in schools?
3. **EXPLAIN** What are three warning signs of being a bully?

>>> **Thinking Critically**

4. **EVALUATE** Find out if your school has a program to deal with preventing bullying. If so, how might students become aware of the program? If not, do research to find a program that might help your school.

>>> **Applying Health Skills**

5. **APPLYING INFORMATION** You want your parent or guardian to learn more about bullying and how he or she can help prevent it. What would you explain to him or her? Write down some basic information on what parents can do to prevent bullying in schools.

↻ Review

🔊 Audio

Teachers who promote tolerance among teens can go a long way toward preventing conflict and bullying at school. *Identify a benefit of tolerance for teens.*

Be Part *of* the Solution

WHAT YOU WILL NEED

* blue construction paper (**Note:** Blue ribbon is the color used to support or create an awareness associated with prevention of bullying and child abuse.)
* scissors
* white Paper 8 ½ x 11
* markers
* glue

Everyone involved in bullying can be affected- the bullies, those being bullied and the people who watch the bullying. In this activity YOU get to make a positive difference by "Stopping the Hate" and promoting and advocating for a school that is safe from bullying.

WHAT YOU WILL DO

1 Take a white sheet of paper and carefully cut it into four equal squares with a scissors.

2 Review lesson one and from the information found in this lesson write a powerful statement on one square of white paper that will advocate and empower others to be part of the solution that ends bullying and harassment. Do the same for lesson two, three and four.

3 Using blue construction paper, trace your hand four times and cut out four handprints. On the fingers of each hand print write: "Be Part of the Solution"

4 Glue one square of paper with your empowering statement written on it to the middle of your handprint.

WRAPPING IT UP

Place the blue "Be Part of the Solution" handprints in your school at the beginning of the week. At the end of one week, write a reflection and identify how your "Be Part of the Solution" project advocated and influenced yourself and/or others to create a school safe from bullying. Discuss other activisms you can create to prevent bullying, harassment and cyberbullying in your school and community.

©Steve Skjold/Alamy

READING REVIEW

FOLDABLES and Other Study Aids

Take out the Foldable® that you created and any graphic organizers that you created. Find a partner and quiz each other using these study aids.

LESSON 1 Bullying and Harassment

BIG IDEA Anyone can experience bullying and harassment, but there are effective ways to stop both.

* Most students have been bullied at one time or another.
* Anyone involved with bullying is affected in negative ways.

LESSON 2 Cyberbullying

BIG IDEA Cyberbullying is a growing problem among teens that causes harm and humiliation.

* Cyberbullying is more difficult to avoid than face-to-face bullying.
* Cyberbullies use several types of technology to attack another person.

LESSON 3 Strategies to Stop Bullying

BIG IDEA Every person can take steps to stop bullying behavior.

* You can stop bullying now and in the future by using strategies.
* By recognizing warning signs, you can help someone who is being bullied.

LESSON 4 Promoting Safe Schools

BIG IDEA Students, teachers, and parents, can promote schools that are safe from bullying.

* Students, teachers, and parents can all take a stand and help prevent bullying in schools.
* At times, the best option for dealing with a bullying is reporting the behavior to an adult.

 Review

 Web Quest

ASSESSMENT

Reviewing Vocabulary *and* Main Ideas

> bullying
> intimidation
> harassment
> cyberbullying
> labeling
> sexual
harassment

» On a sheet of paper, write the numbers 1-6. After each number, write the term from the list that best completes each sentence.

LESSON 1 Bullying and Harassment

1. Purposely frightening another person through threatening words, looks, or body language is called _____.

2. _____ is a type of violence in which one person uses threats, taunts, or violence to intimidate another again and again.

3. Ongoing conduct that offends another person by criticizing his or her race, color, religion, physical disability, or gender is known as _____.

4. _____ is also referred to as name-calling.

5. Uninvited and unwelcome sexual conduct directed at another person is known as _____.

LESSON 2 Cyberbullying

6. Electronic posting of mean-spirited messages about a person, often done anonymously, is called _____.

» On a sheet of paper, write the numbers 7-12. Write True or False for each statement below. If the statement is false, change the underlined word or phrase to make it true.

LESSON 3 Strategies to Stop Bullying

7. <u>Bullying behavior</u> is actions or words that are designed to hurt another person.

8. Using <u>humor</u> can be a strategy to stop bullying when it is happening.

9. Bullies are <u>more</u> likely to target a student who is with his or her friends.

LESSON 4 Promoting Safe Schools

10. A policy that makes no exceptions for anybody for any reason is called a <u>zero tolerance</u> policy.

11. <u>Some</u> bullying is wrong.

12. Any teen who is bullied, or who has a friend who is being bullied, <u>should tell</u> a parent or teacher about the bullying.

 eAssessment

>> Using complete sentences, answer the following questions on a sheet of paper.

🗯 *Thinking* Critically

13. SYNTHESIZE Zoey recently moved from a different part of the country. She speaks with an accent that is different than that of the students in her new school. Kathy, a girl in her class, imitates Zoey's accent, teasing her whenever she speaks. Kathy's teasing really bothers Zoey. What should Zoey do?

14. EVALUATE When Seth walks away from a fight, he hears the bully call him "chicken." What should Seth do? Explain.

🎸 *Write* About It

15. NARRATIVE WRITING Write a fictional story about a bullying incident. Describe the traits of the bully and the form of bullying that occurred.

16. EXPOSITORY WRITING Write a paragraph describing how a zero tolerance policy might stop bullying.

Ⓐ Ⓑ Ⓒ Ⓓ **STANDARDIZED TEST PRACTICE**

Reading
Read the passage below and then answer the questions that follow.

Bullying comes in many shapes and forms. A person might call you names or threaten you with physical violence. A person might tease you or try to keep you from a group. Bullies may even physically attack you. In recent years, technology has also given teens a new way to bully others. They might send a nasty message in an e-mail or through a website. In addition, they might post disrespectful messages about a person or post embarrassing videos.

   If you're being bullied, try to ignore the person and walk away, if possible. Try to remain calm, even if the bully tries to prevent you from leaving. Be forceful and stand up for yourself, but try not to let the confrontation turn physical. It is important to report the incident to a person in authority, such as a teacher, counselor, or other trusted adult.

1. What does *confrontation* mean in this sentence from the passage?
   Be forceful and stand up for yourself, but try not to let the confrontation turn physical.
   A. conflict
   B. agreement
   C. compromise
   D. discussion

2. Which of the following best describes the purpose of the second paragraph?
   A. To explain reasons why people bully others.
   B. To describe how being bullying makes a person feel
   C. To suggest ways to deal with bullying
   D. To give reasons bullies should be tolerated

Unit 3

mental + emotional health

Your Mental + Emotional Health

LESSONS

PREMIUM ONLINE RESOURCES

 Audio

 Fitness Zone

 Videos

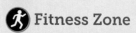 Web Quest

 Bilingual Glossary

 Review

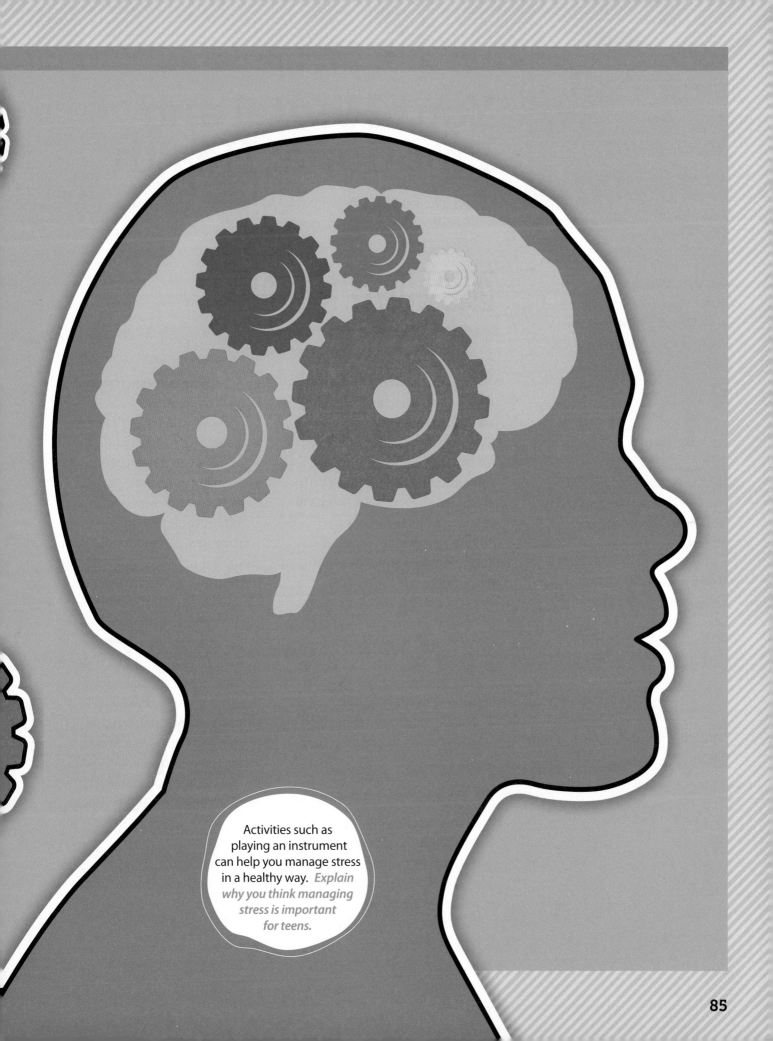

Activities such as playing an instrument can help you manage stress in a healthy way. *Explain why you think managing stress is important for teens.*

Your Mental *and* Emotional Health

BIG IDEA Good mental/ emotional health includes having a positive view of yourself and being resilient.

Before You Read

QUICK WRITE Think about a time when you felt disappointed. Write down how you dealt with your feelings.

▶ Video

As You Read

 FOLDABLES Study Organizer

Make the Foldable® found in the FL pages in the back of the book to record the information presented in Lesson 1.

Vocabulary

› mental/emotional health
› personality
› self-concept
› self-esteem
› confidence
› resilience

🔊 Audio

▶ Video

WHAT IS MENTAL AND EMOTIONAL HEALTH?

MAIN IDEA Mental/emotional health is the ability to handle the stresses and changes of everyday life in a reasonable way.

Raul thinks of himself as a normal teen. He has several good friends and likes his classes in school. Most days, he's in a good mood, but sometimes he feels stressed out. When that happens, he listens to music until he feels better.

Raul shows many signs of good mental/emotional health, or *the ability to handle the stresses and changes of everyday life in a reasonable way.* Other signs of good mental/emotional heath are:

- Having a good attitude and a positive outlook on life
- Recognizing your strengths and working to improve your weaknesses
- Setting realistic goals for yourself
- Acting responsibly
- Being able to relax and have fun, both on your own and in a group
- Being aware of your feelings and expressing them in a healthy way

- Accepting constructive feedback from others without becoming angry or defensive
- Accepting yourself and others
- Adapting to new situations
- Showing empathy—the ability to identify with and share other people's feelings

Your mental/emotional health can affect the other sides of your health triangle. It can also be affected by them. For example, if your physical health is poor, you might often feel tired. This could make it hard for you to feel good about yourself. If your social health is good, by contrast, that means you feel connected to people you trust. That makes it easier for you to deal with your problems, because you always have someone to talk to about them.

> ## Reading Check
>
> **GIVE EXAMPLES** *What are two signs of good mental/emotional health?*

Design Pics/Don Hammond

HOW YOU SEE YOURSELF

MAIN IDEA Your personality and self-concept are two factors that determine your mental/emotional health.

Mentally and emotionally healthy teens can accept themselves as they are. This does not mean that they see themselves as perfect, but they know their own strengths and know how to put them to good use. At the same time, they are aware of their weaknesses and can work to improve them without feeling down on themselves.

Many *factors* shape your **personality**.

Self-Concept

As a teen, you are learning about your own personality. Your personality is *a combination of your feelings, likes, dislikes, attitudes, abilities, and habits.* Many factors shape your personality. Heredity is one of them. For example, if your parents are outgoing, you may be the same.

You also learn some behaviors from other people, such as family members, friends and peers. Your culture, your background, and your role in your family also affect your personality.

Your self-concept, or self-image, is related to your personality, but they are not the same. Your personality is the way others see you, while your self-concept is *the way you view yourself overall.*

> Being able to accept constructive criticism is one sign of good mental/emotional health. *Describe in what areas of your life you receive constructive criticism. Do you respond to it in a positive way?*

Your self-concept affects the way you relate to yourself and the world. Suppose you think of yourself as a person who plays by the rules. Now think about what would happen if you became friends with some teens who are popular at school, but are often getting into trouble. Trying to fit in with this group might go against your self-concept. This might cause you to feel anxious and unhappy.

⟩⟩⟩ Reading Check

IDENTIFY *What two factors influence your self-concept?*

Self-Esteem

A major part of your self-concept is self-esteem, or *how you feel about yourself.* When you have high self-esteem, you like and value yourself. You take pride in your achievements and you face new challenges with confidence—*the belief in your ability to do what you set out to do.*

When you value yourself, you are more likely to take care of your body—for example, by eating right and being physically active.

When you feel confident about yourself, you are more likely to be friendly and outgoing, which will help you make friends.

Many factors can influence your self-esteem. The messages you receive from your friends and family play an important role. Supportive and loving friends and family members can build your self-esteem. Critical or hurtful messages, on the other hand, can damage it.

The media can also affect your self-esteem. It can influence your ideas about how you should look, what you should buy, and how you should act.

However, these messages may not always be realistic. Finally, your own attitude affects your self-esteem. When you think about yourself in a positive way, you boost your own self-esteem.

How Does the *Media* Influence Your *Self-Concept?*

One factor that affects your self-concept is the media. Think about the images you see on television or in movies. They often show attractive people having fun. Some teens try to look and act like the people they see on-screen. They may feel this will improve their self-concept. It's important to recognize the ways in which media messages influence the way you feel about yourself.

With A Group Collect pictures, video clips, or descriptions of images from the media. Analyze the message each sends.

Building Self-Esteem

Self-esteem can always be improved upon. You can develop skills to build your self-esteem and feel good about yourself. The skills you can practice include:

- **Set realistic goals.** Divide your larger goals into smaller goals. You can build on each smaller success to reach your overall goal.

- **Focus on your strengths.** Find something you like to do, such as a hobby, school activity, or sport. Work to improve your skills. Try to enjoy yourself, even when you make mistakes.

- **Ask for help.** Recognize and accept when you might need help. This is especially true when you are learning something new.

- **Remember that no one is perfect.** Everyone has different abilities. Identify your weaknesses without judging yourself and make a solid effort to improve them. If others give you constructive feedback, try to learn from it. Be proud of yourself when you succeed, but know that sometimes failure is out of your control. Mistakes can teach you what doesn't work and push you to grow.

- **Think positively.** A positive attitude can help you be more confident. Being positive also helps you relate better to others. You're likely to be more honest and honorable and to respect others' life experiences.

>>> **Reading Check**

LIST *Name two ways to build self-esteem.*

Roy McMahon/Getty Images

BUILDING RESILIENCE

MAIN IDEA ⟩ Developing resilience can build your self-esteem and improve mental/emotional health.

When you stretch a rubber band, it snaps back into shape as soon as you let it go. Some people are like that too. After a disappointment, they get right back on their feet and keep moving forward. *The ability to recover from problems or loss* is called <u>resilience.</u> Being resilient does not mean that you never have problems. It simply means that you can face them and seek positive solutions.

Resilience is a skill you have to develop and practice throughout your life. Here are some steps you can take to build self-esteem and develop resilience:

- **Focus on your strengths.** If you are a good student, for instance, consider tutoring someone who needs help.
- **Connect with others.** Friends, family members, and others can be an important source of support in difficult times.
- **Motivate yourself.** Set realistic goals and plan out the steps needed to achieve them.
- **Keep a positive attitude.** Focus on what you want to happen, not what you are afraid might happen.
- **Remember that no one is perfect.** It's normal to make mistakes, and it's okay to ask for help when you need it.

⟩⟩⟩ **Reading Check**

DEFINE *What is resilience?* ■

Being resilient means you can bounce back from a difficult experience. *Describe how being resilient helps you achieve your goals.*

REVIEW

⟩⟩⟩ **After You Read**

1. **DEFINE** Define *personality* and use the term in an original sentence.
2. **EXPLAIN** What can happen if you behave in a way that goes against your self-concept?
3. **LIST** What are three steps you can take to build positive self-esteem and resilience?

⟩⟩⟩ **Thinking Critically**

4. **INFER** Give an example of how your mental/emotional health might influence your physical or social health.
5. **ANALYZE** Explain how self-concept and self-esteem are related.

⟩⟩⟩ **Applying Health Skills**

6. **ANALYZING INFLUENCES** What do you think has had the most influence on your personality: your heredity, environment, or behavior? Explain your answer.

ⓒ Review

◆)) Audio

Understanding Your Emotions

BIG IDEA Learning to deal with emotions in healthy ways is important.

Before You Read

QUICK WRITE Think about times when you were upset or angry over something. Try to recall how you dealt with these feelings.

▶ Video

As You Read

STUDY ORGANIZER Make the study organizer found in the FL pages in the back of the book to record the information presented in Lesson 2.

Vocabulary

› emotions
› mood swings
› anxiety
› emotional needs

🔊 Audio

🔤 Bilingual Glossary

🏃 Fitness Zone

Working Out My Emotions

Sometimes my emotions just seem overwhelming. I get angry or upset over little things, and I don't even know why. When that happens, I lace up my track shoes and go out for a run. Focusing on my body for a while helps take my mind off my problems, and when I come back, I always feel more relaxed.

WHAT ARE EMOTIONS?

MAIN IDEA It is normal to experience a variety of emotions.

For two years, Tanesha has played only minor roles in school plays. Now, for the first time, she is being considered for a lead role. As she walks toward the cast list, she feels nervous and excited. When she sees her name at the top, she is so thrilled that she can't resist jumping up and down and clapping her hands.

If you have ever been in a situation like Tanesha's, you can probably imagine her feelings at this moment. In fact, nearly every life experience goes hand in hand with some kind of emotion. **Emotions** are *feelings such as love, joy, or fear.*

It is normal to experience many different emotions, sometimes in just a short period of time. During your teen years, it's normal to have **mood swings,** or *frequent changes in emotional state.*

They can happen because of changing hormone levels in your body, worries over the future, or concerns over relationships. Mood swings are a normal part of growing up.

Mood swings are a **normal** part of *growing* up.

Emotions can sometimes be difficult to recognize. You may not know what you are feeling or why. If you can identify your emotions, you can think of healthy ways to manage them. Remember that emotions are not good or bad. It's how you cope with them that matters.

›››› Reading Check

DEFINE *What are mood swings?*

Health SKILLS ACTIVITY

Practicing Healthful Behaviors

Anger Management

When emotions are running high, they can be difficult to control. It is at these times, though, that self-control is most important. The next time you feel angry, try these steps:

1. Take a deep breath. Try to relax.

2. Identify the specific cause of your anger. If necessary, leave the room so that you can collect your thoughts.

3. When you are calm enough to speak, tell the other person how you feel and what action has caused you to feel this way.

4. Write down your thoughts in a journal.

5. Practice relaxation skills.

6. Do a physical activity. A positive way to manage stress is to keep physically active.

7. Look for opportunities to laugh. Keeping a sense of humor is also a positive skill for managing anger.

 With A Group Role-play a situation in which one or both of you is upset over something. Use anger-management skills to express your anger in a healthy way. Be prepared to perform your role-play for classmates.

EXPRESSING YOUR EMOTIONS

MAIN IDEA The way you express emotions affects your mental/emotional, physical, and social health.

Some emotions, such as happiness and love, can be very pleasant. Others, such as fear and anxiety, can be very unpleasant. However, no emotion is good or bad in itself. All emotions, even unpleasant ones, are normal and healthy. What matters is learning to express them in healthy ways that don't cause any harm to yourself or others. Learning to manage your emotions is an important part of mental and emotional health.

Dealing *with* Anxiety

Have you ever had "butterflies in your stomach" before a big test or a sports match? This is a normal reaction to anxiety, or *a state of uneasiness.* It is a form of stress.

All **emotions** are normal and healthy.

Mild anxiety can be useful because it gives you extra energy. When it builds up too much, however, it can be harmful. It can interfere with necessary functions like eating and sleeping.

When you are feeling anxious, talking with others can sometimes help calm you down. For example, you might try talking through your problems with a family member, friend, or counselor. You can also try writing about your feelings in a journal. Physical activity can also help you feel less tense. One approach that will not help is to run away from the problem. If you meet a challenge directly, you will learn skills that can help you deal with similar challenges in the future.

Dealing *with* Anger

Anger is another emotion that can be difficult to deal with. When you are angry, you may feel like lashing out at the source of your anger. Feelings like these can be helpful in some cases. If a mugger attacks you, for instance, your anger might give you the strength to fight the attacker off and break away. However, if you're angry with your little sister for messing up your room, yelling and hitting will only make the problem worse.

This doesn't mean that anger is a bad emotion that should be hidden from others. You just need to express it in positive ways that do not hurt others. One helpful approach is to think about what is causing your anger and look for a solution. Sometimes, though, you just feel angry for no good reason. In that case, it may help to find ways to relax, such as writing or engaging in physical activity.

Defense Mechanisms

Defense mechanisms are strategies that people use to deal with strong emotions. They are your mind's way of shifting focus away from an emotion that you do not want to face. Everyone uses them sometimes—in most cases, without even realizing that they are doing it.

Some defense mechanisms are more harmful than others. Denial or projection can help protect you from pain in the short term. In the long term, however, they just keep you from dealing with your problems. Sublimation, on the other hand, can turn unwanted emotions into something positive.

❯❯❯ Reading Check

IDENTIFY *Name two defense mechanisms.*

Harry has just found out his parents are getting a divorce. He feels angry and upset, but he doesn't want to recognize those emotions. This chart shows how he uses defense mechanisms to deal with his feelings. *Describe other situations in which a person might use defense mechanisms.*

Defense Mechanisms

Denial

How it works
You refuse to face a feeling you do not want to accept.

Example
Harry tells himself he does not care about the divorce.

Projection

How it works
You pretend someone else is having these feelings.

Example
Harry tells everyone his little sister is really upset about the divorce.

Displacement

How it works
You take your feelings out on someone other than the person who hurt you.

Example
Harry picks a fight with his best friend rather than confront his parents.

Sublimation

How it works
You redirect your feelings into some other, more positive activity.

Example
Harry focuses on his schoolwork to avoid thinking about the divorce.

It is normal to sometimes feel anger. *Explain how to express anger in a positive way.*

©Ingram Publishing/AGE Fotostock

YOUR EMOTIONAL NEEDS

MAIN IDEA All human beings share certain emotional needs.

All people have certain basic physical needs, such as food, water, and shelter. These are needs because people must have them in order to survive. They are much more important than wants, or things people like to have, such as books and DVDs. In the same way, all people share certain emotional needs, which are *needs that affect a person's feelings and sense of well-being.* All people feel a need to love and be loved by others. They also feel a need to belong—for instance, as part of a family, a student at school, or a member of a team. Finally, people feel a need to make a difference in the world. They want to do something that matters to them and to be valued for their achievements.

You can meet your emotional needs in many ways. A healthy way to meet your need for love is to spend time with your family and friends. Similarly, you can meet your need to belong by joining a group or a club. As for your need to make a difference, helping out in your community can be a way to help others. You will earn the respect of your neighbors, and improve your self-esteem.

>>> **Reading Check**

LIST *What are three positive and negative ways to meet emotional needs?* ■

Recognizing your talents is one way of building self-esteem. *Explain how having high self-esteem benefits your health.*

REVIEW

>>> **After You Read**

1. **DEFINE** Define the word *emotions* and give two examples.
2. **DESCRIBE** What are two strategies for dealing with emotions in a healthy way?
3. **IDENTIFY** What emotional needs do all people share?

>>> **Thinking Critically**

4. **INFER** How could the way you express emotions affect your social health?
5. **EVALUATE** Bradley's teacher assigns a presentation to the class. Bradley is very nervous. He is having trouble sleeping. Is Bradley's fear helpful or harmful? Explain why.

>>> **Applying Health Skills**

6. **CONFLICT RESOLUTION** Sophia has asked her mom to pick her up at soccer practice. However, her mom doesn't show up until half an hour after practice is over. Sophia is angry. Write a dialogue between Sophia and her mother showing how she expresses her anger in a healthy way.

Review

Audio

Managing Stress

BIG IDEA Keeping stress under control will improve all aspects of your health.

Before You Read

QUICK WRITE Write a few sentences briefly describing a time when you felt stress. How did you manage your feelings?

▶ Video

As You Read

STUDY ORGANIZER Make the study organizer found in the FL pages in the back of the book to record the information presented in Lesson 3.

Vocabulary

›stress
›stressor
›fight-or-flight response
›adrenaline
›time management

🔊 Audio

🔤 Bilingual Glossary

Myth vs. Fact

Myth: Stress is always bad for you.

Fact: Having a little stress in your life is a good thing. A bit of stress adds excitement and helps you accomplish more. It's only when stress isn't managed well, or when it goes on too long, that it starts to harm your health.

WHAT IS STRESS?

MAIN IDEA Both positive and negative events can cause stress.

Can you remember the last time you were really nervous? You may have had symptoms like sweaty palms, a racing heart, or butterflies in your stomach. These symptoms are all normal responses to **stress**, *the body's response to real or imagined dangers or other life events.* Stress is a normal part of life, and it is not always harmful. For example, you may feel tense while watching a close football game, but that tension is part of the fun. A little bit of stress can also motivate you to work hard and do your best. However, when you face too much stress or the stress lasts too long, it can cause health problems, such as irritability, headaches, or trouble sleeping. *Anything that causes stress* is called a **stressor.** Some stressors are just everyday problems, like an argument with a friend.

> A *little* bit of stress can also **motivate** you to work hard and do *your best.*

Others are serious threats, like experiencing an earthquake, flood, or other natural disaster in your region. Sometimes, even positive events can cause stress. Moving to a new house or adding a new family member is a happy event, but it can still cause stress. Also, different people react to stressors in different ways. One person, for example, might find driving in traffic very stressful, while another might use it as a time to relax and listen to music.

In addition to reacting differently to stressors, each person identifies different stressors. An event that is stressful for you may not be to your friend.

>>> **Reading Check**

EXPLAIN *How can stress be positive?*

HOW YOUR BODY RESPONDS TO STRESS

MAIN IDEA Long-term stress can damage your physical, mental/emotional, and social health.

Suppose that you are riding your bike down the road. Suddenly, a car pulls out right in front of you. As you squeeze the brake, you feel your heart speed up and your stomach clench. What you are experiencing is the fight-or-flight response, which is *the body's way of responding to threats.* When you encounter a stressor, your brain signals your body to begin producing adrenaline, *a hormone that increases the level of sugar in the blood, giving your body extra energy.* The release of adrenaline triggers a process that affects almost your entire body.

The *fight*-or-*flight* response is useful because it helps you **respond** to **threats.**

In some cases, you may feel stress when there is nothing to "fight or flee." In other cases, stress continues for so long that your body can't handle it. High stress levels can affect all three sides of your health triangle.

Physical effects of stress can include headaches, back pain, and upset stomach. Stress can also cause sleep problems. Long term stress can lead to problems such as weight gain, high blood pressure, or heart disease. Stress can also reduce the body's ability to fight off infection.

Mental and emotional effects include feelings of anxiety, anger, or sadness. Stress can affect your mind, making you forgetful and unable to focus.

Social effects of stress can make you irritable. Your anger may become harmful, leading to conflicts in your relationships. In some cases, high levels of stress can lead people to withdraw from friends and family members.

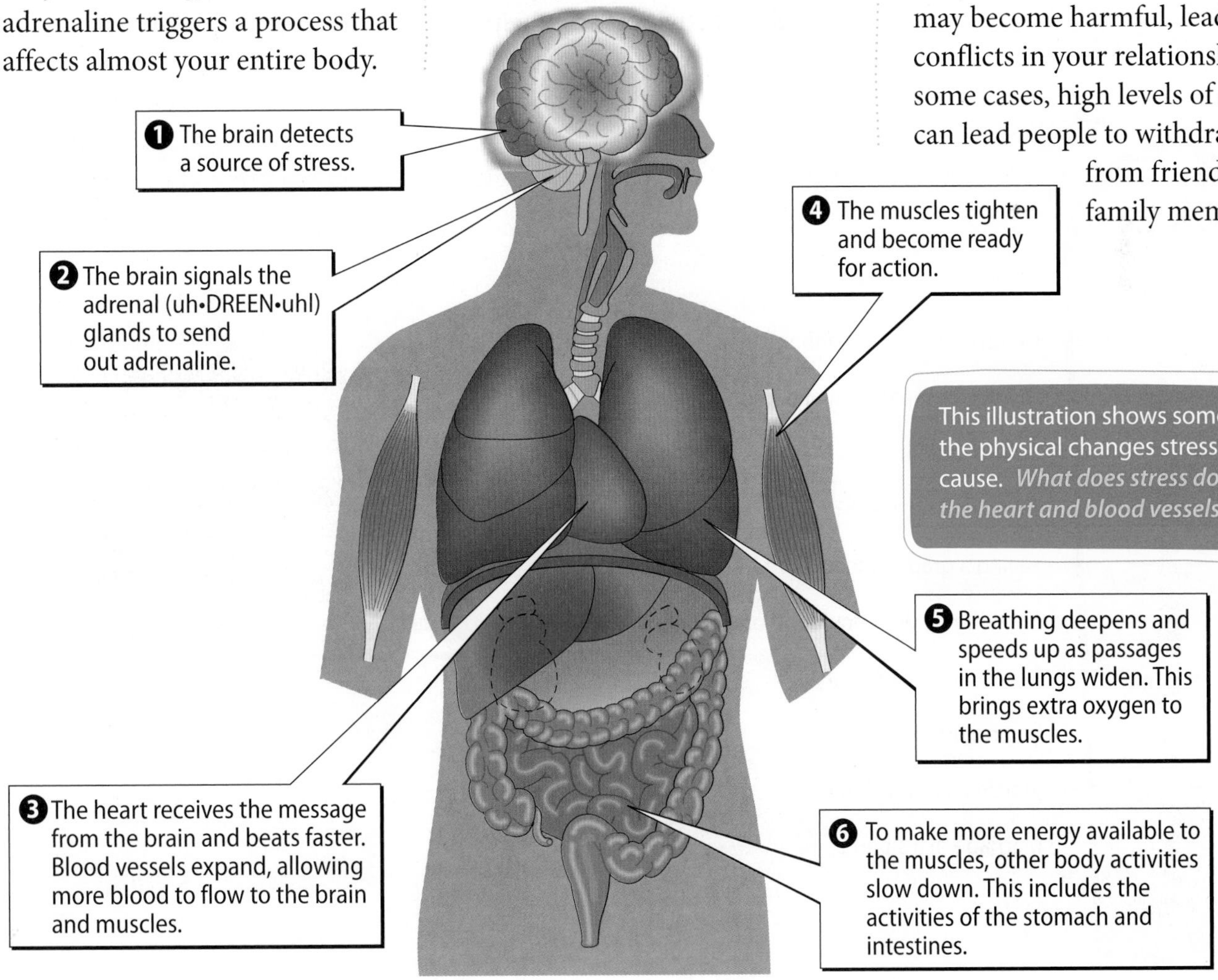

❶ The brain detects a source of stress.

❷ The brain signals the adrenal (uh·DREEN·uhl) glands to send out adrenaline.

❸ The heart receives the message from the brain and beats faster. Blood vessels expand, allowing more blood to flow to the brain and muscles.

❹ The muscles tighten and become ready for action.

This illustration shows some of the physical changes stress can cause. *What does stress do to the heart and blood vessels?*

❺ Breathing deepens and speeds up as passages in the lungs widen. This brings extra oxygen to the muscles.

❻ To make more energy available to the muscles, other body activities slow down. This includes the activities of the stomach and intestines.

MANAGING STRESS

MAIN IDEA You can manage your stress level partly by avoiding stressors and partly by finding ways to cope with stressors you cannot avoid.

In some cases, you can learn to recognize and steer clear of stressful situations. However, it isn't possible to remove all sources of stress from your life. When you can't escape from stress, you can learn to manage your response to it.

Strategies *for* Avoiding Stress

In order to reduce stress in your life, it helps to know what is causing it. Once you know what situations are most likely to be stressful for you, you can sometimes find ways to avoid them.

For example, if riding on a crowded bus makes you feel tense, you could try walking instead.

Another important strategy for reducing stress is **time management,** or *strategies for using time effectively.* If you're like many teens, you may have trouble finding enough time for everything you need to do. Planning ahead can help. If you set aside regular blocks of time for schoolwork and chores, you won't have to rush to get them done at the last minute. Making a day planner—a detailed list of your daily activities—can help you stay organized.

Of course, no matter how carefully you plan your time, there's a limit to how many activities you can cram into your day. Trying to play a sport, act in the school play, and sing in a choir all at once is a recipe for too much stress and too little sleep. That's why you need to learn to set priorities. Decide which activities are the most important to you, and then give yourself the time to focus on them. Learn to say no when you are asked to take on more than you can handle.

≫ Reading Check

IDENTIFY *Give three strategies for avoiding stress.*

- Death of a family member
- Divorce or separation of parents
- Birth of a sibling
- Job loss of a parent
- Failing a grade in school
- Breaking up with boyfriend or girlfriend
- Serious illness or injury
- Beginning to date
- Starting a new school
- Marriage of a sibling

The more of the items on this list you are facing, the more likely it is that stress will affect your health. *Name two other stressors you would add to this list.*

Strategies *for* Coping with Stress

Good general health habits can help you deal with the stressors you can't avoid. Good nutrition and plenty of sleep can give you the energy you need to deal with your problems. It may also help to try some specific stress-relieving strategies, such as:

- **Relaxation.** Try relaxation techniques such as taking deep breaths, stretching, or taking a hot shower to loosen the tension in your muscles. You can also relax simply by doing something you enjoy, such as reading a book or listening to music.
- **Physical activity.** Being physically active releases chemicals in your body that help you feel happy and calm. You can try lifting weights, playing sports, or just going for a walk.
- **Talking it out.** Talking about your feelings, or writing about them in a journal, can help get them off your chest.

It may also help you think of new ways to deal with your problems.

Laughter relieves **tension** and leaves you feeling more *relaxed.*

- **Keeping a positive outlook.** Laughter relieves tension and leaves you feeling more relaxed. It's also helpful to think positively about the events causing you stress, such as a challenging test in the next few days.

>>> **Reading Check**

EXPLAIN *How can budgeting your time help you manage stress?* ■

You can learn strategies to cope with stress. *Describe two healthy ways to cope with stress.*

Tom Merton/OJO Images/Getty Images

>>> **After You Read**

1. **DEFINE** Explain how the words *stress* and *stressor* are related.
2. **GIVE EXAMPLES** Name one positive event and one negative event that could cause stress.
3. **IDENTIFY** What are two healthful strategies for dealing with stress?

>>> **Thinking Critically**

4. **ANALYZE** How does stress affect your physical, mental/emotional, and social health? How are these effects related?
5. **HYPOTHESIZE** Do you think life is more stressful for teens today than it was for your parents? Why or why not?

>>> **Applying Health Skills**

6. **STRESS MANAGEMENT** Give an example of a situation that might cause stress. Then outline strategies you might use for managing stress in this situation.

🔄 Review

🔊 Audio

Coping *with* Loss

BIG IDEA There is no right or wrong way to grieve over a loss.

Before You Read

QUICK WRITE Write a paragraph describing how you would help a friend who has lost a loved one.

▶ Video

As You Read

STUDY ORGANIZER Make the study organizer found in the FL pages in the back of the book to record the information presented in Lesson 4.

Vocabulary

› grief
› grief reaction
› coping strategies

🔊 Audio

🔤 Bilingual Glossary

UNDERSTANDING GRIEF

MAIN IDEA Grief is a normal response when a person suffers a painful loss.

Grief is *the sorrow caused by a painful loss, such as the loss of a loved one.* People may grieve over the death of a relative, a friend, or even a family pet. They might also react this way to other kinds of loss, such as to moving or having a divorce in the family. Grief can cause feelings of sadness, loneliness, or anger. How people experience grief, and how long it lasts, differs for each person.

The Grief Reaction

The process of dealing with strong feelings following any loss is known as the grief reaction. There is no right or wrong way to grieve. However, there are several stages of grief that many people go through. People may not experience all the stages, and they may not happen in a certain order.

Grief can cause *feelings* of **sadness, loneliness,** or **anger.**

DENIAL Immediately after a loss, people may be in a state of shock. They may be unable to believe that the loss occurred.

EMOTIONAL RELEASE When a loss is recognized, it may bring intense emotions, such as crying.

ANGER Some people feel angry with their loved one for leaving them. Some may blame themselves or others.

BARGAINING People think about what they could have done to prevent the loss. They may feel guilt or may wish they had done more for the person.

REMORSE People may cry, feel hopeless or physical symptoms, such as trouble sleeping.

ACCEPTANCE People begin to move on. They begin to accept the loss.

HOPE Thinking about the person becomes less painful. The person who is grieving begins to think about the future.

COPING WITH GRIEF

MAIN IDEA Coping strategies help people deal with grief.

I n most cases, there is no way to undo a loss. However, it is possible to cope with the grief a loss can bring. <u>Coping strategies</u> are *ways of dealing with the sense of loss people feel when someone close to them dies.* One strategy is to give yourself time to grieve and to accept your feelings. Allowing yourself to cry serves as a useful release for strong emotions. It may also help to talk about your feelings with others. Friends and family often want to know what they can do for someone who is grieving, so don't hesitate to let them know what you need.

> *Remember,* though, that each person **grieves** in a *different* way.

> Friends can be a source of comfort in times of grief. *Describe how you can help someone who is grieving.*

Helping Others

When someone close to you has suffered a loss, just being there for that person can be a great help.

Remember, though, that each person grieves in a different way. Some people may need to cry, while others may want to talk about death. Respect the other person's feelings. Physical touch, such as holding hands or hugging, can be very comforting. It is the decision of the grieving person to decide if she or he is comfortable with it, so follow that person's lead.

You can show empathy by acknowledging the pain the person is experiencing. However, try to avoid sharing stories of your own as a way to show that you understand. The grieving person may feel as if you are dismissing his or her pain. Try to listen without judging or giving advice. However, don't force the other person to share feelings if she or he doesn't want to.

Allow the person to decide how much time he or she needs to recover from the loss. Some people recover quickly, while others may not.

> **>>> Reading Check**
>
> **IDENTIFY** *Name two strategies for coping with grief.* ■

LESSON 4

REVIEW

>>> After You Read

1. **DEFINE** What is the *grief reaction?*
2. **LIST** Name five stages that many people experience while grieving.
3. **EXPLAIN** How can you show support for someone who is grieving?

>>> Thinking Critically

4. **EVALUATE** How long should it take to grieve over a death?
5. **ANALYZE** Keiko's parents plan to get a divorce. She is upset and wants them to stay married. She has promised to get straight A's in school from now on if they will stay together. What stage of grief do Keiko's actions most reflect?

>>> Applying Health Skills

6. **STRESS MANAGEMENT** The death of a loved one can cause a great deal of stress. Refer back to the strategies for managing stress. Write a paragraph explaining which of these strategies you think might be useful for someone who is grieving.

🄫 Review

🔊 Audio

Stress Chasers

When you're feeling stress, your whole body is affected. You may feel stiffness in your shoulders or neck. Your mind may be cluttered with troubling thoughts. The table below lists several exercises you can do to help remove this tension from your body and mind.

WHAT YOU WILL NEED

* pencil and paper

WHAT YOU WILL DO

1 Estimate your current level of body tension or stress. Use a scale of 1 to 5, where 1 is "totally calm" and 5 is "very stressed. "Write this number on your paper, along with the words "Starting Stress Level."

2 Perform the first exercise on the list below the photo. When you are done, estimate your stress level again. Write this number down, along with the name of the exercise.

3 Repeat the process for each of the other exercises on the list.

WRAPPING IT UP

Compare your results with those of your classmates. How did each exercise on the list affect your tension level? Which exercises worked best for your classmates in general?

Deep Breathing	Close your eyes and take a deep breath. Hold it for a moment, then slowly exhale. Repeat several times.
Shoulder Lift	Hunch your shoulders up to your ears for a few seconds, then release. Repeat.
Elastic Jaw	Open your mouth and shift your jaw as far to the right as you can without discomfort. Hold for a count of three. Repeat on the left side.
Fist Clench	Make a fist. Tense the muscles in your hand and forearm, then release. Repeat this with your other hand.
Visualization	Close your eyes. Picture a pleasant scene, such as a sunny beach or park. Hold this image in your mind for several seconds.

READING REVIEW

FOLDABLES and Other Study Aids

Take out the Foldable® that you created and any study organizers that you created. Find a partner and quiz each other using these study aids.

LESSON 1 Mental and Emotional Health

BIG IDEA Good mental and emotional health includes having a positive view of yourself and being able to deal with the ups and downs of daily life.

* Your personality, or the way others see you, is related to your self-concept, the way you see yourself.
* Having high self-esteem can improve all three sides of your health triangle.
* Developing resilience can help you recover from problems or losses quickly.

LESSON 2 Understanding Your Emotions

BIG IDEA Learning to deal with emotions in healthy ways is important during your teen years.

* No emotion is good or bad in itself; what matters is learning to express them in healthy ways.
* Defense mechanisms can be helpful for dealing with strong emotions in the short term, but in the long term they prevent you from dealing with your problems.
* All human beings share certain emotional needs, including the need to love and be loved, the need to belong, and the need to make a difference.

LESSON 3 Managing Stress

BIG IDEA Keeping stress under control will improve all aspects of your health.

* Both positive and negative events can cause stress.
* The fight-or-flight response is your body's way of dealing with a real or perceived threat.
* If stress continues for a long time, it can damage your physical, mental/emotional, and social health.
* Strategies for managing stress include avoiding stressors, managing your time, setting priorities, relaxation, physical activity, talking about problems, and keeping a positive outlook.

LESSON 4 Coping with Loss

BIG IDEA There is no right or wrong way to grieve over a loss.

* Many types of loss, including the death of a loved one, can cause a grief reaction.
* Many people go through stages of denial, anger, bargaining, depression, and acceptance while grieving. However, they may not go through all these stages or experience them in order.
* Ways to deal with grief include acknowledging your feelings, crying, and accepting comfort from others.

 Review

 Web Quest

ASSESSMENT

Reviewing Vocabulary *and* Main Ideas

> mood swings

> mental/
  emotional health

> personality

> emotions

> anxiety

> resilience

>> On a sheet of paper, write the numbers 1–6. After each number, write the term from the list that best completes each statement.

LESSON 1 **Your Mental and Emotional Health**

1. The three most important factors that influence your _____ are heredity, environment, and behavior.

2. The ability to recover from problems or loss is _____.

3. The ability to handle the stresses and changes of everyday life in a reasonable way is called _____.

LESSON 2 **Understanding Your Emotions**

4. Frequent changes in emotional state are called _____.

5. _____ are feelings that include happiness, anger, and fear.

6. _____ is a state of uneasiness.

>> On a sheet of paper, write the numbers 7–10. Write *True* or *False* for each statement below. If the statement is false, change the underlined word or phrase to make it true.

LESSON 3 **Managing Stress**

7. The chemical your body produces in response to a stressor is called <u>adrenaline</u>.

8. The tiredness that occurs after a stressful situation is called <u>depression</u>.

LESSON 4 **Coping with Loss**

9. The grief reaction includes denial, anger, <u>time management</u>, depression, and acceptance.

10. One important <u>grief</u> strategy that can help you deal with sorrow over a loss is confronting your feelings head-on.

✔ eAssessment

>> Using complete sentences, answer the following questions on a sheet of paper.

🗨️ *Thinking* Critically

11. SYNTHESIZE Why is being resilient important for mental and emotional health?

12. EXPLAIN Why might people who are often sad become sick more often?

🔑 *Write* About It

13. NARRATIVE WRITING Write a short story about a teen who is stressed out. Be sure to include what is causing stress in the teen's life and the healthful strategies that the teen uses to manage the stress.

14. DESCRIPTIVE WRITING Write a poem about the different activities you can do to manage your emotions.

Ⓐ Ⓑ Ⓒ Ⓓ STANDARDIZED TEST PRACTICE

Math
Use the chart to answer the questions.

Mental and Emotional Disorders in Young People	
Anxiety disorders	As many as 1 in 10 young people
Depression	As many as in 8 young people
Bipolar disorder	About 1 in 100 young people
Conduct disorder	As many as 1 in 10 young people
Eating disorders	1 in 100 to 200 teen females and a much smaller number of males
Schizophrenia	About 3 in 1,000 young people

1. Out of a group of 100 young people, what percentage suffers from anxiety disorders?

2. What is the percentage of young people who have schizophrenia?

3. Of the disorders listed, which is the most common among young people?
   A. Anxiety disorders
   B. Conduct disorder
   C. Depression
   D. Eating disorders

4. Out of a group of 1,000 young people, what percentage suffers from bipolar disorder?

MENTAL/EMOTIONAL HEALTH

Your teenage years bring lots of changes, and those changes also can affect your mental/emotional health.

MOOD SWINGS, ANXIETY AND, ANGER

Mood swings

It's normal to have mood swings. They can happen because of changing hormones, worries over the future, or concerns over relationships.

Anxiety

Have you ever had "butterflies in your stomach" before a big test or an important sports match? Mild anxiety can be actually useful because it gives you extra energy. When it builds up too much, however, it can interfere with functions like eating and sleeping.

Anger

When you are angry, you may feel like lashing out at the source of your anger. Yelling and hitting will only make the problem worse.

STRESS

Both positive and negative events can cause stress.

Anything that causes stress is a stressor. Some stressors include:
- An argument with a friend.
- Test stress
- Natural disasters.

Physical effects of stress can include headaches, back pain, sleep problems, and upset stomach. It can lead to weight gain, high blood pressure, or heart disease.

Mental and emotional effects include feelings of anxiety, anger, or sadness. Stress can make you forgetful and unable to focus

Stress can make you irritable with others, leading to conflicts in your relationships. In some cases, stress can lead people to withdraw from friends and family members.

GRIEF AND LOSS

Grief is a normal reaction to the loss of a loved one, such as a relative, friend or pet, or other kinds of loss, such as changing homes or dealing with divorce. How people experience grief, differs from person to person.

FIVE STAGES OF GRIEF

A widely accepted explanation of how grief affects people includes five stages:

DENIAL

This can't be happening.

ANGER

Why? It's not fair!

BARGAINING

I'll do anything for a few more years.

DEPRESSION

I miss my loved one, why go on?

ACCEPTANCE

It's going to be okay.

MENTAL/EMOTIONAL HEALTH DISORDERS

Mental illness is like any other illness with symptoms and treatments.

1 IN 4
Adults in the U.S. affected by mental and emotional problems

1 IN 10
Children in the U.S. personally affected by mental and emotional problems

ANXIETY DISORDERS

For some people, anxiety never goes away. Doctors do not know exactly what causes these disorders. However, they can still treat them with medication and counseling.

COMMON TYPES OF ANXIETY DISORDERS

Anxiety disorders include panic disorder, social anxiety disorder, obsessive-compulsive disorder, post-traumatic stress disorder, and phobias.

MOOD DISORDERS

Strong emotions that last a long time may signal a mood disorder. The two main types of mood disorders are depression and bipolar disorder.

Depression

Clinical depression causes feelings of:
• Extreme sadness
• Hopelessness
• Worthlessness
• Guilt

Bipolar disorder

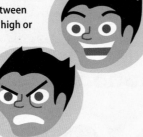

People with bipolar disorder switch between high and low emotional states. During high or manic periods they may:
• Feel either happy or irritable.
• Sleep less.
• Have huge amounts of energy.
• Behave in reckless ways.

PERSONALITY DISORDERS

Three types of personality disorders include paranoid, antisocial, and dependent personality disorders.

PARANOID
People with this disorder are:
• Paranoid.
• Suspicious.
• Hostile.

ANTISOCIAL
People with this disorder may:
• Lie.
• Steal.
• Physically harm a person.

DEPENDENT
People with this personality disorder:
• Cannot make decisions.
• Tend to stay in abusive relationships.

Schizophrenia

Schizophrenia is a severe mental disorder in which a person:
• Loses contact with reality.
• Often has delusions.

GETTING HELP

If you are suffering from depression or another mental or emotional disorder, it's important to remember that you are not alone. Seek help right away for any of the following symptoms:

Feelings of sadness, anger, anxiety, or fear that don't go away

Changes in eating or sleeping patterns

Doing much worse in school all of a sudden

Wanting to be alone all the time

Loss of interest in favorite activities

A feeling of being out of control

Hearing voices

Thoughts of suicide

Mental + Emotional Disorders

LESSONS

PREMIUM ONLINE RESOURCES

 Audio

 Videos

 Bilingual Glossary

 Fitness Zone

 Web Quest

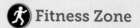 Review

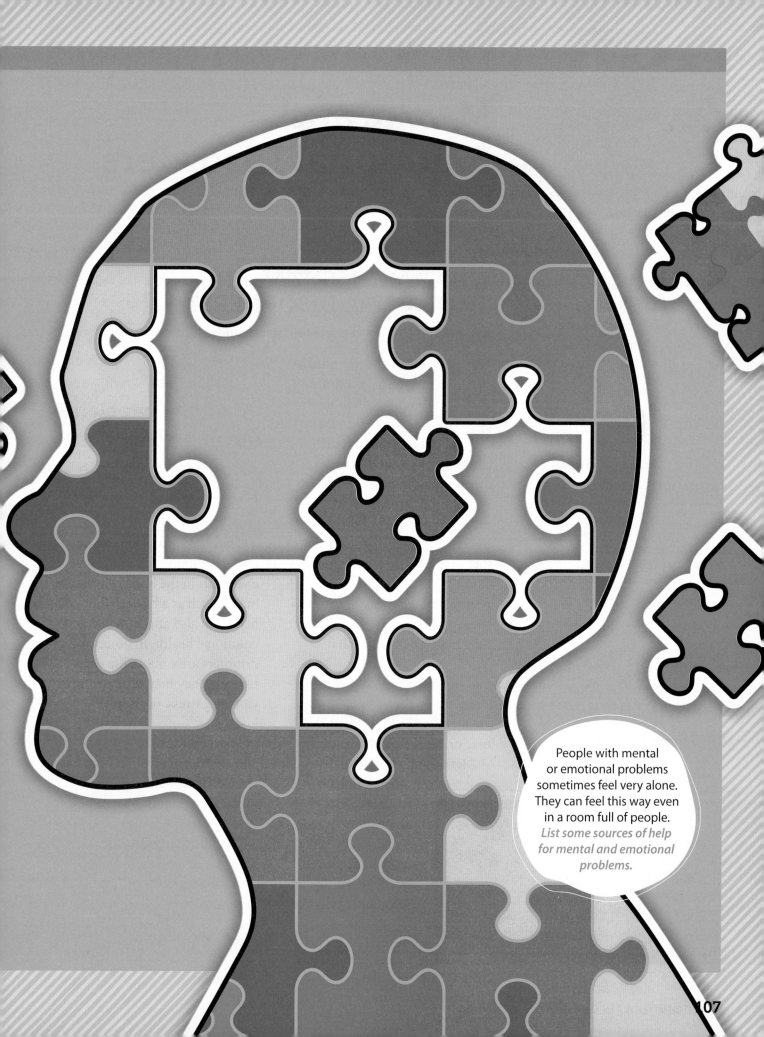

People with mental or emotional problems sometimes feel very alone. They can feel this way even in a room full of people. *List some sources of help for mental and emotional problems.*

Mental *and* Emotional Disorders

> **BIG IDEA** Mental and emotional disorders are real illnesses, just like physical disorders.

Before You Read

QUICK WRITE Think about a time when you were very sad. Tell what you did to overcome your sadness.

▶ Video

As You Read

FOLDABLES® Study Organizer

Make the Foldable® found in the FL pages in the back of the book to record the information presented in Lesson 1.

Vocabulary

› mental and emotional disorders
› anxiety disorders
› phobias
› mood disorder
› personality disorders
› schizophrenia

 Audio

 Bilingual Glossary

Myth or Fact

Myth: People with mental illness are violent and dangerous.

Fact: Most people with mental health problems are no more violent than anyone else. They are much more likely to become victims of crime than they are to harm others.

WHAT ARE MENTAL AND EMOTIONAL DISORDERS?

> **MAIN IDEA** Mental and emotional disorders are health disorders.

Mental and emotional problems are more common than many people realize. Each year, they affect about one out of four adults and one out of ten children in the United States. As common as they are, however, these problems are often misunderstood. Some people think that problems such as depression are just part of someone's personality. They may even see these problems as a character weakness. The truth is that mental and emotional disorders, *illnesses that affect a person's thoughts, feelings, and behavior,* are illnesses, just like heart disease or cancer.

Mental and emotional disorders can have various causes. Some of them may be caused by genetics that are passed from one family member to another.

> Mental and emotional problems are more common than many people realize.

They can result from having too much or too little of certain chemicals in the brain. Physical injury, such as a blow to the head, can also affect mental and emotional health. In other cases, disorders arise out of life experiences. They may result from violence, stress, or the loss of a loved one.

Like other diseases, mental and emotional disorders can be treated. Treatment can include medication, counseling, or both.

> **Reading Check**
>
> **EXPLAIN** *What causes mental and emotional disorders?*

Design Pics/Don Hammond

ANXIETY DISORDERS

MAIN IDEA ⟩ Intense anxiety or fear keeps a person from functioning normally.

Everybody feels anxious sometimes. It's normal to feel nervous when you're trying out for a play or starting an important test. For some people, however, anxiety never goes away. It takes over their lives. These people have anxiety disorders or *extreme fears of real or imaginary situations that get in the way of normal activities.* Doctors do not know exactly what causes these disorders. However, they can still treat them. Usually, treatment involves both drugs and counseling. There are several common types of anxiety disorders.

- **Generalized anxiety disorder** makes people worry all the time, often for reasons that aren't that important. It can also cause physical problems, such as headaches.
- **Panic disorder** causes intense fear called panic attacks.

People with OCD may feel the need to perform certain actions over and over, such as washing their hands. *Predict how this person might feel after turning off the faucet.*

Symptoms include a pounding heart, sweating, or nausea. They make people feel so awful that they will go out of their way to avoid anything that could cause another attack.

- **Phobias** are *intense and exaggerated fears of a specific object or situation.* The following list describes some common phobias.
- **Social anxiety disorder,** also known as social phobia, is a fear of dealing with people. Some people are afraid of specific situations, such as public speaking. Others are afraid to be around other people at all. They may fear that others are watching or judging them.
- **Obsessive-compulsive disorder (OCD)** traps people in a pattern of repeated thoughts and actions. It may force them to do certain tasks over and over, such as washing hands or counting objects.
- **Post-traumatic stress disorder** occurs after a very stressful event. Physical or sexual violence can trigger this disorder. It also affects many war veterans. People with this disorder have flashbacks in which they relive the event. Some people recover quickly from this disorder, but others need more time.

⟩⟩⟩ **Reading Check**

IDENTIFY *What are the symptoms of OCD?*

This table lists several of the most common phobias. *What is the name for fear of flying?*

Acrophobia	Fear of heights
Agoraphobia	Fear of crowded places
Astraphobia	Fear of thunder and lightning
Aviophobia	Fear of flying
Claustrophobia	Fear of enclosed spaces
Hemophobia	Fear of blood
Hydrophobia	Fear of water
Odontiatophobia	Fear of dentists
Trypanophobia	Fear of injections
Zoophobia	Fear of animals (usually spiders, snakes, or mice)

©McGraw-Hill Education/Christopher Kerrigan

MOOD DISORDERS

MAIN IDEA Extreme or inappropriate changes in mood that last a long time may indicate a mood disorder.

The teen years can be a difficult time in life. Friendships may become more complicated. Your relationships with your friends and family are changing. During this time, it's normal to go from feeling happy to feeling sad quite suddenly. However, strong emotions that last a long time may be a sign of a mood disorder. This is *a mental and emotional problem in which a person undergoes mood swings that seem extreme, inappropriate, or last a long time.* The two main types of mood disorders are depression and bipolar disorder.

Depression

You might describe yourself as feeling "depressed" because you did badly on a test or your boyfriend or girlfriend broke up with you. This kind of sadness is normal and happens to everyone sometimes. However, clinical depression is a mental illness that causes feelings of extreme sadness, hopelessness, worthlessness, and guilt. These feelings may last for weeks, months, or even years.

People who are depressed often lose interest in activities they once enjoyed. They may also feel angry or irritable for no reason. They may withdraw from family and friends. Some are unable to sleep, while others sleep all the time.

Depression can also cause physical problems, such as tiredness and headaches. If depression is not treated, people may begin to think about suicide.

> It's normal to go from feeling happy to feeling sad quite suddenly.

If you know someone who seems depressed, encourage him or her to talk to a parent or other trusted adult. If he or she says talking won't help, you should tell an adult about your concerns.

Bipolar Disorder

Bipolar disorder causes extreme mood swings. People switch between low periods and high periods.

Their low periods cause all the symptoms of depression. During their high, or manic, periods, people are in a constant state of excitement. They may feel either happy or irritable. They sleep less and have huge amounts of energy. They may also behave in reckless ways.

However, in between the cycles, people may go through calmer periods. At these times, their behavior is fairly normal.

>>> **Reading Check**

DEFINE *What is a mood disorder?*

Long-term feelings of sadness can be a sign of depression. *Explain why it is important to seek treatment for depression.*

Comstock/Jupiter Images

The highs and lows of bipolar disorder can make a person feel torn in two. *What are some of the symptoms people show during the manic phase?*

PERSONALITY DISORDERS

MAIN IDEA Three types of personality disorders include paranoid, antisocial, and dependent personality disorders.

All people have a basic need to belong and be accepted by others. Some people, however, find it very hard to relate to others. They get stuck in harmful patterns of thinking and acting that cause major trouble in their social lives. These people suffer from **personality disorders**. This term refers to *a variety of psychological conditions that affect a person's ability to get along with others.*

Personality disorders fall into three main groups. The first kind causes people to have a hard time trusting others. One example is paranoid personality disorder.

People with this disorder believe that others want to harm them. They are suspicious and often hostile.

Some people find it very hard to relate to others.

Disorders in the second group cause extremes of emotion and behavior. Antisocial personality disorder is one example. People with this disorder have little respect for others. They may lie, steal, or harm people in other ways, including physically. They often get into trouble with the law.

Finally, some disorders make people feel or act afraid all the time. People with dependent personality disorder, for example, look to others to take care of them. They do not like to make decisions for themselves. Often, they would rather put up with unkind or abusive treatment than be alone. People with this disorder also tend to engage in high-risk behaviors and have poor self-esteem.

⟩⟩⟩ Reading Check

IDENTIFY *Name two personality disorders.*

Fancy Collection/SuperStock

SCHIZOPHRENIA

MAIN IDEA Schizophrenia is a severe mental disorder that is treatable.

One of the most serious mental problems is schizophrenia (skit·zoh·FREE·nee·uh). This is *a severe mental disorder in which a person loses contact with reality.* People with this disorder often have delusions. These are beliefs that have no basis in reality. For instance, they might believe that aliens are attacking their minds. They may also hallucinate, hear voices or see people who are not there.

These symptoms are very frightening and can cause people to behave in strange and unpredictable ways. They may talk without making sense, or they may sit for hours without talking at all. They often have trouble holding a job or caring for themselves.

The strange behavior associated with schizophrenia makes people who have this disorder seem scary to others. They are likely to be feared, avoided, and rejected. They often end up in prison, where they cannot get the treatment they need. Doctors can treat the symptoms of schizophrenia with drugs. However, victims also need therapy and community support to help them learn to live normal lives again.

>>> **Reading Check**

EXPLAIN *How is schizophrenia treated?* ■

The strange behavior associated with schizophrenia makes people who have this disorder seem scary to others.

With therapy and community support, people with schizophrenia can rebuild their lives. *Describe the symptoms of schizophrenia.*

>>> **After You Read**

1. **DEFINE** What are *mental* and *emotional disorders*?
2. **IDENTIFY** Name three kinds of anxiety disorders.
3. **DESCRIBE** What are the symptoms of schizophrenia?

>>> **Thinking Critically**

4. **ANALYZE** How are the symptoms of depression and bipolar disorder similar? How do they differ?

>>> **Applying Health Skills**

5. **ACCESSING INFORMATION** Work with a group to develop a fact sheet about one of the disorders discussed in this lesson. Discuss the symptoms of the disorder and the kinds of treatment available. Provide a list of the sources you use and be prepared to explain why they are reliable.

⟳ Review

◄)) Audio

Brand X Pictures

Suicide Prevention

BIG IDEA Knowing the warning signs of suicide could help you save a life.

⟫⟫⟫ Before You Read

QUICK WRITE Write a paragraph describing what you could do to help a friend who is talking about suicide.

▶ Video

⟫⟫⟫ As You Read

STUDY ORGANIZER Make the study organizer found in the FL pages in the back of the book to record information presented in Lesson 2.

⟫⟫⟫ Vocabulary

› major depression
› suicide

🔊 Audio

🔤 Bilingual Glossary

Ingram Publishing

Myth vs. Fact

Myth: It is dangerous to ask people whether they are thinking about suicide. This might put the idea into their heads.

Fact: Asking directly about suicidal thoughts gives the person a chance to talk about his or her problems. It also shows that there is someone who cares.

DEPRESSION AMONG TEENS

MAIN IDEA Everyone feels depressed at times. Severe or long lasting depression may require treatment.

Julia's friend Logan seemed sad, but she figured he would soon bounce back from whatever was bothering him. Then a month went by and Logan still rarely smiled or spoke to anyone. Julia began to worry that something was really wrong.

Teens face pressure to succeed at school, rejection by peers, and life-changing family events.

The teen years can be stressful. During this time, teens accept new challenges at home and at school. Teens face pressure to succeed at school. They may have to deal with rejection by peers.

In addition, life-changing family events—such as moving, to a new home, divorce, the birth of a new sibling, or the death of a loved one—can feel overwhelming for young people. All these factors can lead to depression among teens.

Major depression is different from ordinary sadness. It is *a very serious mood disorder in which people lose interest in life and can no longer find enjoyment in anything.*

Depressed people feel worthless and empty inside. They lack energy, and they no longer get pleasure from the activities they once enjoyed. Depression can also cause physical symptoms such as headaches or stomachaches, loss of appetite, or trouble sleeping. These problems can last for months. In some severe cases, symptoms of depression can last for years.

⟫⟫⟫ Reading Check

DEFINE *What is major depression?*

SUICIDE

MAIN IDEA Suicide is not a solution to depression.

People suffering from major depression may come to feel that life is not worth living. Some of them begin to think about suicide as an escape from their pain. **Suicide** is *the act of killing oneself on purpose.* It is one of the leading causes of death among teens. This is one reason it is so important for people suffering from major depression to get help. Learning to recognize the warning signs of depression and suicide can help you help others—and maybe even save a life.

Causes *of* Teen Suicide

The biggest risk factor for suicide is mental illness. A teen is at greater risk for suicide if other family members also suffer from depression. The same is true of teens who are living in poverty or those who use alcohol or drugs. Abuse or violence can also be a factor. Depression may also be caused by bullying or cyberbullying. Specific problems may also lead to thoughts of suicide.

Teens who have attempted suicide say that they were trying to escape from a situation that seemed impossible.

Teens may be more likely to think of ending their own lives during times of major stress, such as a divorce in the family. Victims of bullying or cyberbullying are also at greater risk.

Suicide is one of the leading causes of death among teens.

Most teens find ways to cope with stress. They can turn to friends, family members, and others for help when they need it. However, some teens do not have a support network of this kind. Feeling cut off from other people increases their risk for suicide.

All teens sometimes face stressful situations, such as getting a poor grade in school. *Explain why some teens find it harder to cope with stress than others.*

SuperStock

Image Source/Getty Images

Warning Signs *of* Suicide

Teens who are thinking about suicide often show warning signs. They may talk about their plans. They may not use the exact words, "I'm going to kill myself." However, they may give hints, such as "It doesn't matter anymore" or "You won't have me around much longer." Remarks like this should always be taken seriously. Other possible warning signs include:

- Talking about suicide or death
- Talking about feeling hopeless, guilty, or worthless
- Pulling away from family and friends
- Loss of interest in normal activities
- Sudden changes in personal appearance

Teens who are thinking about suicide often show warning signs ahead of time.

- Self-destructive behaviors, such as violence, substance abuse, or running away
- Constant boredom or trouble concentrating

People who may be thinking about suicide need help right away. *Recall what you should do if someone you know shows warning signs of suicide.*

- Sudden drop in grades
- Giving away favorite belongings
- Becoming cheerful after a long period of depression (this may signal that the person has decided on suicide as a way to "solve" his or her problems)

What Teens Want to Know

How do I know if I'm depressed or just sad?
Teens often experience sadness due to school issues, family issues, and mood swings. Depression is a long-term feeling of sadness. A person suffering from depression experiences hopelessness and despair and finds it hard to carry out everyday activities. If you or a friend experience these feelings, talk with your parents or guardians, a school counselor, or a medical professional right away.

Providing Support

People who talk about or attempt suicide may not really want to die. Sometimes, a suicide attempt is a "cry for help." You can provide this help in many ways. If anyone you know shows any warning signs of suicide, it's always a good idea to talk to him or her. Just showing that you care can help the person feel less alone. It can also help the person to see that there are other solutions to his or her problems.

Sometimes, a suicide attempt is a "cry for help."

Let the person know that you care for him or her. Tell the person that you want to help. Sometimes, a person who is considering suicide is hurt or angry about something. As a friend, if you acknowledge the person's pain, you can help your friend.

Many times, the situation that is causing the person to consider suicide may not be an issue in a week or a month. One saying that many people use when referring to suicide is that it is a "permanent solution to a temporary problem."

If your friend is depressed, remind him or her that depression can be treated. Urge him or her to seek help.

Most importantly, never agree to keep someone else's suicide plans a secret. If you suspect that a friend may hurt him or herself, tell a trusted adult immediately. If you think the person might attempt suicide immediately, do not leave him or her alone. Call 911 or a suicide hotline for help.

Remember that life is full of happy and sad moments. At times, each one of us will feel hurt. When those feelings occur, think about the goals that you have set for yourself.

>>> **Reading Check**

IDENTIFY *What are two warning signs of suicide?*

Health SKILLS ACTIVITY

Communication Skills

Giving "Emotional *First Aid*"

People who are severely depressed have an emotional injury. Eventually, they will need professional help for their wounds to heal. Before they can get that help, they require emotional first aid. This can come from anyone with good communication skills. Perhaps you can influence the person to make healthful choices. Here are some tips:

* Let the person speak and express his or her negative emotions.

* Don't challenge or dare the person.

* Remind the person of past successes. Recall challenges that she or he overcame.

* Tell the person how important he or she is to you and others. Provide the person with reasons for living.

 With A Group Make a pamphlet with suggestions for helping troubled teens. Pass out copies of your pamphlet.

DEALING WITH DEPRESSION

MAIN IDEA ⟩ Help is available to manage depression.

Depression can feel like a heavy weight. However, suicide is not the solution; medical help is. With counseling and therapy, people can escape from under the heavy weight of depression and feel like themselves again.

If you are suffering from depression, it's important to remember that you are not alone. There are people who care about you and want to help you. Spend time with family and friends. Talk to someone as soon as possible. A trusted adult can help you get the medical attention you need. Don't wait and hope that the problem will go away on its own. The sooner you tell someone, the sooner you can get help.

Organizations exist that can help teens deal with suicidal thoughts. One is the National Suicide Prevention Lifeline. This free, 24-hour hotline offers help to anyone in crisis.

If you are suffering from depression, it's important to remember that you are not alone.

The National Hopeline Network also has a hotline and a Web site with information about depression. Call 1-800-273-8255 or go online at www.suicidepreventionlifeline.org.

⟩⟩⟩ **Reading Check**

NAME *What are two organizations that can help teens thinking about suicide?* ■

Group counseling is one way for depressed teens to receive emotional support. *List other available sources of help for troubled teens.*

REVIEW

⟩⟩⟩ **After You Read**

1. **VOCABULARY** Define the term *suicide* and use it in an original sentence.
2. **IDENTIFY** What are two symptoms of major depression?
3. **RECALL** What factors can make teens more likely to end their own lives?

⟩⟩⟩ **Thinking Critically**

4. **APPLY** Karen's friend Clay broke up with his girlfriend. Ever since, he has been unhappy and withdrawn. Today, however, he suddenly seemed more cheerful. He gave Karen his new music player, saying he didn't need it anymore. How should Karen respond?

⟩⟩⟩ **Applying Health Skills**

5. **COMMUNICATION SKILLS** Write a dialogue between a teen who has been showing warning signs of suicide and a concerned friend. Show how the second teen uses communication skills to show empathy and concern for the first teen.

 Review

 Audio

Help *for* Mental *and* Emotional Disorders

BIG IDEA Mental and emotional disorders can be treated.

Before You Read

QUICK WRITE Some teens find it very hard to ask for help when they have problems. Is it hard for you to ask for help? In a paragraph, explain why or why not.

 Video

As You Read

STUDY ORGANIZER Make the study organizer found in the FL pages in the back of the book to record the information presented in Lesson 3.

Vocabulary

› therapy
› family therapy
› psychologists
› psychiatrists
› clinical social workers (CSWs)

 Audio

 Bilingual Glossary

WHEN TO GET HELP

MAIN IDEA Mental disorders are treatable medical conditions.

It is not always easy to determine whether a mental health problem is serious. However, certain symptoms may signal a serious problem. If you or someone you know has been experiencing any symptoms, they should not be ignored. Denying a problem exists can prevent the person from seeking treatment and getting help.

Mental and emotional disorders are treatable illnesses. With treatment, most people with mental illnesses—even serious ones—can recover. However, many people with these disorders do not seek help. Some people simply do not realize that they have a mental illness.

> With the right treatment, most people with mental illnesses can recover.

Oftentimes, a person may blame a mental health illness on personal weakness. Others may feel ashamed, embarrassed, or afraid of being labeled as "crazy." Instead of seeking help, they hope that their symptoms will go away on their own. However, denying the problem will not solve it. Understanding that a mental disorder is a sickness—and it can be treated—can make it easier to seek medical help.

Developing Good Character

Respect

One reason many people do not seek help for mental health problems is that they see mental illness as shameful. People often look down on those who are mentally ill in a way they do not do with people who have physical disabilities. Learning to treat mentally ill people with dignity can be a key to helping them. What are some of the harmful labels that society attaches to people who are

©Lisa F. Young/Alamy

Spotting mental illness early is one of the keys to treating it. It isn't always easy to know when you need help with a mental health problem. After all, it's normal for teens to feel moody sometimes. However, some mental problems are too serious to ignore. It's important to seek help right away for any of the following symptoms:

- Feelings of sadness, anger, anxiety, or fear that don't go away
- Changes in eating or sleeping patterns
- Wanting to be alone all the time
- A feeling of being out of control
- Loss of interest in favorite activities
- Doing much worse in school all of a sudden
- Hearing voices
- Thoughts of suicide

>>> **Reading Check**

EXPLAIN *Why are many people unwilling to seek help for mental and emotional disorders?*

WHERE TO GET HELP

MAIN IDEA Trusted adults can help you find a health care professional.

If you think you need help, start by talking to a trusted adult. This could be a parent or guardian, a school nurse, a counselor, or a teacher. Often, just talking about the problem is an important step toward recovery. However, it is only the first step. A trusted adult can help steer you toward sources of professional help. These may include doctors, counselors, and support groups. The Health Skills Activity on the following page has more information on finding mental health resources in your community.

Many teens find therapy helpful at stressful times in their lives.

When you have a problem, talking to a trusted adult can help. *List some people you could talk to if you had a problem.*

Therapy *for* Mental Disorders

Treatment for mental and emotional problems nearly always involves some kind of therapy, or *professional counseling.* Therapy is not for "crazy" people. In reality, people who seek therapy are brave. These people are seeking ways to manage their thoughts and feelings in a way that is not harmful to their health. Many teens find it helpful at stressful times in their lives.

For example, a teen might seek therapy for help in dealing with grief. Therapy can also help people break harmful habits, such as smoking. People may receive therapy in various settings and from various kinds of health professionals. However, the goal is always the same: to help people recognize their problems and find ways to heal.

Tetra Images/Getty Images

Community Resources for *Mental and Emotional* Problems

One of the mental and emotional disorders you have read about in this chapter is obsessive-compulsive disorder (OCD). If you thought that someone you know suffered from OCD, would you know where to seek help? Would you be able to tell them what resources your community offers for counseling, therapy, treatment, or hospitalization? You could provide appropriate, helpful information by researching the topic using reliable sources such as the following:

* **THE INTERNET**

* **SCHOOL AND LOCAL LIBRARIES**

* **TEACHERS**

* **HEALTH CARE PROFESSIONALS**

On Your Own

Create a brochure that lists some community resources for treating OCD. Use several of the sources listed above for your research. Organize your information based on the types of resources and how each resource addresses the disorder. Include descriptions of the disorder as well as tips for how a person could help someone who needs information about treatment.

Therapy can take several different forms. In most cases, a person will meet with a therapist one on one. This is called individual therapy. All conversations with the therapist are private. A person can build a relationship with a therapist over time and discuss her or his feelings openly.

In some cases, a therapist may meet with a group of teens who are all going through the same kind of problems. This is known as group therapy. The group may be called a support group. Group therapy gives people a chance to see that they are not alone in their problems. Group members can learn from and support each other as they all talk about the problems they share.

A special form of group therapy is family therapy. This is *counseling that seeks to improve troubled family relationships.* A therapist can help family members learn to communicate, strengthen their relationships, and solve problems as a group.

In many cases, treatment for mental illness may involve drugs as well as therapy. Medication can help with many mental disorders. However, drugs do not cure mental disorders. They treat the symptoms so that people can function. In many cases, they must be used along with talk therapy to get good results.

>>> **Reading Check**

IDENTIFY *What are two types of therapy?*

Mental Health Professionals

For teens, the first step in finding help is talking with a trusted adult. Once the problem is out in the open, healing may begin. For many emotional problems, professional counseling or therapy is needed.

Mental health professionals have been trained to treat people with mental and emotional problems.

There are several different types of mental health providers. All of them have been trained to treat people with mental and emotional problems. However, they differ in the services they can provide. Which type of mental health professional to see depends on the specific needs of the person getting treatment.

Psychologists (sy•KAH•luh•jists) are *mental health professionals who are trained and licensed by the state to counsel.* They have degrees in the field of mental health, but they are not doctors. Psychiatrists (sy•KY•uh•trists) are *medical doctors who treat mental health problems.* They can provide medication as well as counseling. Clinical social workers (CSWs) are *licensed, certified mental health professionals with master's degrees in social work.* They help people overcome both health problems and social problems.

>>> **Reading Check**

LIST *Name three types of mental health professionals.* ■

Family therapy can help with problems that threaten the health of the family. *Explain how therapists help troubled families.*

Lisa F. Young/Alamy

>>> **After You Read**

1. **VOCABULARY** Define *therapy* and use the word in an original sentence.
2. **IDENTIFY** Name two warning signs of a mental disorder that requires treatment.
3. **EXPLAIN** What is the first step in getting help for a mental or emotional problem?

>>> **Thinking Critically**

4. **INFER** What factors might determine which type of mental health provider a person chooses to see?

>>> **Applying Health Skills**

5. **ADVOCACY** Create a poster about the importance of seeking help for mental and emotional disorders. Include information about signs of mental disorders and ways to treat them.

⟳ Review

🔊 Audio

Hands-On HEALTH ACTIVITY

Interpreting Vocal Stress

Sometimes, the way in which you say something can be more revealing than the words themselves. To better understand this concept, consider this ordinary sentence: "Is he a good catcher?" Read the sentence aloud. Did you notice the sound of your voice rising at the end? This rise in pitch is part of the *intonation pattern* of speech. Different kinds of sentences - statements, questions, and commands – have different intonation patterns.

This activity will give you a chance to work with another aspect of vocal intonation. This time you will be working with *vocal stress*. Also known as "emphasis," vocal stress is saying one word slightly louder than the rest of the words in a sentence. On paper, vocal stress is shown by *italics* or *underlining*. Changing the word that is stressed in a sentence can change that sentence's meaning.

WHAT YOU WILL NEED

* pencil or pen
* paper

WHAT YOU WILL DO

1 As a group, think up and write five sentences on separate sheets of paper. Read each sentence aloud, each time stressing a different word. Notice how the meaning of the sentence changes. Consider these examples:

a. "I told *Ed* to give me the book." (Meaning: I told Ed—not someone else—to give me the book."

b. "I told Ed to give me the *book*." (Meaning: I told Ed what to give me—the book, not something else.)

2 Exchange sentences with another group. Read each sentence with the stress on different words. See how many different meanings you can come up with.

WRAPPING IT UP

After each group has finished, discuss these questions as a class: How does changing the stress change its meaning? Do changes in stress reveal differences in the speaker's feelings? Explain your answer.

READING REVIEW

FOLDABLES and Other Study Aids

Take out the Foldable® that you created and any study organizers that you created. Find a partner and quiz each other using these study aids.

LESSON 1 Mental and Emotional Disorders

BIG IDEA Mental and emotional disorders are real illnesses, just like physical disorders.

* Mental and emotional disorders can result from genetics, chemical imbalances in the brain, physical injury, and life experiences.
* Mood disorders include depression and bipolar disorder. Depression causes feelings of extreme sadness, hopelessness, worthlessness, and guilt. In bipolar disorder, these feelings alternate with periods of constant excitement.
* Anxiety disorders include generalized anxiety disorder, phobias, panic disorder, social anxiety disorder, obsessive-compulsive disorder, and post-traumatic stress disorder.
* Personality disorders are harmful patterns of thinking and acting that affect a person's ability to get along with others. Types include paranoid, antisocial, and dependent personality disorders.
* Schizophrenia is a severe mental disorder in which people lose contact with reality.

LESSON 2 Suicide Prevention

BIG IDEA Knowing the warning signs of suicide could help you save a life.

* Risk factors for teen suicide include mental illness, poverty, substance abuse, violence in the family, stressful situations, bullying, and lack of a support network.
* Warning signs of suicide include talking about death, pulling away from friends, loss of interest in activities, self-destructive behaviors, giving away belongings and becoming suddenly cheerful after a period of depression.
* People who show signs of suicide need to be encouraged to seek professional help.

LESSON 3 Help for Mental and Emotional Disorders

BIG IDEA Mental and emotional disorders can be treated.

* Signs of mental illness include changes in eating or sleeping patterns, loss of interest in people or activities, feeling out of control, hearing voices, and thoughts of suicide.
* If you are concerned about your mental health, talk to a trusted adult.
* Forms of therapy include individual therapy, group therapy, and family therapy.
* Mental health providers include psychologists, psychiatrists, and clinical social workers.

 Review

 Web Quest

ASSESSMENT

Reviewing Vocabulary *and* Main Ideas

› mood disorder › personality › major depression › mental and
› phobias disorders › suicide emotional disorders

≫ On a sheet of paper, write the numbers 1–6. After each number, write the term from the list that best completes each sentence.

LESSON 1 Mental and Emotional Disorders

1. Extreme fears of specific objects or situations are _____.

2. A variety of psychological conditions that affect a person's ability to get along with others is called _____.

3. Illnesses that affect a person's thoughts, feelings, and behaviour are called _____.

4. A _____ is a mental and emotional problem in which a person undergoes mood swings that seem extreme, inappropriate, or last a long time.

LESSON 2 Suicide Prevention

5. _____ is the act of killing oneself on purpose.

6. A very serious mood disorder in which people lose interest in life and can no longer find enjoyment in anything is known as _____.

≫ On a sheet of paper, write the numbers 7-9. After each number, write the letter of the answer that best completes each statement.

LESSON 3 Help for Mental and Emotional Disorders

7. All of the following are signs that a person may need help with an emotional problem *except*
    a. feeling sad or angry for no apparent reason.
    b. loss of interest in favorite activities
    c. treating an injury at home with items from the medicine cabinet.
    d. sleeping too much or waking up too early.

8. Which statement is true of therapy?
    a. It uses the same techniques for everyone.
    b. It may be provided only by psychiatrists.
    c. It is an approach that teaches different ways of thinking or behaving.

9. All of the following are types of therapy for mental/emotional disorders *except*
    a. family therapy.
    b. group therapy.
    c. individual therapy
    d. physical therapy

✔ eAssessment

>> Using complete sentences, answer the following questions on a sheet of paper.

Thinking Critically

10. SYNTHESIZE Do mood swings make teens more likely to become depressed? Explain.

11. APPLY A friend who is depressed considers seeking help a sign of weakness. What do you tell your friend?

12. EXPLAIN What is the difference between sadness and clinical depression?

Write About It

13. PERSUASIVE WRITING Write a paragraph persuading others that people who are mentally ill deserve the same level of respect and care as people who are physically ill.

14. EXPOSITORY WRITING Write an article describing the warning signs of suicide. What are some ways to provide support to someone who may be considering suicide?

15. NARRATIVE WRITING Write a short story about a teen who is depressed. Explain how the teen makes a decision to get help.

Ⓐ Ⓑ Ⓒ Ⓓ STANDARDIZED TEST PRACTICE

Reading
Read the passage below and then answer the questions that follow.

Sigmund Freud was an Austrian doctor and one of the first people to study the human mind. Freud published a study on what he called hysteria in 1893. Freud defined hysteria as a strong emotion such as intense panic or fear. Freud claimed that hysteria was unspent emotional energy connected to disturbing events the person had forgotten.

Freud also studied dreams. He was convinced that dreams told how the mind of a person worked. In 1899, he published a book called *The Interpretation of Dreams.*

Freud revised and added to his theory over the next 30 years. Doctors did not always agree with Freud's theory about dreams. Still, few disagreed with his status as the "father of modern psychology."

1. Which sentence states an opinion?
   A. Sigmund Freud was an Austrian doctor and one of the first people to study the human mind.
   B. In 1899, he published a book called *The Interpretation of Dreams.*
   C. Freud published a study on what he called hysteria in 1893.
   D. Dreams told how the mind of a person worked.

2. What is the theme of the biography?
   A. Different people find different topics interesting.
   B. Freud's investigations led to the founding of the field of modern psychology.
   C. Today, no one disagrees with Freud's ideas.
   D. Sigmund Freud was an Austrian doctor.

Unit 4

conflict resolution + violence prevention

Conflict Resolution

LESSONS

PREMIUM ONLINE RESOURCES >

 Audio

 Videos

Bilingual Glossary

Fitness Zone

Web Quest

Review

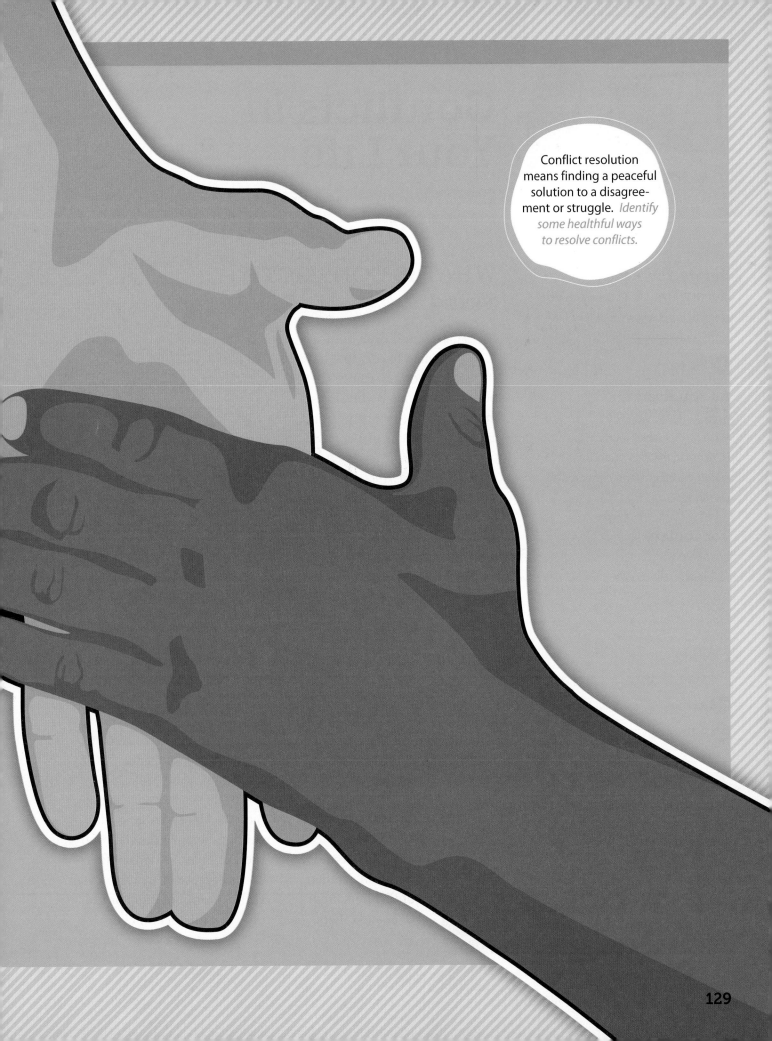

Conflict resolution means finding a peaceful solution to a disagreement or struggle. *Identify some healthful ways to resolve conflicts.*

Conflicts *in* Your Life

BIG IDEA Learning to resolve conflicts in healthful ways can help your overall well-being.

Before You Read

QUICK WRITE List all the words that come to mind when you think of the word *conflict*.

 Video

As You Read

FOLDABLES Study Organizer

Make the Foldable® found in the FL pages in the back of the book to record the information presented in Lesson 1.

Vocabulary

› conflict
› violence
› gang
› revenge
› prejudice
› tolerance
› labeling

 Audio

 Bilingual Glossary

Developing Good Character

The Myth of Positive Prejudice For example, saying that all French people are good cooks may sound like a compliment, but it is really a form of prejudice. Prejudices make assumptions based on race, culture, or group affiliations. It is unfair to assume anything about someone until you get to know that person individually.

WHAT IS CONFLICT?

MAIN IDEA Conflicts are a part of life and can be dealt with in a positive way.

Xavier and Evan both wanted to play first base in a softball game. They both became angry while insisting on playing that one position. This is an example of a conflict, or *a disagreement between people with opposing viewpoints, interests, or needs.* Conflicts happen to everyone. They are part of life.

Conflicts can involve two or more people. They can occur among strangers. Conflicts can arise between close friends or relatives. You can also experience internal conflict, or a conflict in your own mind.

Sometimes conflicts can be helpful. They can allow people to see the consequences of their behavior. When Xavier and Evan argued over who would play first base, the other players asked them to leave the game. Xavier and Evan realized their conflict hurt themselves and others.

Sometimes **conflicts** can be *helpful.*

Conflicts can be very hurtful to all parties involved. In some cases, a conflict may be only an exchange of words. Other times, conflicts can be ongoing and severe. They can damage or even destroy relationships. Conflicts can sometimes result in violence, or *an act of physical force that results in abuse.* Gang confrontations are an example of severe conflicts. A gang is *a group of young people that come together to take part in illegal activities.* Learning how to resolve conflicts in positive ways is an important health skill.

Conflicts occur for many reasons. Understanding why conflicts happen can help you prevent them. Often, conflicts occur because of an act or event. Differences of opinion or jealousy can cause conflicts.

ONOKY - Photononstop/Alamy

CAUSES OF CONFLICT

MAIN IDEA Understanding what causes conflicts can help you learn to prevent them.

You will have many disagreements throughout your life. If those are settled in positive ways, you can avoid escalating conflict. Some of the issues teens argue or disagree about include:

- **Property or territory.** Sometimes teens do not respect others' property or territory. Have you ever had a disagreement with a sibling who went into your bedroom without permission?
- **Hurt feelings.** Teens may feel hurt when a friend pays attention to someone else. They may also feel hurt when they are left out of activities. Insults, teasing, and gossip can result in hurt feelings.

- **Revenge.** When someone is insulted or hurt, that person may take **revenge.** Revenge is *punishment, injury, or insult to a person who is seen as the cause of a strong emotion.* The other person involved may then try to get even.
- **Differing values.** Conflicts can arise when people have different values, cultures, religions, or political views.
- **Prejudice.** Sometimes conflicts are caused by **prejudice,** or *a negative and unjustly formed opinion.* Conflicts can arise when people are not accepted because of differences. Showing respect for all people can help avoid these types of conflicts.

A valuable tool in preventing conflicts is **tolerance,** or *the ability to accept other people as they are* can prevent conflict. Accepting people who are different from you can maintain positive relationships.

>>> **Reading Check**

LIST *What are five causes of conflict?*

Conflicts can arise when people have different **values, cultures, religions,** or **political views.**

COMMON CONFLICTS FOR TEENS

MAIN IDEA Conflicts can be minor or major, interpersonal, or internal.

Conflicts happen all the time. Some conflicts are minor. They may involve a simple exchange of words. Other conflicts can be major and turn violent. Gang confrontations often lead to violence. Violent conflicts that involve weapons can be dangerous or even life-threatening.

The term *conflict* usually refers to conflicts between two or more people, or interpersonal conflicts. Conflicts can also be internal. Imagine that you decide to run for class president.

Later, you learn that your best friend is also running. Campaigning against your friend may cause an internal conflict. You want to avoid competing with your friend, so you are not sure what to do.

Conflicts involving teens can occur at home or at school. For example, you and your sibling might have a disagreement over which TV show to watch at a certain time. At school, a classmate might want to copy your homework, but you know this would be against school rules.

Arguing is a common source of conflict. *Explain some skills you might use to keep an argument from getting out of control.*

You Be the *Judge*

Mike and his friends were having a party and needed a docking station for an MP3 player. Mike borrowed his brother's docking station without asking. As Mike was riding his bike over to his friend's house, he dropped the docking station and broke it. When Mike told his brother what happened, his brother was angry. "How could you take my things without asking? I thought I could trust you," his brother said and walked away. Mike did not know what to do next. How could he mend his relationship with his brother? Analyze how Mike's behavior led to a conflict. Remember the six steps of the decision-making process.

1. State the situation.
2. List the options.
3. Weigh the possible outcomes.
4. Consider your values.
5. Make a decision and act.
6. Evaluate your decision.

With a partner, role-play a scene in which Mike thinks through his decision and then acts on it. Apply the decision-making steps. How would Mike's brother respond to his action?

Conflicts *at* Home

As you begin to grow you will find that you want more independence from your parents or guardians. This may cause conflicts at home. It is important to try to maintain a positive relationship with all family members to reduce the likelihood of conflicts occurring.

Many conflicts that arise between teens and their families occur over limits. Teens usually want fewer limits than their parents or guardians are willing to allow. For example, Juan joined some friends, but promised his parents that he would be home by 10:00 p.m. During the evening, his friends encouraged him to stay out until midnight. When Juan arrived home, his parents were angry that he broke the curfew set for him.

While teens like Juan want greater freedom, their parents want them to be safe. Parents may also require that teens contribute more to the household.

> Conflicts between parents and teens are very common. *Characterize how you might respond when you are told something you do not want to hear.*

This can be a source of conflict. Try to remember that parents and guardians set limits and expectations to help teens grow and develop. In many families, teens can gain greater freedom by showing their parents that they are willing to take on added responsibility within the family.

Siblings can also be a source of conflict at home. Sometimes siblings do not respect one another's space or property. You might have a brother or sister who uses your property without asking for permission. Respecting each other's property and space and discussing these issues regularly can help keep conflicts to a minimum.

> **⟩⟩⟩ Reading Check**
> **IDENTIFY** *What is a common source of conflict among teens and their parents or guardians?*

Conflicts *at* School

Many of the conflicts teens have outside of the home occur at school. Conflicts at school may involve teachers or administrators. Teens might also experience conflicts with friends, peers, or acquaintances at school.

Conflicts often arise from differences in personality, beliefs, and opinions. Conflicts can also be the result of an innocent mistake or other incident. For example, imagine you are standing in line to get into the basketball game. Someone bumps into you, causing you to bump into the person in front of you. The person ahead of you turns around and is upset with you for bumping him or her. That person may say something that upsets you, and before you know it, you are in a conflict. In this type of case, the conflict may begin with a simple exchange of words. However, it could escalate into pushing and shoving or an even more serious form of physical violence.

Bullying is *never* okay.

Conflicts may be one-sided and unprovoked. This can be caused by people not taking the time to understand one another. It can also be caused by people seeking power and attention by putting others down. This type of behavior is called bullying. Bullying is not always physical. It can come in many forms, including words or threats.

Bullying behaviors include:

- Teasing someone or saying hurtful things to a person.
- Labeling, or *name-calling,* based on prejudice.
- Leaving a person out of group activities or events.
- Sending untrue or hurtful e-mails or text messages.
- Becoming physically violent.

Bullying is never okay. No one deserves to be bullied. Use the resources available to you to make your environment as safe as possible. If you are a target of bullying, here are some ways to stop it when it happens:

- **Tell the bully to stop.** Look at the person and speak in a firm, positive voice with your head up. Explain that if the bullying continues, you will report the behavior.
- **Try humor.** This works best if joking is easy for you. Agreeing with a bully in a humorous way could catch that person off guard.
- **Walk away and stay away.** Do this if speaking up seems too difficult or unsafe.
- **Avoid physical violence.** Try to walk away and get help if you feel physically threatened. If violence does occur, protect yourself, but avoid doing anything that might escalate it.
- **Find an adult.** If the bullying is taking place at school, tell a teacher, counselor, or other school official immediately.

> ### >>> Reading Check
> **DEFINE** *What is bullying?* ■

>>> After You Read

1. **DEFINE** What is a conflict?
2. **LIST** What are three causes of conflicts among teens?
3. **IDENTIFY** What are the two main places teens have conflicts?

>>> Thinking Critically

4. **ANALYZE** Ivan and his dad argued about the amount of time Ivan spends playing video games on his computer. Ivan is feeling angry with his dad. What can they do to maintain a positive relationship with each other?
5. **APPLY** Imagine you have a younger sister who gets into your belongings without your permission. This makes you upset and angry. How can you resolve this issue?

>>> Applying Health Skills

6. **ANALYZING INFLUENCES** In what ways does the media misinform teens about how to handle conflicts? Why do you think the media portrays conflicts in these ways?

 Review

 Audio

The Nature *of* Conflicts

BIG IDEA Factors that make conflicts worse include anger, jealousy, group pressure, and the use of alcohol and other drugs.

HOW TO SPOT A CONFLICT

MAIN IDEA Conflicts can often be resolved if the signs of conflict are recognized early.

Conflicts generally occur because you and another person have different viewpoints. However, a disagreement does not need to result in conflict. Recognizing the signs of conflict early is the key to resolving them.

- **Disagreement** If you have a disagreement, be aware that this can lead to conflict.
- **Strong Emotions** Anger, sadness, jealousy and other emotions can be an early sign of conflict. Learn to control your emotions or to walk away to avoid conflict.
  - **Body Language and Behavior** Have you noticed that when some people get angry, they might ball their hands into fists or get a mean look on their face? You might cross your arms or tighten your lips. The other person might start to ignore you. These can be warning signs that a conflict is about to begin.

Managing Your Anger

Sometimes conflicts escalate, or *become more serious.* You can help prevent this by identifying the emotions that cause conflicts to escalate. Anger is a normal emotion, but it can be expressed in healthful ways. Some methods of managing anger include:

- **Step away from the situation** to get your emotions under control. This allows you time to collect your thoughts. It can help to keep you from saying things you will later regret.
- **Share your feelings** with another person, such as a parent or friend. Talking about your anger can help.
- **Engage in physical activity.** Running, biking, or other physical activity can often help to diffuse angry emotions.

Once you feel that your emotions are under control, then let the other person know how you feel. Speak calmly. Remember to use "I" messages, and discuss the problem, not the person.

Handling Feelings *of* Jealousy

Have you ever felt as if someone got better treatment than you did? If you have been in a situation like that, did you feel as if you were treated unfairly? If so, you were probably experiencing jealousy. Like anger, we all feel jealous at one time or another. Perhaps another student seems good at everything. It may appear to you that he or she tries very little to succeed. It is normal to feel jealous every now and then. A feeling of jealousy can sometimes motivate you to work harder to succeed.

Have you ever *felt* as if you were **treated** *unfairly?*

However, jealousy can also result in resentment, or an ongoing feeling of being wronged. Strong feelings of jealousy may lead a person to want to get revenge. Seeking revenge can also turn a minor problem into a major conflict. One or both parties could get hurt—physically, emotionally, or socially. Managing negative feelings such as jealousy in a positive way can help you avoid these situations.

If you have feelings of jealousy, it is important to deal with them in a healthful way. Talking to a good friend or trusted adult can help. Writing in a journal can also help you better understand some of your feelings.

>>> **Reading Check**

RECALL *How can you manage feelings of anger?*

Stress Management

Letting Off *Steam*

It is not always easy to know how or when to deal with anger. Sometimes you may think it is best to say nothing, only to realize later that you are still angry. When you allow anger to build, the emotion becomes like water heating up in a kettle. Eventually, you will need to let off steam. Here are some suggestions that can help you release built-up anger or frustration:

* Close your eyes and focus on relaxing and breathing.

* Find a way to turn your negative energy into positive energy. Write in a journal or work on a hobby. Look for an opportunity to laugh.

* Do some physical activity. Go for a run, bike ride, or walk.

* Talk to a friend, parent, guardian, or another trusted adult.

With A Group Make a list of some of the different techniques each member of your group has used to redirect negative energy. Share your list with other groups in the class.

Talking about feelings of jealousy with a trusted adult is better than seeking revenge. *Identify some situations that might cause a teen to feel jealous of someone else.*

uwe umstÄntter/Getty Images

The Warning Signs of Building Conflict

Physical Signs	Emotional Signs
A knot in the stomach	Feeling concerned
Faster heart rate	Getting defensive
A lump in the throat	Wanting to cry
Balled-up fists	Not feeling valued
Cold or sweaty palms	Wanting to lash out
A sudden surge of energy	Wanting to escape

Some warning signs that a conflict are building are physical. Other signs are emotional. *Identify some other signs of building conflict.*

Group Pressure

Have you seen a public disagreement? A crowd will often form. Sometimes people gather out of curiosity. Some people, however, may begin encouraging the conflict to escalate. This type of behavior is called a mob mentality, or *acting or behaving in a certain manner, often negative, because others are doing it.*

Have you ever noticed that during a **disagreement** in public, a *crowd* will often form?

When mob mentality occurs, some people involved in the conflict might become more aggressive than they would when a conflict occurs in private.

They might get caught up in the crowd's encouragements to do harm. Individuals might forget their own values and instead begin to do what the crowd wants. Situations such as this typically do not end well. Someone is likely to get hurt, either emotionally or physically. Avoid situations such as this. Go and get an adult to help end it.

>>> **Reading Check**

EXPLAIN *How can pressure from a group of peers cause a conflict to escalate?*

Conflicts *and* Substance Use

Using alcohol and other drugs can worsen a conflict. When a person uses alcohol or other drugs, it can affect their emotional state. People under the influence of alcohol and other drugs lose inhibitions and behave in ways that they normally would not. In many cases alcohol and other drugs can cause a conflict to become violent. For teens, alcohol use is illegal. Using drugs, other than for their intended purpose, is illegal for everyone.

Body language can be a sign that conflict is about to begin. *Describe how this teen is showing that she does not agree with what her parent is talking about?*

©Design Pics Inc/Alamy

CONTROLLING CONFLICTS

MAIN IDEA You can often prevent a conflict from building by handling the problem in an appropriate way.

Conflicts are a part of life. However, conflicts can be managed by handling them in ways that help prevent them from escalating. Using the strategies below can help you better handle conflicts.

- **Understand your feelings.** Anger is one of the strongest emotions involved with conflicts. If you take time to understand why you are angry, you may better understand the conflict. Some feelings last only a few minutes. Take time to let your emotions subside before dealing with a conflict.

- **Show respect for others.** It may be difficult to show respect for people who are not kind to you. Sometimes respect can overcome unkindness. It is also important to show respect for yourself. When you avoid letting others force you into doing things you are not comfortable with, you show respect for yourself.

- **See the other person's point of view.** We all have different backgrounds and points of view. If you try to understand these differences, you may be more accepting and considerate of others. This shows that you want to resolve conflicts in a positive way and avoid escalating the conflict.

- **Keep your conflicts private.** Avoid sharing information about your conflicts. If you have a conflict, find a quiet spot to share your differences with the other person.

- **Avoid alcohol and other drugs.** People who use alcohol and drugs may behave in ways that are not normal for them. The alcohol or drugs may control their behavior.

- **Leave the scene if necessary.** At times, it is best to walk away from a conflict. This allows you and the other person to think more rationally.

> ### >>> Reading Check
>
> **EXPLAIN** *Why is it important to show respect for others?* ■

Reaching an agreement that puts an end to a conflict is gratifying. *List some other ways to help prevent conflicts from building.*

>>> After You Read

1. **DEFINE** What does the term *mob mentality* mean?
2. **RECALL** What factors cause a conflict to escalate?
3. **LIST** Name two ways to prevent conflicts from building.

>>> Thinking Critically

4. **ANALYZE** Imagine that you have walked away from a conflict. The other person involved calls you a "chicken." What would you say?

>>> Applying Health Skills

5. **PRACTICING HEALTHFUL BEHAVIORS** Compose a list of factors that can prevent conflicts from escalating. You might list *understanding your feelings, showing respect for others,* and so on. After you have made your list, indicate which items you feel you do well. Then identify which items where you need to improve. For each item where you can use some improvement, write how you can work toward improving in this area.

 Review

🔊 Audio

©Creatas/PunchStock

Conflict Resolution Skills

BIG IDEA You can deal with conflict in constructive ways.

Before You Read

QUICK WRITE List two ways of communicating that could lead to conflict. Next to each, write how the same idea could be expressed in a more positive way.

 Video

As You Read

STUDY ORGANIZER Make the study organizer found in the FL pages in the back of the book to record the information presented in Lesson 3.

Vocabulary

> conflict resolution
> negotiation
> collaborate
> compromise
> win-win solution

 Audio

 Bilingual Glossary

Myth vs. Fact

Myth: It is okay to yell during an argument as long as you do not call someone a name or put another person down.

Fact: Yelling sets everyone on edge. It shows that you are not in control of your emotions. Taking time to calm down can help resolve an issue peacefully.

FINDING SOLUTIONS TO CONFLICTS

MAIN IDEA Conflict resolution involves solving a disagreement in a way that satisfies everyone involved in the conflict.

Sometimes the solution to a conflict is easy to find. Other times, it might be more difficult to find a solution. In cases such as this, you might be tempted to ignore the conflict and not to deal with it. Ignoring or not resolving a conflict can be damaging to a relationship. Avoiding a conflict can often make it worse.

Not resolving a **conflict** can be **damaging** to a *relationship.*

Conflict resolution is *a life skill that involves solving a disagreement in a way that satisfies both sides.* Conflict resolution skills enable two parties to work together toward a positive and healthful resolution to whatever problem they may have.

Using Negotiation Skills

Negotiation is *the process of talking directly to another person in an effort to resolve a conflict.* Negotiation is a powerful skill for addressing disagreements and other conflicts. During negotiation, the two parties involved in a conflict meet and share their feelings, expectations, wants, and the reasons for their wants. The purpose of the negotiating session is to arrive at a peaceful resolution to the conflict.

The T.A.L.K. strategy is often used during negotiations. This strategy allows both parties in the conflict to collaborate, or *work together,* to arrive at a solution. You may find that you are often able to build a better relationship with the other party in a conflict when you are collaborating. The T.A.L.K. strategy, which is explained in detail on the next page, can help you remember the four main steps of the conflict resolution process.

- **T**ake a time-out. Thirty minutes is usually enough time for both sides to calm down and get their emotions under control. This will allow each side to think clearly before they talk.
- **A**llow each person to talk. If both sides are calm, each side should be able to allow the other side to talk uninterrupted. It is important that each person be able to share his or her feelings calmly without being interrupted by the other person. It is also important that the speaker not use angry words or gestures.
- **L**et each person ask questions. Each person should be allowed to ask questions of the other person. Questions should be asked and answered in a calm manner. It is important for both sides to be polite and respectful.
- **K**eep brainstorming. Continue to think of creative solutions until one that satisfies both parties is reached.

It may be difficult to keep your emotions in check during negotiations. However, it is important to do so. Avoid letting your emotions keep you from trying different solutions to your problem. Negotiations can be difficult. It is important to negotiate in private, if possible. It is also important to remember some don'ts for negotiation:
- Don't touch the other person.
- Don't point a finger at the other person.
- Don't call names.
- Don't raise your voice.

There are also some do's to remember:
- Do take a break if either party begins to get angry.
- Do stop negotiations if you are threatened by the other person.
- Do leave and tell a trusted adult if you feel threatened.

Negotiations can be helpful for solving conflicts. However, remember that some issues are not open for negotiation. If something is illegal or potentially harmful to you or others, do not negotiate. Say no or use refusal skills, and if possible, leave the scene of the conflict.

>>> **Reading Check**

IDENTIFY *What are the four steps in the T.A.L.K. strategy?*

These teens are working through a disagreement. *What is the listener doing to demonstrate that he is paying attention?*

Health SKILLS ACTIVITY

Conflict Resolution

Settling a *Disagreement*

Tanisha and Chloe have been friends since elementary school. One Saturday, they agree to meet at the movies. Chloe shows up 30 minutes late. It's the third time in a row that Chloe's been late for something that they were going to do together. Because Tanisha waited for Chloe, they missed the first part of the movie. When Chloe arrives, Tanisha shouts, "You're never on time! You ruin everything!" Remember the four steps of the T.A.L.K. strategy:

* **T**ake a time-out of at least 30 minutes.

* **A**llow each person to tell his or her side uninterrupted.

* **L**et each person ask questions.

* **K**eep brainstorming to find a good solution.

Apply the steps of conflict resolution to Tanisha and Chloe's situation. Write a dialogue that shows them resolving their conflict peacefully.

Possible Outcomes *of* Negotiation

During conflict negotiation, it may be necessary for both sides to **compromise,** or *give up something to reach a solution that will satisfy everyone involved.* Compromise works well when negotiating such things as which television program to watch or what to have for dinner. However, for some issues, such as whether to accept a ride from someone who has used alcohol, compromise is not a solution. Laws, your values, school rules, and limits set by your parents should not be compromised.

Compromise works well when *negotiating.*

Another possible outcome of negotiation is a **win-win solution,** or *an agreement that gives each party something they want.* Often people think that if you negotiate, someone will win and someone will lose. A win-win solution is the best possible outcome. With a win-win solution, both sides come away with something. There are no losers, so each side is satisfied with the outcome. A win-win solution can improve a relationship if both parties feel that the other allowed them to win something.

>>> **Reading Check**

CONTRAST *What is the main difference between a compromise and a win-win solution?* ■

Negotiation is an important tool in conflict resolution. *Describe one way this teen is demonstrating a principle of conflict resolution.*

>>> **After You Read**

1. **DEFINE** What is conflict resolution?
2. **EXPLAIN** Why is it important to take a time-out before beginning a negotiation?
3. **IDENTIFY** List and explain two different outcomes of negotiation.

>>> **Thinking Critically**

4. **EVALUATE** Think of a conflict you have had. Describe how you handled the situation. How might the outcome have been different if you had used the T.A.L.K. strategy to solve the problem?
5. **DETERMINE** Why is it important to brainstorm all the possible solutions to a problem?

>>> **Applying Health Skills**

6. **COMMUNICATION SKILLS** Imagine that you are involved in a conflict with a classmate. What communication skills could you use to help reach a positive resolution?

⟳ Review

🔊 Audio

©BananaStock/PunchStock

Peer Mediation

BIG IDEA Mediation can provide a solution that is acceptable to both parties.

Before You Read

QUICK WRITE Have you ever had a conflict with a friend that you were able to work out? Write two to three sentences describing what you did to help resolve the conflict successfully.

 Video

As You Read

STUDY ORGANIZER Make the study organizer found in the FL pages in the back of the book to record the information presented in Lesson 4.

Vocabulary

› mediation
› peer mediation
› neutrality

 Audio

 Bilingual Glossary

Fitness Zone

Relieving Stress I used to try to please everyone all the time. I would take what people said about me too seriously. I got into conflicts with my classmates because of the stress. All the peer mediators at school knew me by name. Finally, one suggested physical activity to help relieve my stress. I joined the swim team, and guess what? It worked! I have less stress, I'm in better shape , and I haven't been to peer mediation for months.

WHAT IS MEDIATION?

MAIN IDEA Mediation can help the parties in a conflict work together to solve the problem.

At times you are able to resolve a conflict so that both you and the other person are satisfied with the result. You achieve a win-win solution. Knowing that you worked together to arrive at a solution makes you feel good. Unfortunately, not all conflicts are easy to resolve.

Imagine that you and a classmate have teamed up to create a project for your health class. You and your partner disagree on the method you should take to complete the project. Because of the conflict, you have not made any progress on the project.

Sometimes you will not be able to resolve a conflict on your own. You may find you need help to in order to resolve a conflict. A third person who is not involved in the conflict can help people move closer to a solution.

The process of resolving conflicts by using another person or persons to help reach a solution that is acceptable to both sides is called **mediation.** Mediation is similar to negotiation, except that a third party is involved. That person does not resolve the conflict. A mediator helps those involved arrive at a solution.

Sometimes you will need *help* to resolve a **conflict.**

Guidance counselors, school administrators, or other trusted adults may serve as mediators. Mediators can also be students who have been trained in mediation strategies. Some schools have **peer mediation** programs, in which *specially trained students listen to both sides to help resolve conflicts between teens.*

Reading Check

EXPLAIN *Why might you ask a mediator to help resolve a conflict?*

The Mediation Process

Before mediation can begin, both parties must agree to allow a mediator to try to help them. The process begins in a private location. It is important that the only people present are the two parties in conflict and the mediator. If others are present, it could be distracting to the mediation process. It can also be counterproductive if other people start to take sides in the conflict.

A mediation session usually starts with the mediator asking each of the parties involved in the conflict to present his or her side of the conflict. It is important for the mediator to listen carefully. A good mediator does not allow one party to talk while the other party is sharing his or her side. The mediator may interrupt from time to time, but only to ask questions for clarification. This is to make sure that the mediator understands the situation. Asking questions is also important to make sure that the mediator understands the point of view of each of the parties involved in the conflict.

Once the mediator understands the situation, he or she will try to help the parties come up with a solution to the conflict. The mediator and the two parties involved will brainstorm possible solutions. Then they will evaluate the solutions. Some solutions may not be practical. Other possible solutions to the conflict may not be well received by both parties.

> It is **important** for a *mediator* to **listen** carefully.

Then the mediator will encourage the parties to decide on a solution. The mediator does not participate in the selection process. This choice must be made by the parties involved. Both sides of the conflict must be in agreement about the solution. If both parties are not in agreement, the solution will not work.

When following these steps, a mediator can often help the parties in conflict arrive at a solution. Sometimes the parties will have to make compromises.

Often, it may take time to agree on possible solutions. The mediator may have to ask additional questions to find out what might be acceptable to the two sides. The overall solution to the conflict, however, must be acceptable to both parties.

In order to be effective, mediators must be able to both talk and listen. A mediator must be able to help parties communicate without taking sides. Mediators must also be able to help resolve any new conflicts that might arise during the mediation session. Some important traits of a successful mediator include being:

- a good communicator.
- a good listener.
- fair.
- neutral.
- an effective problem solver.

An effective mediator must always be a person who can avoid taking sides in a dispute. A mediator must also respect the privacy of the parties involved and keep all the details secret.

>>> **Reading Check**

IDENTIFY *What are two skills shared by effective mediators?*

A key element to successful mediation is cooperation. Each party must be willing to work with the other and the mediator. *Explain how a mediator helps people in a conflict find a solution.*

Steps in the Mediation Process

1. The parties involved in the conflict agree to seek an independent mediator's help.
2. The mediator hears both sides of the dispute.
3. The mediator and the parties work to clarify the wants and needs of each party.
4. The parties and mediator brainstorm possible solutions.
5. The parties and mediator evaluate each possible outcome.
6. The parties choose a solution that works for each of them.

Being a Peer Mediator

As mentioned before, many schools have peer mediation programs. In these programs, student mediators help other students solve their conflicts. What does it take to become a peer mediator?

Peer mediators must like to help others and be good problem solvers. They must also be able to maintain neutrality, or *a promise not to take sides,* during the mediation process. Another important trait for mediators is to be a good communicator.

During the mediation process, the peer mediator determines that both sides agree to mediation. Then the rules are explained. Generally, each party in the conflict is allowed to tell its side without being interrupted by the other party. The mediator then restates what has been said to make sure that everyone understands. The mediator then identifies what the conflict is and seeks agreement from both parties.

At this point, the mediator tries to determine what each side wants. Sometimes students want to remain friends. Knowing what each side wants, helps the mediator determine what solutions might work. After a solution is reached, the students might be asked to sign a paper that says a solution was agreed upon.

One trait of an effective peer mediator is good listening skills. *Explain why you think it is important for a peer mediator to have these traits.*

What does it take to become a *peer mediator?*

Does the idea of becoming a peer mediator sound interesting to you? If so, you might want to find out whether your school has a peer mediation program. Many schools like to use peer mediators because it allows teens to put problems into words that other teens understand. If you decide to become a peer mediator, you will need to go through training. Peer mediators volunteer their services, so training would likely happen on your own time. Find out whether your school has a peer mediation program. If your school has a program and you are interested, ask how you can get involved.

>>> Reading Check

ASSESS *What do you think is the most important trait of a mediator?* ■

Problem Solver
Easy to Talk to
Enthusiastic
Responsible

Mature
Effective Listener
Decisive
Interested
Alert
Trustworthy
Open-minded
Reliable

REVIEW

>>> After You Read

1. **DEFINE** What is neutrality?
2. **LIST** What are five traits of a good peer mediator?
3. **EXPLAIN** When should the parties of a conflict turn to mediation?

>>> Thinking Critically

4. **APPLY** Think about a conflict you have experienced. Then explain how your conflict resolutions skills helped or could have helped you reach a solution.

>>> Applying Health Skills

5. **CONFLICT RESOLUTION** Imagine you are involved in a conflict with a classmate. Your teacher doesn't seem to understand your problem, so you go the peer mediation program at your school. One of the other students acts as your mediator and your conflict is eventually resolved. Both you and your classmate are satisfied with the result. Why do you think you were able to work out the issue with the peer mediator and not with the teacher?

⟳ Review
🔊 Audio

Hands-On HEALTH ACTIVITY

One Story, Three Endings

WHAT YOU WILL NEED

* The story: Omar and Lou are playing basketball. Peter asks if he can play.
* Ending #1: Omar says, "Sure." Lou doesn't like Peter and would rather he didn't play. However, he just shrugs and starts playing poorly. Peter asks what's bothering him. Lou says, "Nothing."
* Ending #2: Omar says, "Sure." Lou claims that Peter cheats and hogs the ball. Peter replies, "You don't want me to play because I'm better than you." Lou throws the basketball down and walks away.
* Ending #3: Omar says, "Sure." Lou says no because Peter hogs the ball. Omar suggests they trade off after every three shots so no one will have to wait long. They all agree and start to play.

WHAT YOU WILL DO

In a small group, read the story with each of the three endings. Then answer these questions for each of the three endings: Does this solution make someone angry? Do the boys try to listen to one another? Do they understand one another's feelings?

WRAPPING IT UP

You should be able to recognize the three conflict resolution styles in the three endings of the story. The best ending is when problem solving takes place.

Conflicts happen because people have different wants, needs, and opinions. There are at least three ways to handle conflict: denial, confrontation, and problem solving. Denial is when people don't admit they are angry and don't talk about their feelings. Confrontation happens when people are not willing to listen to one another or refuse to compromise. Problem solving occurs when people are able to talk about a problem without insulting or blaming one another.

READING REVIEW

FOLDABLES **and Other Study Aids**

Take out the Foldable that you created and any study organizers that you created. Find a partner and quiz each other using these study aids.

LESSON 1 Conflicts in Your Life

BIG IDEA Learning to resolve conflicts in healthful ways can help your overall well-being.

- Conflicts are a part of life and can be dealt with in a positive way.
- A conflict is a disagreement between people with opposing viewpoints, interests, or needs.
- Understanding what causes conflicts can help you learn to prevent them.
- Common causes of conflict involve disputes over property or territory, hurt feelings, a desire for revenge, differing values, and prejudice.
- Conflicts can be minor or major, interpersonal, or internal.

LESSON 2 The Nature of Conflicts

BIG IDEA Factors that cause conflicts to build include anger, jealousy, group pressure, and the use of alcohol and other drugs.

- Conflicts can often be resolved if the signs of conflict are recognized early.
- Anger and jealousy are normal emotions that can be managed in healthful ways.
- You can often prevent a conflict from building by handling the problem in an appropriate way.
- Strategies for handling conflicts in healthful ways include understanding your feelings, showing respect for others, and seeing other people's points of view.

LESSON 3 Conflict Resolution Skills

BIG IDEA You can deal with conflict in constructive ways.

- Conflict resolution involves solving a disagreement in a way that satisfies everyone involved in the conflict.
- Negotiation is a powerful skill for addressing disagreements and resolving conflicts.
- The T.A.L.K. strategy allows both parties in a conflict to work together to arrive at a solution.
- Possible outcomes of negotiation include reaching a compromise or coming up with a win-win solution.
- Negotiation often requires collaboration in order to reach a positive resolution to a conflict.
- If you are faced with a potentially harmful or illegal situation, do not negotiate—instead, say no, use your refusal skills, or leave the scene.

LESSON 4 Peer Mediation

BIG IDEA Mediation can provide a solution that is acceptable to both parties.

- Mediation is a process that can help the parties in a conflict work together to solve the problem.
- Mediation requires agreement from both parties to work toward a positive solution.
- A good mediator is a good communicator, listens well, is fair and neutral, and is an effective problem solver.
- A mediator can be either a trusted adult or a student who is trained in peer mediation strategies.

 Review

 Web Quest

ASSESSMENT

Reviewing Vocabulary *and* Main Ideas

› conflict	› mob mentality	› revenge	› jealousy
› escalate	› prejudice	› labeling	› tolerance

» On a sheet of paper, write the numbers 1–8. After each number, write the term from the list that best completes each statement.

LESSON 1 Conflicts in Your Life

1. A(n) _____ is a disagreement between people with opposing viewpoints, ideas, or goals.

2. _____ is a negative and unjustly formed opinion, usually against people of a different racial, religious, or cultural group.

3. _____ is punishment, injury, or insult to the person seen as the cause of the strong emotion.

4. Name calling based on prejudice is known as _____.

5. A valuable tool in preventing conflicts is _____, or the ability to accept other people as they are.

LESSON 2 The Nature of Conflicts

6. A conflict that becomes more serious is said to _____.

7. People who act or behave in a certain and often negative manner because others are doing it are expressing a _____.

8. _____ can lead to feelings of anger and resentment or a desire to get revenge.

» On a sheet of paper, write the numbers 9–15. Write *True* or *False* for each statement below. If the statement is false, change the underlined word or phrase to make it true.

LESSON 3 Conflict Resolution Skills

9. The *T* in the T.A.L.K. strategy stands for <u>Tell the other person to cooperate</u>.

10. When a conflict is resolved to the satisfaction of both parties, a <u>win-lose</u> solution has been achieved.

11. <u>Negotiation</u> is the process of talking directly to another person in an effort to resolve a conflict.

12. An arrangement in which each side gives up something to reach a satisfactory solution is known as a <u>resolution</u>.

LESSON 4 Peer Mediation

13. Mediation can help two people unable to reach a compromise on their own.

14. When a mediator promises <u>to take sides</u> in a conflict, that person is maintaining neutrality.

15. Peer mediation involves having specially trained <u>teachers</u> listen to both sides in a conflict to help the parties resolve a conflict.

✔ eAssessment

>> Using complete sentences, answer the following questions on a sheet of paper.

 Thinking Critically

16. EVALUATE Which step in the peer mediation process do you think would be the most challenging? Explain your answer.

Write About It

17. EXPOSITORY WRITING Write a story about an imaginary conflict between two people. Tell how conflict-resolution skills were used to bring about a win-win solution.

Ⓐ Ⓑ Ⓒ Ⓓ STANDARDIZED TEST PRACTICE

Reading

The following are observations made by famous people about conflict. Read the quotes, and then answer the questions.

A. "Conflict is inevitable, but combat is optional." — Max Lucado

B. "The aim of argument, or of discussion, should not be victory, but progress." — Joseph Joubert

C. "A good manager doesn't try to eliminate conflict; he tries to keep it from wasting the energies of his people. — Robert Townsend

D. "You can't shake hands with a clenched fist." — Indira Gandhi

E. "Chance fights ever on the side of the prudent." — Euripedes

1. The two quotes that suggest conflict is a fact of life are
   A. A and B.
   B. B and C.
   C. A and C.
   D. D and E.

2. Which two quotes carry the message that one should be open-minded yet cautious when experiencing a conflict?
   A. The quotes by Max Lucado and Robert Townsend
   B. The quotes by Joseph Joubert and Indira Gandhi
   C. The quotes by Robert Townsend and Indira Gandhi
   D. The quotes by Joseph Joubert and Euripedes

DEALING WITH CONFLICT

Conflicts are a normal part of life, but when you understand why they happen, you can learn ways to prevent them.

COMMON CONFLICTS FOR TEENS

AT HOME
- Limits set by parents
- Sibling disagreements
- Hurt feelings

AT SCHOOL
- Revenge or hurt feelings
- Differences or prejudice
- Bullying

CONFLICT RESOLUTION SKILLS

Use the T.A.L.K. strategy:

TAKE A TIME-OUT

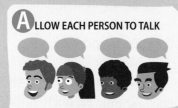

ALLOW EACH PERSON TO TALK

LET EACH PERSON ASK QUESTIONS

KEEP BRAINSTORMING

In the **United States,**
48 states
have **anti-bullying** laws.

More than **25%**
of all **teens** in the **U.S.**
say they have been
bullied at school.

Every year, **8 out of 10**
U.S. schools report
serious crimes
involving students.

In **25** other countries,
1 in 10 students
say they have been
BULLIED at school.

PREVENTING VIOLENCE

Violence is a major health problem in the United States, sending thousands of young people to the hospital each year.

1 IN 3 students report having been in a physical **fight** in the past year.

UNDERSTANDING TEEN VIOLENCE

Most teens do not engage in violence, but teens are influenced by:

PREJUDICE
An unfair negative opinion of a group

MEDIA
Most teens witness 200,000 acts of violence on TV before age 18.

PEER PRESSURE
Negative influences from others

GANGS
Groups of people who take part in illegal activities

DRUGS
Teens who use drugs are more likely to engage in violence.

WEAPONS
People who carry guns are more likely to be hurt by violence.

TEENS ARE TWICE AS LIKELY AS ADULTS TO BE VICTIMS OF VIOLENCE.

Protecting yourself

AVOID EVER PICKING UP A GUN, KNIFE, OR OTHER WEAPON.

MAKE SURE YOUR VALUABLES ARE HARD TO GRAB.

IF THREATENED, GIVE UP YOUR VALUABLES.

MAKE SURE YOUR FAMILY ALWAYS KNOWS WHERE YOU ARE.

OPEN THE DOOR ONLY FOR PEOPLE YOU KNOW.

AVOID GIVING PERSONAL INFORMATION.

Staying safe online

Before meeting an online friend, ask yourself:

- IS YOUR ONLINE FRIEND REALLY ANOTHER TEEN?
- HAVE YOU SHARED YOUR PLANS WITH YOUR FAMILY?
- ARE YOU MEETING IN A SAFE, PUBLIC PLACE?
- ARE YOU SURE IT'S A GOOD IDEA?
- ...

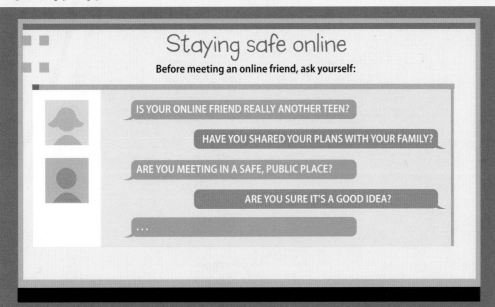

Seek help from:

- **Crisis hot lines**
- **Shelters**
- **Counseling**
- **9-1-1**

CHAPTER 8

Violence Prevention

LESSONS

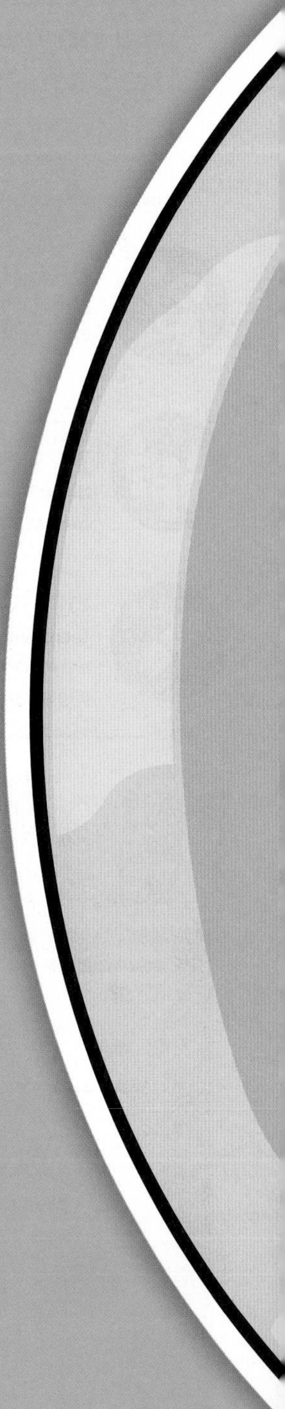

 **PREMIUM ONLINE RESOURCES**

 Audio

 Videos

A BC Bilingual Glossary

 Fitness Zone

 Web Quest

 Review

Violence is an act of physical force that results in injury or abuse. *How might violence also affect the mental/emotional and social sides of your health triangle?*

Understanding Violence

BIG IDEA Violence is a major health problem in our society.

Before You Read

QUICK WRITE What steps do you take to keep yourself safe? Write a paragraph about these strategies.

 Video

As You Read

FOLDABLES **Study Organizer**

Make the Foldable® found in the FL pages in the back of the book to record the information presented in Lesson 1.

Vocabulary

> violence
> intimidation
> bullying
> harassment
> assault
> rape
> homicide
> gang
> victim

 **Audio**

Bilingual Glossary

VIOLENCE IN SOCIETY

MAIN IDEA Many factors contribute to the violence in society.

Have you ever noticed how many news stories are about acts of violence? **Violence** is *any behavior that causes physical or psychological harm to a person or damage to property*. Popular TV shows, movies, music, and video games often feature violent content.

Violence can take many *forms*.

Violence is a major national health concern. A Centers for Disease Control and Prevention study found that violence sends more than 750,000 young people to the hospital each year.

Types *of* Violence

Violence can take many forms. Sometimes words can be violent. Name-calling and teasing can be forms of **intimidation,** or *purposely frightening another person through threatening words, looks, or body language.* Intimidation can also be a form of **bullying,** or *the use of threats, taunts, or force to intimidate someone again and again.*

Spreading rumors can sometimes be a form of bullying. Harassment also uses words as a form of violence. **Harassment** (huh·RAS·muhnt) is *ongoing conduct that offends a person by criticizing his or her race, color, religion, disability, or gender.* Harassing words can be either written or spoken.

The most common violent crimes in the United States are assault, robbery, rape, and homicide. An **assault** is *an attack on another person in order to hurt him or her.* More assaults are reported than any other type of violent crime in the U.S. Robbery involves the taking of another's property by force. **Rape** is *forced sexual activity.* **Homicide,** which is also known as murder, is *a violent crime that results in the death of another person.* In addition to robbery, other property crimes include burglary and vandalism.

> ## Reading Check
>
> **IDENTIFY** *What are the four most common violent crimes in the U.S.?*

UpperCut Images/Getty Images

FACTORS IN TEEN VIOLENCE

MAIN IDEA A variety of factors influence teens toward violence.

One way to defeat violence is through education. *Cite an example of a way to help others learn to avoid homicide, rape, and assault.*

Some teens use violence to get respect from their peers. Others use it to show their independence. Some use it when they feel they are being controlled and think it is their only choice.

Most teens **do not** engage in violence.

Many teens who engage in violence have similar risk factors. Often, teens who use violence have seen violence at home. Teens who commit violence may not have learned to deal with feelings such as anger in healthful ways. They may engage in risk behaviors.

Playing violent video games can affect the way teens think about violence. *Explain how playing violent video games might affect a teen's thinking.*

However, most teens do not engage in violence. Teens who are aware of the risk factors may be able to avoid violent behavior. Other factors that may contribute to violence include prejudice, peer pressure, media influence, gangs, drugs, and weapons.

- **Prejudice** Prejudice is an unfair negative opinion of a group of people. Prejudice can lead to hate crimes. A hate crime is an illegal act that targets a member of a particular group of people.
- **Peer pressure** Pressure from others may cause a teen to go against his or her personal values. In some cases, a group might pressure a teen into committing a violent act.
- **Media influence** Studies show that by age 18, teens will see thousands of violent acts on TV. Violence is also a common theme in movies, music, and video games. Research shows a link between media influences and teen violence.

- **Drugs** Studies show that teens who use drugs are more likely to engage in acts of violence. Drugs and alcohol affect your judgment and your ability to make healthful decisions.
- **Gangs** Some teens join a gang to gain a sense of belonging, for protection, or because a family member or friend has joined. A **gang** is *a group of people who come together to take part in illegal activities.* Gang members may carry weapons, sell drugs, and commit acts of violence.
- **Weapons** In a recent survey, 17 percent of teens admitted to carrying a gun or other weapon at some point. Statistics show that people who carry guns are more likely to be injured as a result of gun violence.

>>> **Reading Check**

EXPLAIN *What is the relationship between gangs, weapons, and drugs?*

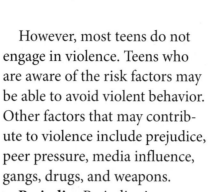

(t)McGraw-Hill Companies, Inc. Ken Karp, photographer, (b)Ingram Publishing

EFFECTS OF VIOLENCE

MAIN IDEA Victims of violence can suffer physical as well as mental and emotional effects.

A teen is twice as likely as an adult to be a victim of violent crime. A **victim** is *any individual who suffers injury, loss, or death due to violence.* Violent crimes always have an effect on the survivors. They may have physical injuries. They may also have emotional injuries.

At times, emotional injuries may be more difficult to handle than physical injuries. Survivors of violent crimes may need help to recover. A trusted adult or mental health professional can help. With the right help, victims of violent crime can recover.

People who experience violent crimes may be reluctant to report them. A rape victim may feel ashamed or embarrassed. That person might even feel partly to blame. A person who has been raped is never to blame for the attack.

However, if you do experience a violent crime, you can take action not only to help yourself but also to protect others.

- **Get medical attention.** You may have injuries you are not aware of. You may be in a state of shock, which can temporarily block out pain.
- **Report the incident to the police.** It is important to file a report. This will help bring the responsible person to justice and help prevent that person from harming someone else.
- **Get treatment for the emotional effects of the crime.** A violent crime is a traumatic experience. Professionals can help you work through the emotional and psychological pain caused by the crime. Counseling can help survivors of crime recover from the experience. They can then move on with their lives.

Reading Check

IDENTIFY *What should you do if either you or a loved one were to experience a violent crime?* ■

A school counselor can be one source of help for people who experience violence. *Identify some other sources of help for survivors.*

REVIEW

After You Read

1. **EXPLAIN** What is violence?
2. **LIST** What are six factors that contribute to violent behavior?
3. **IDENTIFY** What are three actions that the subject of a violent crime should take?

Thinking Critically

4. **ANALYZE** Why should the survivor of a violent crime tell someone about it?
5. **APPLY** Imagine that you have just started attending a new school. Some of the other students tease you because you are new. After a couple of weeks, the teasing continues and becomes more vicious. What should you do?

Applying Health Skills

6. **REFUSAL SKILLS** Your friend's brother has recently joined a local gang. You know that the gang sometimes commits violent crimes. They also use drugs and carry weapons. You are approached about joining the gang. How would you respond?

 Review

 Audio

David Burch/Getty Images

Violence Prevention

BIG IDEA You can help prevent violence.

©McGraw-Hill Education/Ken Karp

Before You Read

QUICK WRITE Find a news article about someone who has experienced crime or violence. Write a one-paragraph summary of the story.

 Video

As You Read

STUDY ORGANIZER Make the study organizer found in the FL pages in the back of the book to record the information presented in Lesson 2.

Vocabulary

› zero tolerance policy
› youth court

🔊 Audio

🔤 Bilingual Glossary

Developing Good Character

Speaking Out About Violence Good citizenship includes helping to keep your school safe. If you know about any violence at school or violent activity that might take place, tell school officials right away. Find out whether your school has a special hotline or other way to report rumors of violent activity. If your school does not, think about how you can feel safe when reporting a potentially dangerous situation. *Write your ideas in a brief paragraph.*

PROTECTING YOURSELF FROM VIOLENCE

MAIN IDEA You can take a number of steps to help protect yourself from violence.

Acts of violence can happen anywhere, including at home, at school, or in your neighborhood. However, you can take a number of steps to help protect yourself. You can learn to avoid unsafe situations that may lead to violence.

- Remember that guns can cause serious injury or death. Avoid picking up a gun, even if you think it is unloaded.
- If a stranger stops his or her car to ask you for help or directions, walk or run in the opposite direction. Do not get close to the car.
- Avoid carrying your backpack, wallet, or purse in a way that is easy for someone to grab.
- If another person threatens you with violence, give that person the money or valuable. Your safety is more important than your possessions.
- Make sure your family always knows where you are, where you are going, and when they can expect you to be home.
- Walk with a friend. After dark, walk only in well-lit areas. Avoid dark alleyways and other places where few people are nearby.
- When you are at home, lock your doors and windows. Open the door only for people you know. If your family has a rule against answering the door, do not open it at all.
- Avoid giving out personal information when you answer the phone or respond to a text message or e-mail. Do not tell anyone you are home alone.

You can *learn* to avoid unsafe **situations** that may *lead* to **violence.**

You can also take steps to protect yourself from gangs. If you live in an area that has gang activity, avoid wearing gang-related colors or clothing. When you are walking on the streets, avoid wearing expensive jewelry. Avoid carrying expensive electronics that might be a target of theft. Another way to help stay safe from gang violence is to choose friends who are not members of a gang.

Practicing safe habits online is another way to protect yourself.

If you are like many teens, you may communicate with many people online. However, avoid meeting an online friend without first learning more about that person.

- Is your online friend really another teen? Sometimes, an adult who could harm you may pretend to be a teen online.
- If you do agree to meet an online friend in person, share your plans with a parent or another trusted adult.

- Make sure you meet your online friend in a public place. First-time meetings are always safer in an area where other people are around in case you need help.

During the teen years, many people form new friendships and start to date. Remember that healthy friendships are built on respect, caring, and honesty. If a friend or dating partner becomes violent or abusive, avoid that person and tell a parent or other trusted adult.

WAYS TO REDUCE VIOLENCE

MAIN IDEA Everyone can make an effort to help reduce the spread of violence.

Violence affects victims, their families, and the rest of society. According to the U.S. government, violence costs the country more than $158 million a year. This figure includes property damage as well as law enforcement and health care costs.

What can you do to *reduce* or *stop* violent behavior?

Acts of violence can damage property, cause injury, and even ruin lives. However, each individual can do his or her part to help reduce violent activity. This often means taking action to stop violence before it starts. What can you do personally to help reduce violent activity?

- Develop your own personal zero tolerance policy regarding violence. A **zero tolerance policy** *makes no exceptions for anyone for any reason.* This might include making a commitment never to fight with, bully, or threaten others.
- Encourage others to resolve conflicts peacefully. By doing this, you can be a role model for nonviolence. You can set a positive example for others.
- Encourage your family to become a member of a Neighborhood Watch program. These programs include volunteers who work closely with law enforcement.
- Report any acts of violence you witness. Talk to your parents, guardians, or a trusted adult and ask what you should do. The adult may help you contact law enforcement.

In addition to the steps listed, you can also become an advocate, or supporter, of safety and victims' rights. One organization you could become involved with is Youth Outreach for Victim Assistance (YOVA). This program is sponsored by the National Center for Victims of Crime. It is a youth-adult partnership that educates teens about what victims experience. This group tells young people where they can go for help if they experience violent crime. You can also search online or contact local law enforcement to learn about other ways you can have a positive impact.

>>> **Reading Check**

EXPLAIN *Why should you tell a trusted adult when you witness an act of violence?*

Many schools today are stepping up measures to prevent acts of violence. *What steps have officials in your school taken to make you feel safer?*

Reducing Violence *in* Schools

You spend a lot of time at school, so it is important to know what your school does to help keep you safe. Most schools are safe places. Each year, however, more than 3 million students report having experienced some type of crime at school. Two million of these crimes are violent. Many schools have taken steps to reduce violence on campus.

It is *important* to **know** what your **school does** to *help keep you safe.*

- **Zero tolerance policies.** Many schools have adopted zero tolerance policies. In schools that have such a policy, any student who brings a weapon to school is expelled. Students who participate in violent acts are also expelled.

- **School uniforms.** Some schools use dress codes or uniforms to help keep students safe. When all students wear similar clothing, it is not easy to tell whether one student is wealthy and another is not. As a result, there may be less violence. Dress codes and uniforms also affect gang members. Dress codes make it difficult for gang members to dress in clothing that shows what gang they are in.

- **Security systems.** Many schools have security systems. They may limit entry to just one door. All of the other doors are kept locked. They may have metal detectors, security cameras, or security guards. These help keep weapons out of schools. Schools may also have a school resource officer from the local police department. Resource officers are usually on campus during school hours. They get to know the students, which helps to prevent problems.

- **Locker searches.** Some schools periodically search students' lockers. Schools may also search students' backpacks. Sometimes trained dogs are brought in by the police. The dogs are used to sniff out drugs and weapons.

- **Conflict resolution programs.** More and more schools are educating students in conflict resolution. A youth court is *a special school program where teens decide resolutions for other teens for bullying and other problem behaviors.* Peer mediation is also popular. These programs involve teens working to help other teens resolve problems in nonviolent ways.

In addition to these efforts, remember some of the actions you can take to help stop violence at school. Become a role model for nonviolence by encouraging others to resolve their differences peacefully. Report any acts or rumors of violence you see or hear about.

NEIGHBORHOOD CRIME WATCH

We immediately report all SUSPICIOUS PERSONS and activities to our Police Dept.

Many communities work together with law enforcement to make their neighborhoods safer. *What is being done in your community?*

Reducing Violence *in* Communities

Students can often influence their peers to make healthful choices. In some schools, students have started their own programs to help reduce violence and criminal activity. They put their ideas into practice in their school and in their community.

Communities are also taking steps to reduce violence. Some communities use their resources to create after-school programs. These programs may be academic, recreational, or cultural. They offer a safe place for teens to spend their afternoons. Teens can stay until their parents or guardians get home from work.

Improved lighting in parks and at playgrounds is another way communities work to reduce violence. Crimes are less likely to happen in well-lighted areas. People who commit crimes are more likely to be seen and recognized if the area has plenty of good lighting.

Protect yourself by avoiding **dangerous** situations.

Neighborhood Watch programs are also popular in many communities. Members of the program watch their neighborhoods for signs of trouble. They report to authorities if they see suspicious activities.

Some communities put their patrol officers on foot, on bicycles, or on horses. This allows the officers to be closer to the people they serve. Police get to know people in the community. This helps law enforcement to prevent criminal activity.

You can protect yourself by avoiding dangerous situations. Walk directly to and from your home. Travel with another person or in a group, especially at night. Avoid taking shortcuts through unfamiliar or unsafe areas of your community.

>>> Reading Check

LIST *What are four ways schools are working to reduce violence?* ■

REVIEW

>>> After You Read

1. **EXPLAIN** What are communities doing to prevent the spread of violence?
2. **RECALL** What can you do to help reduce the spread of violence?
3. **IDENTIFY** What are schools doing to help eliminate violent behavior at school?

>>> Thinking Critically

4. **ANALYZE** Some people claim that school searches violate a person's privacy. How would you respond to this claim?
5. **APPLY** What strategies could you use to keep yourself safe walking home from a mall?

>>> Applying Health Skills

6. **GOAL SETTING** Choose one way you can help to reduce the spread of violence at school or in your community. Use the goal-setting steps you have learned to make an action plan to help reduce violence. Show your plan to your teacher. Follow your action plan for a week, and then write a paragraph discussing your experience.

 Review

 Audio

S. Meltzer/PhotoLink/Getty Images

Image Source/Getty Images

Abuse

BIG IDEA › Abuse affects the physical, mental/emotional, and social health of the person who is abused.

Before You Read

QUICK WRITE Write a paragraph that describes problems that might affect a relationship. Identify healthy ways of dealing with the problems.

 Video

As You Read

STUDY ORGANIZER Make the study organizer found in the FL pages in the back of the book to record the information presented in Lesson 3.

Vocabulary

› abuse
› battery
› domestic violence
› neglect
› sexual abuse

 Audio

Bilingual Glossary

Myth vs. Fact

Myth: Child abuse does not run in families.
Fact: People who abuse children were often abused themselves during childhood. However, not everyone who was abused as a child will become an abuser later in life. Survivors of abuse can make the choice not to continue the pattern. It is often necessary to seek help to learn new strategies for responding when angry.

WHAT IS ABUSE?

MAIN IDEA › Abuse can happen in various ways and take many forms.

Relationships, even close ones, have their good days and bad days. In a healthy relationship, people respect and care for each other. Relationships can be unhealthy, though. Sometimes relationships become unbalanced and difficulties arise. When this happens, abuse can occur. **Abuse** is *the physical, mental, or emotional mistreatment of another person.*

Abuse is never the fault of the person who is being abused. It can affect people of all ages, races, and economic groups. All forms of abuse are wrong and harmful. A person who is being abused should report it to the authorities. Teachers, counselors, nurses, and physicians are required to report suspected abuse to the police. Police investigate reports of abuse to determine whether or not a crime has been committed.

Types *of* Abuse

The four main types of abuse are physical abuse, emotional abuse, neglect, and sexual abuse. You will learn more about these types of abuse as you read the rest of this lesson. In the next lesson, you will also learn strategies for preventing and coping with abuse. Remember, though, that abusive relationships and situations should never be tolerated.

Abuse is never the **fault** of the person who is being *abused.*

Abuse can occur in all kinds of situations. Many times, abuse takes place within a family or other close relationship. For instance, a child may be abused by a parent or guardian. One spouse may be abused by the other spouse. A sibling may abuse a brother or sister. A family member may abuse an older relative. Again, however, abuse is never acceptable behavior.

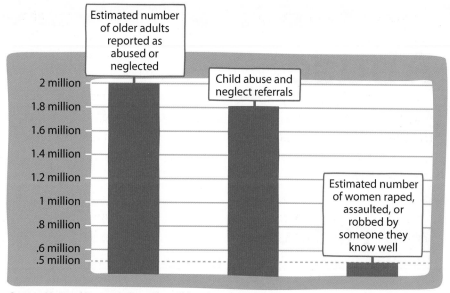

Sources: National Research Council on Elder Abuse, U.S. Department of Justice, U.S. Department of Health and Human Services.

Abuse can happen to people of all ages. *Why do you think people in certain age groups may be more likely to be abused?*

Abuse can also occur between dating partners. Healthy dating relationships are based on each partner respecting the other. When one partner doesn't respect the other, abuse may occur. The abuse may be physical, emotional, or psychological.

Often, abusers will try to make the people they abuse believe that they deserve to be treated harshly. This is never true. Abuse is not the same as discipline. A parent may use discipline, such as a time-out or grounding, to correct behavior or help shape a person's character. On the other hand, abuse causes severe harm to a person—physically, mentally/emotionally, or socially. No one ever deserves to be abused.

>>> **Reading Check**

IDENTIFY *Name some ways in which a person can be affected by abuse.*

Abuse is **not** the same as *discipline.*

Physical Abuse

Physical abuse causes physical harm. A common form of physical abuse is <u>battery,</u> or *the beating, hitting, or kicking of another person.* This is never acceptable behavior. Battery is common in cases of <u>domestic violence,</u> or *physical abuse that occurs within a family.* This situation is where most abuse happens. Domestic violence is about power and control. In such cases, the abuser tries to keep strict control over one or more family members. When physical abuse results from domestic violence, it is considered to be assault. Assault is a crime and should be reported to the police.

Domestic violence includes pushing, slapping, punching, and choking another family member. Sometimes household items are used as weapons. In many cases of domestic violence, physical abuse does not happen all the time. The abuser may engage in other unacceptable behavior in between incidents of assault or battery.

Children who are abused may not get the medical attention they need. Often, when a young victim of abuse is taken for medical attention, the abuser makes excuses for the injuries. For instance, the abuser might claim that the child is clumsy and fell down or ran into something. The abuser might also claim that the young person gets hurt often because he or she is accident prone. Similar excuses are often used when the person who has been abused is an older adult.

Emotional Abuse

Physical abuse often leaves physical signs. It may not be as obvious that a person is being emotionally abused. However, the effects of emotional abuse can be harmful too. The effects of emotional abuse may last even longer than the effects of physical abuse.

The effects of *emotional abuse* may last **even longer.**

Emotional abuse involves words and gestures. These are used to make a person feel worthless, stupid, or helpless. Bullying, yelling, and teasing are all forms of emotional abuse. Insults, harsh criticism, and threats of violence are also forms of emotional abuse. People who have been emotionally abused often feel bad about themselves.

Emotional abuse can occur in the home, at school, or with friends. Sometimes emotional abuse occurs in a dating situation. For example, your partner may be very jealous and want to know where you are all of the time. Your partner may keep you from spending time with family and friends. These are forms of emotional abuse. If you or someone you know has been emotionally abused, get help from a trusted adult.

Neglect

Every person has needs. We all need food, clothing, and shelter. We also need medical care and education. Children need supervision and someone else to provide for them. Sometimes parents or guardians do not meet these needs. **Neglect** is *the failure to provide for the basic physical and emotional needs of a dependent.* This type of abuse sometimes also affects older family members who cannot care for themselves.

People also have emotional needs. People need to feel loved and nurtured. If caregivers do not provide for the needs of the people who depend on them, they may be guilty of neglect. Neglect is against the law and should be reported to the police.

About one out of every ten children in the U.S. is a victim of neglect or child abuse. That means that each year about ten percent of the children are not having their needs met. Neglect can have long-lasting effects on children. Physical injuries can harm a child's growth and development. A neglected child may also have mental/emotional and social problems throughout life.

>>> **Reading Check**

INFER *Why are children more likely to be targets of neglect?*

Emotional abuse can make a person feel isolated or unwanted. *Describe ways you could you reach out to a person who needs help.*

Ray Kachatorian/Photodisc/Getty Images

Sexual Abuse

Every two minutes, someone in the United States is sexually abused. **Sexual abuse** is *sexual contact that is forced on another person.* This can be any type of unwanted touching, kissing, or sexual activity. Photographing a child for sexual materials is another form of sexual abuse. It is also against the law to force a child to look at sexual materials. Sexual abusers often target younger people. A child may be abused by someone who is known and trusted. The abuser may be a family member or a family friend or acquaintance.

Sexual abuse is a type of violence, but abusers may act in other ways toward their victims.

They may try to trick or bribe a child to perform sexual acts. Sexual abuse is never okay. It is damaging to the person who has been sexually abused. Sexual abuse is always a crime. It needs to be reported to the police.

Sexual harassment is another form of sexual abuse. Sexual harassment may happen at school. It includes words, touching, jokes, looks, notes, or gestures with a sexual manner or meaning. Sexual harassment is illegal and should be reported to school personnel immediately.

It can be difficult for a person who has been abused to talk about it. *If you thought someone was being abused, who might you suggest that person turn to?*

Health SKILLS ACTIVITY

Communication Skills

Using "I" Messages

Sometimes a person who makes ethnic jokes or inappropriate remarks may be a friend of yours. He or she may not realize that you find this conduct offensive or abusive. When a friend behaves in a way that offends you, you need to let him or her know that you are offended by these actions. At the same time, you may not want to jeopardize the friendship or create a conflict. You can express your feelings in a positive way by using "I" statements instead of "you" statements. Here are a few examples of "you" statements that can be substituted with "I" statements.

Instead of "You" Statements…	…Use "I" Statements
You're being offensive…	…I am offended by what you are saying.
You're embarrassing me.	…I feel embarrassed.
You aren't being funny.	…I don't think that is funny.

You and the other members of your group should each write three "you" statements. Exchange papers and change the "you" statements to "I" statements.

Ingram Publishing

EFFECTS OF ABUSE

MAIN IDEA People who experience abuse can experience long-term effects.

Physical scars from abuse may go away after time. That is not always the case with the emotional scars. People who experience abuse may blame themselves. They may be afraid or ashamed to get help. Children who suffer abuse or neglect often have a number of problems.

If *you* or someone you *know* experiences **abuse,** seek *help.*

Physical health concerns as a result of abuse include:

- Impaired brain development.
- Impaired physical, mental, and emotional development.
- A hyper-arousal response by certain areas of the brain. This may result in hyperactivity and sleep disturbances.
- Poor physical health, including various illnesses.

Mental/emotional health consequences of abuse include:

- Low self-esteem.
- Increased risk for emotional problems such as depression, panic disorder, and post-traumatic stress disorder.

- Alcohol and drug abuse.
- Difficulty with language development and academic achievement.
- Eating disorders.
- Suicide.

Abuse may also result in various social health consequences:

- Difficulty forming secure relationships.
- Difficulties during adolescence.
- Criminal and/or violent behavior.
- Abusive behavior.

It is important for people who have been abused to deal with their feelings. Help is available from many resources. If you or someone you know experiences abuse, seek help or encourage him or her to get help. Talk to a parent or guardian, a teacher, school counselor, or other trusted adult, or the police. ■

Abuse can affect the physical, social, and mental/emotional health of a person who has been abused. *Name some specific effects of abuse.*

Ingram Publishing

REVIEW

>>> **After You Read**

1. **DEFINE** What is abuse?
2. **DESCRIBE** Name and describe the four major types of abuse.
3. **IDENTIFY** What are four social health concerns for survivors of abuse?

>>> **Thinking Critically**

4. **ANALYZE** Why are children and older adults often the subjects of abuse?
5. **EVALUATE** Do you think it is important for survivors of abuse to get help? Why or why not?

>>> **Applying Health Skills**

6. **ADVOCACY** Locate and research local agencies and organizations that provide help for people who have experienced abuse. For each agency or organization, write down the name, contact information, and whether they specialize in a particular type of abuse. Make copies of the list for your school counselor to give to students.

 Review

 Audio

Preventing *and* Coping with Abuse

BIG IDEA The cycle of abuse can be stopped, but it often requires outside help.

>>> Before You Read

QUICK WRITE Think about adults you trust. Then write one or two sentences that describe what makes those adults trustworthy.

▶ Video

>>> As You Read

STUDY ORGANIZER Make the study organizer found in the FL pages in the back of the book to record the information presented in Lesson 4.

>>> Vocabulary

› cycle of abuse
› crisis hot line

🔊 Audio

🔤 Bilingual Glossary

Developing Good Character

Helping a Victim of Abuse If someone you know has been a target of violence, you can show concern and compassion for that person. You can listen if the person wants to talk. You can help him or her know when to seek help from a parent or other trusted adult. Identify some communication skills that would be helpful in this type of situation. Describe some other ways in which you could show the person that you care.

WARNING SIGNS OF ABUSE

MAIN IDEA Abuse has warning signs and risk factors.

You may notice injuries, such as bruises and burns, on a person who has been abused. The person may have no explanation for the injuries. Sometimes a person who is being abused might appear to be withdrawn or depressed. A person who is being abused may also become aggressive toward others. If you can spot the signs of abuse, you may be able to help.

All families have problems now and then. Usually the problems are dealt with in healthful ways. Good communication skills can help families solve problems in healthful ways. However, some families may not know how to handle problems when they arise. This can lead to extra stress. The stress of family problems can also increase the risk of abusive behavior.

If you can learn to spot *the* **signs** *of* **abuse,** *you may be able to* help.

The Cycle *of* Abuse

Often, abuse in a family may have begun long ago. Health experts have found that patterns of abuse may go back many generations. A generation is a group of people who are born at about the same time. For example, you, your parents, and your grandparents make up three generations. Although it is not always the case, a child who has been abused or who witnessed abuse may grow into an adult who abuses others. This cycle of abuse, or *pattern of repeating abuse from one generation to the next,* does not have to continue forever in a family. However, breaking the cycle of abuse often requires outside help.

>>> Reading Check

DEFINE *What is a cycle of abuse?*

Tetra Images/Getty Images

HELP FOR SURVIVORS OF ABUSE

MAIN IDEA Many resources are available to help people who have been abused.

Abuse in a family affects the whole family. It affects the abuser, the person who has been abused, and other relatives who may not live in the same household. The effects can be long-lasting. Even if a child only witnesses abuse, he or she may grow up thinking abuse is acceptable behavior. In such cases, every member of the family needs help. They need to learn that abuse is not acceptable.

Why Survivors Stay Silent

Many people who have been abused do not tell anyone about it. This prevents them from getting the help they need. Survivors of abuse might choose not to report it for many reasons. Some of the reasons people who have been abused stay silent include:

- Some adults who have been abused think that no one else will believe them.
- A child may fear that adults will think that he or she is lying or exaggerating.
- A person may think that abuse is a private matter.
- Some people may believe that they deserved the abuse because of something they did wrong. However, no one ever deserves to be abused.
- A person may think their abuser will seek revenge.

Men and boys often think that because they are males, they should be able to protect themselves from abuse. This is not true. Abusers usually have an advantage over their victims. They may be stronger or older.

Warning Signs of Abuse

- Illness
- Divorce
- Lack of communication and coping skills
- History of having been abused as a child
- Alcohol or other drug use
- Unemployment and poverty
- Feelings of worthlessness
- Emotional immaturity
- Lack of parenting skills
- Inability to deal with anger

This table presents some factors that can increase the risk of a person becoming abusive. *Explain how identifying and working to improve these factors can help prevent abuse.*

An abuser may be in a position of trust and authority. Abusers may try to make a person feel afraid. A threat can be powerful enough to prevent a victim from reporting the abuse. They may threaten to harm the victim and that person's family if the victim tells.

People who experience abuse should **seek help** for *themselves* and for their *abusers.*

Sometimes, adults who experience domestic violence make excuses for their injuries or for their abuser. This behavior is called enabling. Enabling creates an atmosphere in which a person can continue unacceptable behavior. It can also allow the abuser to avoid taking responsibility for his or her actions. Enabling can establish a cycle or pattern of abuse. If an abuser feels as if he or she will not face consequences for the abuse, it is likely to continue. Sometimes enabling is a result of fear experienced by people who are abused. People who experience abuse should instead seek help for themselves and for their abusers. As you will learn in this lesson, many resources are available to help someone who is being abused. Remember, a person who is abused is never to blame for the abuse.

>>> **Reading Check**

IDENTIFY *What are three reasons that people who are abused often stay silent?*

Sources of Help for Victims of Abuse

If you have a friend you think is being abused, try talking to that person about it. Encourage your friend to seek help. If your friend is too afraid to seek help, go talk to a trusted adult yourself. Let the adult know that you are concerned and worried that your friend is being abused. Remember that child abuse is a crime, and it should be reported to the police.

Talking with a parent or other trusted adult can be one way to help make abuse stop. A crisis hot line, or a *toll-free telephone service where abuse victims can get help and information,* is another resource. Crisis hot lines offer many types of assistance to people who have been abused.

Services provided by crisis hot lines may include help for the abused person, family members, and others who may be affected. People who answer the phones at hot lines have been trained to deal with abuse problems. They know how to help people who have been abused. The caller's identity is kept anonymous.

These organizations help people who have been abused by providing hot lines where people can talk about abuse and get advice. *Identify some others who may be helped by these organizations.*

Crisis hot line conversations are always kept confidential. The figure at the bottom of this page lists information about hot lines that provide help for people who have been abused. Additional hot line numbers can be found online or in the phone book.

If your *friend* is being **abused,** encourage the *friend* to **seek help.**

Breaking *the* Cycle *of* Abuse

Each person has the power to help end the cycle of abuse. People who have been abused—especially children—often feel ashamed. This is especially true if the abuse was sexual. People who have experienced domestic violence may have similar feelings. A person might be afraid that the family will be separated if news of the abuse gets out. A child might also be afraid of getting the abuser in trouble.

However, the cycle of abuse will not end until it is reported. Again, a person who is abused needs help, and so does the abuser. If the abuse is reported to police, the abuser may go to court. As a result, the abuser may be required to seek treatment. If the abuser can learn to manage his or her behavior in more healthful ways, it can help end the cycle of abuse.

Abusers will often threaten their victims or make them promise not to tell anyone. Abuse can also create fear, which can result in enabling. As you have learned, enabling can allow an abuser to continue his or her unacceptable behavior. If you or someone you know is being abused, remember that keeping it secret is not healthful. The only way anyone will get help is if the abuse is reported. Remember that many people have experienced abuse. People who have been abused have many sources of help.

>>> **Reading Check**

RECALL *What types of services do crisis hot lines provide?*

Organization	Whom They Help
Childhelp USA	Child abuse victims, parents, concerned individuals
Youth Crisis Hotline	Individuals reporting child abuse, youth ages 12 to 18
Stop It Now!	Child sexual abuse victims, parents, offenders, concerned individuals
National Domestic Violence Hotline	Children, parents, friends, offenders
Girls and Boys Town	Abused, abandoned, and neglected girls and boys, parents, and family members

NAIC, U.S. Department of Health and Human Services.

Recovering *from* Abuse

Many times, people who have been abused need help in order to recover. Professional counselors are trained to help people overcome emotional trauma. A counselor can help a survivor of abuse by helping the person understand his or her feelings. The person may feel fear or shame. A counselor can also help a survivor learn ways to manage these kinds of emotions. Group counseling is another option. In these sessions, people can meet and talk with others who have also experienced abuse.

Recovery is possible for both the **victim** and the **abuser.**

It is also possible to leave an abusive situation. Children and families who have ongoing problems with abuse may escape to shelters. Shelters are community-run residences where people can feel safe. Families can stay together in shelters, away from the abuser. They can stay while they get help putting their lives back together.

If you have experienced abuse, use what you have learned to get help. If you suspect a friend has been abused, share your knowledge. Encourage that person to get help. Remember that with help, recovery is possible for both the victim and the abuser.

>>> **Reading Check**

IDENTIFY *What are some sources of help for people who have been abused?* ∎

Help is available in any community for people who have been abused and their families. *List some resources that students in your school can turn to if they need to report an abuse problem.*

>>> **After You Read**

1. **IDENTIFY** What are three warning signs of abuse?
2. **EXPLAIN** What is the cycle of abuse?
3. **LIST** Why do some people not report abuse?

>>> **Thinking Critically**

4. **ANALYZE** Why are some abusers older than their victims?
5. **EVALUATE** Why do you think crisis hot lines keep their callers' identities anonymous?

>>> **Applying Health Skills**

6. **COMMUNICATION SKILLS** Write a dialogue between yourself and a person who has been abused. That person wants to report the problem, but doesn't know where to begin.

⟳ Review
🔊 Audio

ZenShui/Alix Minde/Getty Images

Hands-On HEALTH ACTIVITY

Learning to *Manage Threats by Working Together*

WHAT YOU WILL NEED

* Poster board
* Markers or crayons

WHAT YOU WILL DO

1 Your teacher will assign each of four small groups one of the following threatening behaviors: harassment, intimidation, emotional abuse, and bullying.

2 In your group, brainstorm nonviolent ways of responding to the threatening situation. First, imagine an instance in which the threatening behavior takes place. Discuss how a young person might respond in a positive way.

3 Create a colorful poster. On one side, describe the situation your group was assigned. On the other side, write "A Nonviolent Response." Below that, write a nonviolent way of responding.

WRAPPING IT UP

After all the groups have presented their posters, discuss ways teens can help other teens respond to threatening situations in positive ways. Display your posters to help your classmates and other students learn a variety of nonviolent responses.

When faced with a threatening situation, a common reaction is to lash out in anger. When people are pushed, either physically or emotionally, they want to push back. This chapter presents a number of different situations that are threatening, which include harassment, intimidation, emotional abuse, and bullying. When you are faced with a threatening situation, be prepared to respond in a nonviolent way. This activity can help you learn some nonviolent responses.

Blend Images/Getty Images

READING REVIEW

FOLDABLES and Other Study Aids

Take out the Foldable® that you created and any study organizers that you created. Find a partner and quiz each other using these study aids.

LESSON 1 Understanding Violence

BIG IDEA Violence is a major health problem in our society.

* Many factors contribute to the violence in society.
* The most common violent crimes in the U.S. are assault, robbery, rape, and homicide.
* Name-calling, teasing, and making threats are types of violence that can be forms of intimidation.
* A variety of factors influence teens toward violence.
* Victims of violence can suffer physical as well as mental and emotional effects.

LESSON 2 Violence Prevention

BIG IDEA You can help prevent violence.

* You can take a number of steps to help protect yourself from violence, including avoiding unsafe situations.
* Everyone can make an effort to help reduce the spread of violence.
* You can help stop violence by encouraging others to resolve conflicts peacefully and by reporting violent activity.
* Schools and communities have taken action to help stop acts of violence and criminal behavior.

LESSON 3 Abuse

BIG IDEA Abuse affects the physical, mental/emotional, and social health of the person who is abused.

* Abuse can happen in various ways and take many forms.
* The four main types of abuse are physical abuse, emotional abuse, neglect, and sexual abuse.
* People who experience abuse can experience long-term effects.
* If you or someone you know has experienced abuse, seek help or encourage that person to get help.

LESSON 4 Preventing and Coping with Abuse

BIG IDEA The cycle of abuse can be stopped, but it often requires outside help.

* Warning signs of abuse include physical injuries and emotional symptoms.
* Major risk factors for abuse include various types of stress within a family or relationship.
* Each person has the power to do his or her part to help end the cycle of abuse.
* Many resources are available to help people who have been abused.

 Review

 Web Quest

ASSESSMENT

Reviewing Vocabulary and Main Ideas

> assault > youth court > violence > zero tolerance
> homicide > victim policy

>> On a sheet of paper, write the numbers 1–6. After each number, write the term from the list that best completes each statement.

LESSON 1 **Understanding Violence**

1. _____ is any behavior that causes physical or psychological harm to a person or damage to property.

2. The killing of one human being by another is known as _____.

3. A(n) _____ is any individual who suffers injury, loss, or death due to violence.

4. An unlawful threat or attempt to do bodily injury to another person is known as _____.

LESSON 2 **Violence Prevention**

5. A _____ is a policy that makes no exceptions for anybody for any reason.

6. A _____ is a special school program where teens decide resolutions for other teens for bullying and other problem behaviors.

>> On a sheet of paper, write the numbers 7–12. Write *True* or *False* for each statement below. If the statement is false, change the underlined word or phrase to make it true.

LESSON 3 **Abuse**

7. A common type of abuse is <u>zero tolerance</u>.

8. Children, <u>older adults</u>, and people with disabilities all may be targets of neglect.

9. <u>Physical abuse</u> involves words and gestures used to make a person feel worthless, stupid, or helpless.

LESSON 4 **Preventing and Coping with Abuse**

10. The <u>cycle of abuse</u> is a pattern in which children of abuse go on to become abusers.

11. If abuse is ongoing or violent, family members may go to community-run residences known as <u>crisis hot lines</u>.

12. <u>Enabling</u> creates an atmosphere in which a person can continue unacceptable behavior and can establish a pattern of abuse.

✔ eAssessment

>> Using complete sentences, answer the following questions on a sheet of paper.

☁ *Thinking* Critically

13. ANALYZE A teen named Tom lives in a community that has gangs. He often sees gang members in his neighborhood on his way to and from school. Tom does not want to become a victim of gang violence. What are some ways Tom could help keep himself safe?

14. APPLY You have learned that a friend is being abused, but your friend is afraid to seek help. What advice might you give your friend? What could you do to help stop the abuse?

✏ *Write* About It

15. EXPOSITORY WRITING Write a paragraph describing how the media might influence violent behavior.

16. OPINION Use various resources to learn what your community is doing to prevent violence. Think of some additional steps that might help make your community safer. Write a blog post or letter to the editor of your local newspaper offering your suggestions on what more could be done to prevent violence where you live.

Ⓐ Ⓑ Ⓒ Ⓓ STANDARDIZED TEST PRACTICE

Math

The table below contains data about victims of violence for a 10-year period. Use the data to answer the questions that follow.

Violent Victimization Rates by Age, 2000–2010*

Year	12 to 14	15 to 17	18 to 20
2000	5.37	5.19	5.49
2001	5.4	5.24	4.77
2002	4.68	4.82	5.24
2003	5.23	5.79	5.3
2004	4.23	4.42	4.83
2005	5.04	5.5	5.39
2006	5.08	5.12	5.6
2007	5.18	5.03	4.95
2008	5.21	4.66	3.92
2009	4.42	4.38	3.89
2010	3.32	3.37	3.76

* Victimization rates are per 1,000 persons age 12 or older.

1. The two years in which 15- to 17-year-olds experienced a higher rate of crime than 12- to 14-year-olds and 18- to 20-year-olds was

   A. 2000 and 2002.

   B. 2009 and 2010.

   C. 2003 and 2005.

   D. 2001 and 2008.

2. For the years 2008 to 2010, the mean victim rate for 18- to 20-year-olds was

   A. 3.74.

   B. 3.95.

   C. 4.04.

   D. 3.86.

   E. A and C.

   F. D and E.

Unit 5

nutrition + physical activity

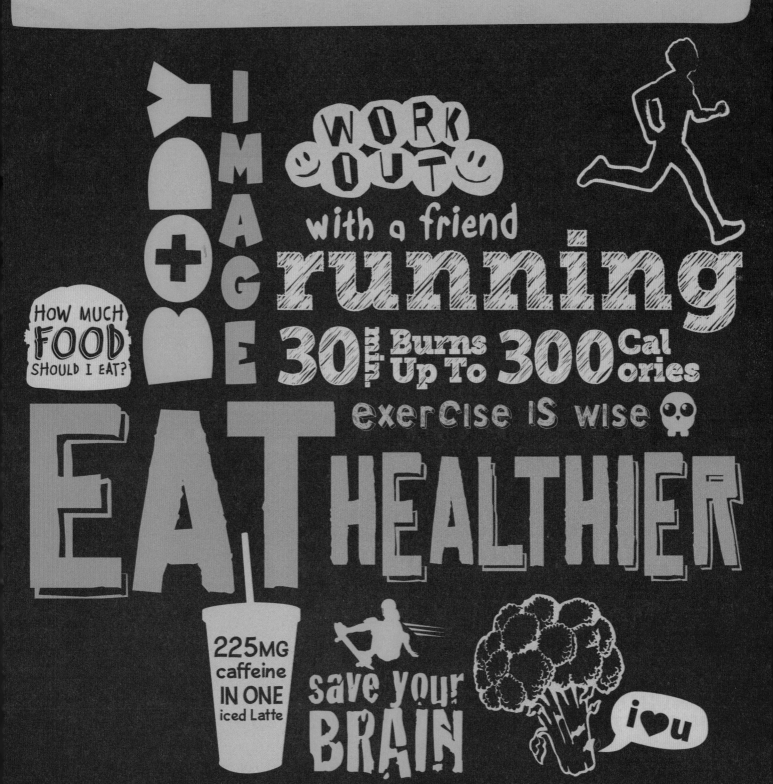

Nutrition

LESSONS

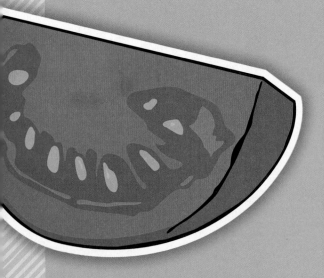

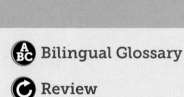

PREMIUM ONLINE RESOURCES

 Audio

 Videos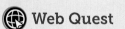

Bilingual Glossary

Fitness Zone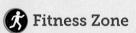

Web Quest

Review

Learning how to
make healthful food
choices helps you meet
your nutritional needs.
*List the healthful foods
that you like to eat.*

Nutrients Your Body Needs

BIG IDEA Each nutrient plays a specific role in keeping your body healthy.

Before You Read

QUICK WRITE List several foods that you think are high in nutrients. After reading the lesson, check to see whether you were right.

 Video

As You Read

FOLDABLES Study Organizer

Make the Foldable® found in the FL pages in the back of the book to record the information presented in Lesson 1 about the types of nutrients.

Vocabulary

> nutrients
> calorie
> nutrition
> proteins
> amino acids
> carbohydrates
> fiber
> fats
> vitamins
> minerals

 Audio

 Bilingual Glossary

UNDERSTANDING NUTRIENTS

MAIN IDEA Nutrients provide energy and help your body function properly.

Imagine you are driving down the road in a car, and it suddenly runs out of gas. The car will stop, because it cannot run without fuel. Your body is similar, but it gets its fuel from food. Nutrients (NOO·tree·ents) are *substances in food that your body needs to function.* Some nutrients give you the energy you need for work and play. A calorie is *a unit of heat that measures the energy available in food.* Other nutrients provide the building blocks your body needs to grow and to repair itself. Since different foods contain a variety of nutrients, you need to eat a variety of foods each day.

Learning about nutrients—what they are, where they come from, and how your body uses them—can help you make more healthful choices about the foods you eat. Nutrition (noo·TRIH·shun) is *the study of nutrients and how the body uses them.* Scientists have identified more than 40 nutrients. They all can be grouped into six main classes: proteins, carbohydrates, fats, vitamins, minerals, and water.

Proteins

Proteins (PROH·teenz) are *the nutrients used to build and repair cells.* They are made of chemical building blocks called amino (uh·MEE·noh) acids. Amino acids are *small units that make up protein.*

Learning about *nutrients* can help you make more **healthful choices**.

Different foods contain different amino acids. Some foods, such as meat, fish, eggs, dairy products, and soybeans, provide all the amino acids your body needs. However, most plant foods are missing at least one of these amino acids. If you choose a vegetarian diet, you can get all the needed amino acids by combining certain plant foods, such as beans and rice. Eating a wide variety of plant foods should give you all the amino acids that your body needs.

Mitch Hrdlicka/Getty Images

Carbohydrates

What does a steamy plate of spaghetti have in common with a ripe peach? Both foods contain large amounts of carbohydrates, your body's main source of energy. Carbohydrates are *the starches and sugars found in foods, especially in plant foods.*

There are two kinds of carbohydrates: simple and complex. *Simple* carbohydrates are sugars. They occur naturally in foods like fruit, milk, and honey. Sugars may also be added when foods are processed.

Complex carbohydrates are starches, or long chains of sugar linked together. Complex carbohydrates are found in foods such as potatoes, seeds, and whole-grain cereals. Your body cannot use these nutrients directly. First, it must break them down through the process of digestion.

Another type of complex carbohydrate, fiber, cannot be digested. Fiber is the *tough, stringy part of raw fruits, raw vegetables, whole wheat, and other grains.*

Although fiber cannot be used for energy, it is still important to your health because it helps carry wastes out of your body. Eating high-fiber foods can lower your risk of developing heart disease. Foods high in fiber can also lower your risk of developing some types of cancer.

A great way to include fiber in your diet is by eating whole-grain cereals. *Explain why it is important to get enough fiber.*

©John Smith/Corbis

Fats

People often talk about fats as something you should always avoid. However, your body actually needs some types of fat in order to function properly. Fats are *nutrients that promote normal growth, give you energy, and keep your skin healthy.*

Your body uses fats to build and maintain cell membranes. These nutrients also carry certain vitamins in your bloodstream and help you feel full after a meal.

What does a steamy plate of **spaghetti** have in **common** with a ripe *peach*?

Not all fats are the same, however. Foods like butter, cheese, and many meats are high in *saturated fats,* which are usually solid at room temperature. Over time, eating too much of these fats can increase your risk of health problems such as heart disease. The same is true of *trans fats,* which start off as liquid oils and are made solid through processing. Stick margarine is one example. However, *unsaturated fats,* which remain liquid at room temperature, can actually lower the risk of heart disease. These fats come mainly from plant foods, such as olive oil, nuts, and avocados.

>>> Reading Check

COMPARE AND CONTRAST
What is the difference between saturated and unsaturated fats?

Myth vs. Fact

Myth: Eating any food with fat in it is unhealthful.

Fact: Some dietary fat is needed to keep you healthy. Those fats should be mostly from the more healthful fats, such as unsaturated fats. To learn more, review the Dietary Guidelines for Americans at *www.usda.gov.*

Vitamins and Minerals

Two other kinds of nutrients that the body needs are vitamins and minerals. Vitamins (VY·tuh·muhns) are *compounds that help to regulate body functions.* Some vitamins help your body fight disease. Others help your body produce energy.

Minerals (MIN·uh·ruhls) are *elements in foods that help your body work properly.* Your body uses only small amounts of these types of nutrients, but they are essential to your health. The table below shows how your body uses several important vitamins and minerals.

>>> **Reading Check**

EXPLAIN *What is the difference between vitamins and minerals?*

Some vitamins, such as vitamins C and B, are *water-soluble.* This means that they dissolve in water. Your body can't store these vitamins, so it needs a fresh supply of them each day. *Fat-soluble* vitamins, such as vitamins A and D, dissolve in fat. Your body can store these vitamins until they are needed.

Water is **essential** to every body *function* you have.

The best way to make sure you get enough vitamins and minerals is to eat a variety of foods. The table below shows which foods are good sources of some important vitamins and minerals. Vitamin supplements can help, but they are not a replacement for healthy eating.

Water

Water is essential to every body function you have. In fact, a person can live for only about a week without water. Your body uses water to carry other nutrients to your cells. Water also helps with digestion, removes waste, and cools you off.

To make sure you are getting enough water, let thirst be your guide. Drink water whenever you are thirsty as well as with your meals. You can also get water from many foods, such as fruits and vegetables, and from beverages such as milk.

Also remember that your body loses water when you sweat in hot weather or during exercise. During these times, you will need to drink extra water to meet your body's needs.

Vitamins and Minerals

VITAMINS	Functions	Food Sources
Vitamin A	Important for good vision and healthy skin	Dark green leafy vegetables (such as spinach), milk and other dairy products, carrots, apricots, eggs, liver
B Vitamins	Helps produce energy; keeps nervous system healthy	Eggs, meat, poultry, fish, whole-grain breads and cereals
Vitamin C	Helps keep teeth, gums, and bones healthy; helps heal wounds and fight infection	Oranges, grapefruits, cantaloupe, strawberries, tomatoes, cabbage, broccoli, potatoes

MINERALS	Functions	Food Sources
Calcium	Helps build strong bones and teeth	Milk and other dairy products, fortified breakfast cereals, oatmeal, dark green leafy vegetables, canned salmon
Iron	Contributes to healthy blood, which helps your body fight disease	Red meat, poultry, dry beans, fortified breakfast cereals, nuts, eggs, dried fruits, dark green leafy vegetables
Potassium	Helps maintain your body's fluid balance	Baked potatoes, peaches, bananas, oranges, dry beans, fish

VITAMINS & MINERALS

TO MEET FDA RECOMMENDED DAILY AMOUNTS OF

CALCIUM
YOU WOULD NEED TO
CONSUME

4 1/3 EIGHT-OUNCE GLASSES OF MILK

OR EAT 2.75 CANS OF

SARDINES

GROSS!

28% 56% 39% 38%

*(%DV) = % OF DAILY VALUE

- FETA
- SWISS
- MOZZARELLA
- PARMESAN

% DAILY VALUE The Nutrition Facts panel shows the % of the recommended daily intake of a particular nutrient in one serving of that food.

A GOOD SOURCE OF VITAMIN B-12 IS CHEESE

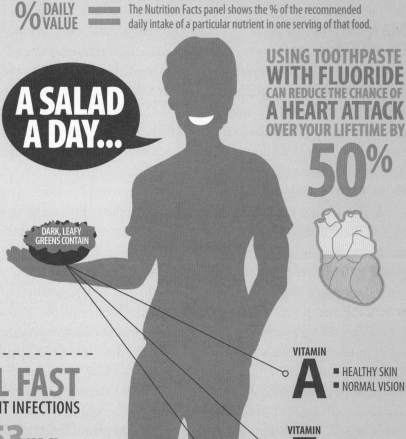

A SALAD A DAY...

DARK, LEAFY GREENS CONTAIN

USING TOOTHPASTE **WITH FLUORIDE** CAN REDUCE THE CHANCE OF **A HEART ATTACK** OVER YOUR LIFETIME BY

50%

VITAMIN A = HEALTHY SKIN / NORMAL VISION

VITAMIN E ANTIOXIDANTS THAT PROTECT CELLS

FOLIC ACID HELPS YOUR BODY FORM AND MAINTAIN NEW CELLS

EAT HEALTHY+HEAL FAST
VITAMIN C HELPS TO HEAL WOUNDS AND FIGHT INFECTIONS

VITAMIN C- RICH FOODS INCLUDE ORANGES, STRAWBERRIES, AND BROCCOLI

53mg

88mg

30mg

YOUR HEART USES
CALCIUM
TO REGULATE RHYTHM

60–75 BPM   60–80 BPM

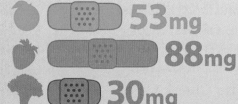

HEALTHY HEART RATE*(BPM)=Beats Per Minute

Guidelines *for* Good Nutrition

The U.S. Department of Agriculture (USDA) and the Department of Health and Human Services (HHS) have developed recommendations for ways to help you eat right. The Dietary Guidelines for Americans offer information about healthy eating and active living. These guidelines are meant for people ages two and up.

Make Smart Food Choices

What can you do to give your body the balance of nutrients it needs? You can start by eating a variety of nutritious foods every day. The Dietary Guidelines for Americans recommend eating:

MORE FRUITS Choose mostly whole fruit rather than drinking a lot of fruit juices. Whole fruit has more fiber.

MORE VEGETABLES Eat more leafy, dark-green vegetables such as broccoli and spinach. Orange vegetables, such as carrots and sweet potatoes are other good sources of nutrients.

MORE WHOLE GRAINS At least half the grains you eat should be whole grains. Try adding oatmeal, whole-wheat bread, and brown rice to your eating plan.

CALCIUM-RICH FOODS Low-fat or fat-free milk, yogurt, or cheese are healthful choices.

PROTEIN-RICH FOODS These include fish, chicken, lean meats, eggs, nuts, seeds, and beans. Go easy on foods that are high in fats, such as fatty meat, butter, and stick margarine.

Avoid Too Much Added Sugars *and* Salt

Some foods are high in added sugars. Choices such as candy, non-diet soft drinks, and sugary desserts are often low in other nutrients. They can fill you up, making you less likely to eat more healthful foods. They can also promote tooth decay. Calories from sugars that are not used by the body for energy are stored as body fat. This can lead to unhealthful weight gain.

Eating too much salt and sodium can also cause problems for some people. Table salt contains sodium, a mineral which helps regulate blood pressure. Too much sodium can increase the risk of high blood pressure. You can avoid consuming too much sodium by cutting down on the amount of salty snacks you eat and avoiding adding salt to your food at mealtimes.

Physical fitness is also important to your health. Try to match how physically active you are with the amount of food you eat. To stay healthy, you need to eat just what your body requires for energy. You will learn more about making healthful food choices in the next lesson. The next chapter focuses on physical fitness.

> ### ⟫⟫ Reading Check
>
> **EXPLAIN** *Why is it healthful to limit the amount of foods you eat that are high in added sugars and salt?* ■

⟫⟫ After You Read

1. **VOCABULARY** Define the term *nutrition*. Use it in an original sentence.
2. **IDENTIFY** Name the six categories of nutrients.
3. **EXPLAIN** How do vitamins help your body?

⟫⟫ Thinking Critically

4. **HYPOTHESIZE** How can the foods you choose to eat today affect your health in the future?
5. **ACCESSING INFORMATION** Research foods that are high in fiber. Identify at least three foods that are high in fiber. Then find all of the high-fiber foods in your house. How many foods did you find? Is your family eating enough high-fiber foods?

⟫⟫ Applying Health Skills

6. **PRACTICING HEALTHFUL BEHAVIORS** Study your school's weekly lunch menu. Find the most healthful food choices. Then make a plan to include these healthful choices in your daily diet.

 Review

🔊 Audio

Creating a Healthful Eating Plan

BIG IDEA Learning to make healthful food choices will help you maintain good health throughout your life.

©Daniel Koebe/Corbis

Before You Read

QUICK WRITE What are your favorite foods? Why do you like them?

▶ Video

As You Read

STUDY ORGANIZER Make the study organizer found in the FL pages in the back of the book to record the information presented in Lesson 2.

Vocabulary

› appetite
› hunger
› MyPlate
› nutrient dense

🔊 Audio

🔤 Bilingual Glossary

What Teens Want to Know

Is a vegetarian diet right for me? People choose vegetarian diets for various reasons. Cost factors, health issues, or religious beliefs may restrict eating meat. Some people simply prefer other foods. Some diets limit all animal products. Others may include dairy, eggs, or fish. Everyone should eat a variety of vegetables, fruits, and grains. See the *Dietary Guidelines for Americans* for more information.

WHAT INFLUENCES YOUR FOOD CHOICES?

MAIN IDEA A variety of factors influence food choices.

Think about the last meal you ate. Perhaps you were grabbing a quick breakfast while rushing to school. Maybe it was lunchtime and you had to choose a meal in the school cafeteria. Or maybe you were just craving a snack.

Many factors affect your food choices, including your own personal preferences. You may simply like the appearance, flavor, or texture of certain foods. These foods appeal to your **appetite,** which is *the emotional desire for food.* Appetite is different from **hunger,** *the body's physical need for food.* Certain foods may trigger your appetite because of their connections to your memories and feelings. If the smell of baking bread reminds you of spending time with your family, for instance, you may feel like eating some even if your body does not really need it.

Other factors that can affect your food choices include:

Family and friends. You may eat more vegetables at home because your family encourages you to eat them. You might choose fast food with your friends because the group enjoys it.

Your culture. Yoshi, who grew up in Japan, likes miso soup and rice for breakfast. His U.S.-born friend Carl prefers cereal and milk in the morning.

Convenience. You might choose a snack from a vending machine because it is handy. If you live near a farmers' market, you may be more likely to eat fresh vegetables and fruits.

Media. Advertisers use many techniques to make you want to buy their foods. You might want to try a new pizza place if the ad shows teens having fun there.

USING MYPLATE

MAIN IDEA MyPlate provides a visual guide to help consumers make more healthful food choices.

As you can see, when mealtime rolls around, there are many different factors leading you in different directions. How can you sort out all the conflicting influences and choose foods that will give your body the nutrients it needs? One tool that can help is MyPlate. This is *a visual reminder to help consumers make healthful food choices.*

When **mealtime** rolls around, there are *many* different *factors* leading you in **different directions**.

The idea behind MyPlate is simple. All the foods you eat are organized into five food groups. Each group provides a different set of nutrients. Choosing a balance of foods from all five groups will give you all the nutrients that your body needs. The MyPlate diagram, shown below, shows how to divide up your plate among these five groups at each meal.

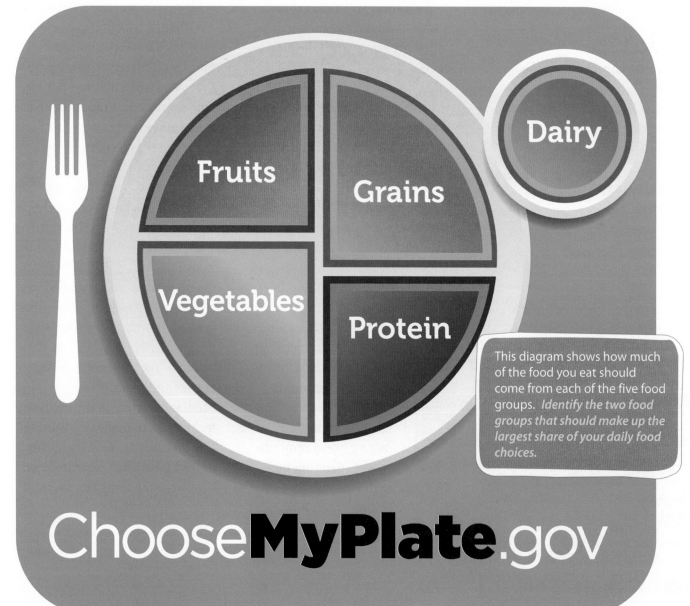

Fruits
Grains
Vegetables
Protein
Dairy

This diagram shows how much of the food you eat should come from each of the five food groups. *Identify the two food groups that should make up the largest share of your daily food choices.*

ChooseMyPlate.gov

US Department of Agriculture

The Five Food Groups

The five food groups highlighted in the MyPlate diagram are:

GRAINS This group includes all grains, such as rice, corn, wheat, oats, and barley. It also includes foods made from these grains, such as bread, cereal, and pasta. In general, teen boys and girls need between 5 and 8 ounces a day of grains. One ounce might be a slice of bread, a cup of cold cereal, or half a cup of rice or noodles. About half the grains you eat should be *whole grains*. Whole grains are foods that contain the entire grain kernel, such as whole-wheat bread, oatmeal, brown rice, and popcorn.

VEGETABLES Teen boys and girls need 2 to 3 cups of vegetables daily. To get all the nutrients you need, you should choose a variety of vegetables. For instance, include some leafy greens, such as spinach, and some starchy vegetables, such as potatoes.

> Teens need 3 cups of *milk* per day to meet their **nutrient needs**.

FRUITS Like vegetables, fruits can be eaten fresh, frozen, or canned. Teens should eat 1½ to 2 cups of fruit per day. Fruits and vegetables together should fill half of your plate.

DAIRY This group includes milk and many other dairy products, such as cheese and yogurt. (Butter or cream, which contain little calcium, are not part of this group.) Teens need 3 cups of milk per day to meet their nutrient needs. Low-fat milk is a healthful choice.

PROTEIN This group includes foods high in protein, such as meat, fish, beans and peas, eggs, and nuts. Teens need between 5 and 6½ ounces of these foods per day. An ounce might be an ounce of meat, one egg, or a tablespoon of peanut butter.

>>> **Reading Check**

LIST *What are the five basic food groups shown in MyPlate?*

Daily Food Plan ChooseMyPlate.gov		Grains	Vegetables	Fruits	Dairy	Protein
12-year-old girl 4'10" tall, 90 lbs.	Active 30 to 60 min.	6 oz.	2.5 cups	1.5 cups	3 cups	5 oz.
12-year-old boy 4'10" tall, 90 lbs.	Active 30 to 60 min.	7 oz.	3 cups	2 cups	3 cups	6 oz.
13-year-old girl 5'1" tall, 100 lbs.	Active 30 to 60 min.	6 oz.	2.5 cups	2 cups	3 cups	5.5 oz.
13-year-old boy 5'2" tall, 105 lbs.	Active 30 to 60 min.	8 oz.	3 cups	2 cups	3 cups	6.5 oz.
14-year-old girl 5'4" tall, 110 lbs.	Active 30 to 60 min.	6 oz.	2.5 cups	2 cups	3 cups	5.5 oz.
14-year-old boy 5'5" tall, 115 lbs.	Active 30 to 60 min.	8 oz.	3 cups	2 cups	3 cups	6.5 oz.

PLANNING HEALTHFUL MEALS

MAIN IDEA ⟩ Use variety, moderation, and balance when choosing foods.

To put the five food groups together into healthy meals and snacks, remember these three words: *variety, moderation,* and *balance.* Choosing a *variety* of foods from the five food groups will help you get all the nutrients you need. It will also help keep your meals and snacks interesting. *Moderation* means keeping your portions to a reasonable size and limiting nutrients that can be harmful, such as fats, sugars, and salt. One way to do this is to drink water instead of sugary drinks. Finally, find the right *balance* between the amount of food you eat and your level of physical activity.

This will help keep your weight under control. You will learn more about managing your weight in the next lesson.

Eating Right at Every Meal

You have probably heard people say that breakfast is the most important meal of the day. When you wake up in the morning, your body needs fuel to get you through the day ahead. Eating a healthful breakfast will boost your energy and help you concentrate in school.

Breakfast
is the most
important
meal of the day.

Good choices for breakfast include foods with complex carbohydrates and protein, such as oatmeal with milk or eggs with whole-grain toast. If you are in a hurry, you can still choose a healthful breakfast. Foods such as fresh fruit, whole-grain bread, and string cheese are all easy to grab and go.

At lunch and dinner, keep the MyPlate diagram in mind. Remember to fill half your plate with vegetables and fruits. If you pack your own lunch, consider a salad to go with—or instead of—your sandwich. A sandwich on whole-grain bread is a good way to get your daily servings of whole grains. Fresh fruits, such as apples and bananas, make an easy-to-carry, healthful dessert.

Eating a variety of foods is the best way to make sure you get all the nutrients you need. *Select which of these healthful foods you enjoy.*

©Tom Grill/Corbis

For growing teens, snacks are also an important part of a day's eating plan. Snacks can help meet your nutrient needs and keep you going through the day. However, many popular snack foods, such as potato chips and cookies, contain a lot of fat, sugar, or salt.

Is a **granola** bar really *better* for you than a **candy** bar?

More healthful snack choices include fresh fruit or dried fruit, air-popped popcorn, whole-grain crackers with cheese, or unsalted nuts. These types of snack foods are nutrient dense, meaning that they *have a high amount of nutrients relative to their number of calories.*

>>> **Reading Check**

IDENTIFY *What are some healthful breakfast foods?*

Eating *Right* When Eating *Out*

Many people eat some meals at restaurants. Part of eating right is making healthful food choices when eating out. When ordering, keep the Dietary Guidelines and MyPlate in mind. Follow these tips:

* **PAY ATTENTION TO THE PORTION SIZES** Many restaurants serve huge portions of food. You may want to eat only part of the meal and take the rest home to eat the next day.

* **STRIVE FOR BALANCE** If you choose to eat the larger portion, eat a smaller meal later.

* **THINK ABOUT WHAT YOU ORDER** Be aware that many restaurants add high-fat sauces or toppings to foods that may already be high in fat.

* **ORDER FEWER FOODS WITH FATS** Choose foods that are baked, grilled, or broiled rather than fried. You can also ask that sauces be served on the side.

* **STAY INFORMED** Some fast-food restaurants make nutritional information available for items on their menus. Ask to see the information before you order.

On Your Own

Think of one of your favorite meals. List the steps you would need to follow in preparing this meal to keep the food safe.

Nutrition Facts

Serving Size 1 cup (226g)
Servings Per Container 2

Amount Per Serving

Calories 250   Calories from Fat 110

	% **Daily Value***
Total Fat 12g	**18**%
Saturated Fat 3g	**15**%
Trans Fat 3g	
Cholesterol 30g	**10**%
Sodium 470mg	**20**%
Potassium 700mg	**20**%
Total Carbohydrate 31g	**10**%
Dietary Fiber 0g	**0**%
Sugar 10g	
Protein 5g	

Vitamin A 4%	•	Vitamin C 2%
Calcium 20%	•	Iron 4%

*Percent Daily Values are based on a 2,000 calorie diet. Your Daily Values may be higher or lower depending on your calorie needs.

	Calories	2,000	2,500
Total Fat	Less than	65g	80g
Saturated Fat	Less than	20g	25g
Cholesterol	Less than	300mg	300mg
Sodium	Less than	2,400mg	2,400mg
Total Carbohydrate		300g	375g
Dietary Fiber		25g	

The Nutrition Facts label on a food package gives you important information about a food's nutritional value. *Determine how many calories a serving of this food contains.*

Getting the Nutrition Facts

Some foods are easy to identify as healthful or less healthful choices. However, with foods that come in a package, it is not always easy to tell. Is a granola bar really better for you than a candy bar? Are pretzels a better choice than chips?

One way to find out is to look at the Nutrition Facts label found on almost all packaged foods. Take a look at the Nutrition Facts label above. It shows the number of calories

and nutrients found in a single serving of the product. Keep in mind that some packages may contain more than one serving. That means that if you eat the whole package, you are taking in more calories and nutrients than are stated on the label. ■

>>> **After You Read**

1. **VOCABULARY** Define *nutrient dense*. Give an example of a nutrient-dense snack.
2. **LIST** Name three factors that can influence your food choices.
3. **EXPLAIN** How many cups of milk or dairy products do teens need each day?

>>> **Thinking Critically**

4. **APPLY** Tom had a peanut butter sandwich and a glass of milk for lunch. Which food groups do these foods represent in MyPlate? What else could Tom eat to make his plate more complete?

>>> **Applying Health Skills**

5. **ANALYZING INFLUENCES** Find an ad for a food or food product. What does the ad tell you about the food? What methods does it use to encourage you to buy the food? Does the ad make you want to try the food? Discuss your findings with your classmates.

 Review

 Audio

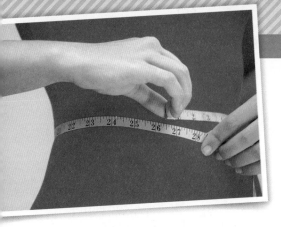

Managing Your Weight

BIG IDEA You can maintain a healthy weight by balancing the food you eat with physical activity.

>>> **Before You Read**

QUICK WRITE Write a paragraph describing what actions you currently take to try to maintain a healthy body weight.

 Video

>>> **As You Read**

STUDY ORGANIZER Make the study organizer found in the FL pages in the back of the book to record the information presented in Lesson 3.

>>> **Vocabulary**

> overweight
> underweight
> Body Mass Index (BMI)
> energy equation.

 Audio

 Bilingual Glossary

What Teens Want to Know

How do I know if I need to lose weight? Teens come in many sizes and shapes. You may not know if you are underweight or overweight. Calculate your Body Mass Index (BMI) using an online tool. Is your BMI near the recommended value for your age? Follow the *Dietary Guidelines for Americans*. If you eat a balanced diet and get at least 60 minutes of physical activity each day, you should reach your ideal weight.

YOUR WEIGHT AND YOUR HEALTH

MAIN IDEA Maintaining a healthy weight can help prevent serious health problems during all stages of life.

One advantage of eating right is that it helps you maintain a healthy weight. Keeping your weight in a range that is right for your body is important for your overall health. Teens who are overweight are at a higher risk for a variety of health problems. Being overweight means *weighing more than is healthy for a person of your gender, height, age, and body type.* As many as one in five children is now considered overweight.

This trend has been linked to lifestyle factors. Some people eat too much. Others eat too many unhealthful foods. Many do not get enough physical activity. Extra pounds put added strain on your heart and lungs. Teens who are overweight are at an increased risk of type 2 diabetes. They are also more likely to develop high blood pressure, heart disease, and cancer later in life.

Concern about overweight teens makes it easy to overlook the issue of teens who are too thin. Being underweight, or *weighing less than is healthy for a person of your gender, height, age, and body type,* is also a problem for many young people.

> Keeping your **weight** in a range that is right for your body is *important* for your **health**.

Underweight teens may not be getting the nutrients their bodies need. They may be at risk of developing anemia. Anemia is a lack of iron in the blood that can make a person feel tired and run down. Teens who are too thin may have a harder time fighting off illness or infection. Finally, they may not have enough stored body fat to help keep them warm and provide them with an energy reserve.

FINDING YOUR HEALTHY WEIGHT RANGE

MAIN IDEA Several factors help to determine your healthy weight range.

Of course, a weight that is healthy for one person won't be the same for another. A six-foot-tall, male football player would not be healthy at the same weight as a five-foot-tall, female ballet dancer. Your "ideal" weight is not a single number but a range that depends on your age, gender, height, and build.

So how can you tell whether your weight is healthy? One way to check is to calculate your Body Mass Index. This is *a method for assessing your body size based on your height and weight.* The chart at the bottom of this page shows BMI ranges for teen males and females.

If your BMI falls into the green range on the chart, it means that you are most likely at a healthy weight for your height. However, it is important to understand that the BMI measurement is not perfect.

For one, it cannot take a person's body build into account. This means that people who are heavily muscled or have a stockier build may be labeled as overweight, even if they do not have a lot of body fat.

It is also important to remember that during your teen years, your body is changing rapidly. Your specific growth pattern may cause you to become overweight or underweight for a time. This is typically normal.

If you are not sure whether your weight is within a healthy range, see a health professional. Do not try to lose or gain weight unless the provider recommends it and suggests a specific eating and physical activity plan.

>>> **Reading Check**

DESCRIBE *What are the limitations of the Body Mass Index as a measure of body size?*

First calculate your BMI using an online tool. Then find your age on the graph. Trace an imaginary line from your age to your BMI to see what range it falls into. *Explain why age and height are important factors in determining BMI.*

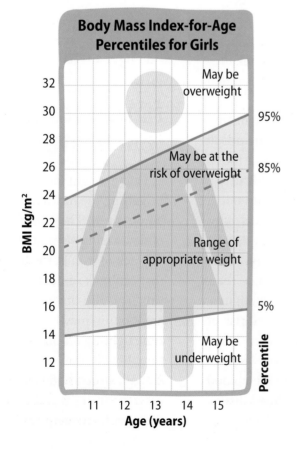

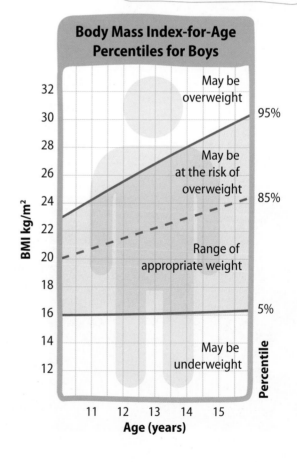

REACHING A HEALTHY WEIGHT

MAIN IDEA Balancing the calories you eat and the ones you burn with activity will help you maintain a healthy weight.

So, if it turns out you are overweight or underweight, what can you do to reach a healthy weight? The key is calories. When you digest food, your body converts food calories into energy. It uses this energy to power your daily body processes and physical activity. When you take in *more* calories than your body uses, it stores the extra calories as body fat. This makes you gain weight. If you eat *fewer* calories than you need, though, your body converts stored body fat into energy. As a result, you will lose weight.

A **chocolate bar** has roughly the *same* number of *calories* as a **turkey sandwich** on whole wheat bread.

How much you eat and what your body needs are related. *The balance between the calories you take in from food and the calories your body uses* is called the energy equation. It takes about 3,500 calories to produce one pound of body fat.

Design Pics Inc./Alamy

To maintain a healthy weight, eat regular meals, choose snacks wisely, and be more physically active. *Explain how planning ahead could help with weight management.*

The energy equation means that if you eat 250 fewer calories than your body burns each day, you can lose one pound of weight after two weeks. The same thing happens if, by exercising, you burn 250 more calories per day than you eat. While 250 calories may not seem like very much, it can make a big difference in your weight over time.

⟩⟩⟩ Reading Check

EXPLAIN *What happens if you take in more calories than your body uses?*

If you are trying to lose weight, choosing nutrient-dense foods will help. A turkey breast sandwich on whole wheat bread, for instance, contains about 250 calories. A chocolate bar has roughly the same number of calories, but it has far fewer nutrients. It will not do as much to fill you up or nourish your body. However, this does not mean that if you need to gain weight, you should eat more candy bars. Instead, choose nutrient-rich foods that are high in calories, such as nuts. If you need to gain weight, you can also try drinking milk or fruit juice with your meals instead of water.

Physical activity is another key to managing your weight. If you need to lose pounds, increasing your level of activity will help. However, teens who want to gain weight still need a certain amount of physical activity to stay healthy. The MyPlate food guidance system recommends that all teens get approximately 60 minutes of physical activity on most days. If you want to gain weight, do not decrease your activity below this level. Instead, increase the number of calories you consume.

Avoiding Unhealthy Diets

Just because you may be a little overweight, it does not mean you have to "go on a diet." In fact, most teens should not diet to lose weight. It is often better just to let your body grow into your healthy weight. Also, many diet plans are ineffective or, worse yet, harmful to your health.

Some unhealthy diet plans require you to eat a lot of certain specific foods, such as grape-fruit. Others require you to avoid some types of food completely, such as high-carbohydrate foods. Some weight-loss plans rely on pills, supplements, or extreme workout programs that promise very fast results.

The **truth** is, when it comes to managing your *weight*, there really are **no** shortcuts.

The truth is, when it comes to managing your weight, there really are no shortcuts. Although you may lose weight on a so-called "fad diet," you will probably gain it back as soon as you start eating normally again. In the meantime, your body will miss out on important nutrients it needs to grow and develop.

Healthful Weight Management

If there are no shortcuts, how can you make sure you reach a healthy weight? You can learn and apply some strategies that will help you in your teen years and beyond. Instead of dieting, try these tips for managing your weight in a healthful way:

- Eat a variety of healthful foods, but reduce the portion sizes. Continue to use MyPlate as your guide, but eat smaller servings from each food group at each meal.
- Drink plenty of water, and avoid high-calorie beverages such as sugary soft drinks.
- Make time for regular meals. Rather than snacking a lot, try to eat only when you're hungry. It is also easier to keep track of how much you eat when you eat at set times.
- Boost your level of physical activity. This will not only help you manage your weight, it will help you in all areas of your health triangle.
- Take your time when you eat. Chew your food thoroughly. This gives your stomach time to signal your brain when you have eaten enough.
- Talk to your health care provider. He or she can recommend a safe, healthy approach that will help you reach your weight goal in a reasonable amount of time. ■

LESSON 3
REVIEW

>>> After You Read

1. **VOCABULARY** Define *Body Mass Index*.
2. **RECALL** Identify one health problem associated with being overweight and one problem associated with being underweight.
3. **IDENTIFY** List the factors that help determine your healthy weight.

>>> Thinking Critically

4. **APPLY** Suppose that a teen takes in 2,000 calories each day and burns 2,300 calories. Over time, what will happen to the teen's weight? Explain your answer.

>>> Applying Health Skills

5. **COMMUNICATION SKILLS** A friend of yours wants to lose some weight. She says she plans to try a new liquid diet that is supposed to take off ten pounds in two weeks. What could you say to convince her that this is not a good idea?

 Review

🔊 Audio

Body Image *and* Eating Disorders

BIG IDEA Teens with a poor body image may develop extreme and harmful eating behaviors.

►►► Before You Read

QUICK WRITE Write a paragraph explaining why you think some teens develop an eating disorder.

► Video

►►► As You Read

STUDY ORGANIZER Make the study organizer found in the FL pages in the back of the book to record the information presented in Lesson 4.

►►► Vocabulary

› body image
› eating disorders
› anorexia nervosa
› bulimia nervosa
› binge eating

 Audio

 Bilingual Glossary

Cultural Literacy

Compulsive Exercising Compulsive exercising is closely related to eating disorders. It is possible to become dependent on or obsessed with exercise. The effects of compulsive exercising are similar to other eating disorders. Use online resources to learn more about compulsive exercising. Find out what to do if you suspect that someone may have this disorder.

BODY IMAGE

MAIN IDEA Your body image is closely related to your weight.

Do I need to drop a few pounds or gain some weight? Many teens ask themselves these questions. Your weight is tied to your **body image,** or *the way you see your body.* Many people have a poor body image. They think they are too thin, too fat, or not muscular enough. People who are unhappy with their bodies may try to change their weight in extreme ways. This can not only damage their health, it may even be life threatening.

Focus on the things you *like* about yourself.

Many factors can influence your body image. Your family and friends may express opinions about the way they think people should look. The media is another influence. Movies and TV show thin models and actors. It is also common to see muscular athletes. However, having a poor body image can hurt your self-esteem, which is the way you feel about yourself inside.

Developing *a* Positive Body Image

Bodies come in all shapes and sizes. Yours depends largely on your gender and what you inherited from your parents. These factors are out of your control. Growth also affects your body shape. Many teens grow in spurts and may carry a few extra pounds or be underweight for a time.

To develop a positive body image, focus on what you like about yourself. Avoid comparing yourself to unrealistic ideals. Spend time with people who like and appreciate you. Show that you value yourself by taking care of your body and mind. Eat well, get lots of rest and exercise, and engage in activities you enjoy.

►►► Reading Check

DESCRIBE *What is the key to having a positive body image?*

EATING DISORDERS

MAIN IDEA An unhealthy body image can lead to an eating disorder.

A person who has an unhealthy body image may be at risk of developing one of a number of eating disorders. These are *extreme eating behaviors that can lead to serious illness or death.* People who feel bad about themselves or are depressed are more likely to develop an eating disorder.

Eating disorders are most common among teen girls and young women. However, males can develop them as well.

Eating disorders are very dangerous. and can even be deadly. Often, though, people with eating disorders may deny that they have a problem. If you think someone you know has an eating disorder, talk about the problem with an adult whom you trust. You can also help a friend who may have an eating disorder by talking about it with that person. Encourage your friend to seek help. An eating disorder is a mental health problem that requires medical treatment.

People who **feel bad** about themselves or are *depressed* are more likely to develop an **eating disorder**.

>>> **Reading Check**

DEFINE *What is an eating disorder?*

Anorexia Nervosa

Anorexia nervosa (an·uh·REK·see·uh ner·VOH·suh) is *an eating disorder in which a strong fear of weight gain leads people to starve themselves on purpose.* People with anorexia eat far fewer calories than they need to stay healthy. They may also exercise excessively. Even after they have become extremely thin, they still see themselves as overweight.

People with this disorder eat so little that their bodies do not get the nutrients they need to grow and repair themselves.

People with anorexia may see themselves as overweight even if they are very slim. *Describe the symptoms of anorexia.*

Photodisc/Getty Images

Their bones may become thin and brittle from lack of calcium. Their blood pressure and body temperature may drop. People with this disorder can starve to death. They may also die from heart failure, kidney failure, or other medical complications. In addition, the depression that often comes with anorexia may lead to thoughts of suicide.

People with *anorexia nervosa* can **starve to death**.

Bulimia Nervosa

Bulimia (byoo·LEE·mee·uh) nervosa is *an eating disorder in which a person repeatedly eats large amounts of food and then purges.* There are several ways that people purge. One way is to vomit, or throw up. Another way is to use laxatives. People with this illness may have a normal weight, but still feel the need to go on an extreme diet. When they can't stay on the diet, they suddenly eat large amounts of food. Then, after eating, they purge. They may also try to burn the calories with constant exercise.

People with this disorder may often go to the bathroom after eating a large meal. While in the bathroom, they may run the water to cover the sound of vomiting. Another sign is swollen cheeks caused by vomiting.

Bulimia robs the body of nutrients and causes **damage**.

Bulimia usually does not lead to extreme weight loss, but it is hard on the body. It robs the body of nutrients and can damage the colon, liver, kidneys, and heart. The coating on the teeth may wear off, because vomiting exposes teeth to the acids from the stomach's contents. The linings of the stomach and esophagus can also be hurt. The body may become dehydrated, meaning it does not have the water levels it needs.

>>> **Reading Check**

IDENTIFY *What are two signs of bulimia nervosa?*

Health SKILLS ACTIVITY

Decision Making

Help for a Friend with an *Eating Disorder*

Kara and Rachel have been friends since kindergarten. Lately, Rachel has begun to notice some changes in Kara. During lunch, Kara hardly eats anything. She claims she is not hungry. She also seems to be getting very thin. Rachel is beginning to worry that Kara may have an eating disorder. Rachel isn't sure whether she should talk to Kara about her concerns or tell someone else.

What Would You Do?

Apply the six steps of decision making to Rachel's problem. Tell what decision you would make if you were Rachel, and why.

1. State the situation.
2. List the options.
3. Weigh the possible outcomes.
4. Consider your values.
5. Make a decision and act on it.
6. Evaluate the decision.

On Your Own

Apply the steps of the decision-making process to Rachel's problem. Explain what you would do if you were Rachel and why.

Percentage of Female High School Students

23%	**15%**	**8%**	**3%**	**50%**	**1%**
are on a diet	binge eat	suffer from bulimia nervosa	use diet pills daily	believe they are overweight	are happy with the way they look

Binge Eating

Binge eating is *a disorder in which a person repeatedly eats too much food at one time.* It is also called compulsive overeating. A *compulsion* is something that you feel you cannot control. People with this disorder may eat even when they are not hungry. They may eat so much food that they feel physically uncomfortable. Binge eaters may hide food they plan to eat. They may also eat alone so that others do not see how much they are eating. In some cases, binge eating may be an attempt to deal with depression.

As you might imagine, binge eating can result in serious weight gain. Binge eaters can develop all the health problems related to being overweight. These include heart disease, diabetes, and some types of cancer. In addition, the guilt that compulsive eaters feel about their problem can lead to depression and low self-esteem.

Males *and* Eating Disorders

As you have learned, eating disorders are linked to body image. People who do not like the way they look may go to extreme measures to change their weight. When you think of anorexia and bulimia, extreme weight loss typically comes to mind. On the other hand, binge eating often leads to extreme weight gain. These eating disorders affect both males and females. Males and females experience similar signs and symptoms.

A related disorder known as *muscle dysmorphia* involves an obsession with adding muscle. It can also affect anyone, but it is more common in males. During the teen years, a girl may mistakenly think she needs to lose weight. A boy the same age may see himself as being too small or not weighing enough. As a result, he may want to gain weight and add muscle. Along with compulsive eating, this disorder often includes using steroids or other dangerous drugs to bulk up.

Eating disorders can affect both males and females. *Explain why extreme weight loss or weight gain can be harmful to a teen's physical, mental/emotional, and social health.*

©McGraw-Hill Education/Robert Manella

Eating disorders may be less noticeable in males and more difficult to diagnose. However, they are just as harmful and require similar treatment.

Treatment for Eating Disorders

An eating disorder is a type of mental health problem. Eating disorders can also cause serious physical harm. People with eating disorders typically need medical treatment to recover.

People with **eating disorders** often *cannot admit* that they have a **problem**.

Health care providers work together to treat eating disorders. Doctors, counselors, and nutritionists can help a person rebuild his or her physical and mental/emotional health. In extreme cases, a hospital stay may be necessary to treat serious physical problems or severe depression.

People with eating disorders often cannot admit that they have a problem. Family and friends can help them recognize the problem and seek treatment. If you are concerned that someone may have an eating disorder, talk to a trusted adult. A person with an eating disorder needs to get help right away. The sooner a person gets treatment, the better the chances of recovering. ■

Getting help with an eating disorder is the first step toward recovery. *Discuss ways that health care providers work to treat eating disorders.*

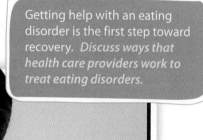

O. Dimier/PhotoAlto

>>> **As You Read**

1. **VOCABULARY** What is another name for *binge eating disorder*?
2. **LIST** What are two health risks associated with anorexia nervosa?
3. **EXPLAIN** How can friends and family members help a person with an eating disorder?

>>> **Thinking Critically**

4. **ANALYZE** How are the symptoms of anorexia nervosa similar to the symptoms of bulimia nervosa? How are they different?

>>> **Applying Health Skills**

5. **ADVOCACY** With a group, write and perform a short dramatic sketch about body image and eating disorders. In it, show how negative body image and eating disorders are linked. Present your sketch for the class.

🔁 Review

🔊 Audio

Hands-On HEALTH ACTIVITY

Jars *of* Sugar

Do you know how much sugar you consume when you grab a quick drink or snack? The table below lists the amount of sugar, in grams, that you might find in several popular foods.

WHAT YOU WILL NEED

* seven empty baby food jars
* a container of sugar
* a set of measuring spoons

WHAT YOU WILL DO

1 Note that 5 grams of sugar is equivalent to 1 level teaspoon of sugar, 1 gram is just under $\frac{1}{4}$ teaspoon, and 2 grams is a little under $\frac{1}{2}$ teaspoon.

2 Calculate how many teaspoons of sugar each product in the list contains.

3 Using the spoons, measure out the amount of sugar in each product and place it in a jar. Label the jar with the name of the product it corresponds to.

WRAPPING IT UP

Evaluate your findings. Which foods contain the most sugar? Which foods are high in other nutrients? What are some ways to reduce your sugar intake?

Grams of Sugar

Cola (12 oz.)	42g
Fat-free fruit yogurt (8 oz.)	35g
Light popcorn (1 c.)	0g
Fruit punch drink (8 oz.)	27g
Sweetened breakfast cereal ($\frac{3}{4}$ c.)	15g
3 reduced-fat chocolate sandwich cookies	14g
Chocolate candy bar (1.55 oz.)	40g

©Ocean/Corbis

READING REVIEW

FOLDABLES and Other Study Aids

Take out the Foldable that you created and any graphic organizers that you created. Find a partner and quiz each other using these study aids.

LESSON 1 Nutrients Your Body Needs

BIG IDEA Each nutrient plays a specific role in keeping your body healthy.

* The six categories of nutrients are proteins, carbohydrates, fats, vitamins, minerals, and water.
* Eating a variety of foods each day will give your body the nutrients it needs.

LESSON 2 Creating a Healthful Eating Plan

BIG IDEA Learning to make healthful food choices will keep you healthy throughout your life.

* Many factors can influence your food choices, including personal preferences, family and friends, culture, convenience, and the media.
* The MyPlate system is designed to help you make healthful food choices. The five main food groups in MyPlate are grains, vegetables, fruits, milk, and protein.
* The keys to planning healthy meals and snacks are variety, moderation, and balance.
* The Nutrition Facts panel on a food package gives you important information about the nutrients that food contains.

LESSON 3 Managing Your Weight

BIG IDEA You can maintain a healthy weight by balancing the food you eat with physical activity.

* Being overweight or underweight can pose risks to your health.
* Body Mass Index is a tool for figuring out whether your weight is in a healthy range.
* It takes 3,500 calories to gain or lose one pound of body fat.
* Weight-loss diets that promise fast results are usually ineffective and often harmful to health.

LESSON 4 Body Image and Eating Disorders

BIG IDEA Teens with a poor body image may develop extreme and harmful eating behaviors.

* Many factors, such as family, friends, and the media, can affect your body image. Developing a healthy body image means accepting yourself as you are.
* Eating disorders are extreme eating behaviors that threaten health. They include anorexia nervosa, bulimia nervosa and binge eating.

 Review

 Web Quest

ASSESSMENT

Reviewing Vocabulary

> appetite   > carbohydrates   > hunger   > nutrients
> calorie   > fats   > minerals   > proteins

Reviewing MAIN IDEAS

>> On a sheet of paper, write the numbers 1–8. After each number, write the term from the list that best completes each sentence.

LESSON 1 Nutrients Your Body Needs

1. Substances in food that your body needs to function are known as _____.

2. A/an _____ is a unit of heat that measures the energy available in food.

3. The nutrients used to build and repair cells are called _____.

4. _____ are the starches and sugars found in foods, especially in plant foods.

5. Your body uses _____ to promote normal growth, give you energy, and keep your skin healthy.

6. Elements in food that help your body work properly are known as _____.

LESSON 2 Creating a Healthful Eating Plan

7. The emotional desire for food is known as _____.

8. _____ is the body's physical need for food.

>> On a sheet of paper, write the numbers 9–13. Write *True* or *False* for each statement below. If the statement is false, change the underlined word or phrase to make it true.

LESSON 3 Managing Your Weight

9. Teens who are <u>underweight</u> have an increased risk of type 2 diabetes, high blood pressure, heart disease, and cancer.

10. The energy equation is the balance between the calories you take in from food and <u>the calories your body uses</u>.

LESSON 4 Body Image and Eating Disorders

11. Having a <u>positive</u> body image can lead to eating disorders.

12. People who starve themselves on purpose because they are intensely afraid of gaining weight have a disorder called <u>bulimia nervosa</u>.

13. <u>Binge eating</u> is a disorder in which people repeatedly eat large amounts of food, but do not purge afterwards.

✔ eAssessment

>> Using complete sentences, answer the following questions on a sheet of paper.

🗨 *Thinking* **Critically**

14. PREDICT What are some short-term and long-term benefits of healthy eating?

15. EXPLAIN Why is it often difficult to tell if a person has an eating disorder?

✏ *Write* **About It**

16. EXPOSITORY WRITING Write an essay that explains clearly how teens can use MyPlate to guide their food choices. Explain how individuals can meet their nutrient needs by choosing wisely from the food groups.

17. NARRATIVE WRITING Write a story about a teen who is concerned about his or her body image. What advice might you give this teen?

Ⓐ Ⓑ Ⓒ Ⓓ STANDARDIZED TEST PRACTICE

Reading and Writing
Read the paragraphs below and then answer the questions.

Most Americans consume too much fat. The Dietary Guidelines suggest that healthy people consume only 20 to 35 percent of their calories from fats. Most of these fats should be unsaturated. Many people, however, eat much more than that. If you look at the Nutrition Facts panel on a food package label, you will see what percentage of fat one label serving contributes to a 2,000-calorie-a-day diet.

People lead busy lives and often depend on fast food when they are away from home or don't have time to prepare a meal. Unfortunately, many fast foods are very high in fat. One burger, for example, may contain more fat than a person should consume in an entire day. However, many fast-food restaurants offer more healthful options, such as salads. Just go easy on the dressing.

1. From the information in the first paragraph, the reader can conclude that the writer thinks Americans
   A. consume too little fat.
   B. consume too much fat.
   C. are poor cooks.
   D. care little about health.

2. Which of the following best describes the purpose of the second paragraph?
   A. To explain reasons why the body needs fats
   B. To rate fats according to health needs
   C. To suggest seeking alternatives to high-fat foods in fast-food restaurants
   D. To rate types of fast-food restaurants

How What You Eat
Affects Your Body

BEING
OVERWEIGHT
PUTS EXTRA STRAIN ON YOUR
HEART & LUNGS

TEENS WHO ARE OVERWEIGHT ARE AT AN INCREASED RISK OF TYPE 2 DIABETES.

DOES YOUR WEIGHT INCREASE YOUR RISK?

HEALTHY WEIGHT	OVERWEIGHT	
NO	YES	Diabetes
NO	YES	Heart disease
NO	YES	High blood pressure
NO	YES	Cancer

IF YOU KNOW YOU ARE OVERWEIGHT, YOU CAN TAKE STEPS TO IMPROVE YOUR OVERALL HEALTH.

WHEN YOU TAKE IN MORE CALORIES THAN YOUR BODY USES, IT STORES THE EXTRA CALORIES AS BODY FAT. THIS MAKES YOU GAIN WEIGHT.

IF YOU EAT FEWER CALORIES THAN YOU NEED, THOUGH, YOUR BODY CONVERTS STORED BODY FAT INTO ENERGY. AS A RESULT, YOU LOSE WEIGHT.

THERE'S NO SINGLE HEALTHY WEIGHT. Genetics, nutrition, physical activity and age all play a role in your body's appearance. So just because you're 12 and feel too thin doesn't always mean you're unhealthy.

Make sure you eat a healthful, balanced diet—that's the first step to physical fitness.

3,500 calories = 1 pound of fat

Eating 250 fewer calories a day for two weeks will produce 1-pound of weight loss.

HOW TO FIND HEALTHFUL FOOD

PROTEINS
WHERE YOU CAN FIND THEM: Meat, fish, soybeans, beans, rice

CARBOHYDRATES
WHERE YOU CAN FIND THEM: Pasta, fruit

FATS
 SOME ARE GOOD olive oil, nuts, avocados

 SOME AREN'T SO GOOD butter, cheese, many meats, stick margarine

VITAMINS AND MINERALS

Vitamin A
spinach, carrots, eggs

B Vitamins
eggs, poultry, whole-grain breads

Vitamin C
orange, tomato, broccoli, potato

Calcium
milk, oatmeal, dark green leafy vegetables, canned salmon
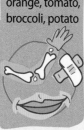

Iron
red meat, poultry, dry beans, nuts, dark green leafy vegetables

Potassium
baked potato, peaches, bananas, oranges, dry beans, fish

Water
To make sure your body is getting enough water, let thirst be your guide. Drink when you are thirsty, as well as with your meals.

GET UP, GET MOVING

Robust physical activity is important to your everyday health.

60 MINUTES
DAILY PHYSICAL ACTIVITY RECOMMENDED FOR TEENS

MY FITNESS SCHEDULE

Day	Activity	Time
Sun	Ride bike	60 min
Mon	Gym class	40+ min
Tue	Soccer practice	60 min
Wed	Gym class	40+min
Thu	Soccer practice	60 min
Fri	Karate class	60 min
Sat	Soccer game	90 min

PHYSICAL FITNESS INCLUDES THESE ELEMENTS

Endurance

Strength

Flexibility

Body composition

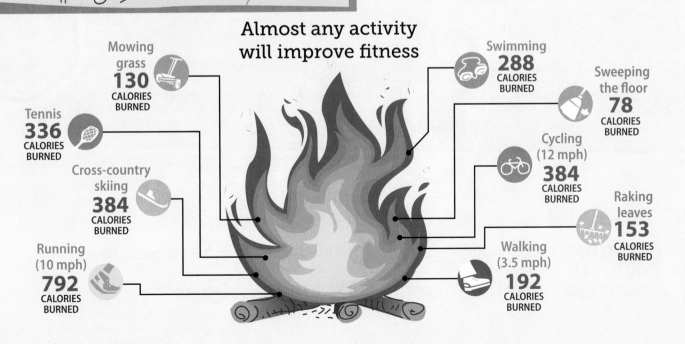

Almost any activity will improve fitness

Mowing grass
130 CALORIES BURNED

Tennis
336 CALORIES BURNED

Cross-country skiing
384 CALORIES BURNED

Running (10 mph)
792 CALORIES BURNED

Swimming
288 CALORIES BURNED

Sweeping the floor
78 CALORIES BURNED

Cycling (12 mph)
384 CALORIES BURNED

Raking leaves
153 CALORIES BURNED

Walking (3.5 mph)
192 CALORIES BURNED

Physical Activity

LESSONS

 **PREMIUM ONLINE RESOURCES**

 Audio

 Videos

 Bilingual Glossary

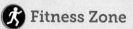 Fitness Zone

 Web Quest

 Review

Finding an activity you are good at helps you feel good about yourself. *Explain how you fit the physical activities you enjoy into your daily life.*

Becoming Physically Fit

BIG IDEA ▶ Being physically active benefits your total health in a variety of ways.

Before You Read

QUICK WRITE Write a short paragraph about the kinds of physical activity you do in a typical day.

▶ Video

As You Read

 Study Organizer

Make the Foldable® found in the FL pages in the back of the book to record the information presented in Lesson 1 about becoming physically fit.

Vocabulary

› physical activity
› physical fitness
› exercise
› strength
› endurance
› flexibility
› body composition
› heart and lung endurance
› muscle endurance
› muscle strength
› joints

 Audio

🔤 Bilingual Glossary

CHOOSING AN ACTIVE LIFESTYLE

MAIN IDEA ▶ An active lifestyle will help to keep you healthy throughout your lifetime.

Anna and Sonja are identical twins. Although they look alike, their habits are different. Anna often plays basketball with her friends. Sonja prefers to stay inside and play video games. Anna would rather take the stairs than wait for the elevator. Can you guess which choice Sonja makes?

Teens need at least *60 minutes* of **physical activity** *every day*.

If you are like Anna, you are getting plenty of physical activity in your daily life. **Physical activity** is *any form of bodily movement that uses up energy.* That should come as great news, because physical activity has many benefits for your health. Being active helps you build strong bones and muscles. It also helps you manage your weight. Regular physical activity keeps your blood pressure at a healthy level and strengthens your heart and lungs.

When you are active, you have more energy. You are also in less danger of developing certain diseases, both now and later in life.

Physical activity is good for your mental and emotional health, too. It helps you sleep better and concentrate better in school. It can also improve your self-confidence and relieve stress. This can help you get along better with others, which improves your social health. In addition, many physical activities, such as team sports, can be a great way to make friends.

Teens need at least 60 minutes of physical activity every day, according to guidelines released by the CDC. However, this doesn't have to mean an hour of activity all at once. For instance, suppose you walk to school several days a week. If it takes half an hour each way, your walking time will add up to 60 minutes. Experts also say teens should get some vigorous activity at least three days a week. Examples include jumping rope, swimming, or playing soccer.

Exercising for Physical Fitness

Leading an active life is the key to **physical fitness,** or *the ability to handle the physical demands of everyday life without becoming overly tired.* When you are fit, you have the energy for everything you want to do. You can work to build physical fitness through **exercise,** or *planned physical activity done regularly to build or maintain one's fitness.* You will learn more in the next lesson about exercises you can use to measure your fitness and develop your own fitness plan.

Raking leaves and *riding* your **bike** can help you **stay fit**.

Some people think of physical activity as either working out or playing sports. However, you can find lots of different ways to be physically active.

Think about things that get your body moving and using up energy. A number of everyday activities can help you stay fit and healthy. Some you may already do from time to time:

- You may walk to and from school every day. Walking is a form of physical activity.
- You probably have various chores to do at home. Cleaning your room, taking out the trash and recyclables, and running the sweeper are also forms of physical activity.
- You might also do yard work in the summer, rake leaves in the fall, or shovel snow in the winter. All of these activities require you to move your body and use up energy.
- If you have a bicycle, you may enjoy riding it just for fun. This is another common daily activity that can help you stay fit.

If you are trying to improve your physical fitness, though, you will also need to learn exercise skills.

>>> **Reading Check**

DESCRIBE *How can physical activity improve all three sides of your health triangle?*

Health SKILLS ACTIVITY

Practicing Healthful Behaviors

Activity + Eating + Sleeping = *Good Physical Health*

In a way, physical health can be viewed as a health triangle within the health triangle. The three sides to physical health—physical activity, good eating habits, and adequate rest—all relate to one another. Teens who are active during the day tend to get more restful sleep during the night. When you eat the right foods, you give your body the fuel it needs for physical activity. Taking care of your physical health can help you maintain a healthy weight and lower your risk of developing serious health problems.

On Your Own

Describe your own physical health triangle. If the sides are not balanced, identify which areas need work. Tell what habits you can adopt that will improve your overall physical health. Develop a plan to practice these habits regularly.

ELEMENTS OF PHYSICAL FITNESS

MAIN IDEA Elements of fitness include strength, endurance, flexibility, and body composition.

Think about someone who is very strong but cannot walk a mile without becoming tired. Would you consider that person physically fit? Probably you would not. True fitness includes **strength,** *the ability of your muscles to use force.* It includes **endurance** (en·DUR·uhnce), *the ability to perform difficult physical activity without getting overly tired,* and **flexibility,** *the ability to move joints fully and easily through a full range of motion.* Fitness also means having a healthy **body composition,** or *proportions of fat, bone, muscle, and fluid that make up body weight.*

Endurance

Endurance is a measure of how long you can last during physical activity. If you can climb several flights of stairs and not feel out of breath, that means you have good heart and lung endurance. **Heart and lung endurance** is *a measure of how efficiently your heart and lungs work when you exercise and how quickly they return to normal when you stop.* This kind of endurance is important for many physical activities, such as running, swimming, and playing team sports. If you can run several miles and your legs don't feel tired, you have good **muscle endurance,** or *the ability of a muscle to repeatedly use force over a period of time.* Activities that build muscle endurance include jumping rope, dancing, and riding a bike.

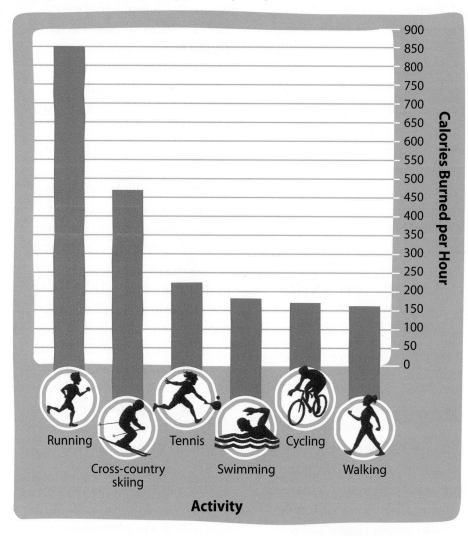

Running — Cross-country skiing — Tennis — Swimming — Cycling — Walking

Activity

Calories Burned per Hour

The **stronger** you are, the **more work** your muscles can do.

Strength

The stronger you are, the more work your muscles will be able to do. Having strong arms and a strong back means that you will be able to move heavy objects more easily. Having strong legs will help you with activities such as running and bicycling. You build **muscle strength,** or *the*

This graph shows how many calories a 100-pound person uses during 60 minutes of different activities. *Identify the activity that burns the most calories. Identify the activity that burns the fewest calories.*

most weight you can lift, when you make your muscles work against a force, such as gravity. Building muscle strength shapes and tones your body and helps you with activities such as sports.

Flexibility

Flexibility allows you to turn, bend, and stretch. Being flexible is especially important for gymnasts and dancers. However, it plays a role in other sports as well. Being flexible makes it easier to change directions quickly. This can help you with everything from stopping and turning to throwing a ball. Also, when your muscles can stretch easily, they are less likely to become injured during sports and other activities. You improve your flexibility every time you stretch your muscles and joints, or *the places where two or more bones meet.* Activities that are good for building flexibility include yoga, swimming, dancing, and karate.

Body Composition

The last element of fitness is body composition. A healthy body generally has more bone, muscle, and fluid than fat. Body composition is not the same as how much you weigh.

Body composition is a measure of how much of your weight is lean tissue instead of body fat. Too much body fat can increase your risk of health problems such as heart disease, diabetes, and certain types of cancer.

Your body composition depends partly on the genes you inherit from your parents. It also depends on how much you eat and how physically active you are. When you are physically active, your body burns up calories from the foods you eat.

Burning extra calories means your body will not store them as fat. The graph on the previous page shows how many calories your body uses during different kinds of activity.

Many of the activities you do every day can contribute to fitness. *Identify which elements of fitness you could improve by raking leaves.*

Being active also improves your body composition by helping you build up your muscles. Physical activity increases the amount of muscle on your body and reduces the amount of fat your body stores. Being active can help you develop a lean, fit, healthy body. In the long run, it will also reduce your risk of many health problems. ■

>>> **Reading Check**

DISTINGUISH *What is the difference between body composition and how much you weigh?*

REVIEW

>>> **After You Read**

1. **DEFINE** What is the difference between *physical activity* and *exercise?*
2. **IDENTIFY** Name one activity that builds muscle endurance and one activity that builds flexibility.
3. **DESCRIBE** What is one way to improve your body composition?

>>> **Thinking Critically**

4. **EVALUATE** In a short paragraph, describe your own thoughts and feelings about the benefits of being physically fit.

>>> **Applying Health Skills**

5. **ANALYZING INFLUENCES** What is your favorite physical activity and why? Did friends or family spark your interest in this activity? Was it something you saw on television or in a magazine? Did social customs play a part? Explain your answer.

 Review

 Audio

Plush Studios/age fotostock

Creating Your Fitness Plan

BIG IDEA You can reach your fitness goals by making a plan to include different kinds of exercise in your life.

Before You Read

QUICK WRITE List at least three ways you can fit more physical activity into your daily life.

▶ Video

As You Read

STUDY ORGANIZER Make the study organizer found in the FL pages in the back of the book to record the information presented in Lesson 2.

Vocabulary

› aerobic exercise
› F.I.T.T. principle
› frequency
› intensity

🔊 Audio

🔤 Bilingual Glossary

🏃 Fitness Zone

Finding Time for Physical Activity Like many teens, I'm very busy. At first, I couldn't find time for exercise. Then I realized that some of my activities, such as dance class, can count toward my 60 minutes of daily physical activity. I also figured out that I don't have to do an hour all at once. Now, on days I don't have dance, I ride my bike to and from school. This, plus gym class three times a week, keeps me in good shape.

MEASURING FITNESS

MAIN IDEA Measuring your fitness level will indicate what areas of fitness you need to improve.

You wouldn't try to build a house without a blueprint, or plan. The same is true for building your physical fitness. The first step in making a physical fitness plan is to figure out how fit you are now. You can use several tests to measure the different elements of fitness.

Measuring Flexibility

You can measure your flexibility with the V-sit reach, or sit-and-reach, test. Mark a line on the floor in inches. Sit with your feet right behind this line with your heels apart. Put one hand on top of the other, palms down. Have a partner hold your legs down while you reach your arms forward as far as you can. Hold the stretch for three seconds. Have your partner note how many inches you could reach past the starting point of the line.

Measuring Heart and Lung Endurance

To test your heart and lung endurance, time how long it takes you to run one mile. If you get too tired while running, you can switch to walking for a while. However, try to cover the distance as fast as you can.

Measuring Muscle Strength and Endurance

You can test the strength and endurance of your abdominal muscles by doing curl-ups. Lie on your back with your knees bent and your arms crossed on your chest. Your feet should be about 12 inches from your buttocks. Have a partner hold your feet down while you raise your body to touch your elbows to your thighs. Then lower your body back down until your shoulder blades touch the floor. Count the number of curl-ups you can do in one minute.

To measure upper-body strength and endurance, you can perform right-angle push-ups.

Frederick Bass/Getty Images

The V-sit reach test measures flexibility. *Explain why flexibility is an important part of physical fitness.*

The curl-up test measures the strength and endurance of your abdominal muscles. *Describe a way to measure upper body strength and endurance.*

The push-up test measures upper-body strength and endurance. *Explain why upper-body strength and endurance is an important part of physical fitness.*

Start out in push-up position: face down with your arms straight and your hands under your shoulders. Your legs should be slightly apart and resting on your toes. Keeping your legs, knees, and back straight, lower your body until your elbows are bent at a 90-degree angle. Your upper arms will be parallel to the floor. Then straighten your arms to raise yourself back up. Do one complete push-up every three seconds and count how many you can do in a row.

Checking Your Results

To see how you did on the fitness tests, look at the table on this page. It shows healthy results for teens of different ages. After you start your fitness plan, you can use these tests as guides to see how your fitness is improving.

>>> **Reading Check**

IDENTIFY *Which fitness test measures heart and lung endurance?*

You wouldn't try to **build** a house *without* a **blueprint**.

This table shows the score you need on each test to do as well as half of all teens of your age and gender. *Identify any areas of fitness you may need to work on.*

	Age	Curl-Ups (# one minute)	OR	Partial* Curl-Ups (#)	Sit and Reach (centimeters)	One-Mile Run (min-sed)	Pull-Ups (#)
BOYS	11	47		43	31	7:32	6
	12	50		64	31	7:11	7
	13	53		59	33	6:50	7
	14	56		62	36	6:26	10
GIRLS	11	42		43	34	9:02	3
	12	45		50	36	8:23	2
	13	46		59	38	8:13	2
	14	47		48	40	7:59	2

SETTING AND REACHING FITNESS GOALS

MAIN IDEA Fitness test results help to determine fitness goals.

Once you know how fit you are right now, you can start to figure out your fitness goals. Your scores on fitness tests might point you toward a fitness goal. For instance, if you did very well on the curl-ups and the one-mile run but not very well on the V-sit reach, you might set a goal to improve your flexibility. You can also set goals based on specific activities you would like to do. If you plan to try out for the track team, for example, then your goals might focus on building better heart-lung endurance.

Choosing Activities

Once you know what your goals are, you can choose activities to help you reach them. Different exercises are good for building up different elements of fitness. The graph above shows the fitness benefits of some popular activities. Consider these other points when choosing exercises.

- **Personal tastes** If you choose activities you enjoy, you are more likely to stick to your plan. If you prefer group activities, you could try a team sport. If you would rather work out alone, you might try jogging or bicycling.

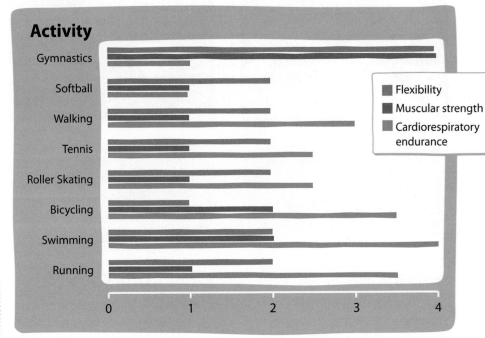

Activity

Legend:
- Flexibility
- Muscular strength
- Cardiorespiratory endurance

Rating scale: 1 = Low   2 = Moderate   3 = High   4 = Very High

Different activities promote different areas of fitness. *List your fitness goals. Identify the activities that can best help you reach those goals.*

- **Requirements** There's no point in choosing swimming as an activity if you don't have a pool you can use. Think about the requirements of different activities. Will you need special equipment? Do you need a partner or teammates? Are there special skills you need to learn first?

- **Time and place** Figure out when you will have time to exercise. Also think about where you will exercise. If you want to start jogging, for example, you need to make sure your route is safe. If you live someplace where it gets very cold in the winter, you may need to run indoors.

A written plan, like the one shown here, will help you stick to your fitness goals. *Identify the activities that are part of this teen's schedule. Determine the number of hours per week that this teen is active.*

Sun	Mon	Tues	Wed	Thurs	Fri	Sat
1 Ride bike (1 hr.)	**2** Gym class (40 min.) Walk briskly home from school (20 minutes)	**3** Soccer practice (1 hr.)	**4** Gym class (40 min.) Walk briskly home from school (20 minutes)	**5** Soccer practice (1 hr.)	**6** Karate class (1 hr.)	**7** Soccer game (90 minutes)

Creating a Schedule

To reach your fitness goals, you will likely need a variety of activities. One is aerobic (ah·ROH·bik) exercise, or *rhythmic activity that uses large amounts of oxygen and works the heart and lungs.* Aerobic exercise is important for improving your heart and lung endurance.

Experts say most of your 60 minutes of daily physical activity should be aerobic activity, done at a moderate to vigorous pace. Moderate activities include hiking, bicycling, or skating. Some vigorous activities are swimming, jumping rope, and team sports. Include vigorous activities at least three times a week.

Some kinds of aerobic exercise, such as swimming, are great for your heart and lungs. Other aerobic exercises, such as running or jumping rope, are good for bone strength. To build up your muscles, however, you will need strength training. Weight lifting, sit-ups, push-ups and climbing are good ways to help make your muscles stronger.

How can you fit all this into your schedule? Write down all the physical activities that are part of your daily routine. Most teens have gym class. If you play a sport, include your practices and games. Also include any outside activities that get your body moving, such as dance lessons. Once you have a written plan, you can see what you are doing now and what other activities you might want to add. Then you can find gaps in your schedule for new activities.

The F.I.T.T. Formula

When you start a new activity, it is best to start small and build up gradually. To increase your activity level over time, remember the F.I.T.T. principle, or *a method for safely increasing aspects of your workout without hurting yourself.* The F.I.T.T. principle stands for:

- **Frequency,** or *the number of days you work out each week.* You may do a new activity two to three times a week. As your fitness improves, you can work out more often.
- **Intensity,** or *how much energy you use when you work out.* Take it easy at first and build up intensity over time. If you lift weights, you can add more weight. If you run, you can go a little faster each time. If you push yourself too hard, you will just quickly tire yourself out. You will also be more likely to get hurt.
- **Time.** Teens need 60 minutes of activity each day. With a new activity, you will not be able to do this all at once. Start with a shorter workout and build up to a full hour.
- **Type.** A complete fitness plan will include a variety of activities. Try to switch among activities so that you can work different muscles on different days. This will also help keep you from getting bored.

> **》》》 Reading Check**
>
> **DESCRIBE** *Cite two types of physical activity that all teens should include in a fitness plan.* ■

》》》 After You Read

1. **DEFINE** Explain what *aerobic exercise* means and give an example.
2. **LIST** Name two factors you should think about when choosing physical activities.
3. **IDENTIFY** What are the four parts of the F.I.T.T. principle?

》》》 Thinking Critically

4. **EVALUATE** Alicia is an inactive teen who wants to get into shape. She has made a fitness plan with three hours of exercise and other physical activity per day. How likely do you think she is to succeed with this plan?

》》》 Applying Health Skills

5. **GOAL SETTING** Identify a personal fitness goal. Then make a fitness plan to meet that goal. Write down which activities you will do and when. Also, note how you can track your progress toward your goal.

 Review

 Audio

Performing *at* Your Best

BIG IDEA Following some basic guidelines will help you get the most out of your workouts.

Before You Read

QUICK WRITE Have you ever felt sore the day after you tried a new physical activity? Explain in a few sentences why you think this happened.

 Video

As You Read

STUDY ORGANIZER Make the study organizer found in the FL pages in the back of the book to record the information presented in Lesson 3.

Vocabulary

› warm-up
› cool-down
› resting heart rate
› target heart rate
› conditioning
› dehydration

🔊 Audio

🔤 Bilingual Glossary

KEYS TO A GOOD WORKOUT

MAIN IDEA Healthy workouts include a warm-up and cool-down.

Jay has decided to start jogging each day before school. He figures that if he gets up 15 minutes early, he can run a mile or so before breakfast. On the first day, he rolls out of bed, puts on his running shoes, and heads out the door. He runs for 15 minutes and then goes home. He feels a bit tired and sore for the rest of the day, but he figures that is normal. The next morning, he is so stiff and sore he cannot even think about getting up to run.

What did Jay do wrong? First of all, he forgot to warm up before his run. Later, he forgot to take the time to cool down after it. Jay also did not drink any extra water after his run. Finally, he tried to do too much too soon. Jay forgot that with a new activity, it is important to start slowly and work your way up to full speed. If you want to avoid Jay's problem, you will need to plan ahead in order to get the most out of your workouts.

Every **workout** should **start with** a **warm-up**.

Warming Up and Cooling Down

Every workout should start with a **warm-up**, or *gentle exercises that get heart muscles ready for moderate-to-vigorous activity.* Warming up increases blood flow and loosens up your muscles so that you are less likely to strain or tear them. Any light activity, such as walking or jogging in place, can make a good warm-up activity. Keep it up for about five to ten minutes. You will know you are warmed up when you begin to sweat and breathe more heavily.

After warming up, it is a good idea to stretch your muscles. This further reduces your risk of injury. Make sure to stretch the muscles that you will be using the most as you exercise.

Stretching should always come after a warm-up. Stretching before your muscles are loose increases the chance of injury.

At the end of a workout, take some time to cool down. A **cool-down,** or *gentle exercises that let the body adjust to ending a workout,* should last five or ten minutes. This will let your heart rate and breathing return to normal. Include light stretching in your cool-down. This will help your muscles relax so they will not feel stiff or sore afterwards.

Monitoring Your Heart Rate

How can you tell whether you are exercising hard enough? One good way is to check your heart rate. You can do this before, during, and after your workout. Your **resting heart rate** is *the number of times your heart beats per minute when you are relaxing.* This is a good measure of your overall heart health.

> Learning the right way to stretch can help prevent injuries. *Try the two stretches shown here. Describe where you feel the "pull."*

In general, a lower heart rate means a healthier heart. When you work out, however, you want to get your heart pumping faster. Your goal should be to reach your **target heart rate,** or *the number of heartbeats per minute you should aim for to help your circulatory system the most during exercise.* This is not a single number but a range. To figure your target heart rate, use the equations below.

$$(220 - \text{your age}) \times 0.5 =$$
bottom of target heart range

$$(220 - \text{your age}) \times 0.85 =$$
top of target heart range

Keeping your heart rate in this range as you exercise will help you work hard enough but not too hard. One way to check your heart rate during a workout is to stop for a minute and take your pulse. If you have been to the doctor recently, you probably remember having your pulse taken. However, you can learn to take your own pulse as part of your workout routine.

To do this, place two fingers against the base of your neck. (Do not use your thumb, which has its own pulse.) You should feel a throbbing sensation. This is the blood pumping through the blood vessels in your neck.

> Measuring your pulse is one way to make sure your heart rate is in the right range for exercise. *Name another way to check your heart rate.*

Use a clock or watch with a second hand to count the number of pulses you feel in ten seconds. Multiply this number by six to get your heart rate.

An easier way to check your heart rate during moderate activity is the "talk test." If you don't have enough breath to talk at all, you are probably working too hard. However, if you have enough breath to sing, you are not working out hard enough.

⟫⟫⟫ Reading Check

EXPLAIN *What is your resting heart rate?*

Karl Weatherly/Getty Images

STAYING IN SHAPE FOR SPORTS

MAIN IDEA › Sports conditioning will help you play to the best of your abilities.

Playing a sport can be a lot of fun. It is also a great way to stay active. However, playing a sport involves a lot more than just showing up for games. Sports put extra demands on your body, and meeting those demands involves proper training. You need to develop the skills and build up the muscles that are necessary for your sport. You also need to eat right so that your body can keep up with everything you are asking it to do.

Playing a sport involves **a lot more** than *just showing up* for games.

Conditioning

Athletes devote many hours to **conditioning,** or *training to get into shape for physical activity or a sport.* This training takes place both on and off the field. Baseball players, for instance, practice batting and fielding before a game. Basketball players do passing drills and take practice shots from different parts of the court. Off the field, athletes may work on shaping up their bodies with running, weight training, or other exercises.

Conditioning is not just important for team sports. Many kinds of physical activity can require training. For example, if you are a dancer, you need to practice regularly to keep your muscles and your skills in shape.

Sports Nutrition

Like other teens, teen athletes have special nutritional needs. They need to eat a variety of foods from all the major food groups. They also need to consume extra calories to replace all the energy they use during their chosen sport. Good food choices will include plenty of complex carbohydrates and healthful fats. Sugary snacks, such as candy bars, are not good choices. They can give you a burst of energy, but that will quickly wear off.

As the saying goes, "Practice makes perfect." *Think of a sport you enjoy. Name one skill you would need to practice for that sport.*

Colin Anderson/Blend Images LLC

Drinking water before, during, and after a game is one good sports nutrition habit. *What is another example of a good sports nutrition habit?*

When you eat is as important as what you eat. It is best not to eat anything in the hour right before a game or a practice. Digesting food takes energy that your body needs to perform at its best. Also, having a full stomach during a workout can make you feel sick.

On the day of a game, it's best to eat a meal with some protein and carbohydrates about two to four hours before you compete. This will give you the energy you need to get through the game, but it will not slow you down by making you feel too full. If you don't have time for a meal, you can choose a light snack, such as fruit or crackers, one to two hours before a game or practice.

When you eat is as important as what you eat.

You should also make sure to get plenty of water to avoid **dehydration,** or *the excessive loss of water from the body.* During any kind of intense activity, your body loses water through perspiration, or sweat. You need to drink more to replace this lost fluid. How much water you need depends on your age and size, the weather, and how hard you are working out.

In general. you should drink water every 15 to 20 minutes during activity. If you wait until you feel thirsty, your body is already low on fluids. It is okay to choose sports drinks if you like them better than water, but they may not be any better for you. If the weather is very hot, or if practice lasts longer than an hour, the extra nutrients in sports drinks may be helpful.

>>> **Reading Check**

GIVE EXAMPLES *List some of the foods an athlete might choose for a snack before a game.* ■

>>> **After You Read**

1. **DEFINE** Explain what *warm-up* and *cool-down* mean.
2. **DESCRIBE** How can you figure out your target heart rate for exercise?
3. **EXPLAIN** Why do you need to drink extra water during physical activity?

>>> **Thinking Critically**

4. **INFER** Why do you think teen athletes should eat a meal 3 to 4 hours before a game?
5. **APPLY** Kerry is planning to start swimming on a regular basis. She says that because you stretch your muscles during swimming, there is no need to cool down. How would you respond?

>>> **Applying Health Skills**

6. **APPLY** Kerry is planning to start swimming on a regular basis. She says that because you stretch your muscles during swimming, there is no need to cool down. How would you respond?

⟳ Review

🔊 Audio

STOCK4B/Getty Images

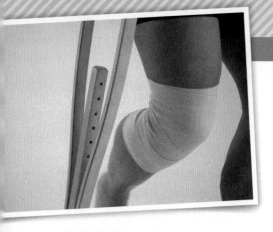

Preventing Sports Injuries

BIG IDEA You can prevent many sports injuries by taking precautions.

QUICK WRITE In a short paragraph, briefly describe how you or someone you know recently got hurt as a result of physical activity.

 Video

As You Read

STUDY ORGANIZER Make the study organizer found in the FL pages in the back of the book to record the information presented in Lesson 4.

Vocabulary

> sprain
> fracture
> dislocation
> stress fracture
> tendonitis
> sports gear
> heat exhaustion
> frostbite
> P.R.I.C.E. formula

 Audio

🔤 **Bilingual Glossary**

COMMON SPORTS INJURIES

MAIN IDEA Some sports injuries can be treated at home, while others require medical care.

Each year, about 3.5 million children and teens are injured while playing a sport. The most common sports injuries are sprains and strains. A **sprain** is *an injury to the ligament connecting bones at a joint.* A strain is a stretch or tear in a muscle or a tendon (the tissue that connects muscles to bones and other muscles). Both sprains and strains can cause pain and swelling. However, they can usually be treated at home.

Two more serious injuries that can occur during sports are fractures and dislocation. A **fracture** is *a break in a bone.* A **dislocation** occurs *when a bone is popped out of its normal place in a joint.* Both of these are very painful injuries that need medical treatment right away.

Some kinds of sports injuries don't happen all of a sudden. Instead, they start out small and get worse over time. These are known as overuse injuries.

A **stress fracture** is *a small fracture caused by repeated strain on a bone.* Athletes in such sports as basketball, tennis, track and field, and gymnastics are commonly at risk for stress fractures. These injuries can even result from suddenly doing too much activity too quickly.

Each year, about 3.5 *million* children and teens are **injured** while playing a *sport.*

Tendonitis, or *the painful swelling of a tendon caused by overuse,* can develop when a muscle is stretched too far, too often. You can help prevent tendonitis by making sure to warm up or stretch your muscles properly before you exercise.

Reading Check

DEFINE *What is a stress fracture?*

216 CHAPTER 10 Physical Activity

WAYS TO AVOID INJURY

MAIN IDEA Staying in shape, using the correct sports equipment, and avoiding bad weather conditions can help you avoid sports injury.

As you can see, sports have their dangers and risk of injury. However, it is possible to make physical activity a lot safer. You have already learned how warming up and cooling down can help you avoid fitness-related injuries.

Teen athletes need to take a few additional precautions as well. They need to have—and use—the right sports gear, or *sports clothing and safety equipment,* for their chosen activity. They should also know their limits and avoid overtraining. Finally, teen athletes need to plan for the weather.

Different sports call for *different types* of safety gear.

Use Proper Gear

Different sports call for different types of safety gear. The right shoes are important for just about any sport. For some sports, such as baseball and soccer, you will need cleats for better traction. Other sports, such as basketball or track, have their own types of shoes. Your coach or doctor can tell you what kind of shoes are best for your sport.

Make sure your shoes fit well, and remember to replace them when they wear out.

Many sports, from football to cycling, require a helmet. Make sure to choose one that is designed for the sport you play. Your helmet should fit snugly but comfortably. Other types of protective gear include pads, mouth guards, athletic supporters, and eye protection. Talk to your coach about what is needed for your sport.

Different kinds of protective gear are used for different sports. *Identify the sports that might require the use of some of the equipment shown here.*

Mouth guards. These soft plastic shields protect your mouth, teeth, and tongue. Wear one for any sport where your mouth could be hit. Examples include baseball, football, and hockey. If you wear a retainer, take it out before you play.

Face and throat protection. A face mask with a throat guard protects the face and throat from being hit by a ball or puck.

Helmets. Always choose a helmet made for the sport you're playing. It should fit snugly but comfortably on your head. Be sure it doesn't tilt backward or forward. Never wear a cap under a helmet.

Chest Protectors. A padded chest protector keeps the torso from being injured in sports such as baseball or softball.

Pads. Pads are used to protect bones and joints from fractures and bruises.

Elbow, knee, wrist, and shin guards. Elbow and wrist guards can prevent arm and wrist fractures. Knee and shin guards can protect these areas during falls.

Know Your Limits

As Vidas dove to stop the ball from entering the net, he came down hard on his right hand. In his excitement, though, he barely noticed it. He did not mention it as he resumed his position as goalkeeper. By the end of the game, he was in real pain. That is when he discovered he had sprained his wrist and would have to sit out the next game.

Vidas's soccer experience shows how important it is to listen to your body. If pain does not stop, seek medical help. Avoid going back to playing too soon after an injury. Your body needs time to heal. How much time will depend on where and how badly you were hurt. A mild ankle sprain may heal in a few weeks, while a broken arm or leg may leave you sidelined for months.

If you *feel* *pain* during a game, you should stop right away and take a *rest*.

It is also important to avoid overtraining. You need to train to get in shape—but not too hard and not too fast. Remember the F.I.T.T. principle. Sudden increases in the frequency, intensity, or duration of exercise can lead to injury. Talk to your coach or doctor about a training program to build your skills safely.

Watch *the* Weather

Both hot weather and cold weather can pose health risks. In hot weather, you sweat more and lose more fluid. This can lead to dehydration. You can prevent this problem by drinking plenty of fluids during physical activity.

If you do not drink enough fluid, you could suffer <u>heat exhaustion,</u> or *an overheating of the body that can result from dehydration.* The symptoms include headache, dizziness, and nausea. Anyone with heat exhaustion should immediately lie down in a cool spot and drink cool water. If the person faints or seems confused, seek medical help right away.

Cold weather also has its dangers. If your skin is exposed to extreme cold, you can develop <u>frostbite,</u> or *freezing of the skin.*

This is most likely to happen to small, exposed body parts like hands, feet, ears, and noses. The skin will turn pale and feel numb. To avoid frostbite, make sure to dress properly for winter sports. Wear layers of clothing, plus a hat, gloves, and boots.

Dressing in layers can protect you from the cold during winter sports. *Describe some precautions you should take while exercising during hot weather.*

Sunburn is a problem that can occur in both hot and cold weather. The skin becomes red and sore, and blisters may form. You probably know that you need to protect your skin with sunscreen when you go to the beach.

However, you may not realize that sunscreen is important for cold-weather sports as well. The ultraviolet (UV) rays that can burn your skin reflect off snow and ice, so sunburn can develop quickly. To avoid sunburn, wear a sunscreen with an SPF of at least 15, and use lip balm as well. Be sure to use sunscreen even on cloudy days—harmful UV rays from the sun can penetrate through cloud cover.

>>> **Reading Check**

DESCRIBE *What are the symptoms of heat exhaustion?*

TREATING SPORTS INJURIES

MAIN IDEA The P.R.I.C.E. procedure is used to treat strains and sprains.

Sometimes, no matter how careful you are, injuries still happen. Minor injuries, such as sprains and strains, can be treated using the P.R.I.C.E. formula, which stands for *Protect, Rest, Ice, Compress, and Elevate.* This procedure works best if you apply it as soon as possible after an injury. It can also help if your muscles feel stiff or sore after a workout. The P.R.I.C.E. formula has five steps.

Sometimes, no matter how careful you are, *injuries* still **happen**.

- **Protect** the body part from further injury. You can tape or splint it to take the strain off it.
- **Rest** the injured body part. Avoid using it for a few days.

- **Ice** the body part. Apply an ice pack for 20 minutes at a time, four to eight times a day, until the swelling goes down (or for two days after the injury). To avoid frostbite, do not put ice directly on the skin.
- **Compress**, or apply pressure to, the injured body part. Wrap it in an elastic wrap or bandage to reduce swelling.
- **Elevate** the injured body part above the level of your heart, if possible. This will help reduce swelling and pain.

>>> Reading Check

LIST *What are the steps of the PRICE formula?* ■

Many minor sports injuries can be treated using the P.R.I.C.E. formula. *Identify the step in the P.R.I.C.E. formula that this teen is demonstrating.*

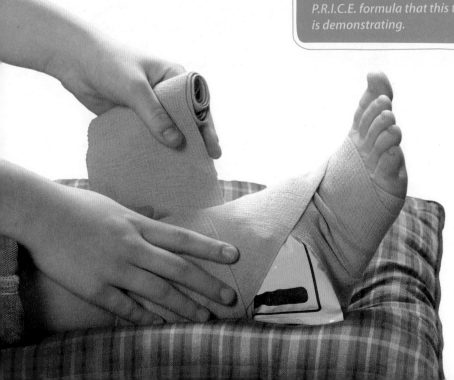

©McGraw-Hill Education/Ken Karp

REVIEW

>>> After You Read

1. **DEFINE** Explain the meaning of the term *sprain*.
2. **DESCRIBE** What are two precautions that can help you avoid sports-related injures?
3. **IDENTIFY** Name one health risk associated with hot weather and one associated with cold weather.

>>> Thinking Critically

4. **ANALYZE** Francis is shopping for a new pair of basketball shoes. He really likes the look of one pair, but when he tries them on, they feel a bit tight. The store does not have them in a larger size. He is thinking of buying them anyway, because they might feel better once he breaks them in. What advice would you give Francis?

>>> Applying Health Skills

5. **INFER** How could being in good physical condition before starting a sport reduce your chances of injury?

🔄 Review

🔊 Audio

Hands-On HEALTH ACTIVITY

Get Up and Get Fit

Now that you know the benefits of fitness, use your knowledge to encourage others. Write a public service announcement (PSA) for a radio show. Do research to learn the physical, mental/emotional, and social health benefits of fitness. The PSA script should persuade others to get up and get fit.

WHAT YOU WILL NEED
* Computers with Internet access
* Recording equipment (optional)

WHAT YOU WILL DO

1 Work in groups of three or four. Identify at least five benefits of fitness, and offer five facts or examples demonstrating the benefits you selected.

2 Write a script featuring at least three examples from your research. Support your position by citing at least one valid resource for each example.

3 Present the PSA to the class as a role-play or a recording.

WRAPPING IT UP

Ask the entire class for feedback on each PSA. Discuss whether your message is clear, whether you offer valid facts or examples, and whether your PSA addresses the audience you are targeting.

READING REVIEW

FOLDABLES and Other Study Aids

Take out the Foldable that you created and any graphic organizers that you created. Find a partner and quiz each other using these study aids.

LESSON 1 Becoming Physically Fit

BIG IDEA Being physically active benefits your total health in a variety of ways.

* An active lifestyle will help to keep you healthy throughout your lifetime.
* Teens should get 60 minutes of physical activity on most days. This can be done all at once or divided into 10- or 15-minute bursts of activity.
* Elements of fitness include strength, endurance, flexibility, and body composition.

LESSON 2 Creating Your Fitness Plan

BIG IDEA You can reach your fitness goals by making a plan to include different kinds of exercise in your life.

* Measuring your fitness level will indicate what areas of fitness you need to improve.
* Fitness test results help to determine fitness goals.
* To reach your fitness goals, you will most likely need to choose a variety of different activities.

LESSON 3 Performing at Your Best

BIG IDEA Following some basic guidelines will help you get the most out of your workouts.

* Healthy workouts include a warm-up and cool-down.
* Conditioning will help you play and perform to the best of your abilities.
* Teen athletes have special nutritional needs. They need to eat a variety of foods and consume extra calories to replace all the energy they use.

LESSON 4 Preventing Sports Injuries

BIG IDEA You can prevent many sports injuries by taking precautions.

* Some sports injuries can be treated at home, while others require medical care.
* Staying in shape, using the correct sports equipment, and avoiding bad weather conditions can help you avoid sports injury.
* The P.R.I.C.E. procedure is used to treat strains and sprains.

 Review

 Web Quest

ASSESSMENT

Reviewing Vocabulary

> cardiovascular endurance
> physical activity
> strength
> flexibility
> exercise
> physical fitness

Reviewing MAIN IDEAS

» On a sheet of paper, write the numbers 1–6. After each number, write the term from the list that best completes each statement.

LESSON 1 Becoming Physically Fit

1. Planned physical activity done regularly to build or maintain fitness is called _____.

2. The ability of your muscles to use force is called _____.

3. The measure of how effectively your heart and lungs work during moderate-to-vigorous physical activity or exercise is called _____.

4. _____ is your ability to move joints fully and easily through a full range of motion.

5. The ability to handle the physical demands of everyday life without becoming overly tired is called _____.

6. _____ is any form of bodily movement that uses up energy.

» On a sheet of paper, write the numbers 7–12. Write *True* or *False* for each statement below. If the statement is false, change the underlined word or phrase to make it true.

LESSON 2 Creating Your Fitness Plan

7. As your fitness level improves, you <u>should not</u> change your goals.

8. <u>Aerobic exercise</u> is any rhythmic activity that uses large amounts of oxygen and works the heart and lungs.

LESSON 3 Performing at Your Best

9. Be sure to eat a <u>large meal</u> one or two hours before physical activity.

10. Stretching is an activity that should be done <u>before</u> warming-up.

LESSON 4 Preventing Sports Injuries

11. Muscle <u>cramps</u> can be relieved by massaging the muscle.

12. A <u>fracture</u> is an injury to a ligament that connects bones at a joint.

✔ eAssessment

>> Using complete sentences, answer the following questions on a sheet of paper.

☁️ *Thinking* **Critically**

13. SUGGEST A friend wants to join a sports team but is worried that she might get hurt. What advice would you give your friend to help her protect herself from injury?

14. APPLY What would you tell someone who said that he does not want to participate in regular physical activity because he does not like team sports?

✏️ *Write* **About It**

15. DESCRIPTIVE WRITING Create a fitness plan that shows how you can be physically active for one hour, 5 to 6 days each week. Use a variety of activities in your plan and include both aerobic and anaerobic exercise. Remember, you can break your hour into 10-15 minute segments if you need to.

16. PERSONAL WRITING Write a journal entry about a fitness goal you would like to reach. What are some realistic steps that you could take to reach this goal?

Ⓐ Ⓑ Ⓒ Ⓓ STANDARDIZED TEST PRACTICE

Math
Answer the questions below.

1. Marie is training for a race. She needs to figure out the low end of her target heart rate.

$T = 0.6 (220 - a)$

In the equation above, T represents the low end of the target heart rate range, and a represents Marie's age. Marie is 16 years old. What is the heart rate Marie should aim for when starting her training program?

2. Marie's younger brother Alex has a target heart rate of 110 beats per minute. His resting heart rate is 50 percent lower than his target heart rate. What is Alex's resting heart rate?

A. 54
B. 55
C. 56
D. 57

Unit 6

health during the life cycle

The Life Cycle

LESSONS

PREMIUM ONLINE RESOURCES ⟩

 Audio
 Videos
 Bilingual Glossary

 Fitness Zone
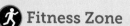 Web Quest
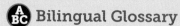 Review

Changes During Puberty

BIG IDEA You will go through many physical, mental/emotional, and social changes during your teen years.

Before You Read

QUICK WRITE Make a list of five ways that you are different now than you were five years ago.

 Video

As You Read

FOLDABLES Study Organizer

Make the Foldable® found in the FL pages in the back of the book to record the information presented in Lesson 1.

Vocabulary

› adolescence
› puberty
› hormones
› community service

 Audio

🔤 Bilingual Glossary

Developing Good Character

Respect Teens can be very sensitive about the changes their bodies are going through, especially if they start to go through puberty earlier than their friends. It is important not to tease others about the way they look. Remember, when your turn comes, you will want to be treated with respect. *Explain what you could do to help a friend who is being teased about his or her appearance.*

CHANGES DURING YOUR TEEN YEARS

MAIN IDEA Adolescence is a time of rapid change and growth.

When you were a baby, you grew and developed very quickly. Now, as a teen, you are going through **adolescence,** or *the stage of life between childhood and adulthood.* During this period, your body will grow faster than it has at any time since you were an infant.

You will also develop in lots of other ways. Along with changes in your body, you will experience changes in how you think and feel as well as changes in your relationships. During your teen years, you will be setting out on a journey of discovery. The object of that discovery is you. You will meet new people, explore new ideas, and take on new responsibilities. The teen years are a chance to practice for life on your own when you become an adult.

>>> **Reading Check**

DEFINE *What is adolescence?*

Physical Changes

Just before she started school this year, Brianna found that none of her fall clothes fit anymore. She had grown several inches in just one year. At the same age, her brother Tyler's voice had gotten deeper, and he developed the beginnings of a mustache.

The changes that Brianna and Tyler are experiencing are normal. The changes are part of **puberty,** or *the time when you develop the physical characteristics of an adult.* During adolescence, your body will go through many changes. Many of these changes are caused by the production of **hormones.** These are the *chemical substances produced in certain glands that help regulate the way your body functions.* Hormones send signals to different parts of your body to prepare for adulthood.

©PhotoAlto/PunchStock

One of the first signs of puberty is often a growth spurt—a sudden, rapid increase in height. As you grow taller, your body shape will also begin to change. Boys' shoulders become wider and their bodies grow more muscular. Girls may notice that their hips are wider and they are developing breasts. Both boys and girls tend to gain weight as their bodies change during puberty, but this weight gain is normal and no cause for concern. The figure below illustrates other changes that affect girls and boys during puberty.

During *adolescence,* your **body** will go through **many changes.**

For most girls, puberty starts sometime between age 8 and age 13. For most boys, it happens a little later—between the ages of 10 and 15. These changes occur at different times for different people. Some 13-year-olds may still look much as they did during childhood. Others are starting to look like adults. You may experience some changes early and others much later. Everyone experiences puberty at their own time and pace.

Mental/Emotional Changes

Adolescence also brings changes in the way you think and feel. Studies have found that teens' brains expand right before puberty. Most of this growth takes place in the part of the brain related to planning and reasoning. During the teen years, your mental skills will improve.

You will grow better at thinking critically and expressing yourself. You will also begin to develop your personal ideals and values.

The hormones that cause changes in your body also affect your emotions. Mood swings may cause sudden shifts from happiness to sadness or anger. This can be confusing, but it is a normal part of adolescence.

During your teen years, you may also begin to develop feelings of attraction toward others. These feelings, combined with changes in your body, may lead you to become interested in dating. However, not everyone is ready for dating at the same age.

> Puberty is the time during which you develop the physical characteristics of an adult. *Identify one change that happens to both males and females during puberty.*

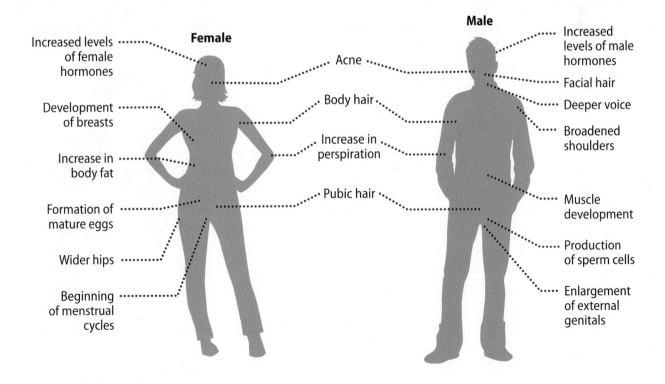

Female
- Increased levels of female hormones
- Development of breasts
- Increase in body fat
- Formation of mature eggs
- Wider hips
- Beginning of menstrual cycles

- Acne
- Body hair
- Increase in perspiration
- Pubic hair

Male
- Increased levels of male hormones
- Facial hair
- Deeper voice
- Broadened shoulders
- Muscle development
- Production of sperm cells
- Enlargement of external genitals

Health SKILLS ACTIVITY · Stress Management

Coping with *Mood Swings*

The mood swings you experience during your teen years can feel like you are on an emotional roller coaster. Below are some tips for dealing with mood swings in healthful ways.

* Do something nice for yourself. For example, listen to your favorite music or get together with a friend.

* Be creative. Write about what you are feeling or express your emotions through painting, music, or another form of artwork.

* Get moving. Physical activities help the body stay in balance. They will also help take your mind off whatever is bothering you.

* Get some help. When you are sad or angry, it is easy to feel alone. Talk to friends, family members, or other trusted adults about what you are feeling.

* Get some rest. With all the changes going on in your body, extra rest can really help. Try taking a nap or just relaxing with a book or magazine.

* Eat well. Help your body through its many changes by eating nutritious foods.

 On Your Own Create a poster, mood board, or digital montage for teens listing healthful ways to cope with mood swings. Include the tips described above and other suggestions you can think of. With permission, hang your creations where other students can see them.

Social Changes

During your teen years, you are likely to go through many changes in the way you relate to others. At home, your relationship with your parents may change as you become more independent. You may want to spend less time with your parents and more with your friends.

You're likely to go through many *changes* in the way you **relate to others.**

You will also become more aware of the world around you. As a child, your world probably centered around your home, your family, and your school.

Listening to music can be a good way to deal with mood swings. *List two other good ways to manage your emotions.*

RubberBall/Alamy

As you get older, you may want to begin to have new experiences and expand your world. One way to do this is by becoming involved in community service. Teens who volunteer learn about the impact that they can make on their community. Volunteering also offers the opportunity to meet new people.

CHANGES IN FAMILY RELATIONSHIPS As a child, you were very dependent on your parents or guardians. They made most of your decisions for you, like what you would eat and when you would go to bed. Now, as a teen, you are learning to act independently and make more decisions for yourself. Your increasing independence can sometimes lead to differences with family members. For instance, you might think that your parents should let you stay out later at night or shop for clothes on your own.

Many times, conflicts can arise when teens and parents or guardians disagree on how much independence a teen should receive. You may have a curfew to be home by 9 P.M. on school nights. Your parents or guardians set rules to help keep you safe. Other rules come in the form of laws that apply to the whole community. If you follow the rules your parents set and avoid breaking any laws, you show your parents that you respect them. They will recognize that you are trustworthy. They may allow you to make more of your own decisions.

CHANGES IN PEER RELATIONSHIPS During your teen years, your peers are likely to play a growing role in your life. You may spend more time with friends and less with family members. Often, your peers will influence your ideas and your behavior. For example, you may change the way you dress or the music you prefer in order to be more like your friends. Changes like this are typical, but in other cases, peer influences can be unhealthy. For example, peers may urge you to engage in harmful behaviors, such as drinking or smoking. On the other hand, peer influence can also be positive, as in the case of friends who study hard or do volunteer work.

YOUR RELATIONSHIP WITH THE COMMUNITY You are part of several communities. They include your neighborhood, your school, and the city or town where you live. Your teen years can be a good time to make a positive contribution. Community service involves *volunteer programs aimed at improving the community and the life of its residents.* Community service gives you a chance to help people and develop your independence at the same time. You will be able to make your own decisions about when and where to work. You can also learn new skills and develop a sense of responsibility.

>>> **Reading Check**

EXPLAIN *In what ways can peers have a positive influence on teens?* ■

REVIEW

>>> **After You Read**

1. **DEFINE** What are hormones?
2. **IDENTIFY** Name three ways in which teens' bodies change during puberty.
3. **DESCRIBE** How are relationships with parents likely to change during the teen years?

>>> **Thinking Critically**

4. **INFER** Why do you think teens go through puberty at different rates?
5. **EVALUATE** How do you think the changes you experience during adolescence help you prepare for adulthood?

>>> **Applying Health Skills**

6. **COMMUNICATION SKILLS** Write a dialogue between a teen and a parent. In the dialogue, the teen should explain his or her need for more independence in a way that shows consideration for the parent's feelings.

 Review

 Audio

The Male Reproductive System

BIG IDEA The male reproductive system produces cells that can join with cells from a female's body to produce a child.

Before You Read

QUICK WRITE Do you plan to have children when you become an adult? How do you picture your family life when you are older?

▶ Video

As You Read

STUDY ORGANIZER Make the study organizer found in the FL pages in the back of the book to record the information presented in Lesson 2.

Vocabulary

› reproduction
› reproductive system
› sperm
› testes
› semen
› hernia

🔊 Audio

🔤 Bilingual Glossary

🏃 Fitness Zone

Playing It Safe I never used to see the point of wearing a cup when playing soccer. I'd do it because the coach said we had to, but it made me feel silly. Then one time during a game I got hit with a ball—right there. It hurt, but it also made me realize just how much worse it would have been if I hadn't had the cup on. Now I make sure never to go out on the field without it.

REPRODUCTION

MAIN IDEA The reproductive system makes it possible for people to produce children.

Romi's older sister has recently come home with a new baby. Romi enjoys holding his niece and feeling her tiny hand squeezing his finger. Romi even helps change and bathe the baby. Playing with his niece and helping his sister and her husband take care of their new arrival has made Romi start to think about whether he would like to have children of his own some day.

Reproduction is one of the most **important functions** of all living things.

Reproduction is *the process by which organisms produce others of their own kind.* Reproduction is one of the most important functions of all living things.

The reproductive system is made up of *the body systems and structures that make it possible to produce children.* All individuals will eventually grow old and die, but through the process of reproduction, new ones are produced to keep the species going.

In human beings, the process of reproduction involves the joining of two cells that come from the reproductive systems of a male and a female. Every human being on earth begins life in this way. Puberty and adolescence is the time during which your body goes though a series of physical, mental/emotional, and social changes. These changes prepare your body to be able to produce children during adulthood. In this lesson and the next one, you will learn more about how the male and female reproductive systems work and how to take care of them.

⟫ Reading Check

DEFINE *What is the reproductive system?*

©A. Chederros/Onoky/Corbis

THE MALE REPRODUCTIVE SYSTEM

MAIN IDEA The main purpose of the male reproductive system is to produce and release sperm.

The male reproductive system has two main jobs. The first is to produce **sperm**, or *male reproductive cells.* The second is to deliver sperm to the body of a female.

The **testes** are *the pair of glands that produce sperm* and male hormones. The testes, or testicles, are located in a pouch called the scrotum. The scrotum keeps the testes at the proper temperature. Before sperm leave the body, they mix with fluids from the prostate gland and seminal vesicles.

The **male reproductive system** has two main jobs.

Semen is *a mixture of fluids that protect sperm and carry them through the male reproductive system.* Semen travels to the urethra and is released by the penis. This happens in a series of muscle contractions known as ejaculation. An ejaculation may contain 500 million sperm.

>>> Reading Check

IDENTIFY *What are the two primary functions of the male reproductive system?*

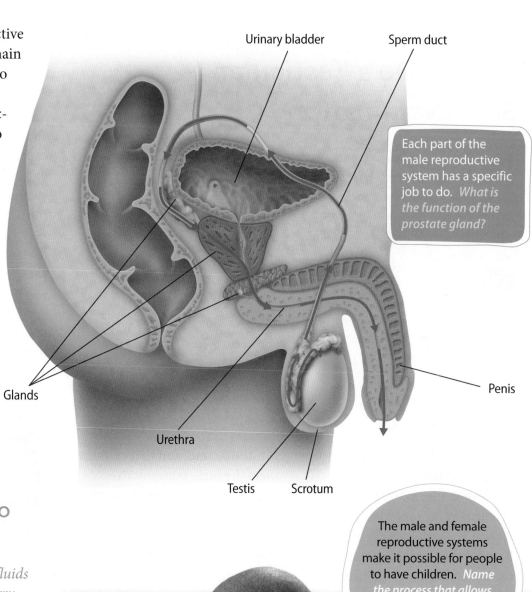

Urinary bladder

Sperm duct

Each part of the male reproductive system has a specific job to do. *What is the function of the prostate gland?*

Penis

Glands

Urethra

Testis   Scrotum

The male and female reproductive systems make it possible for people to have children. *Name the process that allows living things to produce others of their own kind.*

Ingram Publishing

How to Do a *Testicular Self-Examination*

When a male goes for his yearly physical checkup, the doctor will usually examine him for testicular cancer. This can help detect the disease before it becomes very serious or difficult to treat. Testicular cancer can grow and spread very quickly if it is not detected. It is important for males to do a testicular self-examination every month.

On Your Own
Go online or to the library to find instructions for how to do a testicular self-examination. Write out the instructions. Make a list of at least four signs of testicular cancer to watch out for. Then write two questions that a male might have for a health care provider about testicular self-examinations.

Male Health Problems

For teen males, a common problem of the reproductive system is injury to the testes. Because the testes hang outside the body, they can be injured if they are hit, kicked, or crushed. This can easily happen during sports activities. That is why many sports require a protective cup to cover the scrotum and testes. Another type of injury to the male reproductive system is testicular torsion. This happens when the cord that holds the testes together twists around. Testicular torsion cuts off the flow of blood to a testicle, which results in pain and swelling.

A third condition that can affect the male reproductive system is a **hernia,** which occurs when *an internal organ pushes against or through a surrounding cavity wall.* For instance, part of the intestine can push into the groin area or the scrotum. Hernias are typically caused by muscle weakness and strain. A hernia may appear as a bulge or swelling in the groin area. This can be corrected with surgery.

It is *important* for males to do a **testicular self-exam** every month.

Diseases can also affect the male reproductive system. Some of them are sexually transmitted diseases (STDs), which spread through sexual contact. One of the most serious diseases that can affect this system is cancer of the testes. This is the most common form of cancer in males aged 15 to 35. If it is not treated, the cancer may spread to other parts of the body. If it is caught early, it can almost always be cured. If the testicles are removed, however, the man will become sterile, or unable to produce offspring. Men may also become sterile as a result of drug use or being exposed to radiation or certain types of chemicals. Also, some STDs can cause sterility if they are not detected and treated.

The American Cancer Society recommends that all males perform a testicular self-exam once a month. *Explain why it is a good idea for teen males to start performing testicular self-exams.*

Jack Hollingsworth/Photodisc/Getty Images

Caring *for the* Male Reproductive System

Like any other body system, the male reproductive system needs care. Taking good care of your reproductive system during your teen years will help keep it in good working order. This will help ensure that you can have healthy children later in life.

Take **good care** of your *reproductive system.*

Here are a few steps male teens can take to keep their reproductive systems healthy:

- **Practice good hygiene.** Take a bath or shower daily, and be sure to clean the penis and scrotum carefully.

- **Wear protective gear.** For contact sports such as baseball and football, a protective cup is needed. Talk to a parent or your coach about the proper way to wear a cup. Other sports that involve running require an athletic supporter.

- **Conduct self-exams.** Once a month, check the testes for any lumps or swellings that could be a sign of cancer. See the Health Skills Activity for more information.

- **Get regular checkups.** Your doctor or other health care provider can screen for reproductive health problems and catch them before they become serious.

>>> Reading Check

IDENTIFY *Name two problems that can affect the male reproductive system.* ■

Injuries to the testicles can occur during sports activities. *Name what the boy in the picture can do to protect himself from injury.*

©Ben Blankenburg/Corbis

>>> After You Read

1. **VOCABULARY** What is a hernia?
2. **IDENTIFY** What are the two main functions of the male reproductive system?
3. **NAME** What are two types of gear that protect the reproductive system of a male athlete?

>>> Thinking Critically

4. **EVALUATE** Why is it important to take care of the male reproductive system?
5. **APPLY** Anthony has noticed that one of his testicles is swollen. He does not feel comfortable talking about the problem, but it is not going away. What steps would you suggest that Anthony take?

>>> Applying Health Skills

6. **ACCESSING INFORMATION** Some teens and adults use anabolic steroids to change their bodies. Use online resources to learn about the ways steroids affect the male reproductive system. Summarize your findings in a fact sheet about the dangers of illegal steroid use.

⟳ Review

◗⟩ Audio

The Female Reproductive System

BIG IDEA The female reproductive system contains the organs that nurture a baby before birth.

Before You Read

QUICK WRITE What do you think is the most important thing teens can do to care for their reproductive systems?

 Video

As You Read

STUDY ORGANIZER Make the study organizer found in the FL pages in the back of the book to record the information presented in Lesson 3.

Vocabulary

› fertilization
› ovaries
› uterus
› ovulation
› menstruation
› gynecologist

 Audio

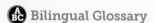 Bilingual Glossary

Myth vs. Fact

Myth: Females can practice good hygiene by douching, or rinsing out the vagina with a solution of water and vinegar.

Fact: Douching can be harmful. The body has its own natural system for flushing bacteria out of the vaginal area. Douching is more likely to cause irritation or infections than to prevent them.

HOW THE FEMALE REPRODUCTIVE SYSTEM WORKS

MAIN IDEA The female reproductive system has three main functions: to produce and store eggs, to create and nurture offspring, and to deliver a baby.

The first function of the female reproductive system is to produce, store, and release egg cells. Eggs, or ova, are the female reproductive cells. The second function is to create offspring. This process begins with **fertilization,** which is *the joining of a male sperm cell and a female egg cell to form a fertilized egg.* The female reproductive system must also nourish the developing baby until it is born. Giving birth is the system's third function.

Parts *of the* Female Reproductive System

The **ovaries** are *the female endocrine glands that release mature eggs and produce the hormones estrogen and progesterone.* These hormones control the female reproductive system and sexual development.

Once a month, the ovaries release an egg cell into the fallopian tubes. If an egg is fertilized, it continues along the tube and attaches itself to the wall of the **uterus,** *a pear-shaped organ, located within the pelvis, in which the developing baby is nourished and protected.* In the uterus, the fertilized egg continues to develop for about nine months. At birth, the baby moves out of the uterus and passes out of the body through the vagina.

The Menstrual Cycle

The menstrual cycle is the series of events that prepares the female reproductive system to produce a baby. The term *menstrual* comes from the fact that the cycle usually takes about a month, but that length can vary. For some females, a cycle may be as short as 23 days, while it may be as long as 35 days for others. Menstrual cycles begin when a girl reaches puberty. However, it may take a couple years for a girl's cycle to become regular.

Image Source/Getty Images

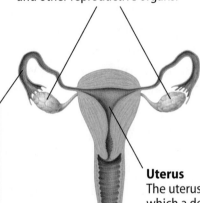

This diagram shows the parts of the female reproductive system. *Identify the functions of the ovaries.*

Ovaries
The ovaries hold the female's eggs. The ovaries also make the hormones estrogen and progesterone. These control female sexual development and other reproductive organs.

Cervix
This is the narrow part of the bottom of the uterus. The opening of the cervix enlarges to allow a baby to leave the uterus during birth.

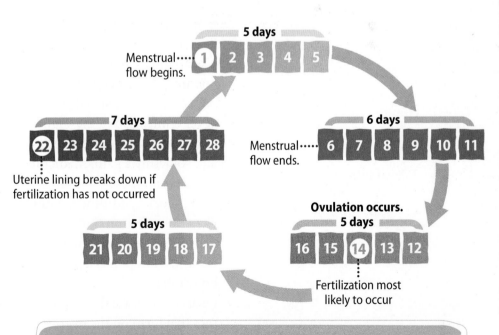

Uterus
The uterus is the organ in which a developing child is nourished.

Fallopian tubes
Eggs travel from the ovaries to the uterus through the fallopian tubes. Eggs are usually fertilized in these tubes.

Labia
Labia are folds of skin that cover the opening of the vagina.

Vagina
The vagina is the passageway that leads from the cervix to the outside of the body. Menstrual flow leaves the body through the vagina. Sperm enter the female reproductive system through the vagina. During birth, a baby leaves the mother's body through the vagina.

The cycle begins with <u>ovulation,</u> or *the process by which the ovaries release mature eggs, usually one each menstrual cycle.* When the ovaries release an egg, it moves through the fallopian tubes toward the uterus. At the same time, the lining of the uterus begins to thicken. The lining thickens as the uterus prepares to receive the egg if it is fertilized. However, if this does not happen, <u>menstruation</u> occurs with *the flow from the body of blood, tissues, and fluids that result from the breakdown of the lining of the uterus.* The lining of the uterus then leaves the body through the vagina. A menstrual cycle typically lasts from three to five days. Then, about two weeks later, the ovaries release another egg, and the cycle starts over again.

5 days
Menstrual flow begins. | 1 | 2 | 3 | 4 | 5 |

6 days
Menstrual flow ends. | 6 | 7 | 8 | 9 | 10 | 11 |

Ovulation occurs.
5 days
| 16 | 15 | 14 | 13 | 12 |
Fertilization most likely to occur

5 days
| 21 | 20 | 19 | 18 | 17 |

7 days
| 22 | 23 | 24 | 25 | 26 | 27 | 28 |
Uterine lining breaks down if fertilization has not occurred

The menstrual cycle is usually around 28 days. However, it may be shorter or longer for some females, especially during the first two years of menstruation. *What process begins the menstrual cycle?*

Advocacy

Promoting *Breast Self-Examinations*

Doctors usually examine a female's breasts for lumps or signs of breast cancer. Breast cancer can grow and spread very quickly if not detected early. Females can detect signs of cancer by doing a breast self-examination every month. It should be done about seven days after menstruation.

Go online or to the library to find instructions for how to do a breast self-examination. Make a brochure that features these instructions. In your brochure, include a list of at least four signs of breast cancer to watch out for.

FEMALE HEALTH PROBLEMS

MAIN IDEA The female reproductive system can be affected by a wide range of problems, ranging from minor to life-threatening.

Before their menstrual cycles, many females experience a set of symptoms known as premenstrual syndrome, or PMS. Some of these symptoms are physical. For example, females may develop headaches or acne. They may also feel tired or crave certain foods. Emotional symptoms can also occur, such as depression or irritability. This is one of the most common issues affecting the female reproductive system. Other problems are less common, but may be more serious.

Females can detect signs of cancer by doing a *breast self-exam* every month.

- Many females have abdominal cramps during the first few days of their menstrual cycles.
- A yeast infection in the vagina can cause itching, discharge, and sometimes pain. Medicine can clear up these symptoms.
- Toxic shock syndrome is a rare bacterial infection that causes fever and vomiting. It may occur if a tampon is left in the body too long.
- Ovarian cysts are fluid-filled sacs that form on the ovaries. These are usually harmless, but they may become painful if they grow too large.

- Sexually transmitted diseases, or STDs, can harm the female reproductive system.
- Women can get cancers of the reproductive system. Cancer may affect the breasts, ovaries, uterus, or cervix. Certain STDs can increase the risk of cervical cancer.
- Infertility is the inability to become pregnant. Some STDs can cause this problem if they go undetected or are not treated. Other causes include blocked fallopian tubes and fibroids, which are noncancerous growths in the uterus.

Keeping track of her menstrual cycles can help a female monitor her health. *Identify the length or duration of a typical menstrual cycle.*

©age 100/Corbis

Caring *for the* Female Reproductive System

Care of the reproductive system is important to a female's overall health. Good hygiene, including a daily shower or bath, is especially important. So are regular medical checkups. In addition to a regular doctor, females should also see a **gynecologist,** or *a doctor who specializes in the female reproductive system.* Medical professionals recommend that females make their first visit to a gynecologist at around age 14 and then once a year after that.

Take **good care** of your *reproductive system.*

In addition to medical exams, females should examine their own breasts once a month for signs of cancer. Breast cancer is rare among teens, but self-examinations can help girls become familiar with both the look and feel of their developing breasts.

These self-exams allow females to learn what is normal for them. Some changes to look for include lumps, swelling, or irritation of the skin. If a female does notice anything unusual or different during a breast self-examination, it is important to see a doctor.

Females should make sure that they change their tampons or sanitary napkins at least every four hours during their menstrual cycles. This practice helps maintain good hygiene. Changing tampons regularly can also help prevent potentially serious infections such as toxic shock syndrome. Finally, females can keep track of their menstrual cycles. This will help them notice problems such as missed cycles. A female should see a doctor if she misses several menstrual cycles in a row. She should also see a doctor if she experiences severe pain or bleeding in between cycles.

>>> Reading Check

LIST *What are four ways to care for the female reproductive system?* ■

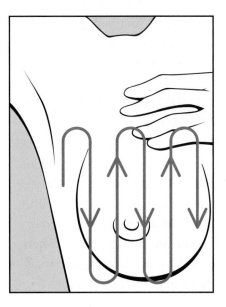

In a vertical pattern, check from the underarm to the sternum and from the collarbone to the ribs. *Explain why it is important for females to start performing regular breast self-exams during the teen years.*

LESSON 3

REVIEW

>>> After You Read

1. **VOCABULARY** Define *fertilization.*
2. **LIST** What are the three functions of the female reproductive system?
3. **RECALL** How often should females perform breast self-exams?

>>> Thinking Critically

4. **INFER** How does fertilization affect the menstrual cycle?
5. **SYNTHESIZE** How are the male and female reproductive systems similar? How are they different?

>>> Applying Health Skills

6. **ADVOCACY** By practicing abstinence as a teen, you can avoid STDs that could damage your reproductive system. Design a bumper sticker or button with a slogan that promotes abstinence. You may also use graphics or illustrations to support your ideas.

 Review

 Audio

Infant *and* Child Development

BIG IDEA Each human being passes through several stages of development before reaching adolescence.

Before You Read

QUICK WRITE Describe a time when you or someone you know played with a baby or helped care for one. What could the baby do without your help? What did you have to help the baby do?

▶ Video

As You Read

STUDY ORGANIZER Make the study organizer found in the FL pages in the back of the book to record the information presented in Lesson 4.

Vocabulary

› cell
› tissues
› organs
› body system
› embryo
› fetus
› developmental task
› infancy
› toddler
› preschooler

 Audio

 Bilingual Glossary

HOW LIFE BEGINS

MAIN IDEA Every human being starts out as a single cell that develops into a baby over the course of nine months.

As strange as it may seem, your whole body began as a single **cell,** *the basic unit of life.* This cell formed when an egg cell from your mother joined with a sperm cell from your father. During the next few days, this single cell divided many times before reaching your mother's uterus. There, over the course of nine months, this tiny clump of cells continued to divide and form new structures. Eventually, these structures developed into all the parts of the body you have now. To understand how this process took place, first you need to know a little more about how the human body is constructed.

Building Blocks *of* Life

Have you ever planted seeds and watched a new plant grow? You are excited when the first little shoot emerges from the ground. As the plant grows, the stem grows longer and thicker. Then leaves form, and the plant may develop flowers or produce fruit.

Have you ever thought about how a computer or a household appliance works? If you have, you know that it contains many different parts, each with its own job to do. However, all these different parts work together to make the object function properly. In the same way, your body contains many different parts that work together to enable you to breathe, eat, walk, speak, and do everything else you need to do in the course of a day.

Your **whole body** began as a *single cell.*

Machines can have hundreds or thousands of parts. The body is made up of trillions of tiny parts known as cells. Cells that have similar functions in the body work together in larger units called **tissues,** which are *a group of similar cells that do a particular job.* Some examples include muscle tissue, brain tissue, and connective tissue.

Andersen Ross/Getty Images

Tissues, in turn, can form an organ, *a body part made up of different tissues joined to perform a particular function.* Your heart, brain, liver, and lungs are all organs. When *a group of organs work together to carry out related tasks,* they are a body system.

Development *before* Birth

The process of a single cell dividing to form all the body's tissues, organs, and systems is long and complex. Fertilization usually occurs inside the mother's body. The egg cell and the sperm cell join in one of her fallopian tubes. As soon as the sperm cell enters the egg, the cell begins to divide, preparing to form a new person.

How does a *single cell* divide to form all the **tissues, organs, and systems** that make up your *body?*

The fertilized egg cell makes its way down the tube, continuing to divide along the way. By the time it reaches the uterus, it has become a hollow ball of cells about the size of a pinhead. This ball attaches itself to the lining of the uterus, which will nourish it as it grows. From this point on, it is called an embryo, or *the developing organism from two weeks until the end of the eighth week of development.*

Once an embryo is attached inside the uterus, the cells continue to multiply and change, forming tissues that do specific jobs. Those tissues combine to form organs. After about eight weeks, all the major organs of the body have begun to take shape, including the heart, brain, and lungs. From this point, the embryo is called a fetus, which is *the developing organism from the end of the eighth week until birth.* The average time from conception to birth is about nine months, or 280 days.

> Cells work together in the body to form tissues, organs, and systems. *Name some of the body's organs.*

The fetus floats in a fluid-filled sac that cushions and protects it. It gets its nourishment from the mother's blood through the placenta, a thick, rich tissue that lines the walls of the uterus. A tube called the umbilical cord connects the mother's placenta to the fetus. Through it, the fetus can take in oxygen and nutrients. However, substances such as alcohol or tobacco can also reach the fetus through the placenta. Women need to be careful to avoid these harmful substances.

>>> **Reading Check**
DEFINE *What are tissues?*

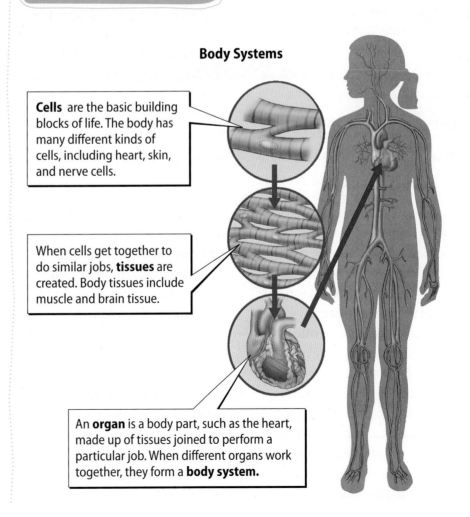

Body Systems

Cells are the basic building blocks of life. The body has many different kinds of cells, including heart, skin, and nerve cells.

When cells get together to do similar jobs, **tissues** are created. Body tissues include muscle and brain tissue.

An **organ** is a body part, such as the heart, made up of tissues joined to perform a particular job. When different organs work together, they form a **body system.**

1 Infancy
Description: The child depends on others to meet his or her needs.
Developmental task: learning trust

2 Early Childhood (Toddler Stage)
Description: The child is learning to walk, talk, and do other things on his or her own.
Developmental task: developing autonomy (the ability to do things for yourself)

3 Middle Childhood (Preschool Years)
Description: The child begins to come up with ideas on his or her own and to develop a sense of right and wrong.
Developmental task: developing initiative (the ability to plan and act on your own)

4 Late Childhood (Early School Years)
Description: The child starts school and learns new skills.
Developmental task: developing a sense of self-worth

5 Adolescence
Description: The teen explores many different roles.
Developmental task: developing identity (a sense of who you are)

6 Early (Young) Adulthood
Description: The young adult tries to establish close personal relationships.
Developmental task: developing intimacy (a strong connection with another person)

7 Middle Adulthood
Description: The adult seeks a sense of satisfaction through career, raising a family, and being active in society.
Developmental task: making a contribution to the world

8 Late Adulthood (Maturity)
Description: The adult looks back on the achievements of his or her life.
Developmental task: to be satisfied with the life one has lived.

Each stage of life has a different developmental task. *Name the developmental task during adolescence.*

STAGES OF LIFE

MAIN IDEA A human life can be divided into a series of stages, each with its own tasks to be mastered.

A newborn baby can do very little by itself. For the first several months of life, a baby depends on others for tasks as simple as eating and getting from place to place. Newborns also rely on others for basic hygiene. However, a baby can eventually grow into an adult who can do complicated tasks such as driving a car or programming a computer. To reach adulthood, a baby will go through a series of stages or steps with different needs.

The psychologist Erik Erikson divided the human life cycle into eight separate stages. Erikson's stages of life include infancy, early childhood, middle childhood, late childhood, adolescence (or the teen years), early adulthood, middle adulthood, and late adulthood. At each stage, Erikson explained, a person needs to master a **developmental task,** which is *an event that needs to happen in order for you to continue growing toward becoming a healthy, mature adult.*

The *human life cycle* can be **divided into eight stages.**

With each new task someone masters, that person builds character and improves his or her social and emotional health. The developmental tasks for a toddler or preschooler are fairly simple. The tasks that an adolescent or a young adult need to master are far more complex.

Infancy

During **infancy,** or *the first year of life,* babies learn about their own bodies. Tasks such as focusing the eyes or sitting up are big achievements for a baby. Infants also begin to observe the world around them. Babies learn to recognize familiar people and objects. They begin to learn language, making sounds and recognizing others' voices. During this first stage of life, infants also develop bonds of love and trust with their parents and others close to them.

Early Childhood

A **toddler** is *a child between the ages of one and three.* During the second year of life, toddlers become more aware of themselves and their surroundings. Toddlers learn to do things on their own, such as feeding themselves and using the toilet. They can walk and are interested in exploring new objects and people. Young toddlers may also speak in short sentences and be able to follow simple directions.

After age 2, toddlers start to learn more complicated skills. Physically, they may be able to run, jump, or kick a ball. Mentally, toddlers can usually recognize different shapes and follow directions that have two or three steps. Socially, toddlers will often imitate the behavior of their parents or siblings. However, they are also becoming more interested in being independent. Many parents of toddlers will say that their child's favorite word is "no."

Toddlers will often **imitate** the **behavior** of *parents* or *siblings.*

Middle Childhood

A **preschooler** is *a child between the ages of three and five.* If a toddler's favorite word is "no," a preschooler's favorite word might be "why?" At this stage of life, children tend to ask a lot of questions and use their imaginations. Preschoolers may enjoy telling stories or singing songs.

Children between the ages of three and five become more interested in people outside their immediate families. They learn to cooperate with other children. Preschoolers' physical skills also improve. They can do exercises like hopping on one foot or riding a tricycle. They may be able to dress and undress themselves or choose foods at mealtimes.

>>> **Reading Check**

RECALL *What is the first stage of life?*

In early childhood, children learn to walk on their own. *What are some other skills children learn at this age?*

Late Childhood

From ages six to eleven, school plays an important part in children's lives. It brings them into regular contact with other children, as well as adults. Children become more independent from their families, and friends become more important to them. Through schoolwork, friends, and activities, children at this stage develop increased confidence in themselves.

During your school years, friendships become more important. *Describe some other differences between middle and late childhood.*

By the time you reach your teen years, you have already learned and accomplished quite a lot in your life. You can do physical tasks such as riding a bike or throwing a ball. You can read and write, and you have learned to recognize and express your emotions. You can make decisions and solve problems on your own, but you have also formed new friendships and learned to work with others. All these skills will help you as you move through adolescence.

Adolescence

During the time between ages 12 and 18, you are no longer a child, but you are not yet an adult. During adolescence, you experience physical, mental/emotional, and social changes that prepare you for adulthood. ■

▶▶▶ After You Read

1. **DEFINE** What is a developmental task?
2. **IDENTIFY** What are the parts that make up a body system, from smallest to largest?
3. **LIST** What are the eight stages of life?

▶▶▶ Thinking Critically

4. **ANALYZE** What is the difference between an embryo and a fetus?
5. **INFER** What might happen to an infant who did not receive proper care or attention from adults?

▶▶▶ Applying Health Skills

6. **GOAL SETTING** Vanessa's older sister, Kayla, has tried several times to quit smoking without success. Now she and her husband want to have a baby. She has asked Vanessa for her advice. Use the steps of the goal-setting process to show how Kayla could reach her goal of becoming tobacco free before she gets pregnant.

⟳ Review

🔊 Audio

Kablonk/Purestock/SuperStock

Staying Healthy *as* You Age

BIG IDEA Practicing good health skills will allow you to remain healthy throughout early, middle, and late adulthood.

››› Before You Read

QUICK WRITE Pretend that you are 50 years older than you are now. Write a letter to your teen self. What kinds of stories or advice would the older you want to share with the younger you?

 Video

››› As You Read

STUDY ORGANIZER Make the study organizer found in the FL pages in the back of the book to record the information presented in Lesson 5.

››› Vocabulary

› chronological age
› biological age
› social age

 Audio

 Bilingual Glossary

Developing Good Character

Respect Their bodies may not move as fast as yours, but older adults have a lot of wisdom and experience to share. You can show respect for older adults by listening and speaking in a polite manner. *Make a list of topics you think would be interesting to discuss with older adults. Describe some other ways to show respect to older adults.*

STAGES OF ADULTHOOD

MAIN IDEA People continue to learn and develop throughout adulthood.

Right now, as a teen, you are going through a lot of changes as you move toward adulthood. However, this does not mean that you will stop changing once you reach become an adult. Your body will change gradually as you age. Your mental abilities will continue to change as you learn and develop new skills throughout your life. Your social life will also change as you form new relationships, which might include marriage or parenthood.

Just like childhood, adulthood can be divided into several stages. Each stage may involve milestones such as starting a career, getting married, or raising a family. However, these milestones will not be the same for everyone. Some people, for instance, marry late in life, or they may never marry at all. Some adults choose to have many children, while others have none.

Some people work in one career throughout adulthood, while others may change careers many times. Each person's path through adulthood is unique, because no two people in the world are exactly alike.

You will *never* stop **changing.**

Early Adulthood

Many people, as they leave their teen years, are still working on their education. Others have finished school and are starting their first jobs. Young adults may try several different jobs before settling on a career path. People this age also work on forming close relationships with others. This may include marrying or starting a family. However, some adults do not marry until they are older, while others never marry. It is also important for people in early adulthood to continue habits that will contribute to good health later in life. These positive steps include a healthful diet and staying active.

Middle Adulthood

In middle adulthood, many people have settled down in their work and personal lives. At this point, they may start to think about what they want to contribute to the world. People in middle adulthood often feel a need to achieve and be recognized. They may focus on advancing in their careers, raising their families, or helping improve their communities.

At some point in middle adulthood, people often think about what they have done with the first half of their lives and what they would like to do next.

Some people decide to make major changes in their lives at this time. Others focus on getting the most out of the paths they have already chosen. Many adults at this stage of life also renew their commitment to a healthful diet and remaining physically active.

Late Adulthood

Late adulthood begins around age 65. People at this age often decide to retire from their jobs and pursue new interests. Many become more active than when they were younger. Retirement may allow older adults to travel.

Someone who has always lived and worked in a city may start visiting parks. In late adulthood, it is also important to maintain good physical health.

Some older adults pursue new careers or volunteer work. Late adulthood can also be a time of reflection. Older adults may ask whether they have fulfilled their potential. If they are satisfied, they can be content at this stage.

>>> **Reading Check**

COMPARE *What are some of the differences between life as an early and middle adult?*

WAYS TO MEASURE AGE

MAIN IDEA Your age in years may be different from the age suggested by your physical health or your social behavior.

According to one saying, "You're only as old as you feel." Many people in late adulthood feel young for their age because they have maintained good health. Because of this, there are several different ways of measuring age.

- **Chronological age** is *age measured in years.* This is how long you have been alive. It does not reflect your health. People who are the same age may be very different physically, mentally, and socially.
- **Biological age** is *your age determined by how well your various body parts are working.* This measure of age depends primarily on the physical condition of a person's body along with his or her overall health.

For instance, a 50-year-old with arthritis would have an older biological age than a 50-year-old in good health. Heredity, environment, and behavior all play a role in this type of age. People who have made healthful choices may have a younger biological age.

- **Social age** is *your age measured by your lifestyle and the connections you have with others.* This has to do with the role you play in the world. Consider two young adults, both 22 years old. One of them is still attending college and living at home.

The other has taken a job, gotten married, and started a family. The second person might have an older social age.

Buying a first home is one milestone many people reach in middle adulthood. *What are some other changes people may go through during this time?*

GOOD HEALTH FOR OLDER ADULTS

MAIN IDEA Staying active can help older adults maintain good health.

Wellness, or balanced health, is as important for older people as it is for teens. Adults who take good care of their physical health may have a biological age that is younger than their age in years. Being physically active and eating healthy foods are two ways to do this.

Staying *active* is **important** for *social health.*

Good mental health is also a part of biological age. Older people may be at risk for mental problems such as depression.

Getting regular medical care can be a key to finding and treating these problems. Being involved in social and physical activities can also help.

Staying active is important for social health as well. Older people will enjoy better total health when they stay involved with friends and continue to engage in activities they enjoy. They may even live longer as a result. Their social age, as well as their biological age, may be much younger than their actual years.

>>> Reading Check

EXPLAIN *How do health habits affect biological and social age?* ■

>>> After You Read

1. **DEFINE** Explain what the term *social age* means.
2. **GIVE EXAMPLES** Name three things that people may focus on during early adulthood.
3. **EXPLAIN** Why is regular medical care important for mental health in late adulthood?

>>> Thinking Critically

4. **ANALYZE** How are chronological age and biological age related? How are they different?
5. **ANALYZE** Ophelia retired last year. She was looking forward to spending lots of time on her favorite hobby. However, she soon began to feel dissatisfied. She missed the daily social contact of her job. What could Ophelia do to bring her health triangle back into balance?

>>> Applying Health Skills

6. **ANALYZING INFLUENCES** Sylvester is 65 years old and recently retired. He has just decided to get an Internet connection in his home for the first time. How might this affect his total health?

🔄 Review

🔊 Audio

Fit and active people tend to be healthier throughout their lives. *Predict how your choices as a teen may affect your biological age later in life.*

Hands-On HEALTH ACTIVITY

Looking Ahead

WHAT YOU WILL NEED
* Pencil or pen and paper

WHAT YOU WILL DO
Respond "always," "sometimes," or "rarely" to each of the statements below. Write your responses on a separate sheet of paper.

1 I try to think through problems. I think about the consequences before I act. I think about how my actions affect other people.

2 I am able to communicate well with my parents or other adults. I do some things alone or with friends that I used to do with my family.

3 I am able to list my four most important beliefs. I like who I am; I don't try to be something I'm not.

4 I listen to other people's ideas even when they are different than mine. I am concerned about problems in the world today.

5 I have one or two close friends with whom I can talk about almost anything.

6 I do some things alone or with friends that I used to do with my family.

Preparing for adulthood involves behaving in a more mature and responsible way. This activity will give you a chance to do some thinking like an adult and see how you feel about questions that you may face as an adult.

Odilon Dimier/PhotoAlto/Getty Images

WRAPPING IT UP
When you are finished, look at how many questions you answered "always" and "sometimes." This shows that you are already beginning to think and act in a mature and responsible way. As a class, brainstorm some other ways you can demonstrate mature and responsible behavior.

READING REVIEW

FOLDABLES and Other Study Aids

Take out the Foldable® that you created and any study organizers that you created.
Find a partner and quiz each other using these study aids.

LESSON 1 Changes During Puberty

BIG IDEA You will go through many physical, mental/emotional, and social changes during your teen years.

* Adolescence is a time during which the body grows and changes faster than at any time since infancy.
* Puberty is characterized by the production of hormones that send signals to different parts of your body to prepare for adulthood.
* Teens experience many physical, mental/emotional, and social changes during adolescence.

LESSON 2 The Male Reproductive System

BIG IDEA The male reproductive system produces cells that can join with cells from a female's body to produce a child.

* The reproductive system makes it possible for people to produce offspring, or children.
* The main purpose of the male reproductive system is to produce and release sperm.

LESSON 3 The Female Reproductive System

BIG IDEA The female reproductive system contains the organs that nurture a baby before birth.

* The female reproductive system has three main functions: to produce and store eggs, to create and nurture offspring, and to deliver a baby.
* The female reproductive system can be affected by a wide range of problems, ranging from minor to life-threatening.

LESSON 4 Infant and Child Development

BIG IDEA Each human being passes through several stages of development before reaching adolescence.

* Every human being starts out as a single cell that develops into a baby over the course of nine months.
* A human life can be divided into a series of eight stages, each with its own tasks to be mastered.
* Psychologist Erik Erikson's stages of development include infancy, early childhood, middle childhood, late childhood, adolescence, early adulthood, middle adulthood, and late adulthood.
* Childhood is the period of time from infancy through adolescence or the teen years.

LESSON 5 Staying Healthy as You Age

BIG IDEA Practicing good health skills will allow you to remain healthy throughout early, middle, and late adulthood.

* People continue to learn and develop throughout the various stages of adulthood.
* Your age in years may be different from the age suggested by either your physical health or your social behavior.
* Your age can be measured in at least three ways—in years, by how well your various body parts are working, or by your lifestyle and the connections you have with others.
* Staying active and involved can help older adults maintain good physical, mental/emotional, and social health in their later years.

 Review

 Web Quest

ASSESSMENT

Reviewing Vocabulary *and* **Main Ideas**

> ovaries
> reproduction

> puberty
> ovulation

> testes
> fertilization

> reproductive
> system

> adolescence
> hormones

›› On a sheet of paper, write the numbers 1–9. After each number, write the term from the list that best completes each statement.

LESSON 1

1. The stage of life between childhood and adulthood is _____.

2. _____ is the time when a person develops the physical characteristics of his or her gender.

3. During adolescence, certain _____ increase and send signals to your body to prepare for adulthood.

LESSON 2

4. Without the process of _____, humans could not produce children.

5. The pair of endocrine glands that produce sperm are the _____.

6. The body systems and structures that make it possible to produce children is called the _____.

LESSON 3

7. The release of one egg cell each month is _____.

8. _____ is the joining of a male sperm cell and a female egg cell to form a fertilized egg.

9. The _____ are the female glands that release mature eggs and produce hormones that control female sexual development.

›› On a sheet of paper, write the numbers 10–17. Write *True* or *False* for each statement below. If the statement is false, change the underlined word or phrase to make it true.

LESSON 4

10. A fetus differs from an embryo in that a fetus is <u>less</u> than 8 weeks old.

11. A <u>tissue</u> is a group of similar cells that do a particular job.

12. A <u>toddler</u> is an event that needs to happen in order for you to continue growing toward becoming a healthy, mature adult.

13. During <u>infancy</u>, you experience many physical, mental/emotional, and social changes that prepare you for adulthood.

LESSON 5

14. All adults age at <u>the same</u> times and in <u>the same</u> ways.

15. <u>Middle</u> adulthood is when most people retire from their jobs.

16. <u>Chronological age</u> is your age measured by your lifestyle and your connections with others.

17. The age determined by how well your body parts are working is your <u>biological</u> age.

✓ eAssessment

≫ Using complete sentences, answer the following questions on a sheet of paper.

Thinking Critically

18. INTERPRET How might a pregnant woman's health behaviors affect the health of her baby?

19. ANALYZE What are some ways an older adult can keep his or her health triangle balanced?

Write About It

20. EXPOSITORY WRITING Find out what organizations in your community pair teens and older adults. Then, write a short letter to the editor of your school paper describing these projects.

STANDARDIZED TEST PRACTICE

Reading/Writing
Read the prompts below. On a separate sheet of paper, write an essay that addresses each prompt. Use information from the chapter to support your writing.

1. Imagine that you are going to a party with friends. When you arrive, you notice that some of the guests are using alcohol and tobacco. How does using alcohol and tobacco introduce risk in a person's life? Write an essay describing how tobacco and alcohol use can damage your health and some ways to avoid them.

2. A person's reproductive system develops the ability to produce a child when they reach puberty. However, he or she is not ready to be a parent. Explain why an adolescent is unable to care for a child. What steps can an adolescent take to reduce the risk of becoming a teen parent?

Find out how much time you have to write your essay. Use this time wisely. Spend some of the time at the beginning of the test organizing your thoughts and planning out your essay. Save some time at the end of the test so you can revise and review your essay.

THE CYCLE OF LIFE

You experience change during every stage of your life.

ERIKSON'S STAGES OF DEVELOPMENT

 Feed me.

 I can do it!

 I'll wear this.

 Check it out!

 Who am I?

 Together!

 How can I help?

I rock!

Age	0–2 years	2–4 years	4–5 years	5–12 years	13–19 years	20–24 years	25–64 years	65+ years

ADOLESCENCE

The teen years are a time of rapid growth and development.

Physical changes

- Hormones
- Hair growth
- Acne

Mental/emotional changes

- Emotions
- Mood swings
- Attraction

Social changes

- Independence
- Awareness
- Relationships

STAGES OF LIFE

Infancy — Mom

Early childhood

Middle childhood

Late childhood

Late adulthood

Middle adulthood

Early adulthood

Adolescence

- Infants need to be fed and sheltered as they observe the world.

- Toddlers become more aware, start to explore, and learn to speak.

- Preschoolers ask questions, use their imaginations, and improve their physical skills.

- Between the ages of 6 and 11, the focus is on schoolwork, friends, and activities.

- During the teen years, the physical characteristics of adulthood begin to emerge.

- Young adults typically focus on education, careers, and forming close relationships.

- People in middle age often feel a need to achieve and be recognized.

- Older adults may pursue new activities and interests.

HYGIENE AND PERSONAL CARE

Hygiene is about keeping your body clean and healthy. Personal care involves practicing healthful behaviors that help you look and feel good.

HYGIENE

Keeping it clean

 Use deodorant

 Wash your hands

Healthy hair

 Wash your hair

 Brush or comb your hair

Teeth and gums

 Brush your teeth

 Floss your teeth

PERSONAL CARE

Healthy eyes

 Never look directly at the sun

 Wear sunglasses

Healthy ears

 Do not insert anything inside

 Avoid loud noises

Skin care

 You can use acne medication

Utilize deodorant soap

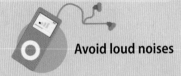 Use sunscreen

Nails

 Cut your nails

 File your nails

Facial and body hair

 You may choose to shave your body hair

Personal Health Care

LESSONS

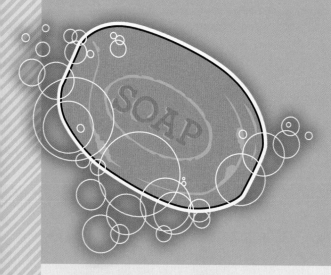

 **PREMIUM ONLINE RESOURCES** 〉

 Audio

 Videos

 Bilingual Glossary

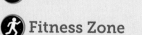 Fitness Zone

 Web Quest

 Review

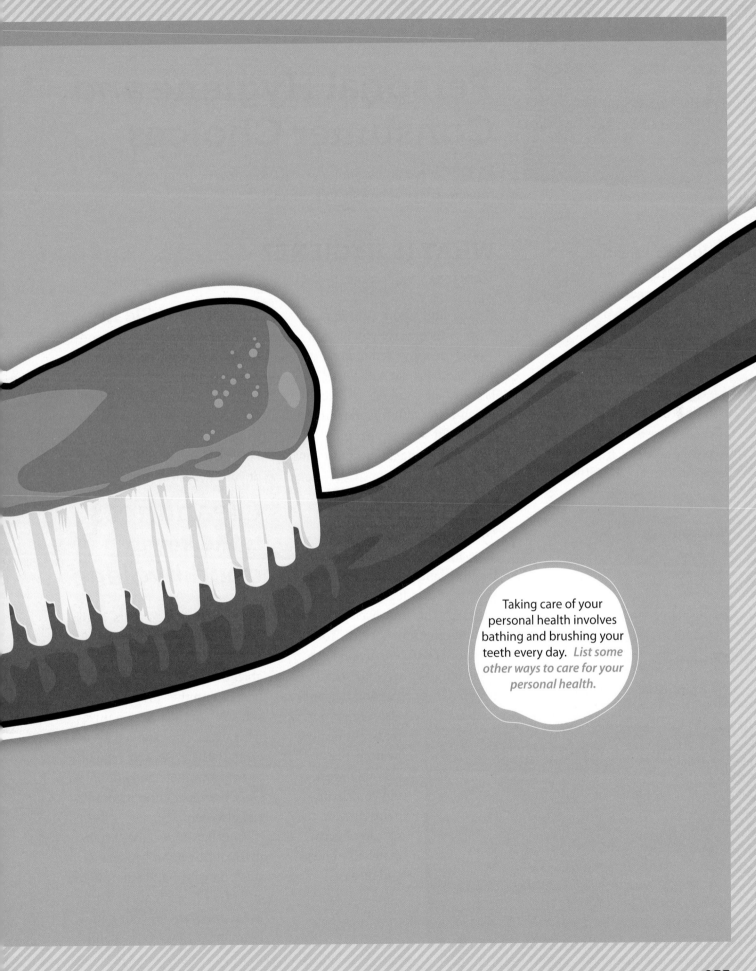

Taking care of your personal health involves bathing and brushing your teeth every day. *List some other ways to care for your personal health.*

Personal Hygiene *and* Consumer Choices

BIG IDEA It is important for you to know how to maintain good hygiene as well as be a smart consumer.

Before You Read

QUICK WRITE Write two ways that washing your hands with soap helps keep you and others healthy.

▶ Video

As You Read

 FOLDABLES Study Organizer

Make the Foldable® found in the FL pages in the back of the book to record the information presented in Lesson 1 about personal hygiene and consumer choices.

Vocabulary

› hygiene
› body odor
› consumer
› comparison shopping
› warranty

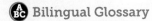 Audio

🄰🄱🄲 Bilingual Glossary

Developing Good Character

Responsibility Skilled consumers understand what they are buying. They also understand why they make certain choices. Being aware of advertising is an important skill. Ads make products seem appealing to get people to buy them. It is also important to be aware of a product's impact on the world around you. You can show responsibility by learning more about the things you buy.

WHAT IS HYGIENE?

MAIN IDEA Your personal hygiene affects all sides of your health triangle.

Think about your appearance. Have you bathed or showered today? Are your clothes neat and clean? Is your hair combed? Did you brush your teeth this morning? Caring for your appearance includes paying attention to your personal **hygiene,** or *cleanliness.*

Keeping your body clean is an example of good hygiene. Your hygiene and your appearance affect all three sides of your health triangle. When you look your best, you feel good about yourself. This improves your mental/emotional health. You are more confident around others, strengthening your social health. Of course, good hygiene also keeps your body physically healthy. For example, washing your hands helps prevent the spread of germs that cause illness and disease.

Keeping It Clean

Cleanliness is the key to good hygiene. Bathing or showering every day helps keep your skin and body clean. Use soap to wash away dirt, sweat, oils, and bacteria that collect on your skin. Having clean, healthy skin is part of your overall appearance.

Cleanliness is **the key** to *good hygiene.*

During the teen years, **body odor,** or a *smell from the body,* may become more noticeable. Body odor is caused by bacteria on the surface of your skin. When the bacteria mix with sweat, it creates a bad odor. Bathe or shower regularly to help eliminate body odor. Deodorants or antiperspirants can also help.

Another important part of maintaining good hygiene is washing your hands thoroughly and often. Use plenty of warm water and soap on your hands. Remember to wash the palms

P. Ughetto/PhotoAlto

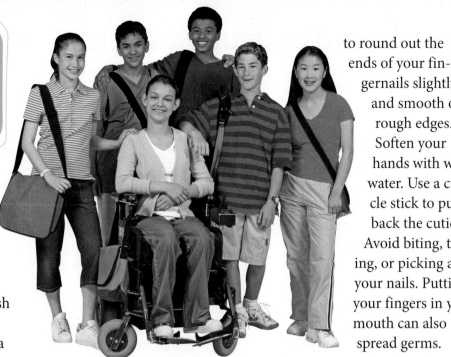

The changes your body undergoes during adolescence make hygiene especially important. *Explain how having good hygiene can make you feel more confident.*

and backs of your hands and between your fingers. Washing your hands helps prevent illness and the spread of germs. Always wash your hands before you prepare and eat food. Remember to wash your hands after you use the bathroom, play with pets, visit a sick person, or touch garbage or another source of germs.

⟩⟩⟩ Reading Check

IDENTIFY *What are two ways body odor can be eliminated?*

Facial *and* Body Hair

As you get a little older, you will develop facial and body hair. The growth of hair in new areas of your body, such as under your arms or on your face, is caused by hormones. As facial and body hair becomes thicker, males may choose to shave their facial hair. Females may choose to shave the hair on their legs or under their arms. Shaving your body hair is a personal choice.

Caring *for* Your Nails

Nails protect sensitive fingertips and the tips of toes. Your fingernails and toenails are made of a tough substance called keratin. Around the nail is a nonliving band of skin called the cuticle. Sometimes, the cuticle can

become torn or cut. Nail care includes caring for the cuticle. Common problems with fingernails and toenails include:

- **Hangnails.** A hangnail is a split in the cuticle near the fingernail's edge. You can treat hangnails by carefully trimming the skin. The cuticle should heal in a few days.
- **Ingrown toenails.** An ingrown toenail occurs when the toenail pushes too far into the skin along the side of the toe. This can result from trimming the nail on a curve rather than straight across or wearing shoes that are too tight. See a doctor if your toe becomes inflamed and sore, as it may be infected.

To care for your nails, wash your hands regularly. Use hand lotion to keep nails and skin moist. Trim your nails using a nail clipper or small scissors. Cut your nails straight across, so the nail is at or just beyond the skin. Use an emery board or nail file

to round out the ends of your fingernails slightly and smooth out rough edges. Soften your hands with warm water. Use a cuticle stick to push back the cuticle. Avoid biting, tearing, or picking at your nails. Putting your fingers in your mouth can also spread germs.

Washing your **hands** helps **prevent** illness.

Proper nail care keeps your nails looking clean and healthy. *Identify some steps you can take to improve the appearance of your nails.*

DEVELOPING CONSUMER SKILLS

MAIN IDEA > Knowing how to evaluate products and services will help you become a smart consumer.

Have you ever shopped for personal care items? Soap, shampoo, deodorant, toothpaste, shaving cream—you have many different sizes, styles, and brands of products to choose from. As a **consumer,** or *a person who buys products and services,* you have likely walked through store aisles many times. How do you decide which items to buy? How do you know what will be best for you?

Many factors can influence your choices as a consumer. Cost is likely to be a factor. Another is personal taste. You may simply prefer one brand over another. Environmental impact can also be an influence. You may prefer a product that is all natural, organic, or recyclable.

Another major factor is the media. The Internet, TV and radio stations, newspapers, and magazines feature advertising.

Consumers are not always aware of the influence of advertising. Advertisements are designed to make products seem more appealing to consumers. You may not realize it, but you might be tempted to buy an item just because it has a catchy ad.

>>> **Reading Check**

LIST *What are some factors that influence your choices as a consumer?*

Health SKILLS ACTIVITY

Analyzing Influences

Persuasive Advertising

Advertisers often use the following positive and negative techniques to persuade you to buy their items.

* Hidden messages. Sometimes messages are in the form of pictures. A picture may show attractive people smiling when they use a product. This is telling you that the product will make you happier and healthier.

* Comments by previous users. These ads show people who claim to have used the company's product and gotten great results. These people may be paid actors.

* Celebrity backing. Popular actors, athletes, or celebrities promote some products, making them seem glamorous. Remember that these people are paid to promote these products.

Find an example of an ad that uses one of the techniques described here. How might these ads influence your decision to try the product? Discuss your findings with the class.

With A Group

©McGraw-Hill Education/Robert Manella

Shopping Smart

The first step in becoming a smart shopper is to understand what you are buying. With many products, this means reading the product label. By law, labels for foods, medicines, and many health products must include certain kinds of information. A label tells you the product name and what the product is intended to do. The label also gives directions, lists the ingredients, and may feature warnings.

> The first step in becoming a *smart shopper* is to **understand what you are buying.**

Directions tell you how much of the product to use and how often to use it. Use a product only as directed. A product may not be effective if it is not used as directed. If any problems occur when you use a product, stop using it immediately. The product may contain an ingredient that is causing the problem.

Product labels can also help you compare similar products. Different brands of the same type of product may contain slightly different ingredients. The same brand may feature a range of products designed for slightly different uses. Also, product labels may make similar claims, but they may not do exactly the same thing.

Consumers can sometimes be persuaded to try products that have fancy packaging with catchy phrases. Packages are designed to make you want to buy the product. They may feature bright colors, attractive logos, and certain words designed to attract your attention in an ad or on a store shelf. Companies spend lots of money on designing packaging that will help sell their products.

When you compare two or more similar products, you are comparison shopping. Comparison shopping involves *collecting information, comparing products, evaluating their benefits, and choosing products with the best value.* As you compare products, consider the benefits of one brand over another. Which brand offers more of what you are looking for? Does one brand fit your needs better than another? You should also consider the brand's reputation. Do you know anyone else who has used and liked it? How about the cost? Some brands may have the same volume and ingredients but different prices.

Finally, check to see if the item carries a warranty. A warranty is *a written agreement to repair a product or refund your money if a product does not function properly.* A warranty shows that a store or company stands behind its product.

>>> Reading Check

LIST *What factors are evaluated in comparison shopping?* ∎

>>> After You Read

1. **DEFINE** What does hygiene mean?
2. **EXPLAIN** How does good hygiene affect all parts of your health triangle?
3. **IDENTIFY** What are two kinds of information found on health product labels?

>>> Thinking Critically

4. **APPLY** What can you do to care for your nails and cuticles?
5. **SUGGEST** Turtles and other reptiles carry salmonella. This bacterium often makes people sick if they ingest it. What would be a good way to make sure you do not get sick from your friend's pet turtle?

>>> Applying Health Skills

6. **ANALYZING INFLUENCES** Imagine that you are selecting a deodorant. Compare your wants and needs to the product's claims. What other factors would influence your decision?

 Review

 Audio

Taking Care *of* Your Skin *and* Hair

BIG IDEA › Caring for your skin and hair is important to your overall physical health.

>>> **Before You Read**

QUICK WRITE Write a short description of how you care for your skin and hair each day.

 Video

>>> **As You Read**

STUDY ORGANIZER Make the study organizer found in the FL pages in the back of the book to record the information presented in Lesson 2.

>>> **Vocabulary**

› epidermis
› dermis
› subcutaneous layer
› acne
› dermatologist
› ultraviolet (UV) rays
› melanin
› dandruff

 Audio

 Bilingual Glossary

HEALTHY SKIN

MAIN IDEA › The skin is an organ with several important functions.

When you think of an organ, you might think of a body part such as your heart or your stomach. You might overlook your body's largest organ, which is not found beneath your skin—it is your skin! The average person's skin has a surface area of 2 m² and weighs about 3 kilograms.

Your skin is the **largest organ** in your body.

Your skin has many jobs. It acts as a waterproof shield that defends against germs and keeps them from getting into your body. Your skin helps maintain your body temperature. When you sweat, your skin gets rid of water and salts and cools your body. Your skin allows you to feel and sense pressure, temperature, and pain. Your skin also uses energy from sunlight to make vitamin D, which helps keep bones and teeth healthy.

The skin is composed of three main layers. The epidermis is *the outermost or top layer of the skin.* The palms of your hands and the soles of your feet are where the epidermis is thickest on your body. This outer layer of skin is at its thinnest where it covers the outside of your eyelids.

The dermis is *the skin's inner layer,* underneath the epidermis. The dermis contains nerve endings, blood vessels, and oil and sweat glands. The dermis is the part of your skin that enables your sense of touch. This layer is also the part of your skin that helps keep you cool by getting rid of waste when you sweat.

The innermost layer of skin, the subcutaneous layer, is *a layer of fat cells.* This layer connects your skin to muscle and bone. The subcutaneous layer is also the part of the skin that holds the roots of your hair follicles.

>>> **Reading Check**

NAME *What are the three layers of the skin?*

©McGraw-Hill Education/Ken Karp

CARING FOR YOUR SKIN

MAIN IDEA You can take several steps to help keep your skin healthy.

Keeping your skin healthy can help you feel, look, and smell good. Having healthy skin is also an important part of your overall physical health. Try these tips for taking care of your skin:

- Bathe or shower every day. Use soap to wash away dirt, sweat, oils, and bacteria that collect on your skin.
- If you have dry or delicate skin, you may find it helpful to moisturize it with lotion. Dry or cracked skin can itch or become irritated.
- Limit the amount of time you spend in the sun, especially between the hours of 10:00 a.m. and 4:00 p.m. This is when the sun's UV rays are strongest and most direct.

Whenever you do spend time in the sun, wear protective clothing and apply sunscreen with a sun protection factor (SPF) of 15 or higher. Reapply sunscreen about every two hours and after swimming.

Healthy skin helps you **feel, look,** and **smell good.**

- Avoid tanning beds, which can be harmful or damaging to the skin in ways similar to exposure to the sun.
- Avoid tattoos and piercings. Permanent body decoration can put you and your skin at risk for disease and scarring.

You will see advertisements for many types of skin care products. Your family and friends may make recommendations. Entire store aisles are filled with various cleansers, lotions, creams, and other items that promise to make your skin healthier and look better.

You may be tempted to try many of these products. You may find some that work as promised, while others appear to have no effect. However, remember to use your skills as a wise consumer. You can make informed choices about the types of products to buy and use on your skin.

> Your skin is a very complex body organ. It has many parts. *Describe the function of the sweat glands.*

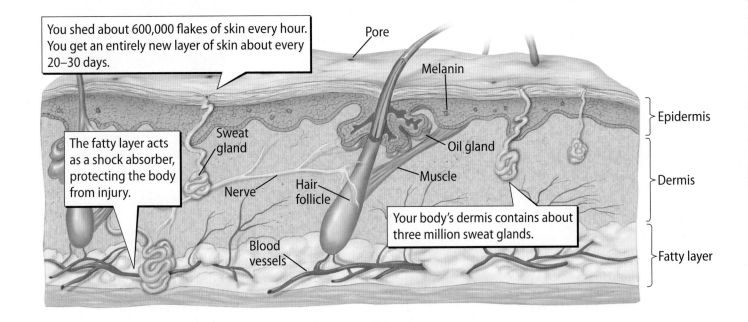

You shed about 600,000 flakes of skin every hour. You get an entirely new layer of skin about every 20–30 days.

The fatty layer acts as a shock absorber, protecting the body from injury.

Your body's dermis contains about three million sweat glands.

Pore

Melanin

Sweat gland

Oil gland

Muscle

Nerve

Hair follicle

Blood vessels

Epidermis

Dermis

Fatty layer

Skin Problems

Did you ever wake up, look in the mirror and see a huge pimple? One skin problem experienced by many teens and some adults is <u>acne</u>, which is *a skin condition caused by active oil glands that clog hair follicles.* The openings of hair follicles onto the skin are called pores. Bacteria gather in the clogged pores, making them swell up. The figure on this page shows how acne forms.

Did you ever **wake up,** look in the mirror, and **see** a *huge pimple?*

If you get acne, avoid picking it or try to pop the pimple. In fact, try not to touch your skin at all. Instead, wash the area with a mild soap and warm water. Be gentle, and don't scrub too hard. Washing helps remove the dirt, oil, sweat, or makeup that can cause pores to clog. Over-the-counter acne medications can often help clear up breakouts.

You will find many different acne medications on the market. Keep in mind the guidelines for being a good comparison shopper you learned in the previous lesson. If you have a bad case of acne, you may need to see a <u>dermatologist,</u> *a doctor who specializes in problems of the skin.*

Viruses can also affect the skin. Cold sores and warts are both caused by viruses. Both can be treated but are contagious. Contagious means the virus can be spread to others. A person with a wart or cold sore should avoid skin contact with others. Washing the hands thoroughly and often is also important.

> **》》 Reading Check**
>
> **DESCRIBE** *How can bacteria and viruses affect the skin?*

Body Piercing *and* Tattooing

For some adults, piercings or tattoos are ways to express themselves. However, unlike different clothes or hair styles, body piercings and tattoos are permanent.

Both carry potential health risks because they break the physical barrier of the skin. This can allow germs to enter the body. Germs that are passed along by dirty tattoo or piercing needles can cause infection.

If the needles used for tattooing are not sterile, they can spread bacteria and viruses. Piercings can also spread disease and cause scarring. An oral piercing can damage your mouth and teeth. A tattoo or piercing may also impact your social health. They can limit future opportunities and relationships.

Additionally, many states in the U.S. have limits on the age at which a person can get a tattoo or body piercing. In many states, teens under the age of 18 need parental approval to get a tattoo or body piercing. This typically also applies to ear piercings.

Whiteheads, blackheads, and pimples are all forms of acne that commonly affect teens. *Identify the structure in the skin that becomes clogged to form acne.*

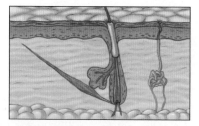

Whiteheads
A whitehead forms when a pore gets plugged up with the skin's natural oils and dead skin cells.

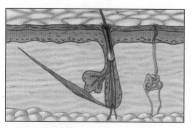

Blackheads
When a whitehead reacts to the air and darkens, it becomes a blackhead.

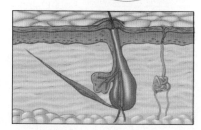

Pimples
A clogged follicle can burst, which lets in bacteria. The follicle becomes infected and a pimple forms.

Sun Damage

Factors in the environment can affect your personal health. For example, the sun gives off **ultraviolet (UV) rays,** *an invisible form of radiation that can enter skin cells and change their structure.* Sunburn happens when UV rays damage skin cells. You might think soaking up lots of UV rays will give you a healthy-looking tan. In reality, a tan indicates that skin damage has occurred. One way that the result of tanning will show up on your skin is through wrinkles, or premature aging of the skin. Even worse, too much time in the sun increases the risk of skin cancer.

In **reality,** a **tan** indicates that *skin damage* has occurred.

You can protect yourself from UV rays by limiting sun exposure and using sunscreen. Check the weather forecast for the UV Index. This is a scale that categorizes levels of UV radiation reaching Earth's surface.

Special cells in the epidermis make **melanin,** *the substance that gives skin its color.* Darker skin has more melanin than paler skin. Melanin can block some, but not all, UV rays from reaching the lower layers of skin.

What about indoor tanning beds? Many people use these devices to stay tan all year round. However, tanning beds also emit UV rays. They can damage the skin, lead to skin cancer, and injure the eyes and immune system.

Some people may choose spray tans to make the skin look darker. Spray tans are safer than being exposed to UV rays through direct sunlight or tanning beds. However, a spray tan does not offer any additional protection from the sun's UV rays.

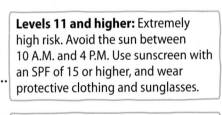

UV Index

Levels 11 and higher: Extremely high risk. Avoid the sun between 10 A.M. and 4 P.M. Use sunscreen with an SPF of 15 or higher, and wear protective clothing and sunglasses.

Levels 8 through 10: Very high risk. Avoid the sun between 10 A.M. and 4 P.M. Use sunscreen with an SPF of 15 or higher, and wear protective clothing and sunglasses.

Levels 6 and 7: High risk. Cover up, wear a hat with a wide brim, and use sunscreen with an SPF of 15 or higher.

Levels 3 through 5: Moderate risk. Stay in the shade around midday.

Level 2 or below: Low risk. Wear sunglasses on bright days, cover up, and use sunscreen.

11+
10
9
8
7
6
5
4
3
<2

Ultraviolet rays from the sun can harm your skin. *How can you protect yourself from ultraviolet rays and reduce your risk for developing skin cancer?*

Health SKILLS ACTIVITY

Communication Skills

Is *Tanning* Worth the *Risk*?

Kaitlyn walks toward the pool. She has always taken pride in her deep tan. Lately, however, she's been thinking about her aunt. After years of sun exposure, her aunt is undergoing skin cancer treatments. Now Kaitlyn doesn't think tanning is such a good idea.

"C'mon, Kaitlyn!" says her friend Sabrina, motioning to Kaitlyn. "The pool opens in ten minutes. Let's grab the best spots in the sun. School starts again in a month, so we have to make the most of every sunny day!"

Kaitlyn hesitates. She'd rather find a spot in the shade. She wants to convince Sabrina to do the same, but she thinks her friend may not agree.

In a paragraph or two, write your thoughts about what Kaitlyn might say to change her friend's mind about tanning.

HEALTHY HAIR

MAIN IDEA ⟩ Hair grows from follicles in the dermis and is made of keratin.

You might spend a lot of time and money to make your hair look thick or to keep it from getting frizzy. The roots of hairs are in the dermis of your scalp. The scalp is the skin beneath the hair on your head. Hair grows from follicles in the dermis. As new hair cells are formed, old ones are forced out. Your hair contains a protein called keratin, the same substance that forms fingernails and toenails. Keratin is the substance that gives hair strength and allows it to bend and blow in the wind without breaking.

The part of the hair that you can see is the shaft. The shape of the hair shafts determines whether your hair is wavy, curly, or straight. As is the case with living skin, hair gets its color from the pigment melanin. The color of your hair is determined by heredity. You inherited genes from your parents that determined your hair color.

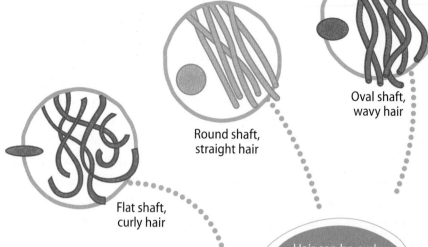

Flat shaft, curly hair

Round shaft, straight hair

Oval shaft, wavy hair

Hair can be curly, straight, or wavy depending on the shape of the hair shaft. *Identify the type of hair a person will have if his or her hair shafts have a round shape.*

Hair *and* Scalp Problems

Two conditions that can affect the health of your hair are dry or oily scalp. Either problem can be corrected by choosing the right shampoo. Read the label carefully. Different products are made for dry, oily, and normal hair. Chlorine in swimming pools can dry out your hair and even cause it to change color. Special shampoos can remove chlorine.

Another common scalp problem is dandruff, *a skin condition with many dead skin cells flaking off the outer layer of the scalp.* Symptoms of dandruff include itchy scalp and flaking of the scalp. Dandruff can be caused by dry skin or certain skin diseases. Washing your hair regularly will help control itching of the scalp and flakes of dandruff.

If regular hair washing does not work, using a special medicated shampoo can also help control and prevent dandruff.

Sometimes, an itchy scalp is caused by head lice. Lice are tiny insects that live in hair and on the scalp and feed on human blood. Having lice is not an indication that a person is unclean. However, lice spread easily from one person to another.

To help prevent lice from spreading, avoid sharing hats, combs, and brushes. If you get head lice, use a medicated shampoo that is developed for treating lice. You will also need to wash all your bedding, towels, combs, brushes, and clothing. Since lice spread easily, everyone else in your home will need to take these same steps.

Myth vs. Fact

Myth: Head lice can hop or fly from one person to another.

Fact: The head louse (Pediculus humanus capitis) cannot fly or hop. It moves by crawling. The most common way that head lice are transmitted from one person to another is through direct contact, such as head-to-head contact. Less commonly, transmission occurs through clothing, such as hats, scarves, and coats, or personal items, such as combs, brushes, and towels.

Caring *for* Your Hair

Keep your hair healthy by washing it regularly with a gentle shampoo and using conditioner if needed. Keep in mind that when shopping for shampoos and conditioners, you will find many different types and brands on store shelves. Use the comparison shopping skills you have learned to choose the hair care products that are best for you. If possible, let your hair dry naturally. If you use a blow dryer, use low heat. Styling irons and high heat from hair dryers can make hair dry, brittle, and faded.

Brushing or combing your hair helps you improve your appearance. It also helps keep your hair healthy and whole. Brushing and combing removes dirt from your hair and scalp in between shampooing. It also helps spread natural scalp oils down along the hair shaft. The natural oils produced in the scalp help keep your hair strong and in good condition.

>>> **Reading Check**

DESCRIBE *What are two problems that can affect the scalp and hair?* ■

THINKSTOCK /age fotostock

Brushing or combing your hair regularly can keep it looking healthy. *List some other ways you can take care of your hair.*

>>> **After You Read**

1. **DEFINE** Explain the meanings of the terms epidermis and dermis.
2. **IDENTIFY** Name three functions of the skin.
3. **EXPLAIN** What protein does hair contain? How does hair grow?

>>> **Thinking Critically**

4. **APPLY** A friend of yours is planning to get a pierced lip. What potential risks does this plan involve?
5. **EVALUATE** Your friend has told you that for the last several days her scalp has been itchy. This is not normal for her. What are some possible causes for this symptom? For each cause, suggest a course of action your friend could take.

>>> **Applying Health Skills**

6. **PRACTICING HEALTHFUL BEHAVIORS** Tisha's friends have invited her to join them at the beach. They plan on lying out in the sun to tan. How can Tisha protect her skin from the sun's UV rays?

⟳ Review

⏽ Audio

Caring *for* Your Mouth *and* Teeth

BIG IDEA Your teeth perform important functions and keeping them healthy is part of responsible healthful behaviors.

 Before You Read

QUICK WRITE Describe the steps you take to keep your teeth and gums healthy.

▶ Video

 As You Read

STUDY ORGANIZER Make the study organizer found in the FL pages in the back of the book to record the information presented in Lesson 3.

 Vocabulary

› plaque
› tartar
› gingivitis
› fluoride
› orthodontist

🔊 Audio

🅰🅱🅲 Bilingual Glossary

YOUR TEETH AND GUMS

MAIN IDEA Your teeth have several important functions.

Your teeth are vital to your health and appearance. They allow you to chew and grind food. They aid in forming certain speech sounds. Your teeth help give structure to your mouth.

Each tooth has a root that goes into your jaw. Roots are surrounded by pink flesh called gums. About three-fourths of each tooth is located below the gum line. The top of the tooth is the crown, which is covered with a layer of hard, white enamel. The neck of a tooth connects the crown to the root.

> *Your teeth* are vital to your **health** and **appearance.**

Tooth *and* Gum Problems

Proper care of your teeth and gums can prevent tooth decay. The process of tooth decay begins with **plaque,** which is *a thin, sticky film that builds up on teeth, leading to tooth decay.*

Bacteria in plaque feed on the carbohydrates—sugars and starches—in the foods you eat. These bacteria produce acids that can break down tooth enamel and lead to cavities, or holes in your teeth. The germs in plaque can also cause bad breath.

If plaque is not removed, it hardens and becomes **tartar,** *a hardened plaque that hurts gum health.* You cannot brush away tartar. A dentist must remove it with special tools that clean and polish the teeth.

The gums serve as anchors for your teeth. The teeth you can see in a mirror are only the top parts of your teeth. Most of each tooth is rooted in your gums. A common gum problem is **gingivitis,** which is *a common disorder in which the gums are red and sore and bleed easily.* If it is left untreated, gingivitis can lead to tooth loss or infection.

 Reading Check

EXPLAIN *How does tartar form?*

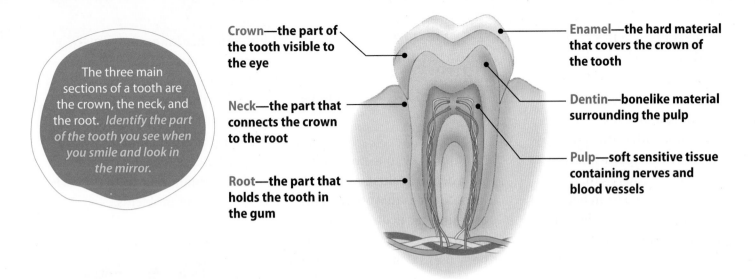

The three main sections of a tooth are the crown, the neck, and the root. *Identify the part of the tooth you see when you smile and look in the mirror.*

Crown—the part of the tooth visible to the eye

Neck—the part that connects the crown to the root

Root—the part that holds the tooth in the gum

Enamel—the hard material that covers the crown of the tooth

Dentin—bonelike material surrounding the pulp

Pulp—soft sensitive tissue containing nerves and blood vessels

KEEPING TEETH AND GUMS HEALTHY

MAIN IDEA ⟩ Ways to keep your teeth and gums healthy include brushing and flossing every day and having dental checkups twice a year.

Two important keys to having healthy teeth and gums are proper brushing and flossing. Brush after every meal (or at least twice a day) with toothpaste and a soft-bristled toothbrush. Brushing cleans the teeth and stimulates the gums. Brushing your teeth after eating also removes plaque from the surface of the teeth, before bacteria can produce acid that harms the teeth. It is important to brush to remove plaque before it becomes tartar, which only a dentist can remove.

Whenever you do brush your teeth, be sure to use a toothpaste or mouthwash that contains **fluoride,** which is *a chemical that helps prevent tooth decay.*

In many areas, fluoride is added to tap water. The fluoride in the water also protects your teeth. However, fluoride from water is not a replacement for brushing or flossing. Remember to use the consumer skills you have learned to choose the type and brand of toothbrush, toothpaste, and mouthwash that best fit your needs.

Brushing your teeth after eating can also help prevent halitosis, or bad breath. Using mouthwash can also help kill bacteria and control bad breath. Halitosis can be caused by eating certain foods, poor oral hygiene, tobacco or alcohol use, bacteria on the tongue, decayed teeth, and gum disease.

⟩⟩⟩ **Reading Check**

EXPLAIN *Why should you brush your teeth two to three times each day?*

Stage 1
The bacteria in plaque combine with sugars to form a harmful acid. This acid eats into the enamel, the hard outer surface of the tooth.

Stage 2
Repeated acid attacks on the enamel cause a cavity, or hole, to form.

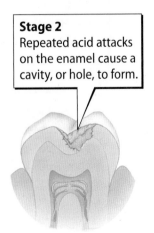

Stage 3
If the cavity grows and reaches the sensitive inner parts of the tooth, it can cause a toothache.

Tooth decay occurs when bacteria in the mouth convert food into acids that dissolve enamel. *Explain what can happen if tooth decay goes untreated.*

In addition to brushing your teeth, flossing is also important. Dental floss is a thin plastic, nylon, or coated silk thread. When you slide it between each tooth after brushing, floss removes food particles from the sides of your teeth. It also removes plaque from between the teeth and under the gum line that a toothbrush cannot reach. Flossing keeps gums healthy and prevents gum disease. Flossing also helps to clean underneath braces. The figure at the bottom of this page shows how to brush and floss to prevent tooth decay.

Flossing keeps gums **healthy** and prevents **gum disease.**

A third key to maintaining your dental health is your diet. Choose foods that are high in the mineral calcium, such as yogurt, cheese, and milk. Your body uses calcium to build teeth and bones. You should also limit foods that are high in sugar, which can cause tooth decay. If you do eat sugary foods, brush your teeth as soon as you can.

Another important way to protect your teeth and gums is to have dental checkups twice a year. The dentist or dental hygienist will clean your teeth to help prevent tooth decay and gum disease. The dentist will also examine your teeth for cavities or other problems.

Brushing and flossing regularly helps prevent tooth decay and other health problems. *Explain why you should brush your teeth at least three times a day.*

Brush the outer surfaces of your upper and lower teeth. Use a combination of up-and-down strokes and circular strokes.

Thoroughly brush all chewing surfaces with a soft-bristle brush to protect your gums.

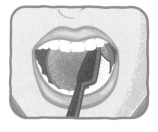

Brush the inside surfaces of your upper and lower teeth.

Brush your tongue and rinse your mouth.

Take about 18 inches of floss and wrap each end around the middle finger of each hand.

Grip the floss firmly between your thumb and forefinger.

Slide the floss back and forth between teeth toward the gum line until it touches the gum line.

Curve the floss around the sides of each tooth. Keep sliding the floss back and forth gently as you move it up and down.

Other Dental Issues

Other problems of the teeth and mouth include misalignment and impacted wisdom teeth. Improperly aligned upper and lower teeth are also know as having a "bad bite." This can be caused by crowded teeth, extra teeth, thumb sucking, injury, or heredity. If left untreated, a bad bite can result in lost teeth.

Your dentist will also check for wisdom teeth, or extra molars in the back of your mouth. These typically appear in your late teens. An impacted wisdom tooth can cause swelling and be very painful. Wisdom teeth may crowd or push on other teeth or become infected.

If you have problems when your wisdom teeth come in, your dentist may have to do surgery to remove them.

If your teeth need straightening, your dentist may refer you to an orthodontist, who is *a dentist who prevents or corrects problems with the alignment or spacing of teeth.* Orthodontists specialize in correcting irregularities of the teeth and jaw. An orthodontist may apply braces in order to straighten your teeth.

Braces can help improve your appearance and make your teeth easier to clean. Some teens may not like having braces on their teeth, because they do change your appearance for a while. Braces can be made of either metal or a clear material. Clear braces have become popular because they are less noticeable. Many adults choose this option if they did not have their teeth straightened earlier in life.

BananaStock/Alamy

> ### ⟩⟩⟩ Reading Check
>
> **EXPLAIN** *How is your diet related to the health of your teeth and gums?* ▪

Your smile is an important part of your appearance. *List some steps you can take to help keep your teeth healthy.*

⟩⟩⟩**After You Read**

1. **VOCABULARY** What is plaque? Why should plaque be removed?
2. **LIST** What are the functions of the teeth?
3. **DESCRIBE** Name two healthful behaviors that keep your teeth and gums healthy.

⟩⟩⟩**Thinking Critically**

4. **HYPOTHESIZE** What can happen to your teeth and gums if you do not floss regularly?
5. **EVALUATE** Why is good gum care important for the health of your teeth?

⟩⟩⟩**Applying Health Skills**

6. **ADVOCACY** Create a booklet that explains the importance of proper tooth and gum care. Include original art, if you like, with step-by-step instructions. Distribute copies to students in other classes.

 Review

 Audio

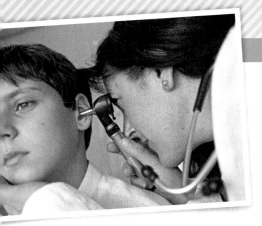

Protecting Your Eyes *and* Ears

BIG IDEA Caring for your eyes and ears will keep them healthy.

> ## Before You Read

QUICK WRITE Make a list of five ways you used your sense of hearing today. Describe how these activities would be different if you could not hear very well.

 Video

> ## As You Read

STUDY ORGANIZER Make the study organizer found in the FL pages in the back of the book to record the information presented in Lesson 4.

> ## Vocabulary

> farsightedness
> nearsightedness
> astigmatism
> optometrist
> ophthalmologist
> decibel
> tinnitus

 Audio

Bilingual Glossary

Myth vs. Fact

Myth: Sun exposure will not damage my eyes.
Fact: Long-term exposure to UV rays can lead to the development of cataracts, an eye disease in which the lens becomes cloudy. Cataracts can impair vision and cause blindness. Surgery can replace the affected lens with an artificial lens.

HEALTHY EYES

MAIN IDEA The function of your eyes is to focus light on the retina, which sends information to your brain about what you see.

Your eyes are your windows to the world. Like a camera, your eyes focus light in order to give your brain a picture of the world around you. Eyes allow you to see shapes, colors, and motion. Each part of the eye plays an important role in how you see. Light enters the eye through the pupil and the lens focuses the light on the retina. The retina contains nerve cells that send signals to the brain though the optic nerve.

Vision Problems

Two common vision problems are farsightedness and nearsightedness. **Farsightedness** is *a condition in which faraway objects appear clear while near objects look blurry.* If someone is farsighted, the words on this page may look unclear. That person might need reading glasses to see this page clearly.

However, if the same person looked at a sign across the room, the words would be in focus. The opposite is true for a person who is nearsighted. **Nearsightedness** is *a condition in which objects that are close appear clear while those far away look blurry.* A nearsighted person can clearly see objects only if they are up close.

Your eyes are your **windows to the world.**

A third common eye problem is astigmatism, *an eye condition in which images appear wavy or blurry.* Eye problems are usually corrected with eyeglasses or contact lenses. Both help the lens of the eye focus light on the retina. An eye doctor can determine if you need corrective lenses.

> ## Reading Check

IDENTIFY *What are the two most common vision problems?*

Pixtal/AGE Fotostock

CARING FOR YOUR EYES

MAIN IDEA You can keep your eyes healthy by taking care of them.

Your eyes need light to see. However, too much light can hurt them. The sun can be very hard on your eyes. For example, you should never look directly at the sun. In addition, UV rays can damage eye cells. You can solve this problem by wearing sunglasses that block UV rays.

Your eyes need light to see.

Sitting too close to the television can cause eyestrain and headaches. Sit at least six feet away. If you get eyestrain while sitting at a computer, change the monitor position to cut down on the glare. Reading in dim light can also cause eyestrain. Be sure you have adequate lighting while you read, work, or watch TV.

Light should come from above your reading material. Taking frequent breaks while using your computer, watching TV, or reading can help prevent eyestrain.

Protect your eyes when you are doing yard work, handling power tools, or playing sports that involve flying objects. Wear glasses or goggles designed for these activities. Wear goggles when you work with strong chemicals, such as when you are working in a science laboratory.

If something gets in your eye, do not rub it. Instead, flush particles out with clean water or eye drops. Sharing eye makeup or eye care products can spread germs. Diseases such as conjunctivitis, or pinkeye, can make eyes red and painful. If your eyes hurt for more than a short time, see a doctor right away.

Keeping your eyes healthy includes getting regular exams or vision screenings. If you already wear glasses or contacts, you should regularly visit an **optometrist,** *a health care professional who is trained to examine the eyes for vision problems and prescribe corrective lenses.* Schedule a visit at least once a year. An optometrist may do initial tests for eye diseases. He or she may then send you to an **ophthalmologist,** *a physician who specializes in the structure, functions, and diseases of the eye.* An ophthalmologist also fits people with corrective lenses.

⟩⟩⟩ Reading Check

EXPLAIN *Why is it important to protect your eyes while in the sun?*

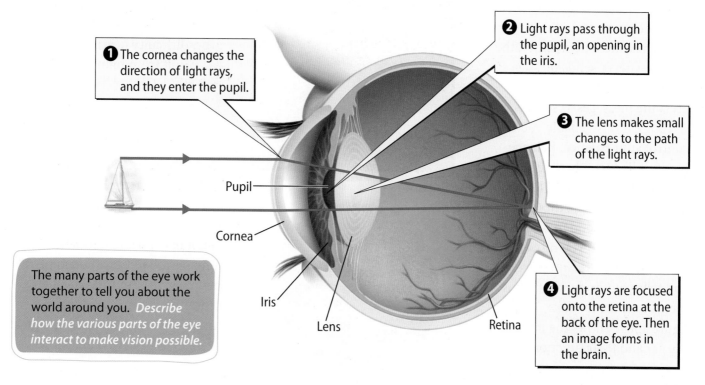

❶ The cornea changes the direction of light rays, and they enter the pupil.

❷ Light rays pass through the pupil, an opening in the iris.

❸ The lens makes small changes to the path of the light rays.

❹ Light rays are focused onto the retina at the back of the eye. Then an image forms in the brain.

Pupil

Cornea

Iris

Lens

Retina

The many parts of the eye work together to tell you about the world around you. *Describe how the various parts of the eye interact to make vision possible.*

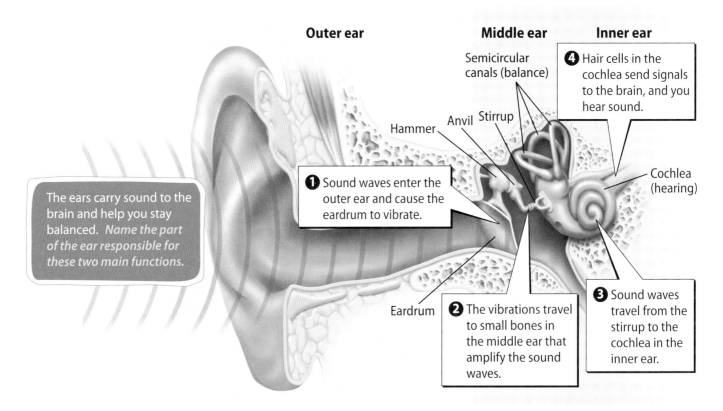

Outer ear | **Middle ear** | **Inner ear**

Semicircular canals (balance)

❹ Hair cells in the cochlea send signals to the brain, and you hear sound.

Hammer Anvil Stirrup

❶ Sound waves enter the outer ear and cause the eardrum to vibrate.

Cochlea (hearing)

The ears carry sound to the brain and help you stay balanced. *Name the part of the ear responsible for these two main functions.*

Eardrum

❷ The vibrations travel to small bones in the middle ear that amplify the sound waves.

❸ Sound waves travel from the stirrup to the cochlea in the inner ear.

HEALTHY EARS

MAIN IDEA Ears gather sound and help you stay balanced.

Like your eyes, your ears allow you to receive information. Your ears gather sound. Sound waves travel through the ear canal and strike the eardrum, causing it to vibrate. The vibrations cause tiny hairs to move in the inner ear. When these hairs move, the auditory nerve sends messages to the brain. Your brain interprets the messages as sounds.

Your ears also enable you to control your balance. Balance is controlled by tube-like structures in the inner ear. Fluid and tiny hair cells inside these tube-like structures send messages to your brain when you move. The brain interprets the messages and tells your body what adjustments it needs to make. The figure above shows the parts of the ear and their functions.

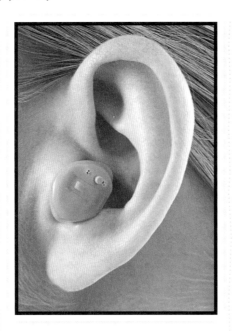

People with hearing loss may wear hearing aids. These increase the loudness of sounds. *Identify a common way that people with hearing loss can communicate.*

Ear Problems

Ear infections are the most common ear problems. Germs can spread into the ear from the nose or throat. Ear infections can be treated by a doctor.

The most serious ear problem is hearing loss. Deafness, or great difficulty hearing sounds, can result from injury, disease, or birth defect. *The unit for measuring the loudness of sound* is a **decibel.** Normal conversation is about 60 decibels. Noises above 85 decibels can damage the ear and cause hearing loss.

A constant ringing in your ears is called **tinnitus.** This is often a result of exposure to noise for a long period of time. For some people with tinnitus, the ringing sensation is constant. Ongoing tinnitus is an early warning sign of inner-ear nerve damage.

Image Source./Getty Images

CARING FOR YOUR EARS

MAIN IDEA Protecting your ears from loud sounds is the best way to care for them.

To care for your ears, wash and dry them regularly. Your ears produce earwax, which helps trap dirt and carry it out of the ear opening, or auditory canal. Use a wet washcloth to wipe off dirt and earwax on the outside of your ears. Do not insert anything inside your auditory canal. If you get water in your ears, you can use special eardrops that will help dry out the water and prevent an infection. You can protect your ears from the cold by wearing a hat or scarf that covers your ears, or earmuffs. See a doctor if any parts of your ears hurt or become infected.

Protect your ears from loud sounds.

The best way to care for your ears is to protect them from loud sounds. Repeated exposure to sounds above 85 decibels is can cause hearing loss. When listening to music, especially with earbuds or headphones, keep the volume down. How can you tell how loud is too loud? If other people can hear sound coming from your headphones or earbuds, you may have your music turned up too loud.

>>> Reading Check

EXPLAIN *Why is it important to limit your exposure to loud sounds?* ■

Music at a rock concert might be 140 decibels or louder. Earplugs let you enjoy loud music without damaging hearing. *List some other factors that might damage your hearing.*

©McGraw-Hill Education/Ken Karp

>>> After You Read

1. **VOCABULARY** What is the difference between an optometrist and an ophthalmologist?
2. **DESCRIBE** List three habits that you would recommend to promote eye health and protect vision.
3. **EXPLAIN** What happens when sound waves reach your outer ear?

>>> Thinking Critically

4. **ANALYZE** Tracy is nearsighted. What type of vision correction do you think might be best for Tracy? Explain your reasoning.
5. **EVALUATE** A woodshop produces noise that reaches 100 decibels. What can you do to reduce the risks of injury to your ears? Explain.

>>> Applying Health Skills

6. **PRACTICING HEALTHFUL BEHAVIORS** Imagine that you are planning a backpacking trip in the mountains. The weather will be sunny but very cold. What kind of gear or clothing would you pack to protect your eyes and ears?

⟳ Review

🔊 Audio

Hands-On HEALTH ACTIVITY

Observing *the Eye*

WHAT YOU WILL NEED

* Mirror
* Pencil or pen
* Paper

WHAT YOU WILL DO

1 Turn off the lights. Sit in the dark for two to three minutes.

2 Turn the lights back on, and quickly look at your eyes in the mirror. Watch what happens in the center of your eyes. Record what you notice about your eyes.

3 Once your eyes have adjusted to the light, do the color vision test. Look at the circle shown on this page. Can you see a number in the circle? If not, you may have trouble distinguishing between the colors red and green.

WRAPPING IT UP

As a class, make a chart or graph that compares the results for all students. What do your findings show?

Your eyes can adjust very quickly to different levels of light. The muscles inside the eye change so that more or less light comes in. Most people can also distinguish colors with their eyes. Some, however, are born without the ability to see certain colors. Try this activity to observe how your eyes react to light and color.

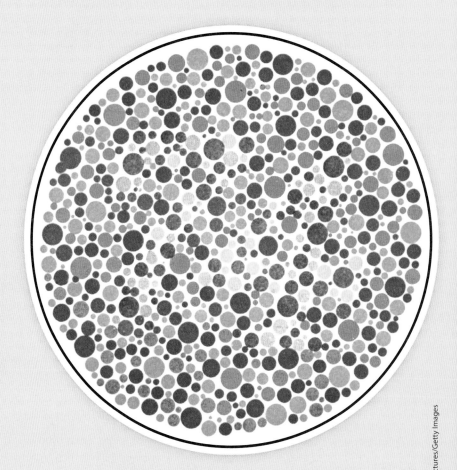

Steve Allen/Brand X Pictures/Getty Images

READING REVIEW

FOLDABLES and Other Study Aids

Take out the Foldable® that you created and any study organizers that you created. Find a partner and quiz each other using these study aids.

LESSON 1 Personal Hygiene and Consumer Choices

BIG IDEA It is important for you to know how to maintain good hygiene as well as be a smart consumer.

* Your personal hygiene affects all sides of your health triangle.
* The term *hygiene* refers primarily to cleanliness, or keeping your body clean.
* Knowing how to evaluate products and services will help you become a smart consumer.
* Part of being a smart consumer is learning skills that will help you become a good comparison shopper.

LESSON 2 Taking Care of Your Skin and Hair

BIG IDEA Caring for your skin and hair is important to your overall physical health.

* The skin is the body's largest organ, and it has several important functions.
* Your skin is made up of three main layers: the epidermis, the dermis, and the subcutaneous layer.
* You can take several steps to help keep your skin and hair healthy, starting by keeping it clean.
* Hair grows from follicles in the dermis and is made of keratin.

LESSON 3 Caring for Your Mouth and Teeth

BIG IDEA Your teeth perform important functions and keeping them healthy is part of responsible healthful behaviors.

* Your teeth have several important functions, including chewing food, helping to form certain speech sounds, and giving structure to your mouth.
* Tooth and gum problems include the development of plaque and tartar and gingivitis.
* Gingivitis is a gum disease that can cause tooth loss.
* Having healthy teeth and gums requires brushing and flossing every day and regular dental checkups.

LESSON 4 Protecting Your Eyes and Ears

BIG IDEA Caring for your eyes and ears will help keep them healthy.

* The function of your eyes is to focus light on the retina, which sends information to your brain about what you see.
* You can keep your eyes healthy by taking care of them.
* Ears carry sound to the brain and help you stay balanced.
* Protecting your ears from loud sounds is the best way to care for them.

 Review

 Web Quest

ASSESSMENT

Reviewing Vocabulary *and* Main Ideas

- melanin
- consumer
- ultraviolet (UV) rays
- acne
- hygiene
- comparison shopping
- body odor
- dermatologist

>> On a sheet of paper, write the numbers 1–8. After each number, write the term from the list that best completes each statement.

LESSON 1

1. An important part of maintaining good _____ is washing your hands thoroughly and often.

2. Factors that influence your choices as a(n) _____ include cost, your likes and dislikes, and the media.

3. _____ involves evaluating the benefits of two or more similar products.

4. During the teen years, _____, which is caused by bacteria on the skin, becomes more noticeable.

LESSON 2

5. Skin gets its color from _____.

6. _____ is a skin condition caused by active oil glands that clog hair follicles.

7. Sunburn is caused by _____ from the sun damaging skin cells.

8. If you have a bad case of acne or another serious skin problem, you may need to see a _____.

>> On a sheet of paper, write the numbers 9–16. Write *True* or *False* for each statement below. If the statement is false, change the underlined word or phrase to make it true.

LESSON 3

9. Gingivitis is a chemical that helps prevent tooth decay.

10. Tooth decay begins with the formation of plaque.

11. The two most important keys to having healthy teeth and gums are braces and flossing.

12. A dermatologist is a dentist who specializes in problems with the alignment or spacing of teeth.

LESSON 4

13. Nearsightedness is an eye condition in which images or objects appear wavy or blurry.

14. A condition in which a person has constant ringing in the ears is called tinnitus.

15. The best way to care for your ears is to listen to loud sounds.

16. An obstetrician is a physician who specializes in the structure, functions, and diseases of the eye.

✔ eAssessment

>> Using complete sentences, answer the following questions on a sheet of paper.

🗨️ Thinking Critically

17. SYNTHESIZE How can awareness of the influence of advertising claims help you when you are shopping for personal care items?

18. ANALYZE How can taking care of your teeth, hair, skin, and nails affect all sides of your health triangle?

19. PREDICT If you don't treat a hearing problem, how might it affect other areas of your health?

20. EVALUATE Are consumer skills helpful only for saving money? Explain your answer.

🖊️ Write About It

21. OPINION Imagine you are teaching a younger sibling about personal care. Write a dialogue between you and your sibling in which you explain how to care for the eyes, teeth, ears, and skin.

22. DESCRIPTIVE WRITING You have never worn glasses or contact lenses, but lately you have found that it is becoming harder to read the board in class. You are also having trouble seeing the ball when you are playing for your school team. Describe what steps you might take to help address this situation.

Ⓐ Ⓑ Ⓒ Ⓓ STANDARDIZED TEST PRACTICE

Writing
Read the prompts below. On a separate sheet of paper, write an essay that addresses each prompt. Use information from the chapter to support your writing.

1. A person is experiencing headaches and eye-strain. Explain what might be the causes of the problem and how to fix them. Then describe different ways that the person can protect their eyes from injury.

2. Imagine that you have a company that makes shampoo. How would you convince buyers to choose your product? Write a script for a television or radio advertisement that sells your product. What important information such as directions for use and warnings would you include?

3. You have noticed that a close friend has bad breath. What can that person do to help control or eliminate this problem? Explain the causes of bad breath and the steps someone could take to help avoid it.

Unit 7

your body systems

Your Body Systems

PREMIUM ONLINE RESOURCES >  Audio    Videos    Bilingual Glossary

 Fitness Zone    Web Quest   Review

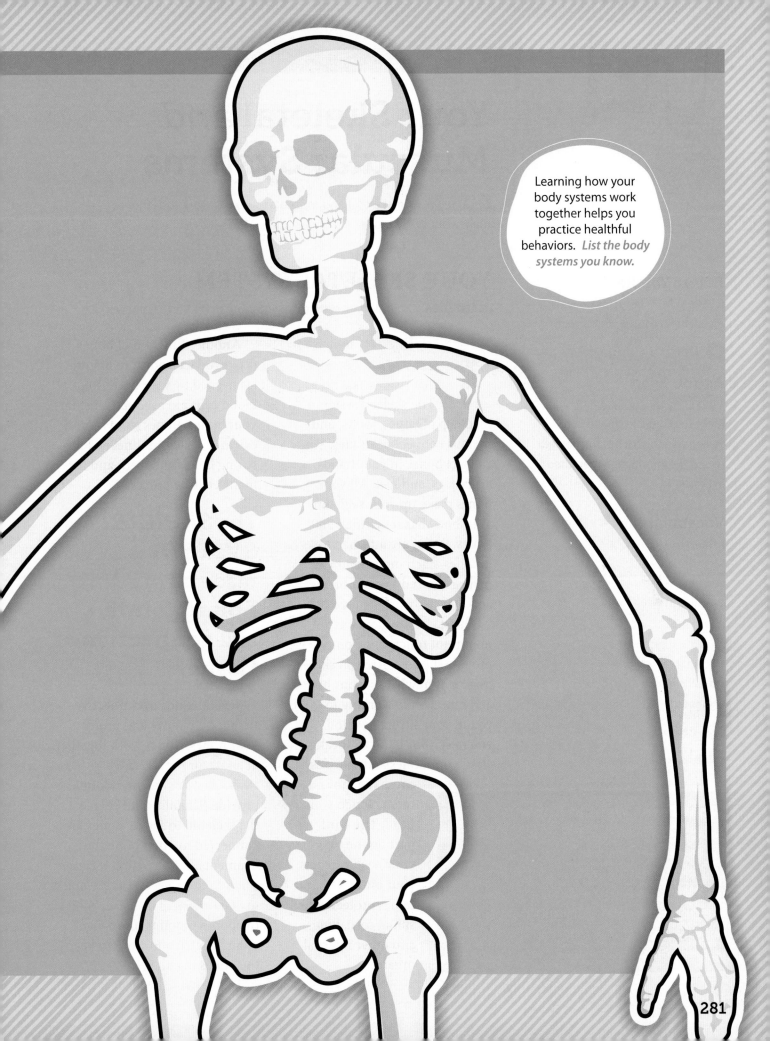

Your Skeletal *and* Muscular Systems

BIG IDEA Your skeletal and muscular systems work together to make your body move.

Before You Read

QUICK WRITE List some bones and muscles that you are familiar with.

 Video

As You Read

FOLDABLES Study Organizer

Make the Foldable® found in the FL pages in the back of the book to record the information presented in Lesson 1.

Vocabulary

› skeletal system
› joints
› ligament
› cartilage
› tendons
› muscular system
› skeletal muscle
› cardiac muscle
› smooth muscle

 Audio

 Bilingual Glossary

What Teens Want to Know

What causes paralysis? The nervous system can be affected by injury or disease. A spinal cord injury causes the nervous system to lose the ability to send messages to the muscles. Muscular disorders such as muscular dystrophy cause the muscular system lose the ability to respond to messages from the nervous system.

YOUR SKELETAL SYSTEM

MAIN IDEA Your skeletal system provides your body with a framework.

Your **skeletal system** is *the framework of bones and other tissues that supports the body.* The skeletal system is made up of bones, joints, and various connective tissues. You can feel bones in your hands, arms, legs, and feet. All your bones make up your skeleton. Your body has more than 200 bones. Bones are attached to muscles. Each movement you make is caused by your bones and muscles working together.

The skeletal system has important functions. Bones provide support. Are you sitting right now? If so, your bones and muscles are working to hold you in your sitting position. If you raise your hand to speak in class, your bones and muscles cause the movement. Touch your head. The hard part on the top of your head protects your brain. Other bones protect your spinal cord, lungs, and other internal organs.

Another function of bones is to produce and store materials needed by your body. Red blood cells are produced inside your bones. Bones store fat and calcium. Calcium is needed for strong bones and teeth and for many cellular processes.

The *skeletal system* is made up of **bones, joints,** and **connective tissue.**

Several kinds of connective tissues help move and protect your bones. Bones work together at **joints,** or *the places where two or more bones meet.* Joints provide flexibility and enable the skeleton to move. Bones are connected to other bones by ligaments. A **ligament** is *a type of connecting tissue that holds bones to other bones at the joint.* When the bones in joints move, ligaments stretch and work to keep the bones together.

image100/Getty Images

Ligaments connect bones but do not protect them. Bones are protected by **cartilage,** which is *a strong, flexible tissue that allows joints to move easily, cushions bones, and supports soft tissues.* Cartilage protects bones, such as those connected by your knee joint. Your elbows and shoulders are also protected by cartilage.

Your bones are also protected by **tendons,** *a type of connecting tissue that joins muscles to bones and muscles to muscles.* Tendons help to stabilize joints and keep them from moving out of place.

Your skeletal system contains two types of joints—immovable joints and movable joints. Immovable joints do not move.

For example, your skull contains several immovable joints. Movable joints allow you to move your hands and feet and bend parts of your body such as your knees and elbows.

>>> **Reading Check**

COMPARE *Identify two types of joints and give an example of each.*

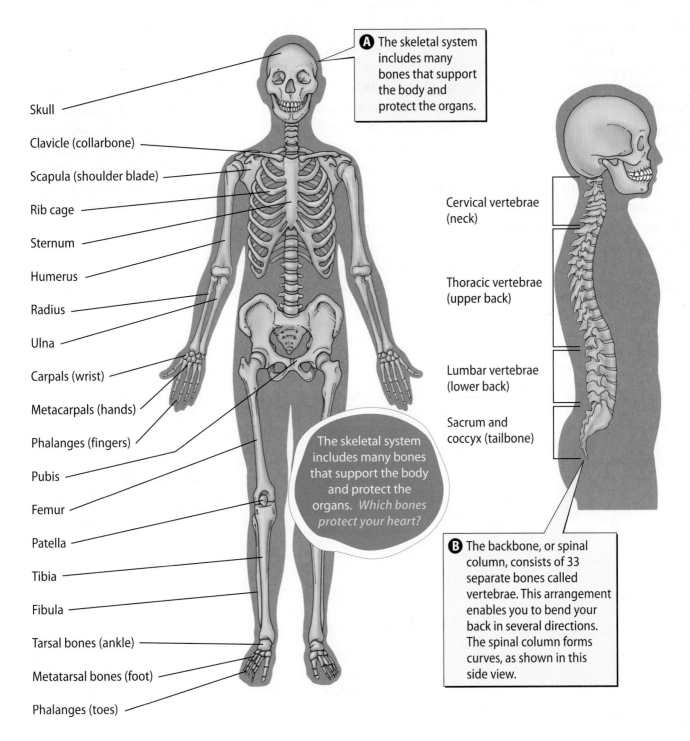

A The skeletal system includes many bones that support the body and protect the organs.

Skull

Clavicle (collarbone)

Scapula (shoulder blade)

Rib cage

Sternum

Humerus

Radius

Ulna

Carpals (wrist)

Metacarpals (hands)

Phalanges (fingers)

Pubis

Femur

Patella

Tibia

Fibula

Tarsal bones (ankle)

Metatarsal bones (foot)

Phalanges (toes)

The skeletal system includes many bones that support the body and protect the organs. *Which bones protect your heart?*

Cervical vertebrae (neck)

Thoracic vertebrae (upper back)

Lumbar vertebrae (lower back)

Sacrum and coccyx (tailbone)

B The backbone, or spinal column, consists of 33 separate bones called vertebrae. This arrangement enables you to bend your back in several directions. The spinal column forms curves, as shown in this side view.

YOUR MUSCULAR SYSTEM

MAIN IDEA Your muscular system allows your body to move and helps keep it stable.

You use muscles when you walk and stretch. Movement is an important part of the muscular system. The muscular system is made up of *tissues that move parts of the body and control the organs.* Muscles also provide your body with stability and protection.

Muscles attached to bones support your body to provide stability and balance. If you stumble and lose your balance, your muscles pull you back to a stable position. Muscles cover most of your skeleton like a layer of padding. Muscles cover your abdomen, chest, and back to protect your internal organs. Muscles also work to maintain your body at its normal temperature of around 37°C.

Muscles provide **stability, protection,** and maintain body **temperature.**

When you are cold, your muscles contract quickly and cause you to shiver. When you are too warm or have exercised, your body may sweat. In either case, when your body is too cold or too warm, your muscles work to turn chemical energy into thermal energy to keep your body at a safe temperature.

Many of your muscles are attached to bones by tendons to enable your skeleton to move.

The movement of your body can be fast, such as when you run, or slow, such as when you stretch. Muscles are made of strong tissue that can contract in an orderly way. When a muscle contracts, the cells of the muscle become shorter. When the muscle relaxes, those cells return to their original length.

Many of your muscles are not attached to bones. As these muscles contract, they cause blood and food to move through your body. These are the muscles that cause your heart to beat. These types of muscles also make the hair on your arms stand on end when you get goose bumps.

>> **Reading Check**

EXPLAIN *How do muscles work?*

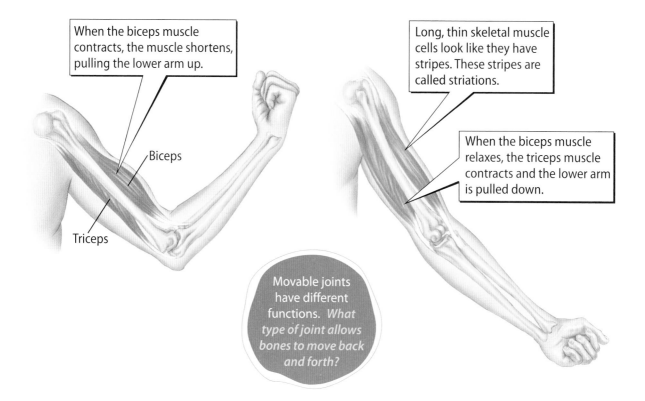

When the biceps muscle contracts, the muscle shortens, pulling the lower arm up.

Biceps

Triceps

Long, thin skeletal muscle cells look like they have stripes. These stripes are called striations.

When the biceps muscle relaxes, the triceps muscle contracts and the lower arm is pulled down.

Movable joints have different functions. *What type of joint allows bones to move back and forth?*

Types *of* Muscles

Your body has three different types of muscles: skeletal, cardiac, and smooth. Each of these muscle tissues has a specific function. A skeletal muscle is a type of *muscle attached to bones that enables you to move your body.* Skeletal muscles are voluntary muscles. This means that you control the skeletal muscles to make your body move.

Your heart is made of cardiac muscle. Cardiac muscle is the *muscle found in the walls of your heart.* Cardiac muscles are involuntary muscles. They work on their own, without your control. When cardiac muscles contract and relax, they pump blood through your heart and blood vessels throughout your body.

Your body has *three* different **types** of *muscles:* **skeletal, cardiac,** and **smooth.**

A smooth muscle is a *type of muscle found in organs and in blood vessels and glands.* Smooth muscles are involuntary muscles named for their smooth appearance. Blood vessels in your body are lined with smooth muscles.

These are the major skeletal muscles and their functions. *What muscle is used when you straighten your arm?*

Your stomach, bladder, and intestines also contain smooth muscles. Contraction of the smooth muscles controls the movement of blood through the vessels. They also move other materials through the body, such as food in the stomach.

How Muscles Work Together

Skeletal muscles work by pulling. They do not push on your bones. Each movement that you make involves your muscles pulling on your bones. Muscles often work together to help your body move.

Muscles contract and expand to pull on bones and create movement at a joint. The process of two muscles working together is called paired movement. For example, when you pull up your lower arm at the elbow, your biceps muscle contracts. When you then lower your arm, your biceps muscle relaxes and the triceps muscle contracts in order to pull the arm down.

>>> **Reading Check**

DIFFERENTIATE *Explain the difference between voluntary and involuntary muscles.*

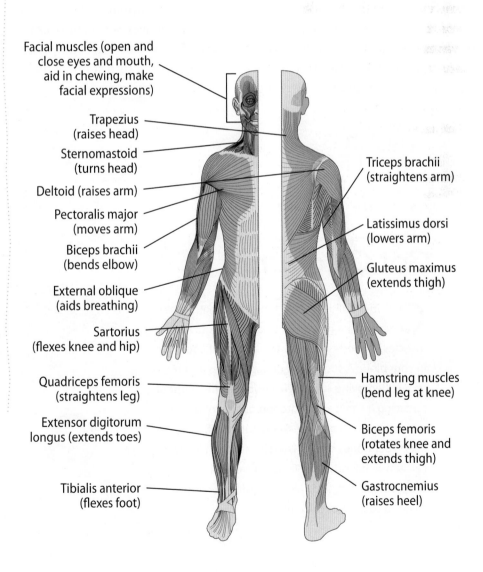

Facial muscles (open and close eyes and mouth, aid in chewing, make facial expressions)

Trapezius (raises head)

Sternomastoid (turns head)

Deltoid (raises arm)

Pectoralis major (moves arm)

Biceps brachii (bends elbow)

External oblique (aids breathing)

Sartorius (flexes knee and hip)

Quadriceps femoris (straightens leg)

Extensor digitorum longus (extends toes)

Tibialis anterior (flexes foot)

Triceps brachii (straightens arm)

Latissimus dorsi (lowers arm)

Gluteus maximus (extends thigh)

Hamstring muscles (bend leg at knee)

Biceps femoris (rotates knee and extends thigh)

Gastrocnemius (raises heel)

PROBLEMS WITH BONES AND MUSCLES

MAIN IDEA Your bones and muscles can develop problems.

Your bones, muscles, and connective tissues are strong, but they need your care. Problems can develop because of injury, infection, poor posture, and lack of nutritious foods. Some problems of the skeletal system can include:

- **Fracture.** A fracture is a break in a bone caused by an injury.
- **Dislocation.** This occurs when a bone is pushed out of its joint. Dislocation can stretch or tear a ligament.
- **Sprain.** A sprain is an injury to the ligament connecting bones at a joint. This occurs when a ligament is stretched or twisted and causes swelling.

*Your **bones, muscles,** and **connective tissues** are strong, but they need your **care.***

- **Strain.** This is a small tear in a muscle or tendon. Strains can occur when a muscle has been overstretched. A strain may be referred to as a pulled muscle.
- **Overuse injuries.** Injuries as a result of overuse occur over a period of time. An example of overuse is a shin splint, which can develop in runners.

- **Osteoporosis.** This condition results in brittle or porous bones. Osteoporosis can be caused by long-term lack of nutrition or exercise.
- **Scoliosis.** This is a curving of the backbone. The spine curves to one side of the body in an S-shape or C-shape.
- **Muscular dystrophy.** This disorder weakens muscles over time. It is usually inherited and causes skeletal muscle tissue to gradually waste away.

>> **Reading Check**

SUMMARIZE *What are the most frequent causes of injuries to bones and muscles?*

Health SKILLS ACTIVITY

Practicing Healthful Behaviors

Got *Calcium?*

Calcium builds healthy bones and teeth. It also keeps your heartbeat steady and your nerves and muscles in good condition. If the amount of calcium in your blood is too low, your body draws it from your bones. Removing calcium from your bones without replacing it can lead to osteoporosis, a condition in which bones become weak and can break easily. To keep your bones strong, exercise and eat plenty of calcium-rich foods. Low-fat or fat-free dairy products, broccoli, spinach, and calcium-fortified orange juice are all good sources of calcium.

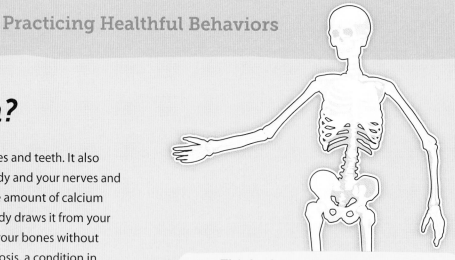

 On Your Own

Think of ways to incorporate calcium-rich foods into meals and snacks. For example, include some spinach leaves in a salad or melt some reduced-fat cheese on top of cooked broccoli. Create a list of your ideas and share them with the class.

CARING FOR YOUR BONES AND MUSCLES

MAIN IDEA › You can help keep your bones and muscles healthy.

Your good health habits can keep your bones and muscles strong and healthy. Bone and muscle health requires energy from the foods you eat. A diet that is rich in nutrients such protein, fiber, and potassium, and vitamin C can help keep your muscles strong.

Physical activity also helps keep muscles healthy and strong. Decreased muscle strength can increase the risk of heart disease and injury and make joints less stable. Do flexibility exercises so you can move more easily and work out more safely. Choose activities that strengthen your muscles and bones. Improve your cardiovascular endurance to give your heart and lungs more power. Warm up before and cool down after physical activity. If you feel pain, stop and give your body time to recover.

Exercise and **physical activity** can help keep your muscles *strong.*

Also important is your posture, or the way you hold your body. Good posture means the bones and joints in your back stay in place and your muscles are used properly. To prevent too much strain on your back, avoid carrying a heavy backpack. Bend and use your legs, not your back, when you lift something heavy. Running, walking, cycling, and swimming can help you keep your back strong and healthy.

> ››› **Reading Check**
> EXPLAIN *How can good posture contribute to bone and muscle strength.* ■

››› **After You Read**

1. **VOCABULARY** Define the skeletal system.
2. **EXPLAIN** Describe two functions of the skeletal system.
3. **COMPARE AND CONTRAST** What is the difference between voluntary and involuntary muscles?

››› **Thinking Critically**

4. **SYNTHESIZE** Imagine you are going to start running to prepare for a marathon in six months. What steps can you take to protect your bones and muscles?
5. **ANALYZE** Why do you think poor posture can cause backaches?

››› **Applying Health Skills**

6. **PRACTICING HEALTHFUL BEHAVIORS** List the physical activities in which you participate. Evaluate your activities to determine how well you are strengthening your bones and muscles. Do you think you should add any activities? Tell how you might change your physical activities to improve bone and muscle strength.

⟳ Review

◆⟩ Audio

These runners exercise to keep fit. *How are these teens helping their bones and muscles?*

Your Nervous System

BIG IDEA Your nervous system controls and sends messages throughout your body.

Before You Read

QUICK WRITE Describe some ways that your brain and your nerves tell your body what to do.

 Video

As You Read

STUDY ORGANIZER Make the study organizer found in the FL pages in the back of the book to record the information presented in Lesson 2.

Vocabulary
› nervous system
› brain
› neurons
› central nervous system
› spinal cord
› peripheral nervous system

 Audio

 Bilingual Glossary

Developing Good Character

Citizenship You can demonstrate good citizenship by sharing what you learn about protecting your health. For example, encourage family members to protect their brains by always wearing a helmet when riding a bike. *What are some other ways you could promote healthy choices in your family or neighborhood?*

PARTS OF THE NERVOUS SYSTEM

MAIN IDEA Your movements and body processes are controlled by the nervous system.

Your nervous system carries messages back and forth between your brain and the rest of your body. The **nervous system** is *the body's message and control center.* Your **brain** is *the command center, or coordinator, of the nervous system.* The nervous system gathers, processes, and responds to information received by the brain. This happens very quickly. Your nervous system can receive information, process it, and respond to it in less than a second.

The nervous system controls all body processes, such as digestion, breathing, and blood flow through the body. Your nervous system processes your physical and emotional feelings, and reactions to stimuli. A stimulus is a change in environment that causes a response. Examples of stimuli include heat, cold, and actions such as seeing and catching a thrown ball.

The nervous system receives messages from your five senses—vision, hearing, smell, taste, and touch. Your senses send a message to your brain, and your body reacts. For example, think about how you react when you touch something hot—you quickly sense the heat and pull away.

Your *nervous system* **carries messages** throughout your **body.**

The nervous system is made up of neurons. **Neurons** are *cells that make up the nervous system.* Neurons are also called nerve cells. Neurons are the message carriers that help the different parts of your body communicate with one another.

> **Reading Check**
>
> DEFINE *What is another phrase for neurons?*

Tom Merton/OJO Images/Getty Images

The Central Nervous System

Your nervous system has two parts. The central nervous system (CNS) includes *the brain and the spinal cord.* The brain controls your thoughts, speech, memory, and muscle movement. The brain controls voluntary muscle movement, or things you have to think about doing, such as standing, running, waving, and speaking. The brain also controls involuntary muscle movement, or things you do without thinking, such as your heartbeat, swallowing, blinking, coughing, and sneezing.

The largest and most complex part of the brain is the cerebrum.

It controls memory, language, and thoughts. The part of the brain that controls voluntary muscle movement is called the cerebellum. It also stores muscle movements you have learned, such as tying your shoes or riding a bike. The brain stem is the area of the brain that controls involuntary muscle movement.

Your spinal cord is also part of your central nervous system. The spinal cord is *a long bundle of neurons that sends messages to and from the brain and all parts of the body.* The neurons in the spinal cord reach out to other parts of the body. The brain responds to information through the neurons in the spinal cord.

The Peripheral Nervous System

The peripheral nervous system (PNS) includes *the nerves that connect the central nervous system to all parts of the body.* If a bee stings you, it hurts because your nerves sense pain. They can send a message through the peripheral nervous system to the central nervous system and on to the brain—in under a second!

The peripheral nervous system handles movements you control and those you do not control. Raising your hand is a voluntary movement, or one you control. The beating of your heart is an involuntary movement, or one you do not control.

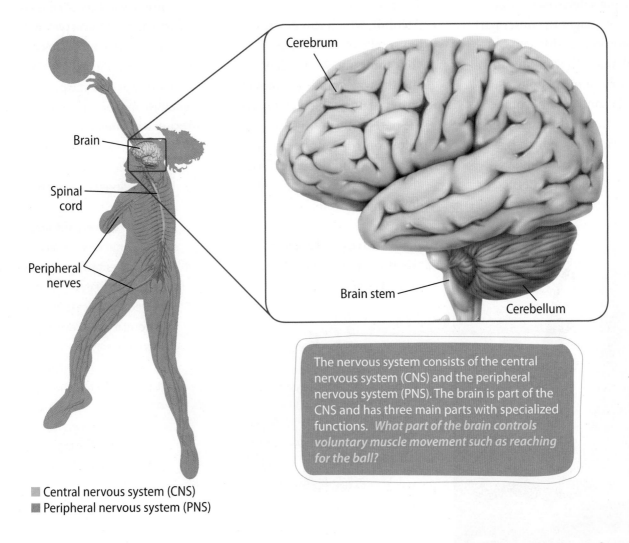

Brain

Spinal cord

Peripheral nerves

Cerebrum

Brain stem

Cerebellum

The nervous system consists of the central nervous system (CNS) and the peripheral nervous system (PNS). The brain is part of the CNS and has three main parts with specialized functions. *What part of the brain controls voluntary muscle movement such as reaching for the ball?*

■ Central nervous system (CNS)
■ Peripheral nervous system (PNS)

PROBLEMS AFFECTING THE NERVOUS SYSTEM

MAIN IDEA Injury or disease can harm the nervous system.

The nervous system can become injured or be affected by diseases or disorders. One of the most common causes of damage is injury to the head, neck, or back. For example, an injury to the spinal cord could lead to paralysis. This means the loss of feeling in or being unable to move some body parts. Since the brain is your control center, an injury can cause memory loss, brain damage, or the loss of some physical abilities. Other issues that affect the nervous system include:

- **Multiple sclerosis,** or MS, which damages the outer part of some nerves. MS can cause problems with thinking and memory. Some people lose muscle control or become unable to walk because of MS.
- **Cerebral palsy,** which is a disease of the nervous system that is either inherited or caused by brain damage.

- **Alzheimer's disease,** which most often affects older adults, harms the brain and causes loss of memory.
- **Parkinson's disease,** which is a brain disorder that causes shaking and stiffness of the arms and legs.
- **Epilepsy,** which occurs when signals in the brain do not send messages in the normal way. Epilepsy can cause a person to briefly lose muscle control or have seizures.
- **Viruses,** such as polio, rabies, meningitis, encephalitis, and West Nile virus.
- **Alcohol,** which can destroy brain cells. Alcohol also affects your thinking, your balance, and the way your body moves.
- **Other drugs,** which can harm the part of the brain that helps control your heart rate, breathing, and sleeping. Drugs also either speed up or slow down the nervous system.

>>> **Reading Check**

EXPLAIN *How can alcohol affect the nervous system?*

Garrett has cerebral palsy, a nervous system disorder. He gets good grades and is active in his community. *Name two other nervous system disorders.*

Practicing Healthful Behaviors

Avoiding *Repetitive Motion Injuries*

Many people use computers for long periods of time, either at work, home, or school. Some people may develop a nervous system problem called *repetitive motion injury.* This is an inflammation of the nerves in the wrists caused by prolonged, repeated movements. The following strategies can help you prevent repetitive motion injuries when you are using a computer:

1 Keep your wrists relaxed and straight.
2 Use only finger movements to strike the keys.
3 Press the keys with the least pressure that is necessary.
4 Move your entire hand to press hard-to-reach keys.
5 Take frequent breaks, which are also good for your eyes.

On Your Own Practice these strategies each time you use a computer.

Realistic Reflections

CARING FOR YOUR NERVOUS SYSTEM

MAIN IDEA Healthy behaviors can protect the nervous system.

A healthful lifestyle will help protect your nervous system. Make healthful food choices, drink plenty of water, and get plenty of sleep. Stay physically fit and maintain a healthy weight. You can also guard against illnesses that affect the nervous system. Protect yourself against insects and avoid animals that may carry disease. Wash your hands thoroughly and often.

Make **healthful food choices**, *drink* plenty of **water**, and *get* plenty of **sleep.**

Use safety equipment when you participate in physical activities. Protect your brain by wearing a safety helmet when riding a bike or skating. If you participate in gymnastics, have a person nearby who can spot you.

If you lift weights, it is also helpful to have a spotter on hand. You can protect your back and spinal cord by lifting properly.

Be sure to follow basic safety rules. Always wear a seat belt when riding in a car. Pay attention to signs telling where you may or may not ride a skateboard. When you skate or ride a bike, watch for traffic, people, and animals in your path.

You practice positive health behaviors when you decide not to use alcohol or other drugs. These destroy brain cells and affect your thoughts, emotions, and judgment. You want to keep your brain cells healthy!

>>> Reading Check

IDENTIFY *Name two other ways to protect you nervous system.* ■

These teens wear a protective helmet when they ride their bicycles. *Why is a protective helmet important when riding a bike?*

REVIEW

>>> After You Read

1. **VOCABULARY** What are neurons?
2. **DEFINE** Define CNS and PNS.
3. **DESCRIBE** What are stimuli?

>>> Thinking Critically

4. **SYNTHESIZE** List five voluntary muscle movements that you make each day.
5. **ANALYZE** Choose a sports figure or athlete that you know about. It could be a member of a team, a dancer, or a skateboarder. Describe how that person practices safe habits to protect the body's health.

>>> Applying Health Skills

6. **ACCESSING INFORMATION** Epilepsy is a nervous system disorder in which a person has seizures. During a seizure, the person may lose consciousness, twitch, and shake. Use online and library resources to investigate what happens in the brain of a person who has epilepsy. Write a paragraph describing what you find.

🔄 Review

🔊 Audio

Daniel Simon Westend 61/Getty Images

Your Circulatory *and* Respiratory Systems

BIG IDEA Your heart is the center of your circulatory system, and your lungs are the center of your respiratory system.

Before You Read

QUICK WRITE In a few sentences, explain what you already know about how the heart works and what it does.

 Video

As You Read

STUDY ORGANIZER Make the study organizer on page 42 to record the information presented in Lesson 3.

Vocabulary

› circulatory system
› cardiovascular system
› heart
› arteries
› veins
› capillaries
› blood pressure
› respiratory system
› lungs
› larynx
› trachea
› bronchi
› diaphragm

 Audio

🔤 Bilingual Glossary

YOUR CIRCULATORY SYSTEM

MAIN IDEA Your circulatory system is like a transportation system inside your body.

The body works all the time, even when you are asleep. Your circulatory system keeps your body working. The circulatory system is *the group of organs and tissues that carry needed materials to cells and remove their waste products.* The circulatory system includes the heart, different types of blood vessels, and the blood. It is also called the cardiovascular system. This includes *organs and tissues that transport essential materials to body cells and remove their waste products.*

The body **works** *all the time,* even when you **are asleep.**

Cardio refers to the heart, and *vascular* refers to the blood vessels. The circulatory system moves blood to and from tissues in the body. The blood delivers oxygen, food, and other materials to cells. It also carries waste products away from cells.

Think about a network of busy roads, all traveling in different directions. The circulatory system is like a highway system, and your blood cells are like the semi trucks that travel the roads, carrying materials around, through, and out of your body.

> ## Reading Check
>
> **RECALL** *What are the main parts of the circulatory system?*

Parts *of the* Circulatory System

Your heart is always at work. The heart is *the muscle that acts as the pump for the circulatory system.* It pushes blood through tubes called blood vessels. There are three different types of blood vessels. Arteries *carry blood away from the heart to various parts of the body.* Veins *carry blood from all parts of the body back to the heart.* Between the arteries and veins are tiny blood vessels called capillaries, which *carry blood to and from almost all body cells and connect arteries and veins.*

Sean Justice/Corbis

The heart has four chambers. The top chambers are called atria, or "rooms." Blood enters the heart through the two atria. The lower chambers are called ventricles. Blood leaves the heart through the two ventricles.

Blood pressure is *the force of blood pushing against the walls of the blood vessels.* Blood pressure is highest when the heart contracts, or pushes out blood. It is lowest between heartbeats, when the heart relaxes.

Blood supplies all parts of **your body** with materials needed to survive.

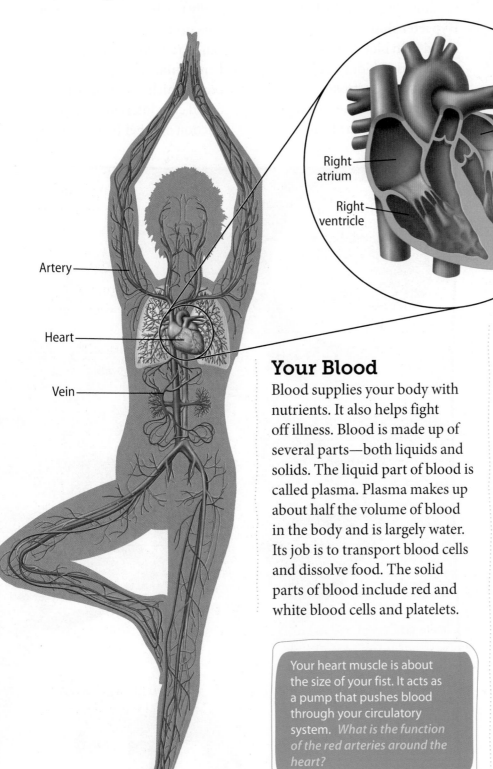

Artery

Heart

Vein

Right atrium

Right ventricle

Left atrium

Left ventricle

Your Blood

Blood supplies your body with nutrients. It also helps fight off illness. Blood is made up of several parts—both liquids and solids. The liquid part of blood is called plasma. Plasma makes up about half the volume of blood in the body and is largely water. Its job is to transport blood cells and dissolve food. The solid parts of blood include red and white blood cells and platelets.

Your heart muscle is about the size of your fist. It acts as a pump that pushes blood through your circulatory system. *What is the function of the red arteries around the heart?*

- **Red blood cells** carry oxygen to all other cells in the body. They also carry away some waste products.
- **White blood cells** help the various body systems destroy disease-causing germs.
- **Platelets** are small, disk-shaped structures that help your blood clot. Clotting helps keep you from losing too much blood when you have a cut or other injury. Red blood cells are one of four specific types: A, B, AB, or O. Your blood type is inherited from your parents and remains the same throughout your life.

It is good to know your blood type. Some blood types are compatible. This means they can be safely mixed if a person needs blood. Mixing blood types that are not compatible can be harmful or even fatal.

People with any blood type can receive type O. As a result, people with type O blood are called "universal donors." People with type AB blood can receive any blood type but can only give to others with type AB. They are known as "universal recipients."

Blood is given during surgery or when a person needs blood due to a serious injury or illness. Blood may also carry an Rh factor, or a protein found on the surface of red blood cells. The Rh factor is another inherited trait. Blood is either Rh-positive or Rh-negative. People with Rh-positive blood can receive blood from donors who are either Rh-positive or Rh-negative. People with Rh-negative blood can only receive blood from donors who are also Rh-negative. Both the blood type and the Rh factor must be compatible in order for blood to be received safely.

>> **Reading Check**

RECALL *Which blood cells carry oxygen from the lungs to all parts of the body?*

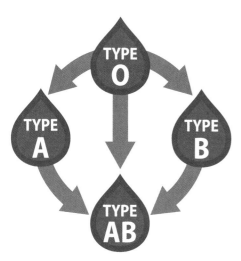

Donated blood saves many lives each year. *Which blood type is compatible with the other three blood types?*

Problems Affecting *the* Circulatory System

Some circulatory problems affect the heart or blood vessels. Others mainly affect the blood, while some affect other body systems. Circulatory problems include:

- **Hypertension,** which is also called high blood pressure. It can lead to kidney failure, heart attack, or stroke.
- **Heart attack,** or the blockage of blood flow to the heart.
- **Stroke,** which usually results from blood clots in the brain or from a torn blood vessel.
- **Arteriosclerosis,** or a condition in which arteries harden and reduce blood flow.
- **Anemia,** which is an abnormally low level of hemoglobin. It's a protein that binds to oxygen in red blood cells.
- **Leukemia,** or a type of cancer in which abnormal white blood cells interfere with production of other blood cells.

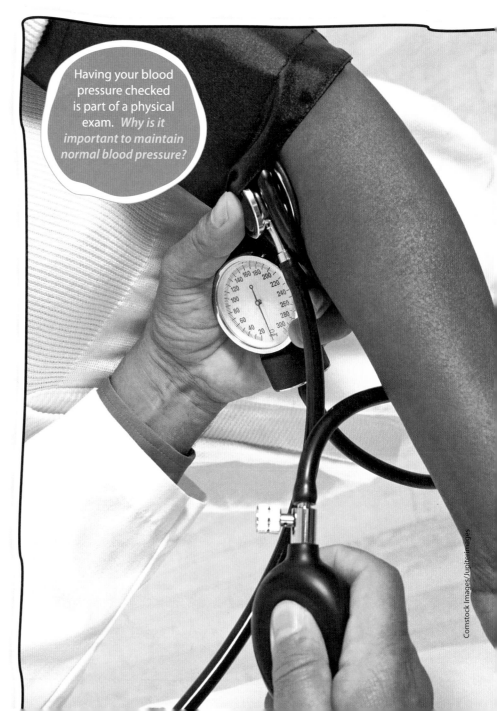

Having your blood pressure checked is part of a physical exam. *Why is it important to maintain normal blood pressure?*

YOUR RESPIRATORY SYSTEM

MAIN IDEA Your respiratory system controls your breathing.

Oxygen is essential to the body for survival. You get oxygen by breathing. Breathing is the movement of air into and out of the lungs. Breathing enables your respiratory system to take in oxygen and eliminate carbon dioxide. Your respiratory system contains *the organs that supply your blood with oxygen.*

The lungs are *two large organs that exchange oxygen and carbon dioxide.* Air moves in and out of your lungs through the respiratory system. Breathing in, or inhaling, brings oxygen into your lungs. Your blood circulates through your lungs, exchanging carbon dioxide for oxygen. Exhaling, or breathing out, is the action of your lungs getting rid of carbon dioxide and other waste materials from your body.

Parts *of the* Respiratory System

When you breathe in, air enters through the nose and mouth. In the nose, air is warmed and moistened. Hairs and sticky mucus in the nose help track dust and dirt from the air. Air passes through the nose and mouth into the throat.

Oxygen is **essential** to the body for **survival.**

The pharynx is a tube-like passageway at the top of the throat that receives air, food, and liquids from the mouth or nose. The epiglottis is a flap of tissue at the lower end of the pharynx. It keeps food and liquids from entering the respiratory system.

Air passes from the pharynx into a triangle-shaped area called the voice box or the larynx. This is *the upper part of the respiratory system, which contains the vocal cords.* Two thick folds of tissue in the larynx—the vocal cords—vibrate and make sounds as air passes over them. The tissues of the larynx allow a person to speak. Air then enters the trachea, *a passageway in your throat that takes air into and out of your lungs.* The trachea branches into two narrower tubes called bronchi, *two passageways that branch from the trachea, one to each lung.* Inside the lungs, the bronchi continue to branch off into even smaller tubes.

> **⟫⟫⟫ Reading Check**
>
> **RECALL** *Name the main organs of the respiratory system.*

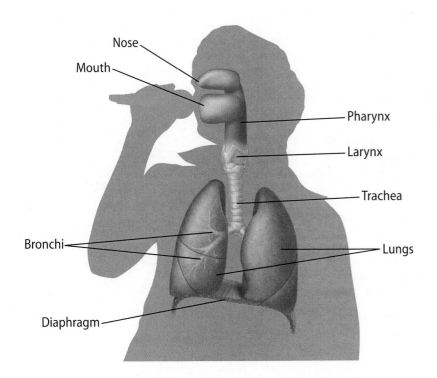

Nose
Mouth
Pharynx
Larynx
Trachea
Bronchi
Lungs
Diaphragm

Air moves into and out of the lungs through the respiratory system. *Which part of the respiratory system contains bronchi?*

How You Breathe

When carbon dioxide is in your blood, your nervous system signals your body to let it out, or exhale. The change in air pressure inside your chest causes breathing to occur. Breathing begins with the **diaphragm**, which is *a large, dome-shaped muscle below the lungs that expands and compresses the lungs, enabling breathing.*

When you breathe in, the diaphragm contracts. This allows the lungs to expand and fill with air. When you breathe out, the diaphragm expands. As it gets larger, it pushes on the lungs, forcing out the air.

Breathing involves both voluntary and involuntary muscle movements. You do not have to think about breathing. However, you can hold your breath or control the rate of your breathing.

Problems Affecting *the* Respiratory System

Tobacco smoke, chemicals, germs, and air pollution are harmful to your health because they can damage the many parts of your respiratory system. Tobacco use may not cause all the problems that can affect your respiratory system, but it can make many problems worse. Some respiratory illnesses can make breathing difficult. Others can become life-threatening.

Tobacco smoke, *chemicals,* germs, and air pollution are *harmful* to your *health.*

Coughing and difficulty breathing are symptoms of respiratory problems. *Name other symptoms of a cold.*

Respiratory Illnesses

Illness	Causes	Symptoms
Colds, flu	viruses	congestion, runny nose, watery eyes, coughing, sneezing
Bronchitis (brahn KI tus)	viruses, bacteria	coughing and fatigue due to mucus blocking the bronchi and bronchioles slows air movement
Pneumonia (noo MOH nyuh)	viruses, bacteria	difficulty breathing due to fluid in the alveoli that slows gas exchange
Asthma (AZ muh)	dust, smoke, pollen, pollution	difficulty breathing due to swollen airways and increased mucus
Emphysema (em fuh SEE muh)	smoking	coughing, fatigue, loss of appetite, and weight loss due to destruction of alveoli
Lung cancer	smoking	coughing, difficulty breathing, and chest pain

Respiratory problems can be prevented. *Which of these respiratory illnesses are caused by risk behaviors?*

Getty Images

KEEPING YOUR CIRCULATORY AND RESPIRATORY SYSTEMS HEALTHY

MAIN IDEA You can help keep your heart, blood vessels, and lungs healthy and strong.

The health of your circulatory system has a major effect on your current and future health. Your blood carries oxygen to your cells and carbon dioxide away from them. Your blood also carries vital nutrients to your organs, muscular system, and skeletal system.

Caring *for* your Circulatory System

The best way to help keep your heart healthy is to be physically active. Teens should set a goal of getting 60 minutes of physical activity each day. Regular activity strengthens your heart muscle and allows it to pump more blood with each heartbeat.

What you eat can have either a positive or negative effect on your circulatory system. Try to limit the amount of fat you eat.

Fats, especially saturated and trans fats, can cause deposits to form in your arteries. These deposits increase blood pressure. As you lower your fat intake, increase your intake of dietary fiber. Whole grains and fresh vegetables are a great source of fiber. Whole-grain cereals, raw vegetables, and breads made with whole grains also make filling and satisfying snacks.

Set a goal to get *60 minutes* of **physical activity** each day.

Another way to keep your heart and blood vessels healthy is to avoid tobacco. Tobacco use can cause lung cancer, emphysema, and other lung diseases.

The nicotine in tobacco can constrict your blood vessels. When your blood vessels are constricted, or narrower, your heart has to work harder. This can result in high blood pressure and lead to heart disease.

Finally, learn to manage the stress in your life. Stress can cause high blood pressure, which puts a strain on the entire cardiovascular system. You can learn strategies to deal with stress in healthful ways. Regular physical activity is a very effective way to help relieve stress. Staying active can also help you maintain a heart-healthy weight.

> **》》 Reading Check**
>
> **RECALL** *What is one good way to help keep your circulatory system in good health?*

Marcus knows how to enjoy the outdoors. *How is Marcus caring for his body on his walk?*

Paul Edmondson/Corbis

Caring *for* Your Respiratory System

Your whole body depends on having a healthy respiratory system. However, you can take positive action to help keep your lungs breathing strong. Here are some things you can do to benefit your respiratory system:

- **Avoid tobacco use.** Smoking can cause cancer. All tobacco products contain substances that can cause cancer.
- **Stay away from people who smoke.** Avoid places where the air is smoky. Breathing secondhand smoke, or air that has been contaminated by others' tobacco use, can be just as harmful as smoking.
- **Take care of your body.** Give your body a chance to heal and recover when you have a cold or the flu. See a health professional if an illness does not go away.
- **Drink plenty of fluids.** Whether you feel ill or healthy, you always need plenty of fluids. Drink more water when you participate in physical activity or exercise.
- **Be physically active on a regular basis.** Keep your body systems active and strong.

- **Eat a healthful diet.** As with all body systems, your respiratory system needs a proper balance of nutrients.
- **Pay attention to weather alerts for your area.** Allergy, ozone, and pollution alerts can prepare you for when outside air may be less healthful.

Your **whole** body *depends on* having a **healthy respiratory system.**

- **Manage stress.** Use strategies to maintain your stress levels. As you have learned, stress can have an impact on all sides of your health triangle.
- **Protect yourself from infections.** Wash your hands thoroughly and frequently with soap and water. Keep your body covered and protected when walking in the woods. Eat a healthful diet and get plenty of sleep.

> **>> Reading Check**
>
> **RECALL** *What is one good way to protect your respiratory system?* ■

Myth vs. Fact

Myth: The hiccups are a mystery.

Fact: We know what causes hiccups. The hiccups are caused in the diaphragm. Occasionally, the diaphragm has spasms that cause air to be taken in or pushed out rapidly. The sound of a hiccup is caused by the sudden rush of air being stopped by the vocal cords. Some hiccups can be caused by eating a big meal, swallowing air, stress, or excitement. The mystery is how to get rid of the hiccups!

>>> **After You Read**

1. **VOCABULARY** What are the main organs of the respiratory system?
2. **EXPLAIN** How are sounds made when a person speaks?
3. **DESCRIBE** Tell how the diaphragm helps you breathe.

>>> **Thinking Critically**

4. **SYNTHESIZE** Think about the movement of your chest as your lungs take in air. Is this voluntary or involuntary movement? What changes when you do deep-breathing exercises?
5. **ANALYZE** When Nora's father went to donate blood, he was asked his blood type. He wasn't sure. How can Nick learn his blood type? Why is it important?

>>> **Applying Health Skills**

6. **ANALYZING INFLUENCES** A number of factors in the environment might influence respiratory health. Make a list of these factors and discuss their role in the health of the community.

Review

Audio

Your Digestive *and* Excretory Systems

BIG IDEA Your digestive and excretory systems process the food you eat for use by your body.

>>> **Before You Read**

QUICK WRITE Have you ever had a stomachache or pain? Write one or two sentences to describe how it felt.

 Video

>>> **As You Read**

STUDY ORGANIZER Make the study organizer on page 42 to record the information presented in Lesson 4.

>>> **Vocabulary**

› digestive system
› saliva
› enzymes
› digestion
› small intestine
› liver
› gallbladder
› pancreas
› excretory system
› kidneys
› colon

 Audio

 Bilingual Glossary

YOUR DIGESTIVE SYSTEM

MAIN IDEA Digestion is the first step in the way your body processes the food you eat.

Do you know what happens to food after you eat it? As soon as food enters your mouth, it begins its journey through your digestive system. Your digestive system is *the group of organs that work together to break down foods into substances that your cells can use.* No matter what you eat, your food goes through four steps— ingestion, digestion, absorption, and elimination. The first step, ingestion, is the act of putting food in your mouth, or eating.

The Process *of* Digestion

The digestive system begins in your mouth. With your first bite of food, your teeth begin to smash and grind the food into small bits. The food mixes with your saliva, *a digestive juice produced by the salivary glands in your mouth.* Chemicals called enzymes, are in your saliva. Enzymes are *proteins that affect the many body processes.* Digestion is *the process by which the body breaks down food into smaller pieces that can be absorbed by the blood and sent to each cell in your body.* In other words, digestion allows the body to get energy and nutrients from the food you eat.

Do you know *what happens* to food **after you eat it?**

Once you chew and swallow something you eat, the food first enters your throat. Throat muscles contract and expand to push the food down the esophagus into the stomach. The esophagus is a muscular tube that connects the mouth to the stomach. Waves of muscle contractions allow food to move through the esophagus and the rest of the digestive tract. Once the partially digested food leaves the esophagus, it enters the stomach. The stomach is a large, hollow organ that stores food temporarily.

The stomach also aids in chemical digestion. Chemical digestion is when chemical reactions in the body break down pieces of food into small molecules. In the stomach, food mixes with gastric juices until it forms a watery liquid. During this digestive process, the food may stay in the stomach up to four hours. The food then passes from the stomach to the small intestine. The **small intestine** is *a coiled tube between 20 and 23 feet long, in which about 90 percent of digestion takes place.*

The small intestine is directly connected to the stomach. Most chemical digestion occurs in the small intestine. Nutrients in the small intestine enter the blood through blood vessels. This is known as absorption: the body begins to absorb the nutrients from the food.

Organs *that* Aid *in* Digestion

The liver and pancreas are both organs that produce substances that enter the small intestine and help with chemical digestion.

The **liver** is *a digestive gland that secretes a substance called bile, which helps to digest fats.* The **gallbladder** is *a small, saclike organ that stores bile until it is needed in the small intestine.* The **pancreas** is *a gland that helps the small intestine by producing pancreatic juice, a blend of enzymes that breaks down proteins, carbohydrates, and fats.*

>>> **Reading Check**

PARAPHRASE *What is the first step of the digestive system?*

YOUR EXCRETORY SYSTEM

MAIN IDEA Waste from the food you eat is processed by your excretory system.

The **excretory system** is *the group of organs that work together to remove wastes.* The main organs of this system are the **kidneys**, or *organs that filter water and dissolved wastes from the blood and help maintain proper levels of water and salts in the body.* Other main organs include the bladder, which stores urine until it is ready to be passed out, or removed, from the body. Another main organ is the **colon**, or *the large intestine.*

Foods that are not absorbed in the coiled small intestine move into the shorter but wider large intestine, which is also called the colon. The materials that pass through the large intestine are the waste products of digestion. The waste products become more solid as water is absorbed.

The waste products are pushed into the final section of the large intestine, or the rectum. Muscles in the rectum and anus control the release of solid waste, or feces.

The excretory system controls your body's water levels. Your skin and lungs also help to remove waste from your body. Your skin gets rid of some wastes in the form of sweat. Your lungs get rid of carbon dioxide when you exhale, or breathe out.

Kidney

Ureter

Bladder

Urethra

Many wastes are excreted through the kidneys and bladder. *What body system sends a signal that the bladder is full?*

DIGESTIVE AND EXCRETORY PROBLEMS

MAIN IDEA Some problems can occur in the digestive and excretory systems.

Have you ever had a stomachache or felt uncomfortable after eating? Several factors can cause problems of the digestive system. As you might imagine, many issues may be related to what you eat, but other problems can be symptoms of serious illness.

Have you ever had a *stomachache* after eating or felt **uncomfortable?**

A common warning from your digestive system is indigestion. The mildest symptom of indigestion is often a bloated or unusually full feeling after eating. It can also include belching, painful gas, nausea, or a burning sensation in the stomach area. Indigestion is often your digestive system's way of telling you to eat more slowly and healthfully.

Another common problem is heartburn, or a burning sensation in the chest or throat. It is caused by stomach acids flowing back into the esophagus. Heartburn can be caused by diet. However, if it lasts too long, medical attention is required.

Contaminated food or water is often the cause of diarrhea, or watery feces. Diarrhea might also be a symptom of a disease of the colon. A person with severe diarrhea for a long time should also see a medical professional.

Pain in the stomach area can be caused by an ulcer, or *an open sore in the stomach lining.* Ulcers can be caused by bacteria, and alcohol use can also be a factor.

Sometimes a person can feel severe pain caused by mineral crystals, or stones, that develop in the digestive system. Gallstones, which form in the gall bladder, and kidney stones often require medical care.

Appendicitis is inflammation of the appendix. The appendix is a tube about four inches long, located near where the small intestine and large intestine meet. If your appendix becomes inflamed, you feel pain in the lower right side of your body. Appendicitis is very serious and requires emergency surgery.

Hemorrhoids are swollen veins at the opening of the anus. They may be itchy or painful and sometimes cause bleeding.

⟩⟩⟩ Reading Check

PARAPHRASE *What is a common sign that you might need to eat more healthfully?*

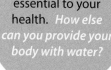

Drinking water is essential to your health. *How else can you provide your body with water?*

Myth vs. Fact

Myth: All forms of bacteria are bad.

Fact: Your digestive system contains between 10 and 100 trillion bacteria. That's ten times the number of cells in your body! Certain bacteria are necessary for the digestion of food. Without "friendly" bacteria, you could eat all you wanted, but the food could pass through your intestines mostly undigested. The trick is to stay away from "unfriendly," or harmful, bacteria.

CARING FOR YOUR DIGESTIVE AND EXCRETORY SYSTEMS

MAIN IDEA A healthful diet and lifestyle are important to digestive and excretory health.

Your digestive and excretory systems are important to your overall physical well-being. As with all body systems, a healthy lifestyle can keep your digestive and excretory systems healthy. Here are some steps you can take:

- **Eat a healthful diet with plenty of fiber.** Choose low-fat and high-fiber foods from all food groups. Include plenty of fresh fruits and vegetables.
- **Take time to eat and chew food thoroughly.** Avoid rushing your meals, which can overload your digestive system. Eating and chewing slowly will also help prevent you from eating too much.

A *healthy* lifestyle can keep your **digestive** and excretory systems **healthy**.

- **Drink plenty of water.** Your digestive system needs water to work properly. Drink six to eight 8-ounce glasses of water each day. Unsweetened fruit juices, low-fat milk, soup, and many fruits and vegetables are also sources of water.

- **Take care of your teeth and gums.** Your teeth begin the digestive process by helping to chew, mash, and grind your food into small pieces to swallow. Brush your teeth at least twice a day with fluoride toothpaste and floss daily. Get regular dental checkups.
- **Wash your hands.** Make a habit of washing your hands thoroughly with soap and water. This is especially important before preparing or eating foods. Regular hand washing will help prevent the spread of bacteria that can upset the digestive system.
- **Avoid risk behaviors.** Alcohol use can interfere with the way your digestive system absorbs nutrients. It can also contribute to ulcers and indigestion. Tobacco use has been linked to ulcers and other digestive problems such as heartburn, gallstones, and kidney stones.
- **Be physically active.** As with all body systems, keeping your body fit and maintaining a healthy weight will have positive effects on your digestive and excretory health.

>>> **Reading Check**

RECALL *Why are your teeth important to the digestion process?* ■

REVIEW

>>> **After You Read**

1. **VOCABULARY** Define the digestive system.
2. **EXPLAIN** In which body part does most of your digestion take place?
3. **DESCRIBE** What role does the stomach play in the digestive process?

>>> **Thinking Critically**

4. **SYNTHESIZE** Describe the path of food from the mouth to the colon.
5. **ANALYZE** Imagine you just ate a huge meal from a fast-food restaurant. Now you have a stomachache. What could be the problem and the cause?

>>> **Applying Health Skills**

6. **GOAL SETTING** Identify a behavior that can promote digestive health but which you are not currently practicing. Use the skill of goal setting to help you make this behavior a habit. Share the steps in your action plan with your classmates.

 Review

 Audio

Your Endocrine *and* Reproductive Systems

BIG IDEA ⟩ Your body has glands and organs that allow it to function and reproduce.

Before You Read

QUICK WRITE Think about when you may have experienced a growth spurt or seen one in someone else. Write one or two sentences to describe the growth spurt.

▶ Video

As You Read

STUDY ORGANIZER Make the study organizer found in the FL pages in the back of the book to record the information presented in Lesson 5.

Vocabulary

› endocrine system
› gland
› metabolism
› diabetes
› reproductive system
› fertilization

 Audio

🔤 Bilingual Glossary

YOUR ENDOCRINE SYSTEM

MAIN IDEA ⟩ Your endocrine system produces chemicals that regulate body functions.

When you start back to school each year, do you notice that many of your classmates have grown? The body system responsible for growth and other changes is the endocrine system. The endocrine system is *the system of glands throughout the body that regulate body functions.*

Do you **notice** that many of **your classmates** have *grown?*

The endocrine system sends messages to the body through the blood in the form of hormones. Hormones are chemical substances produced in glands that help to regulate the way your body functions. A gland is *a group of cells, or an organ, that secretes a chemical substance.* For example, the thyroid gland controls metabolism. Metabolism is *the process by which the body gets energy from food.*

Glands *and* Hormones

The major glands of the endocrine system include the pituitary, thyroid, parathyroid, adrenals, hypothalamus, thymus, and the pancreas. The glands of the reproductive system are also part of the endocrine system. Glands produce specific hormones, which travel through the bloodstream to cells that need them. Some hormones are produced continuously, while others are produced at certain times.

When you feel nervous or stressed, your heart rate and blood flow to the brain may increase. Your blood sugar and blood pressure may rise. Sweat production increases and air passages expand. Digestion and other bodily processes may slow down to conserve energy. Your adrenal glands release the hormone adrenaline, which allows your body to respond to stress.

⟩⟩⟩ Reading Check

RECALL *What is metabolism?*

Problems *of the* Endocrine System

The most common problem of the endocrine system is **diabetes,** *a disease that prevents the body from converting food into energy.* Diabetes can cause heart disease, kidney failure, blindness, and circulatory problems. Diabetes is the seventh leading cause of death in the U.S.

People with diabetes have too much sugar in their blood. Type 1 diabetes occurs when the pancreas cannot produce enough insulin. People with type 1 diabetes need to take extra insulin to help control their blood sugar. Type 2 diabetes occurs when the body cannot properly use the insulin it produces. However, type 2 diabetes is preventable.

Maintaining a healthy weight and staying physically active can help prevent type 2 diabetes.

Another endocrine system problem is an overactive or underactive thyroid. With an overactive thyroid, the gland makes too many hormones. Symptoms of hyperthyroidism include swelling in front of the neck, nervousness, increased sweating, and weight loss.

An underactive thyroid gland is also known as hypothyroidism. In the case of an underactive thyroid, the gland is not making enough hormones to regulate metabolism. Symptoms of an underactive thyroid gland include tiredness, depression, weight gain, hair loss and pain in the muscles and joints.

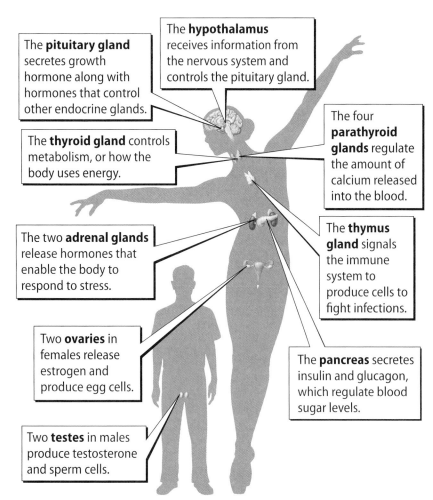

The **pituitary gland** secretes growth hormone along with hormones that control other endocrine glands.

The **hypothalamus** receives information from the nervous system and controls the pituitary gland.

The **thyroid gland** controls metabolism, or how the body uses energy.

The four **parathyroid glands** regulate the amount of calcium released into the blood.

The two **adrenal glands** release hormones that enable the body to respond to stress.

The **thymus gland** signals the immune system to produce cells to fight infections.

Two **ovaries** in females release estrogen and produce egg cells.

The **pancreas** secretes insulin and glucagon, which regulate blood sugar levels.

Two **testes** in males produce testosterone and sperm cells.

Health SKILLS ACTIVITY

Advocacy

Managing *Diabetes*

People with diabetes must carefully keep track of the types and amounts of foods they eat. If they eat foods with too much sugar, they can become ill. If they don't eat enough food, or wait too long to eat, their blood sugar levels can become dangerously low.

On Your Own

Use reliable online resources or library materials to find valid information about diabetes. Create a brochure that encourages teens who have diabetes to manage their condition carefully. Be sure to include information that explains why managing diabetes is so important.

The endocrine system is made up of glands that secrete hormones to control body systems. *Which gland signals the immune system to produce cells to fight infection?*

YOUR REPRODUCTIVE SYSTEM

MAIN IDEA Male and females have different reproductive systems.

One of the body's organ systems is responsible for the survival of the human species. The **reproductive system** includes *the body organs and structures that make it possible to produce children.* The reproductive system is the only body system that is different in males and females. Males produce sperm, or male reproductive cells. About once a month, a female produces an egg cell, or the female reproductive cell that joins with a sperm cell to make a new life.

>>> **Reading Check**

EXPLAIN *What is the purpose of the reproductive system?*

The Male Reproductive System

The main purpose of the male reproductive system is to produce sperm. Each sperm can join with a female reproductive cell and make another human.

The *reproductive system* is the **only** body system that is *different* in **males** and **females.**

The testes are the two male reproductive glands that produce sperm. During puberty the testes produce testosterone, a male hormone. The testes also produce sperm. A male's two testes are located in the scrotum. Next to the testes is a collection of tubes called epididymis. Mature sperm are stored in the epididymis. The scrotum keeps the testes at the right temperature to produce sperm.

Once sperm develop, they move to a tube called the sperm duct and are stored. Mature sperm mix with fluids produced by the prostate and other glands. This mixture is called semen. Semen leaves the body from the urethra through the penis. The urethra is a small tube that runs from the bladder along the length of the penis. Both semen and urine pass through the urethra but at different times. Semen exits the penis through ejaculation, which is a series of forceful muscle contractions.

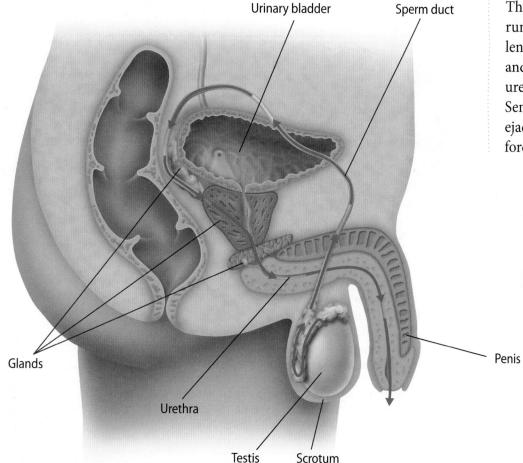

Urinary bladder

Sperm duct

Glands

Urethra

Testis

Scrotum

Penis

Each part of the male reproductive system has a job to do. *What part produces sperm?*

Problems *of the* Male Reproductive System

For male teens, a common reproductive-system problem is injury to the testes. One way to avoid this type of injury is to wear appropriate protective gear. A more serious problem is cancer of the testes, which can occur in males in their teens and young years. Males can also contract sexually transmitted diseases (STDs) or become sterile, which means the reproductive system is unable to produce offspring.

A hernia occurs when an internal organ pushes against or through a surrounding muscle. A hernia appears as a lump or swelling in the groin or lower abdomen. Hernias are caused by muscle weakness or strain. They can be corrected with surgery.

Testicular torsion occurs when a cord that holds the testes together becomes twisted around a testicle. This cuts off blood flow and causes pain and swelling. Testicular torsion is a serious condition that requires immediate medical attention.

The Female Reproductive System

The female reproductive system has three main functions. The first function is to produce and store eggs. Eggs, or ova, are the female reproductive cells. Ovaries are the female endocrine glands that store and release mature eggs. The second function is to create offspring, or babies, through the process of fertilization. The third function is to give birth to a baby.

Fertilization is *the joining of a male sperm cell and a female egg cell to form a fertilized egg.* After fertilization, the fertilized egg travels from the fallopian tube to the uterus. The egg attaches to the wall of the uterus and begins to grow. During the first eight weeks, the fertilized egg is called an embryo. The cells of an embryo divide as it grows. These cells will come together as tissue and form body systems. After eight weeks, the embryo becomes a fetus. The fetus continues to grow in the uterus, and in about nine months, a baby is ready to be born.

Cervix
This is the narrow part of the bottom of the uterus. The opening of the cervix enlarges to allow a baby to leave the uterus during birth.

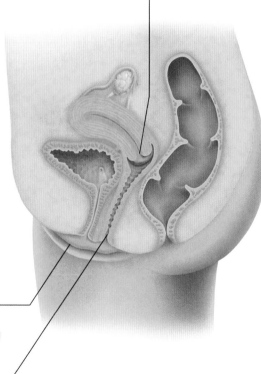

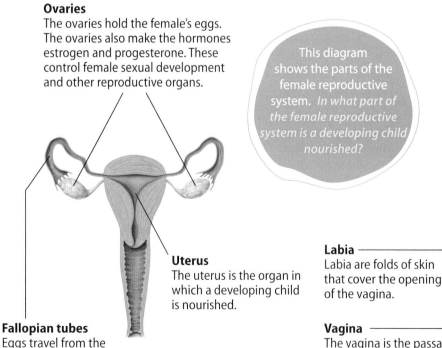

Ovaries
The ovaries hold the female's eggs. The ovaries also make the hormones estrogen and progesterone. These control female sexual development and other reproductive organs.

This diagram shows the parts of the female reproductive system. *In what part of the female reproductive system is a developing child nourished?*

Uterus
The uterus is the organ in which a developing child is nourished.

Fallopian tubes
Eggs travel from the ovaries to the uterus through the fallopian tubes. Eggs are usually fertilized in these tubes.

Labia
Labia are folds of skin that cover the opening of the vagina.

Vagina
The vagina is the passageway that leads from the cervix to the outside of the body. Menstrual flow leaves the body through the vagina. Sperm enter the female reproductive system through the vagina. During birth, a baby leaves the mother's body through the vagina.

Maintaining a healthy reproductive system can help you have a healthy baby one day. *What is the best way to protect yourself from sexually transmitted disease?*

Brand X Pictures/Jupiterimages

During puberty, egg cells mature and are released by the ovaries in a process called ovulation. Ovulation is the release of one mature egg cell each month. Just before one of the ovaries releases an egg cell, the lining of the uterus thickens. The uterus is getting ready to receive and nourish a fertilized egg.

If an egg is not fertilized, the lining of the uterus breaks down and is shed by the body through menstruation. Menstruation is when the lining material, the unfertilized egg, and some blood flow out of the body. Menstruation usually lasts from five to seven days and happens about every 28 days. This is called the menstrual cycle. The menstrual cycle results from hormonal changes that occur in females from the beginning of one menstrual cycle to the start of the next cycle.

Problems *of the* Female Reproductive System

Infertility and STDs can also occur in the female reproductive system. In females, infertility means the inability to become pregnant. The most serious female reproduction problems are cancers. Cancer can occur in the breasts, ovaries, uterus, or cervix.

Ovarian cysts may also become a problem in the female reproductive system. Ovarian cysts are growths on the ovary. Symptoms of ovarian cysts can include a feeling of heaviness in the abdomen and abdominal pain, swelling, and bloating.

Developing Good Character

Advocacy Your friend Tim has noticed a swelling in one of his testes. He is uncomfortable discussing this problem with his parents, but the swelling is not going away. You want to show Tim that you care and advocate for some positive health practices. *What advice can you give Tim to help him prevent a serious problem?*

Caring *for* Your Reproductive System

It is important to maintain the health of your reproductive system. This is true for both males and females. Keeping your reproductive system healthy starts with good hygiene.

- **Shower or bathe daily.** Cleaning all parts of the body regularly is an important aspect of good health.
- **Abstain from sexual activity.** Choosing abstinence will protect you from STDs and unplanned pregnancy.
- **Regular self-exams.** It is recommended that females do breast self-examinations every month. Males should check their testes for lumps, swelling, or soreness.
- **Regular physical check-ups.** See a health care provider for regular check-ups. Females should see a gynecologist, a doctor who specializes in the female reproductive system.

It is important to maintain the health of your reproductive system.

- **Wear protective gear (males).** Males should wear an athletic supporter or protective cup when participating in sports activities. Talk to your coach to find out the proper method of wearing these items.
- **Track menstrual cycles (females).** Your period may not be regular for the first one or two years. Missed periods, heavy bleeding, or severe cramps may require seeing a health care provider.

Reading Check

IDENTIFY *What are the most serious problems that can occur in the male and the female reproductive systems?* ■

LESSON 5

REVIEW

After You Read

1. **VOCABULARY** Define hormones.
2. **DESCRIBE** What does the endocrine system do?
3. **EXPLAIN** What is diabetes?

Thinking Critically

4. **ANALYZE** Why are the male and female reproductive systems different?
5. **APPLY** Anthony will play catcher this year on his baseball team. What special precautions does Anthony need to take to prevent injuries to his reproductive system?

Applying Health Skills

6. **ACCESSING INFORMATION** Use print or online resources to research why females should do regular breast self-exams. Report your findings in a short paragraph.

◯ Review

◉ Audio

Staying hydrated will help keep all your body systems healthy. *Why is it important to drink plenty of water?*

©JGI/Blend Images LLC

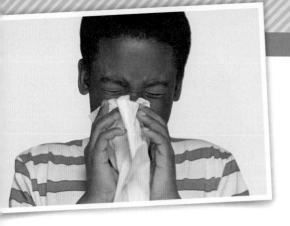

Your Immune System

BIG IDEA Your immune system helps your body defend itself against infections.

Before You Read

QUICK WRITE Write one or two sentences describing the last time you had a cold or fever.

▶ Video

As You Read

STUDY ORGANIZER Make the study organizer found in the FL pages in the back of the book to record the information presented in Lesson 6.

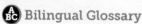

Vocabulary

› immune system
› immunity
› inflammation
› lymphatic system
› lymphocytes
› antigens
› antibodies

🔊 Audio

🔤 Bilingual Glossary

What Teens Want to Know

Do teens need any vaccines? The CDC recommends that teens be vaccinated for chicken pox, hepatitis B, measles-mumps-rubella, tetanus-diphtheria, and human papillomavirus (HPV). Teens should get the first three vaccines if they did not receive them in childhood or if they have not had the diseases.

YOUR IMMUNE RESPONSES

MAIN IDEA Your body has several defenses against pathogens.

Pathogens, germs that cause diseases, are every-where. They are in the air you breathe, the water you drink, and on the objects you touch. Most bacteria, viruses, and other pathogens do not make you sick. Your body has natural barriers between you and pathogens.

Your first line of defense against infection is your body's natural barriers in other body systems. Natural barriers are your skin, the saliva and stomach acid in the digestive system, and mucous membranes in the respiratory system. The circulatory system and nervous systems also work together to raise the body's temperature, killing the pathogens with fever.

When pathogens get past the body's natural barriers, your immune system responds. The immune system is *a combination of body defenses made up of all the cells, tissues, and organs that fight pathogens that enter the body.*

Your immune system has two main responses—nonspecific and specific. Together these responses provide immunity, or *the ability to resist the pathogens that cause a particular disease.*

Your **first line of defense** against infection is your body's *natural barriers.*

If a pathogen gets through, the immune system takes action with a specific response. It recognizes specific pathogens that have attacked before. Once your immune system has created a specific response, those respnse cells remain in your body. When the same pathogen attacks again, your immune system is prepared to fight it and reacts right away.

Reading Check

RECALL *What parts of the digestive system provide barriers to pathogens entering the body?*

Inflammation

White blood cells flow through the circulatory system. Their job is to fight germ-causing diseases and pathogens. They do most of their work attacking pathogens in the fluids outside your blood vessels. They fight infection in several different ways.

Some white blood cells can surround and destroy bacteria. Other white blood cells release chemicals that make it easier to destroy pathogens. Some white blood cells can produce proteins that destroy viruses and other foreign substances in the body.

White blood cells *fight* germ-causing diseases and pathogens.

When pathogens get past the body's first-line defenses, your immune system reacts with what is called a nonspecific response.

A nonspecific response typically begins with inflammation. **Inflammation** is *the body's response to injury or disease, resulting in a condition of swelling, pain, heat, and redness.* The brain sends signals that tell white blood cells to rush to the affected area and destroy the pathogens. Circulation to the area slows down. The symptoms of inflammation are caused by white blood cells surrounding pathogens and destroying them.

When you have an infection, the body starts producing a protein to stimulate the body's immune system. If pathogens spread, your body temperature may rise and cause a fever. A higher body temperature makes it harder for pathogens to reproduce. A fever also signals the body to produce more white blood cells to destroy pathogens. When you have a fever, you know that your white blood cells are fighting pathogens.

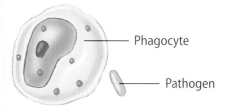

A phagocyte destroys a pathogen by surrounding it and breaking it down. *How does a phagocyte know to go to an inflammation?*

Phagocyte

Pathogen

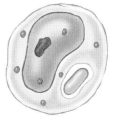

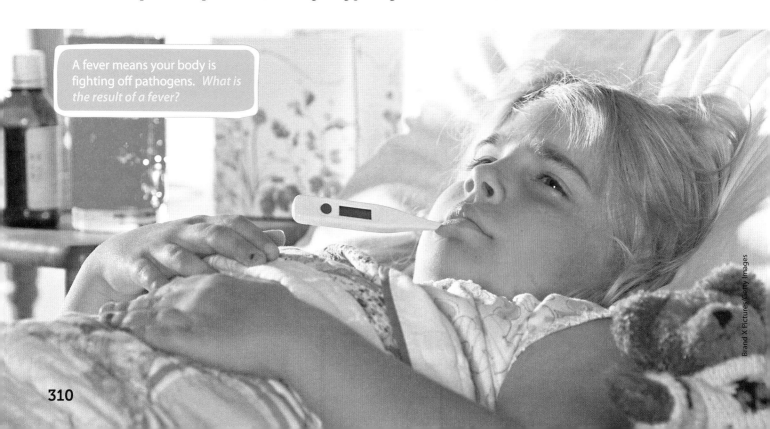

A fever means your body is fighting off pathogens. *What is the result of a fever?*

Brand X Pictures/Getty Images

310

The Lymphatic System

If the body's first-line and second-line defenses do not destroy all the invading pathogens, another type of immune response occurs. This is a specific immune response that calls on the lymphatic system.

The **lymphatic system** is *a secondary circulatory system that helps the body fight pathogens and maintains its fluid balance.* The fluid that circulates through the body's lymphatic system is known as lymph. The *special white blood cells in the lymphatic system* are called **lymphocytes.**

Two main lymphocytes are B cells and T cells. Macrophages are found in the fluid, or lymph. Macrophages attach themselves to invading pathogens and destroy them. Macrophages help lymphocytes recognize the invading pathogens and prepare for future attacks.

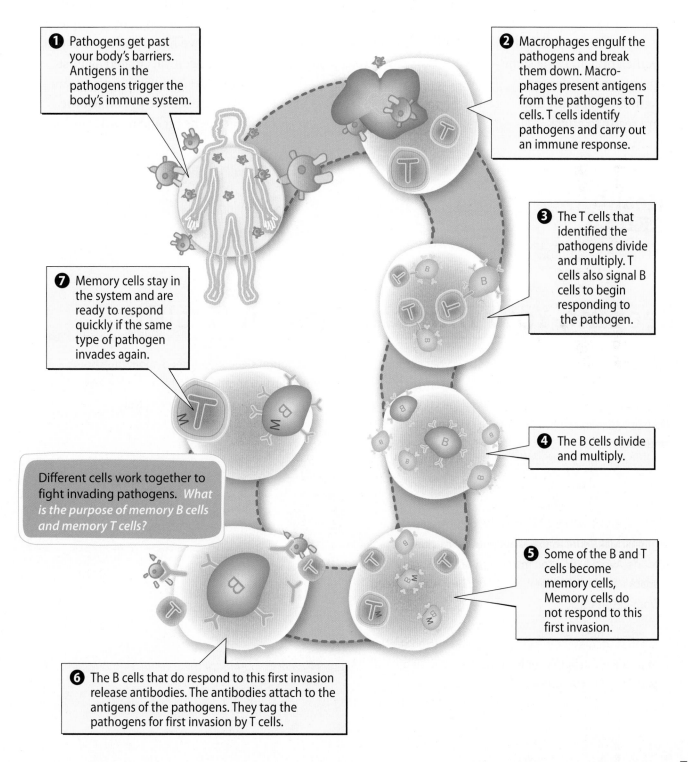

❶ Pathogens get past your body's barriers. Antigens in the pathogens trigger the body's immune system.

❷ Macrophages engulf the pathogens and break them down. Macrophages present antigens from the pathogens to T cells. T cells identify pathogens and carry out an immune response.

❸ The T cells that identified the pathogens divide and multiply. T cells also signal B cells to begin responding to the pathogen.

❼ Memory cells stay in the system and are ready to respond quickly if the same type of pathogen invades again.

❹ The B cells divide and multiply.

Different cells work together to fight invading pathogens. *What is the purpose of memory B cells and memory T cells?*

❺ Some of the B and T cells become memory cells, Memory cells do not respond to this first invasion.

❻ The B cells that do respond to this first invasion release antibodies. The antibodies attach to the antigens of the pathogens. They tag the pathogens for first invasion by T cells.

Antigens *and* Antibodies

Lymphocytes react to antigens. Antigens are *substances that send the immune system into action.* The immune system responds to these antigens by producing antibodies. These are *proteins that attach to antigens, keeping them from harming the body.*

Lymphocytes known as B cells produce a specific antibody for each specific antigen. If the same type of pathogen invades the body again, these specific antibodies are ready to attack.

T cells in the lymphatic system can do two things. T cells known as helper cells stimulate the production of B cells to produce antibodies. T cells called killer cells attach to invading pathogens and destroy them.

Many kinds of **cells** work **together** in your immune system to *fight* **invading pathogens.**

Some of the new B cells and T cells do not react to pathogens immediately. These B cells and T cells wait and are ready to react if the same kind of pathogen invades the body again. These cells are called memory B cells and memory T cells.

These cells help your immune system stop diseases that have attacked before. For example, if you have had measles, or if you have been vaccinated against measles, your immune system remembers. It will attack the antigens for the measles virus.

Health SKILLS ACTIVITY

Practicing Healthful Behaviors

Keeping Your *Immune System* Healthy

You can play an active role in keeping your immune system healthy. A healthy immune system means your body will be better able to fight off infection. Follow these tips to help keep your immune system in top condition.

✳ Get plenty of physical activity and regular exercise.

✳ Eat plenty of vitamin-rich fruits, vegetables, and whole grains. Make sure that you get enough calcium-rich foods. Avoid eating too many high-fat and sugary snacks. A healthful diet is one of the best ways to keep from getting sick.

✳ Learn strategies for managing stress. Reducing stress in your life can help your immune system fight off pathogens more successfully.

✳ Get enough sleep. Teens need about nine hours a night. Rest strengthens the body's defenses and reduces your chances of becoming ill.

Use online or library resources to find articles that explain how vitamins and minerals can affect the immune system. Create a chart that describes at least five vitamins and minerals that strengthen the immune system. List foods that are rich in these vitamins and minerals.

IMMUNITY

MAIN IDEA The body develops immunity in different ways.

Everyone is born with natural immunity. Even before a mother gives birth, antibodies pass from her body to her developing fetus. However, these immunities last only a few months. The baby's immune system becomes active and produces antibodies on its own to fight pathogens.

Everyone is born with *natural immunity.*

Vaccination causes the immune system to produce antibodies for certain diseases. A **vaccine** is *a preparation of dead or weakened pathogens that is introduced into the body to cause an immune response.* This process is called immunization.

Vaccines have been developed for many diseases, such as polio, measles, mumps, chicken pox, hepatitis, and strains of the flu.

Some vaccinations, or shots, are given in a series over several months. Others must be given repeatedly over a lifetime. If you cut yourself on a piece of rusty metal, a doctor may ask when you got your last tetanus shot. Rusty metal can introduce harmful pathogens into your body. To fight them, your immune system will need antibodies. To stay healthy, it is important to keep your vaccinations current. If you have not had a certain vaccination for a while, you may need what is called a booster shot.

> ### >>> Reading Check
>
> **PARAPHRASE** *How does a baby gain natural immunities?* ■

> Your skin is a first line of defense against germs. *Why do you think it is important to wash you hands frequently?*

>>> After You Read

1. **VOCABULARY** Define immune system.
2. **RECALL** What does the lymphatic system do in the body?
3. **IDENTIFY** Name three ways the body achieves immunity against diseases.

>>> Thinking Critically

4. **EXPLAIN** What is the difference between a nonspecific immune response and a specific immune response? Which kind of response does the immune system "remember"?
5. **APPLY** Why should you avoid drinking from the same container as a friend who has a cold?

>>> Applying Health Skills

6. **ACCESSING INFORMATION** Each of the 50 states determines which vaccines they require for students entering school. Research the vaccine requirements in your state. Why do you think states require certain vaccines before students enter school?

⟳ Review

🔊 Audio

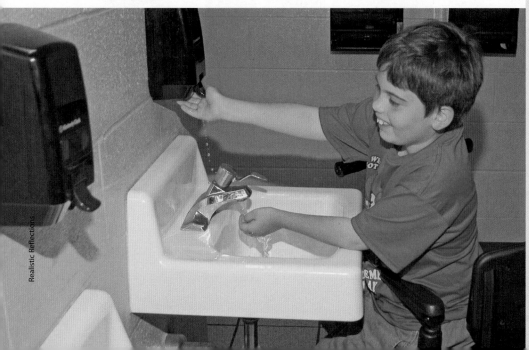

Realistic Reflections

Hands-On HEALTH ACTIVITY

How Muscles and Bones Work Together

WHAT YOU WILL NEED

* 2 strips of stiff poster board, 2" x 10"
* Hole punch
* Metal fastener
* 2 long balloons
* String

WHAT YOU WILL DO

1 Round off one end of each strip. Punch holes about 1 inch from both ends of one strip. This strip will represent the upper arm. In the other strip, punch one hole about 1 inch from the rounded end. Punch another hole 4 inches from that same end. This strip will be the forearm.

2 With a fastener, loosely join the rounded ends of the two strips. This represents the elbow joint. Slightly inflate the balloons. Tie knots in both ends of each balloon.

3 Tie the end of one balloon to each end of the upper arm. This balloon represents the triceps muscle on the back of the arm.

4 Tie the second balloon between the top of the upper arm and the second hole in the forearm. This represents the biceps muscle.

5 Experiment with moving the upper arm while the elbow joint rests on a surface. Observe what happens to the balloons and to the forearm.

The muscles, bones, and joints in the body work like living levers. They use the same principles used in lifting and moving machines such as cranes. The small movement of one arm of a lever causes a larger movement in the other arm. This activity will show you how to make a model arm. Your finished arm will demonstrate how muscles and bones work together in a system of joints and levers.

WRAPPING IT UP

What happens to the biceps when the forearm is extended? What happens when it is closed? How do the triceps and biceps work together? Observe how far the forearm moves when the upper arm is moved a short distance. Compare these distances.

READING REVIEW

FOLDABLES and Other Study Aids

Take out the Foldable® that you created and any study organizers that you created.
Find a partner and quiz each other using these study aids.

LESSON 1 Your Skeletal and Muscular Systems

BIG IDEA Your skeletal and muscular systems work together to make your body move.

* Your skeletal system provides your body with a framework.
* Your muscular system allows your body to move and helps keep it stable.
* You can help keep your bones and muscles healthy.
* Your bones and muscles can develop problems.

LESSON 2 Your Nervous System

BIG IDEA Your nervous system controls and sends messages throughout your body.

* Your movements and body processes are controlled by the nervous system.
* Injury or disease can harm the nervous system.
* Healthy behaviors can protect the nervous system.

LESSON 3 Your Circulatory and Respiratory Systems

BIG IDEA Your heart is the center of your circulatory system, and your lungs are the center of your respiratory system.

* Your circulatory system is like a transportation system inside your body.
* Your respiratory system controls your breathing.
* You can help keep your heart, blood vessels, and lungs healthy and strong.

LESSON 4 Your Digestive and Excretory Systems

BIG IDEA Your digestive and excretory systems process the food you eat for use by your body.

* Digestion is the first step in the way your body processes the food you eat.
* Waste from the food you eat is processed by your excretory system.
* A healthy diet and lifestyle are important to digestive and excretory health.

LESSON 5 Your Endocrine and Reproductive Systems

BIG IDEA Your body has glands and organs that allow it to function and reproduce.

* Your endocrine system produces chemicals that regulate body functions.
* Males and females have different reproductive systems.
* Good hygiene is important to keeping your reproductive system healthy.

LESSON 6 Your Immune System

BIG IDEA Your immune system helps your body defend itself against infections.

* Your body defends itself against pathogens, or germs.
* The body develops immunity in a variety of ways.
* You can help your body become better able to fight off infection.

 Review

 Web Quest

ASSESSMENT

Reviewing Vocabulary *and* Main Ideas

> neurons
> trachea

> central nervous
> system

> skeletal muscles
> veins

> joint

>> On a sheet of paper, write the numbers 1–6. After each number, write the term from the list that best completes each statement.

LESSON 1 Your Skeletal and Muscular Systems

1. The point at which two bones meet is called a _____.

2. The muscles attached to bones that enable you to move your body are called _____.

LESSON 2 Your Nervous System

3. Your nervous system is made up of _____.

4. The _____ is made up of the brain and spinal cord.

LESSON 3 Your Circulatory and Respiratory Systems

5. _____ carry blood back to the heart.

6. The _____ is the tube in your throat that allows air into and out of the lungs.

>> On a sheet of paper, write the numbers 7–12. Write True or False for each statement below. If the statement is false, change the underlined word or phrase to make it true.

LESSON 4 Your Digestive and Excretory Systems

7. Saliva is a digestive juice produced by the <u>liver</u>.

8. The <u>excretory system</u> is the group of organs that work together to remove wastes.

LESSON 5 Your Endocrine and Reproductive Systems

9. The chemicals secreted by the endocrine glands are called <u>hormones</u>.

10. The release of one egg cell each month is called <u>fertilization</u>.

LESSON 6 Your Immune System

11. The immune system is a combination of body defenses made of up of the cells, tissues, and organs that fight <u>vaccines</u>.

12. <u>Immunity</u> is the body's response to injury or disease, resulting in a condition of swelling, pain, heat, and redness.

 eAssessment

☁ *Thinking* Critically

13. ANALYZE Which body systems do you think benefit the most from healthful eating habits? Explain your answer.

14. ASSESS What factor do you think poses the biggest risk to the health of your respiratory system? Explain your answer.

15. PREDICT Describe the possible consequences for the rest of the body if the digestive system is not working properly.

16. INTERPRET Sometimes after you get a vaccination, you later have to get more of the same vaccine. Why do you think you might need this "booster shot"?

●◆ *Write* About It

17. EXPOSITORY WRITING Choose one of the body systems discussed in this chapter. Write an article or blog post for teens on ways they can protect this body system. Include information on how this body system functions and how it can affect other body systems.

18. ADVOCACY Diabetes is a disease that has been termed an epidemic in the U.S., which means that it is out of control. In this chapter, you have learned some of the risk factors for type 2 diabetes. Use online or library resources to learn more about this disease. Write a blog post or letter to the editor recommending healthful actions people can take to help prevent type 2 diabetes.

Ⓐ Ⓑ Ⓒ Ⓓ STANDARDIZED TEST PRACTICE

Reading
Read the passage below and then answer the questions that follow.

Some people believe that cracking your knuckles is detrimental to your health. In fact, many people believe this habit can even cause arthritis, a skeletal-system disease that causes pain and loss of movement in the joints. However, this is just a myth. The sharp popping noise you hear when you flex your knuckles is actually gas bubbles escaping inside the joints in your fingers.

Have you heard the urban legend that people who can bend their fingers far back are double-jointed? This is another myth. There is no such thing as being double-jointed. People who can bend their fingers back have the same number of joints that everyone else does. The real difference is that the ligaments surrounding their joints stretch more than normal.

1. The word *detrimental* in the first paragraph seems to mean
  A. good.
  B. indifferent.
  C. resourceful.
  D. harmful.

2. Which statement best captures the main idea of the passage?
  A. Arthritis is a skeletal-system disease that causes pain and loss of movement.
  B. A number of myths exist about the skeletal system.
  C. People who can bend their fingers far back are not really double-jointed.
  D. Cracking your knuckles is not really harmful.

YOUR BODY SYSTEMS

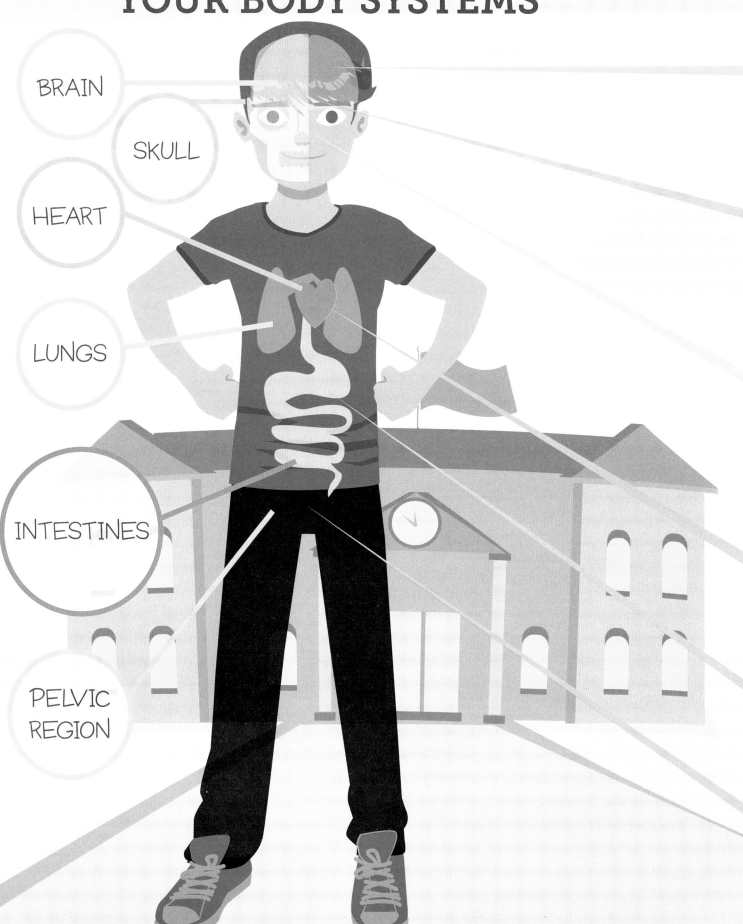

BRAIN

SKULL

HEART

LUNGS

INTESTINES

PELVIC REGION

WHAT'S GOING ON IN THERE?

NERVOUS SYSTEM

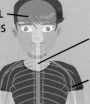

BRAIN
NEURONS

CENTRAL NERVOUS SYSTEM
SPINAL CORD
PERIPHERAL NERVOUS SYSTEM

FUNCTIONS	PROBLEMS	CARE
Controls body processes	Injury	Healthful choices
Physical sensations and reactions	Diseases	Plenty of sleep
Thoughts, language, and memory	Epilepsy	Good hygiene
Movement	Alcohol and drugs	Basic safety

SKELETAL AND MUSCULAR SYSTEMS

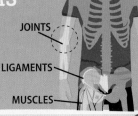

BONES
JOINTS
LIGAMENTS
MUSCLES

FUNCTIONS	PROBLEMS	CARE
Support	Fracture	Diet
Movement	Dislocation	
Protection	Sprain/strain	Physical Activity
Storage of red blood cells, fat, and calcium	Overuse	

ENDOCRINE SYSTEM

GLANDS
HORMONES

FUNCTIONS	PROBLEMS	CARE
• Produces hormones • Controls metabolism • Manages blood pressure and blood sugar	• Diabetes • Overactive glands • Underactive glands	• Maintain healthful diet and weight • Ensure good personal hygiene to prevent infection

CIRCULATORY AND RESPIRATORY SYSTEMS

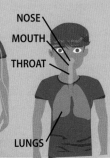

NOSE
MOUTH
THROAT
HEART
BLOOD VESSELS
CAPILLARIES
LUNGS

FUNCTIONS	PROBLEMS	CARE
• Deliver nutrients • Remove wastes • Transport oxygen • Remove carbon dioxide	• High blood pressure • Heart attack • Stroke • Arteriosclerosis • Anemia • Leukemia • Environmental factors • Illnesses	• Physical activity • Healthful diet • Avoid tobacco • Stress management • Good hygiene

DIGESTIVE AND EXCRETORY SYSTEMS

LIVER
STOMACH
BLADDER
PANCREAS
GALLBLADDER
KIDNEYS
COLON

FUNCTIONS	PROBLEMS	CARE
• Break down food • Process nutrients • Remove waste • Manage body's water levels	• Stomachache • Heartburn • Indigestion • Ulcers • Gallstones and kidney stones • Appendicitis	• Fiber-rich diet • Eat slowly • Plenty of water • Good personal hygiene

REPRODUCTIVE SYSTEM

REPRODUCTIVE CELLS
SPERM
EGG

FUNCTIONS	PROBLEMS	CARE
• Produces eggs (female) • Produces sperm (male) • Enables fertilization • Protects and nourishes fetus	• Injury • Cancer • Hernia • Infertility • Ovarian cysts	• Good hygiene • Breast self-exams (females) • Testicular self-exams (males)

Unit 8

tobacco, alcohol + other drugs

Tobacco

LESSONS

PREMIUM ONLINE RESOURCES 〉 Audio Videos Bilingual Glossary Fitness Zone  Web Quest  Review

Tobacco use can pose serious risks to your health. *What are some of the consequences of using tobacco?*

Facts About Tobacco

BIG IDEA The substances contained in tobacco products are very harmful to your health.

>> **Before You Read**

QUICK WRITE Make a list of as many tobacco products as you can think of prior to reading this lesson. Briefly describe each product.

▶ Video

>> **As You Read**

 Study Organizer

Make the Foldable® found in the FL pages in the back of the book to record the information presented in Lesson 1.

>> **Vocabulary**

> nicotine
> addictive
> smokeless tobacco
> snuff
> tar
> bronchi
> carbon monoxide

 Audio

 Bilingual Glossary

Myth vs. Fact

Myth: Cigarettes advertised as "all natural" are safer to smoke than regular cigarettes.
Fact: All cigarettes are harmful to the human body. The smoke from any cigarette contains tar and carbon monoxide. There is no evidence that any type of cigarette is less harmful than another.

WHAT IS TOBACCO?

MAIN IDEA Tobacco is a harmful and addictive substance in all its forms.

Tobacco is a woody, shrub-like plant with large leaves. It is grown throughout the world. Tobacco contains harmful substances that are released when a person smokes or chews it. Tobacco companies add more harmful ingredients when they prepare tobacco to be sold. It is estimated that there are more than 4,000 chemicals in tobacco—many of which have been proven to cause cancer. Some of the same ingredients found in cleaning products or pest poisons are added to tobacco products.

Forms *of* Tobacco

Tobacco companies harvest leaves from tobacco plants. The leaves are then prepared for smoking or chewing. Tobacco products come in many forms. The most common include cigarettes, cigars, pipes, specialty cigarettes, and smokeless tobacco.

CIGARETTES These are the most commonly used forms of tobacco products. Cigarettes contain shredded tobacco leaves. They may also have filters intended to block some harmful chemicals. However, filters do not remove enough chemicals to make cigarettes less dangerous. Also, some tobacco users may try not to inhale the smoke, but smoking in any form is not safe for your body.

There are more than *4,000* **chemicals in tobacco.**

Tobacco smoke contains many harmful chemicals that put smokers at risk for lung diseases, heart disease, and various types of cancer. Tobacco also contains **nicotine,** *an addictive, or habit forming, drug.* **Addictive** means *capable of causing a user to develop intense cravings.* These cravings make it hard to quit.

Flora Torrance/Life File/Getty Images

Nicotine is found in all types of cigarettes, cigars, pipe tobacco, and smokeless tobacco. The nicotine in tobacco products is so addictive that most users find it very difficult to stop using tobacco once they start. You will learn more about nicotine addiction and ways to avoid tobacco later in this chapter.

CIGARS AND PIPES As is the case with cigarettes, the tobacco used in cigars and pipes is made up of shredded tobacco leaves. However, one large cigar can contain as much tobacco and nicotine as an entire pack of 20 cigarettes. Pipes and cigars also cause some of the same serious health problems that cigarettes do. Cigar smoke contains up to 90 times more cancer-causing chemicals than those found in cigarette smoke. People who smoke cigars or pipes are more likely to develop mouth, tongue, or lip cancer than people who do not use tobacco. Cigar and pipe smokers also face an increased risk of dying from heart disease compared to nonsmokers.

SPECIALTY CIGARETTES

This category includes flavored, unfiltered cigarettes and clove cigarettes. Flavored cigarettes are typically imported from other countries. In addition to clove, some have added flavors such as cherry, strawberry, or cinnamon. The U.S. has tried to ban sales of flavored cigarettes in an effort to discourage young people from trying them. Flavored tobacco is also smoked in special water pipes called hookahs. These forms of tobacco contain higher concentrations of harmful chemicals than do regular cigarettes.

ELECTRONIC CIGARETTES

These products look like regular cigarettes. "E-cigarettes" do not contain tobacco, but they still have nicotine inside. Liquid nicotine is heated, and users inhale the vapor. These "smokeless cigarettes" are sometimes marketed as a way to help smokers quit or as an alternative to regular cigarettes. However, the nicotine in electronic cigarettes has the same harmful addictive effects as that found in regular cigarettes.

SMOKELESS TOBACCO

Smokeless tobacco is *ground tobacco that is chewed, placed inside the mouth along the gum line, or inhaled through the nose.* It comes in two forms. Chewing tobacco is often called "dip" or "spit tobacco." Snuff is *finely ground tobacco that is inhaled or held in the mouth or cheeks.* While it does not affect the lungs the way smoking does, smokeless tobacco is not a safe alternative to cigarettes. It contains addictive nicotine and other chemicals and causes gum disease and oral cancer.

>>> **Reading Check**

IDENTIFY *What are the most common health risks associated with cigars and pipes?*

Sports leagues have restricted the use of smokeless tobacco by athletes. *Describe the influence this might have on sports fans.*

Design Pics Inc./Alamy

CHEMICALS IN TOBACCO

MAIN IDEA Tobacco contains many chemicals that can harm your body.

armful chemical compounds exist in all forms of tobacco. These are released when a person smokes or chews tobacco. Chemicals in tobacco can also affect nonsmokers who inhale others' smoke. Most of these chemicals hurt your body's ability to work properly.

Harmful chemicals exist in all forms of tobacco.

- **Nicotine** is one of the harmful substances found in tobacco leaves and in all tobacco products. Nicotine is an addictive, or habit-forming, drug. A person begins to depend on it after it has been in the body regularly. A person can become addicted to nicotine very quickly. Nicotine has other effects, too. It makes your heart beat faster and raises your blood pressure. It causes dizziness and an upset stomach and reduces the amount of oxygen your blood carries to the brain.

- **Tar** is *a dark, thick liquid that forms when tobacco burns.* When a person smokes, the smoker inhales this dangerous substance. When it is inhaled, tar covers the airways and the **bronchi**, which *are passageways that branch from the trachea to each lung.* Lungs covered with tar can become diseased and cause serious breathing trouble.

- **Carbon monoxide** is *a colorless, odorless, poisonous gas that is created when tobacco burns.* Carbon monoxide harms the brain and the heart by reducing the amount of oxygen available to these organs. If too much carbon monoxide enters your body, it can kill you.

Reading Check

NAME *What are three harmful substances found in tobacco smoke?* ■

Smoking tobacco in cigarettes is one way people are at risk of getting cancer or having other health problems. *Explain some of the healthful decision can you make about tobacco.*

REVIEW

After You Read

1. **VOCABULARY** What does the term *addictive* mean? Use it in a complete sentence.
2. **SUMMARIZE** Why is tobacco harmful?
3. **IDENTIFY** Name three substances in tobacco smoke that are harmful to the body.

Thinking Critically

4. **ANALYZE** If many cigarettes have filters, why are they still not safe?
5. **EXPLAIN** What is harmful about the chemicals in tobacco?

Applying Health Skills

6. **COMMUNICATION SKILLS** Ryan has learned that his friend Spencer smokes cigars and uses smokeless tobacco. Spencer tells Ryan that this is not as bad as smoking cigarettes. What facts might you suggest that Ryan share with Spencer to convince him that his tobacco use is still harmful?

  Review

🔊 Audio

Photodisc/Getty Images

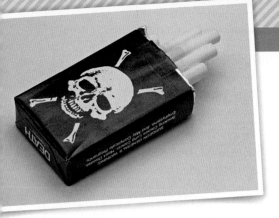

Health Risks *of* Tobacco Use

BIG IDEA Tobacco is a dangerous drug with serious health consequences.

Before You Read

QUICK WRITE List two future goals. Then write a short paragraph telling how a tobacco-related illness could affect each of these goals.

 Video

As You Read

STUDY ORGANIZER Make the study organizer found in the FL pages in the back of the book to record the information presented in Lesson 2.

Vocabulary

› alveoli
› emphysema

 Audio

 Bilingual Glossary

What Teens Want to Know

Will smoking make my skin look worse? Smoking cigarettes causes changes in the skin that speed up the aging process. Only sun exposure does more damage to the skin than smoking. A research study found microscopic wrinkles developing on the skin of smokers as young as 20 years old. In addition, a smoker's skin tends to develop a yellowish or grayish look.

TOBACCO USE IS HARMFUL TO YOUR HEALTH

MAIN IDEA Health experts have been warning about the dangers of tobacco for many years.

The message that smoking is bad for your health is not new. In 1964, the Surgeon General issued a report saying that smoking may be hazardous. A year later, tobacco companies were ordered to add health warnings to cigarette packages. Since then, the warning labels have become more prominent. Other tobacco products now also carry similar warnings.

Tobacco use has serious consequences. The chemicals in tobacco and tobacco smoke can cause damage to most of the body's systems. In the United States, more than 400,000 people die every year from smoking-related illnesses. Tobacco use is especially damaging to teens because their bodies are still growing. The chemicals in tobacco interfere with this process of growth and development.

You have already learned that nicotine raises the heart rate and blood pressure. Tobacco users often cannot run as long or as fast as they did before they started smoking. They get sick more often and tend to stay sick longer. Tobacco use can cause diseases of the mouth and lungs.

400,000 people *die* every year from **smoking-related illnesses.**

Tobacco use also damages the rest of the body. It can cause diseases of the circulatory system, respiratory system, nervous system, digestive system, and excretory system. Many of these illnesses can be prevented if you choose the positive health behavior of staying tobacco free.

Reading Check

RECALL *Why is smoking especially hazardous to teen health?*

Respiratory System

Tobacco smoke damages the air sacs in the lungs. This damage can lead to a life-threatening disease that destroys these air sacs. Smokers are also between 12 and 22 times more likely than nonsmokers to develop lung cancer.

Digestive System

All forms of tobacco increase the risk of cavities and gum disease. Tobacco dulls the taste buds and can cause stomach ulcers. Tobacco use is linked to cancers of the mouth, throat, stomach, esophagus, and pancreas.

Nervous System

Tobacco use reduces the flow of oxygen to the brain, which can lead to a stroke.

Excretory System

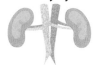

Smokers have at least twice the risk of developing bladder cancer as nonsmokers. Smokeless tobacco can also put users at risk of developing bladder cancer.

Circulatory System

Tobacco use is linked to heart disease. It increases the chances of a heart attack. Smoking also raises blood pressure and heart rate.

Respiratory System

Tobacco smoke contains tar, which coats the inside of the lungs. Smoke damages the alveoli, or *tiny air sacs in the lungs.* When this happens, your lungs are less able to supply oxygen to your body. This damage can cause emphysema, *a disease that results in the destruction of the alveoli in the lungs.* When this disease affects a large part of the lungs, it can cause death.

The chemicals in tobacco also put smokers at a greatly increased risk of developing lung cancer. Lung cancer is the leading cause of death among people who smoke. A person who quits smoking completely can greatly reduce the risk of lung cancer.

A person who **quits smoking** can *reduce the risk* of **lung cancer.**

Circulatory System

Tobacco use affects the circulatory or cardiovascular system, which includes the heart and blood vessels. As nicotine enters the circulatory system, blood vessels constrict, or squeeze together. Over time, the blood vessels can harden. When this happens, the blood vessels cannot carry enough oxygen and nutrients to all the parts of the body that need them.

Using tobacco harms many body systems, causing many health problems or diseases. *Explain the effects of tobacco use on the digestive system.*

Tobacco use also raises blood pressure and heart rate. Blood vessels narrow and harden due to nicotine and other factors. As a result, the heart has to work harder to move blood, oxygen, and nutrients through the body. When the heart has to work harder and blood vessels are narrower, blood pressure goes up. High blood pressure puts more stress on the heart and blood vessels. This increases the chance of a heart attack, stroke, or heart disease.

Nervous System

Your brain needs oxygen. The carbon monoxide in tobacco smoke can cut down the amount of oxygen that the blood can carry to the brain. Nicotine reaches the brain in only a few seconds and attaches to special receptors in brain cells. The brain then adapts by increasing the number of nicotine receptors. Tobacco users then have a strong need for more tobacco.

Digestive System

Smoking can damage your digestive system. It can lead to mouth and stomach ulcers, which are painful, open sores. People with ulcers may not be able to eat certain foods. They may not get all the nutrition they need. Smoking also harms teeth and gums, causing teeth to yellow. Smokers are also more likely to get cavities and gum disease.

Excretory System

Tobacco can also harm your excretory system. Smokers and tobacco users are much more likely to develop bladder cancer than are nonsmokers. Chemicals in tobacco smoke are absorbed from the lungs and get into the blood. From the blood, the chemicals get into the kidneys and bladder. These chemicals damage the kidneys and the cells that line the inside of the bladder and increase the risk of cancer. Smoking tobacco is also a factor in the development of colorectal cancer, a cancer that affects the colon and the rectum.

>>> **Reading Check**

SUMMARIZE *List two harmful effects of smoking tobacco.*

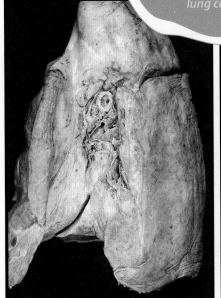 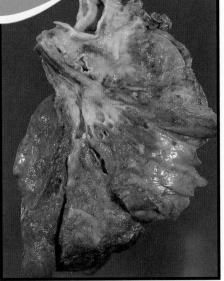

The first photo is a healthy lung. The second photo is a cancerous lung. Healthy lungs provide oxygen to your body. *Evaluate the choices you can make to decrease your risk of getting lung cancer.*

Health SKILLS ACTIVITY

Decision Making

Fresh-Air *Friend*

Mike gets a ride to school with his friend Ashley and her mom. Mike is concerned because Ashley's mom smokes in the car. He hasn't mentioned it because he doesn't want to sound ungrateful. He has tried opening a window, but Ashley says it makes her cold. Mike doesn't know what to do or say. Use the decision-making process to help Mike make a decision.

* State the situation.

* Consider your values.

* List the options.

* Make a decision and act.

* Weight the possible outcomes.

* Evaluate the decision.

 On Your Own

Apply the six steps of the decision-making process to Mike's problem. What are Mike's options? Show how Mike makes a healthful decision.

EFFECTS OF TOBACCO USE

MAIN IDEA Tobacco use causes both short-term and long-term damage to the body.

Tobacco use causes changes in the body. Some of the effects of tobacco use are immediate. These short-term effects can often be felt right away.

- **Cravings** Nicotine is a very addictive drug, which means it causes the body to want more of it. A person who uses tobacco may feel a need for more very soon after using it.
- **Breathing and heart rate** For a smoker, it becomes harder to breathe during normal physical activity. It is more difficult for a tobacco user to work out for a long period of time. Nicotine also causes the heart to beat faster than normal.

Tobacco use causes **changes** in **the body.**

- **Taste and appetite** Tobacco use dulls taste buds and reduces appetite. Tobacco users may lose much of their ability to enjoy food. However, when a person quits using tobacco, taste buds will heal.
- **Unpleasant feelings** Tobacco users may experience dizziness. Their hands and feet may also feel colder than normal.
- **Unattractive effects** Tobacco use causes bad breath, yellowed teeth, and smelly hair, skin, and clothes. It also ages the skin more quickly.

As you have learned, the chemicals in tobacco cause damage to many body systems. The impact of tobacco use is not limited only to smokers and other tobacco users. Simply being around other people who smoke can also cause health problems. Some of the long-term damage caused by tobacco use can even be life-threatening.

- **Bronchitis** Tobacco smoke can damage the bronchi, or the passages through which air travels to the lungs. Also, a buildup of tar in the lungs can cause a smoker to have fits of uncontrollable coughing.
- **Emphysema** This disease can make a person use most of his or her energy just to breathe. Emphysema is a common cause of death for smokers.
- **Lung cancer** Nearly 90 percent of lung cancer deaths are caused by smoking.
- **Heart disease** The nicotine in tobacco greatly increases the risk of heart attack or stroke.
- **Weakened immune system** Long-term tobacco use harms the body's defenses against various diseases. Tobacco users are also more likely to get common illnesses such as coughs, colds, and allergies.

> ### Reading Check
>
> **DESCRIBE** *What are two long-term and short-term effects of tobacco use?* ■

After You Read

1. **DEFINE** What does *emphysema* mean? Use it in an original sentence.
2. **SUMMARIZE** Describe the ways in which smoking harms the systems in the body.
3. **IDENTIFY** What is the leading cause of death among people who smoke?

Thinking Critically

4. **ANALYZE** Which of the health risks associated with tobacco use do you consider the most serious? Explain your answer.
5. **APPLY** Bethany is at a party where another girl lights a cigarette. When Bethany points out that smoking is bad for her health, the other girl shrugs. "I'm a strong person," she says. "I can quit any time I want to." How might Bethany reply?

Applying Health Skills

6. **ACCESS INFORMATION** Conduct more research into the harmful effects of tobacco. Use health journals, magazines, and web sites of national organizations. Write a short report about the information you find.

 Review

 Audio

Tobacco Addiction

BIG IDEA Tobacco contains nicotine, which is an extremely powerful and addictive drug.

Before You Read

QUICK WRITE Make a list of habits you know are hard to break. Choose one that interests you and write a paragraph describing what you think it would be like to change that habit.

 Video

As You Read

STUDY ORGANIZER Make the study organizer found in the FL pages in the back of the book to record the information presented in Lesson 3.

Vocabulary

› addiction
› psychological dependence
› physical dependence
› tolerance
› withdrawal
› relapse

 Audio

🆎 Bilingual Glossary

Cultural Literacy

Smokeless Tobacco Smokeless tobacco includes chewing tobacco and a finely ground tobacco known as snuff. The CDC has found that about 11 percent of teen boys and 2.2 percent of teen girls have used smokeless tobacco. However, it is not a safer alternative to smoking. Smokeless tobacco causes a variety of health problems, including white patches in the mouth that can lead to oral cancer.

A POWERFUL DRUG

MAIN IDEA Tobacco contains strong substances that make it difficult to stop using once a person has started.

Many people know the dangers of tobacco use, but they continue to use it over many years. As you have learned, tobacco contains nicotine. Nicotine is a powerful drug that causes addiction, or *a mental or physical need for a drug or other substance.* Scientific studies have shown that nicotine is as addictive as powerful drugs such as cocaine or heroin. This addiction is both psychological and physical.

The Path *to* Addiction

When nicotine enters the body, it interacts with receptors in the tobacco user's brain. The brain sends a message to the body to speed up heart and breathing rates. As heart and breathing rates return to normal, the user wanting more. Tobacco use soon becomes a habit, and a user can quickly become addicted.

Nicotine is as **addictive** as powerful drugs such as **cocaine** or **heroin.**

Studies have shown that as many as 90 percent of adult smokers began using tobacco before the age of 18. Teens are more likely to develop a severe level of addiction than people who begin smoking at a later age. Teens who use tobacco are also much more likely to use drugs such as marijuana, cocaine, and alcohol. For example, a recent national survey shows that more than 90 percent of cocaine users smoked cigarettes before they started using cocaine. Another study has shown that nicotine addiction may lead to other addictions.

Reading Check

DESCRIBE *Why is it especially risky for teens to try tobacco?*

Tobacco use leads to nicotine addiction. Once a person is addicted to the nicotine in tobacco, it becomes very difficult to quit. Here are some facts about how people become addicted to tobacco use:

- The government has found that tobacco companies market to young people. Some people start using tobacco as early as age 11 or 12.
- Research has shown that every day in the United States, more than 6,000 teens and preteens try their first cigarette or other form of tobacco.

- Teens can feel symptoms of nicotine addiction only days or weeks after they first start using tobacco. The symptoms of addiction are felt even before teens start to use tobacco regularly. This can be especially harmful since the teen years are a time of rapid growth and development.
- The earlier in life someone tries tobacco, the higher the chances that person will become a regular tobacco user. Early tobacco use also lowers the chances a person will ever be able to quit.

The best way to prevent tobacco addiction is to never start using tobacco. *Explain why it is difficult to stop smoking once a person has started.*

Health SKILLS ACTIVITY

Accessing Information

Quitting *Tobacco Use*

A person's physical dependence on nicotine makes it very hard to quit using tobacco. Some people cannot quit on their own. Sometimes they need help to overcome the physical addiction to nicotine. Two common methods for quitting are nicotine replacement and certain medications.

* Nicotine replacement may involve nicotine gum, patches, or lozenges for tobacco users over age 18. Adults don't need a prescription for these treatments. You do need a prescription for nicotine inhalers and nasal sprays.

* People who want to quit smoking may join support groups. They can find suggestions about how to stop from organizations such as the American Heart Association or the American Lung Association.

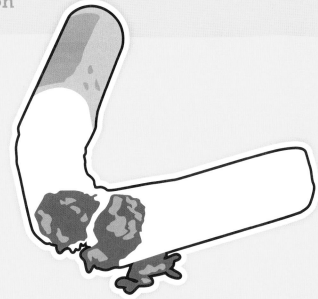

On Your Own

Use various resources to find valid information about a method of quitting tobacco use. Create an informative brochure covering how the method works and how it helps people deal with cravings for nicotine. Present your brochure to the class.

Physical *and* Psychological Dependence

Psychological dependence is *a person's belief that he or she needs a drug to feel good or act normally.* In other words, the desire for tobacco becomes greater than the fear of its dangers. *Physical dependence* is *an addiction in which the body develops a chemical need for a drug.* This can happen quickly for a person who uses tobacco. Teens can develop a physical dependence for nicotine even more easily than adults.

The *desire* for tobacco **becomes greater** than the *fear* of its dangers.

As you have learned, nicotine is a drug. As with any drug, the body will develop a *tolerance,* or *a need for larger and larger amounts of a drug to produce the same effect.* This grows over time and causes a tobacco user to crave nicotine. Anyone who quits tobacco goes through *withdrawal.* This is *a series of symptoms a person experiences when he or she stops using an addictive substance.* When a person goes through withdrawal from tobacco, cravings for nicotine typically increase. Someone who is trying to quit may also have mood changes, feel nervous or irritable, or be extra hungry.

The body undergoes physical changes when a person no longer uses tobacco. For example, the nicotine and carbon monoxide from tobacco are replaced by more oxygen in the blood. The extra oxygen is healthier for the body, but it may affect the brain and result in headaches or dizziness for a while. The extra oxygen in the blood may also cause a person who has quit to feel tingling in the fingers or toes. Since nicotine acts as a stimulant, someone who has stopped using tobacco may suddenly feel extra tired or sleepy.

Sometimes, the symptoms of nicotine withdrawal are so bad that a person starts using tobacco again. Many people who stop smoking have a *relapse,* which is *a return to the use of a drug.* Relapses are common during the first few months after a tobacco user quits.

The physical and psychological effects of withdrawal may lead a person to return to nicotine. However, if someone starts smoking again after working hard to quit tobacco, it can leave that person frustrated and angry. As a result, many people try to quit several times before they finally break the habit. Fortunately, many resources are available to help people either say no to tobacco before they start or to help them kick the habit for good.

>>> **Reading Check**

DESCRIBE *What are the symptoms of nicotine withdrawal?* ■

LESSON 3

REVIEW

>>> **After You Read**

1. **DEFINE** What does the term *addiction* mean? Use it in a complete sentence.
2. **IDENTIFY** What is the substance in tobacco that causes addiction?
3. **RECALL** Explain the difference between physical dependence and psychological dependence.

>>> **Thinking Critically**

4. **SYNTHESIZE** Explain how a person becomes addicted to tobacco.
5. **ANALYZE** Why do you think it is important for a teen never to try tobacco?

>>> **Applying Health Skills**

6. **REFUSAL SKILLS** Some teens try tobacco for the first time because of peer pressure. With a small group, brainstorm effective ways to say no when peers offer or suggest that you use tobacco. Make a list of the best ideas and share them with your class.

 Review

 Audio

Costs *to* Society

BIG IDEA Tobacco has consequences in addition to the harm it causes tobacco users.

Before You Read

QUICK WRITE Make a list of the costs associated with tobacco use. As you read, add costs to your list.

 Video

As You Read

STUDY ORGANIZER Make the study organizer found in the FL pages in the back of the book to record the information presented in Lesson 4.

Vocabulary

› secondhand smoke
› mainstream smoke
› sidestream smoke
› passive smoker

 Audio

 Bilingual Glossary

Developing Good Character

Citizenship Good citizens look out for the welfare of the community. The term *community* includes more than just your neighborhood. It also includes the environment and the air you breathe. Obeying laws that regulate smoking is one way of showing good citizenship. Brainstorm ways of showing good citizenship when it comes to tobacco.

TOBACCO'S MANY COSTS

MAIN IDEA Tobacco use is costly to society.

You have learned about how harmful it is to use tobacco. Unfortunately, tobacco use includes many more costs. Society as a whole pays a price, and people who do not use tobacco may also experience health problems as a result. For example, tobacco companies spend more than $34 million each day on marketing to encourage people to use tobacco products. Whether or not they use tobacco, U.S. taxpayers pay billions of dollars each year in federal taxes to treat the many health problems caused by tobacco use. In addition, many more costs and consequences are associated with tobacco use.

Costs *to* Smokers

Tobacco use is expensive. Researchers have determined that the average cigarette smoker uses one-and-a-half packs each day. The average price per pack in most states has risen to more than $5.

This means the typical smoker spends more than $8 per day on cigarettes. That adds up to $250 over the course of one month, or $3,000 per year. In ten years' time, the average smoker in the U.S. will have spent more than $30,000 on cigarettes.

Society as a whole *pays a price* as a result of **tobacco use.**

Tobacco users also spend more money on health care. They pay higher health insurance rates than nonsmokers. Tobacco users can also expect to live shorter lives and have more health problems than people who do not use tobacco. This is especially true for females. Tobacco use shortens a woman's life by an average of five years. Fires are another health risk posed by smoking. Tobacco left burning is often the cause of fires at home. Carelessly discarded cigarettes or matches can also spark wildfires.

Costs to Nonsmokers

Even if you do not smoke, being around those who do can be harmful. Whenever people smoke near you, you breathe *air that has been contaminated by tobacco smoke* called **secondhand smoke** or environmental tobacco smoke (ETS). Secondhand smoke is filled with nicotine, carbon monoxide, and other harmful substances.

Secondhand smoke comes in two forms: mainstream smoke and sidestream smoke. **Mainstream smoke** is the *smoke that is inhaled and then exhaled by a smoker.* **Sidestream smoke** is *smoke that comes from the burning end of a cigarette, pipe, or cigar.* The U.S. Environmental Protection Agency (EPA) has labeled secondhand smoke as a human carcinogen. This means it causes cancer.

Even if you **do not smoke,** being around those who do can be harmful.

A **passive smoker** is *a nonsmoker who breathes in secondhand smoke.* Adults who do not use tobacco but who regularly breathe secondhand smoke can get sick from it. They can have the same health problems as smokers. An estimated 46,000 nonsmokers die each year from heart disease. About 3,000 additional nonsmokers die of lung cancer each year.

Secondhand smoke is especially harmful to young children and people with asthma. When children are exposed to secondhand smoke, they are more likely to have respiratory and other health problems. Allergies, asthma, ear infections, bronchitis, pneumonia, and heart problems have all been linked to exposure to secondhand smoke.

Secondhand smoke is dangerous to everyone's health. *Evaluate why smoke-free restaurants are considered to be healthier for customers and restaurant workers.*

Costs to Unborn Children

Women who use tobacco while pregnant put their unborn children in serious danger. Their babies could be born too early or too small or even die. The lower a baby's birth weight, the higher the chances that the baby will have health problems. Sudden infant death syndrome (SIDS) is linked to babies whose mothers smoked during or after pregnancy. Babies who are born to mothers who used tobacco while pregnant may also grow and develop more slowly than other children. This can cause health issues throughout childhood.

Costs to Society

Tobacco not only harms the body, it costs the user and society a lot of money. Smoking and other forms of tobacco use have hidden costs. Two additional costs involved in tobacco use are:

- **Lost productivity.** Productivity is how much a person is able to finish in the time he or she works. People who use tobacco have lower productivity levels on the job. They are sick more often than nonsmokers and get less done. Lost productivity costs businesses a lot of money. The nation as a whole pays a large price too. The government estimates that smoking costs the U.S. economy $96 billion per year in lost productivity.
- **Health care costs.** People who use tobacco tend to need more medical treatment than those who do not. If tobacco users have health insurance, it may help pay for some of their treatment. However, because health insurance companies face more costs to cover tobacco users, they charge higher rates for their insurance. If a tobacco user has no health insurance, the government helps cover the costs. This means that every family in the U.S. pays for tobacco use through their taxes.

>>> **Reading Check**

IDENTIFY *Name three groups of people affected by tobacco use.*

C Squared Studios/Getty Images

COUNTERING THE COSTS OF TOBACCO USE

MAIN IDEA Laws and education protect nonsmokers and lower the cost of tobacco use to society.

According to the U.S. Centers for Disease Control and Prevention, each pack of cigarettes sold results in more than $10 in added costs for health care and lost productivity. Some public action groups and Congress have investigated ways to lower these costs. New laws have been proposed that would restrict or even ban the manufacture and sale of tobacco products. Many smokers, however, claim that such restrictions would interfere with their constitutional rights.

Tobacco Taxes

One important action has been taken to reduce the cost of tobacco use to society. Taxes are now added to each pack of cigarettes sold in the United States and in certain states. For example, there is a federal excise tax that adds $1 to the price of each pack of cigarettes.

Additional taxes make it more costly to buy tobacco products and give the government more money to educate people about the dangers of tobacco use.

> **Each pack of cigarettes** sold results in more than *$10* in **added costs** for **health care** and **lost productivity.**

Smoke-Free Environments

Many state and local governments have tried to lower the cost to nonsmokers through laws that ban smoking in public spaces. This protects nonsmokers from dangerous secondhand smoke. It has become common to find businesses and restaurants where smoking indoors or even outdoors is not allowed. These laws protect the health of all restaurant patrons.

Some towns and cities have even made it illegal to smoke in certain outdoor locations, such as beaches, playgrounds, and gardens. The federal government has also passed laws to protect the rights of nonsmokers. For example, since 1989, it has been illegal to smoke on all airplane flights in the United States.

Labeling Laws *and* Advertising Limits

Laws now control the ways that tobacco companies are allowed to market and sell tobacco products. Cigarette packages must have clear warning labels, or disclaimers. The disclaimers state clearly that smoking is harmful. Cans and pouches of smokeless tobacco must also display similar disclaimers. The same laws apply to advertisements for tobacco.

Limits on tobacco advertising are more extensive than ever before. In the United States, laws protect young people from tobacco advertisements. For example, tobacco companies cannot place outdoor ads within 1,000 feet of schools and playgrounds or advertise on television or radio. They cannot make or sell promotional hats, T-shirts, and other items. This is why tobacco companies try to place their products in the media being used by celebrities. It allows them to work around the laws in place.

SURGEON GENERAL'S WARNING: Smoking Causes Lung Cancer, Heart Disease, Emphysema, And May Complicate Pregnancy.

Federal law requires that cigarette packages have one of four different warning labels. *Describe the goal of tobacco warning labels.*

Antismoking Campaigns

Anti-tobacco ads are helping create awareness about the dangers of tobacco use. The goal is to send a message to people of every age that using tobacco is a risk behavior. Antismoking ads also explain that tobacco use has many negative consequences. Smokers who see these ads may recognize the dangers of tobacco. As a result, they may quit or seek treatment. The ads can help non-smokers recognize the benefits of remaining tobacco free.

Antismoking ads are helping **create awareness** about the **dangers of tobacco use.**

State and local governments are also making efforts to stop tobacco use. Most communities enforce laws against selling tobacco to minors. Smoking is often prohibited on school grounds. States have sued tobacco companies to recover costs to the public related to tobacco use. Part of the money awarded in these cases has helped fund anti-tobacco campaigns. Many communities plan activities to promote healthy lifestyles. Individuals can also help to stop or limit tobacco use. You can do your part by encouraging others to avoid tobacco use.

>>> Reading Check

SUMMARIZE *What actions have been taken to lower the cost to society of tobacco use?* ■

Smoke free areas in the community help promote healthy lifestyles. *Describe how no smoking signs placed throughout the community can help promote healthy lifestyles.*

C. Zachariasen/PhotoAlto

>>> After You Read

1. **DEFINE** Explain what the terms *sidestream smoke* and *mainstream smoke* mean. Use both terms in complete sentences.

2. **IDENTIFY** What health care costs are involved with tobacco use?

3. **DESCRIBE** Explain the effects smoking can have on unborn babies and children.

>>> Thinking Critically

4. **ANALYZE** How can laws to protect you from secondhand smoke help to protect your health?

5. **EXPLAIN** What solutions have been offered for the problems of tobacco use? Explain how these solutions seem to help lower the costs of tobacco use to society.

>>> Applying Health Skills

6. **ACCESS INFORMATION** Research the latest government restrictions on tobacco companies or advertisements. Write a paragraph describing your findings.

⟳ Review

🔊 Audio

Saying No *to* Tobacco Use

BIG IDEA It is important to have strategies to resist the strong influences around tobacco use.

Before You Read

QUICK WRITE Write a short dialogue between yourself and a peer who wants you to try tobacco. Show how you can politely but firmly refuse.

 Video

As You Read

STUDY ORGANIZER Make the study organizer found in the FL pages in the back of the book to record the information presented in Lesson 5.

Vocabulary

> target audience
> product placement
> point-of-sale promotion
> cold turkey
> nicotine replacement therapies

 Audio

ABC Bilingual Glossary

⚡ Fitness Zone

Fitness Calendar. To stay healthy and tobacco free, it helps me to have an activity calendar. I can set goals for how much physical activity I want to get each day. When I choose new activities to try, it makes it more interesting. I also like to use an online diary to keep track of how well I am doing with my goal.

WHY DO TEENS USE TOBACCO?

MAIN IDEA Many sources can influence teens to try tobacco.

With all the risks you have read about in this chapter, you may wonder why any teen would use tobacco. Many influences pressure teens to try tobacco or make it look appealing. In this lesson, you will learn more about those factors and about ways to stay tobacco free.

Influence *of* Peers *and* Family

Peer pressure is one of the major reasons that teens try tobacco. People are more likely to use tobacco if their friends use it. Some teens use tobacco in order to fit in with their friends. Others use tobacco because they think it makes them seem cooler or more mature. Others think smoking will help them feel more confident around others. It is important to choose friends who encourage you to make healthful choices, such as committing to be tobacco free.

> It is **important** to choose friends who *encourage* you to make **healthful** choices.

It is also common for teens to start smoking when tobacco is used in their homes. Similar to peer pressure, families can pressure teens into using tobacco even if family members do not directly encourage smoking. For example, parents may warn against the dangers of tobacco use, but they may smoke. Having a sibling who smokes may also encourage tobacco use. Other reasons teens try tobacco include wanting to rebel, being curious, or thinking that the health risks do not apply to them.

> ## Reading Check
>
> **DESCRIBE** *Explain how teens are influenced by their peers to use tobacco.*

Influence *of* the Media *and* Advertising

Another negative influence on teen smoking is the media. Although efforts have been made to reduce this influence, it is still common. Television shows, movies, and video games often show characters having fun while smoking. An estimated one-third of popular movies made for children and teens still show images of people smoking.

Tobacco companies spend millions of dollars to advertise their products. Colorful ads show happy, attractive people smoking. Nine out of ten smokers start smoking by age 18.

> Here is what teens across the United States said in response to statements about tobacco use. *Evaluate how you would respond to these statements.*

Eighty percent of underage smokers use the three most-advertised brands. Tobacco companies see teens as a **target audience,** or *a group of people for which a product is intended.*

One key strategy tobacco companies use is **product placement,** which is *a paid arrangement a company has made to show its products in media such as television or film.* If you see a favorite celebrity smoking, it could have a strong influence on you. This would be especially true if you did not know how harmful tobacco products are to your health. Tobacco companies use another strategy to get shoppers' attention as they wait to pay for other items. **Point-of-sale promotions** are *advertising campaigns in which a product is promoted at a store's checkout counter.*

	Agree	Disagree	No Opinion or Do Not Know
Seeing someone smoke turns me off.	67%	22%	10%
I would only date people who don't smoke.	86%	8%	6%
It is safe to smoke for only a year or two.	7%	92%	1%
Smoking can help you when you're bored.	7%	92%	1%
Smoking helps reduce stress.	21%	78%	3%
Smoking helps keep your weight down.	18%	80%	2%
Chewing tobacco and snuff cause cancer.	95%	2%	3%
I strongly dislike being around smokers.	65%	22%	13%

Source: Centers for Disease Control and Prevention.

Refusing *Tobacco*

Sindhu and Andrea have been friends since the third grade. Now they are older and go to different schools. Andrea spends much of her time with her new friends. One day they meet after school, and Andrea offers Sindhu a cigarette. Sindhu wants to keep her friend, but she does not want to smoke. What should she say to Andrea? Remember the S.T.O.P. strategy:

* **S**ay no in a firm voice

* **T**ell why not.

* **O**ffer another idea.

* **P**romptly leave.

On Your Own Role-play how Sindhu reacts when Andrea asks her to have a cigarette. How can Sindhu use the S.T.O.P. formula in this situation?

RESISTING TOBACCO USE

MAIN IDEA Knowing how to resist tobacco will help you stay tobacco free.

You can protect your health now and in the future if you make a commitment to stay tobacco free. If you make that choice, you will enjoy better physical, mental/emotional, and social health. Ninety percent of adult smokers start smoking before age 18. If you avoid tobacco use as a teen, you will greatly decrease the chance that you will smoke as an adult.

Reasons *to* Say No

Being tobacco free is a safe behavior that includes many benefits. A healthier body is just one of those benefits. Read the list below to learn more.

- **Overall health.** People who smoke get sick more easily and more often than nonsmokers.

- **Clear, healthy skin.** If you use tobacco, your skin cells are less able to take in oxygen and other important nutrients.
- **Fresh breath.** Cigarettes and smokeless tobacco products cause bad breath.
- **Clean, fresh-smelling clothes and hair.** Smokers usually smell like smoke. Stinky cigarette odors stick to clothes and hair and are not easy cleaned.
- **Better sports performance.** Nonsmokers are usually better athletes than smokers because they can breathe more deeply and are more healthy overall.
- **Money savings.** Tobacco is expensive. Increased taxes on tobacco mean the costs will keep going up. Teens who do not buy tobacco will have more to spend on other items such as clothes and music.

- **Environmental health.** By staying tobacco free, you help reduce secondhand smoke. You are also protecting the people around you.

Ways *to* Say No

You take responsibility for your health when you choose not to use tobacco. Choose to spend time with others who are tobacco free. Be prepared to be asked if you want to try tobacco. Practice your refusal skills to help you make the best decision. You can practice saying no in an assertive style that shows you are serious but also that you respect others. Speak in a firm voice with your head up. This will tell others you mean what you say.

Your Rights *as* a Nonsmoker

You have the right to breathe air that is free of tobacco smoke. Many laws are in place to protect nonsmokers. It has become easier to find smoke-free places. As a nonsmoker, you can ask people not to smoke around you.

> **Reading Check**
> **GIVE EXAMPLES** *Identify three ways you can say no to tobacco.*

"I dislike the smell of smoke."

"I can't. I'm on the soccer team and need to keep my lungs in shape."

"My parents would ground me if they found out I was smoking."

"I can't afford to spend all my money on tobacco."

"Smoking is bad for you."

"It's against the law for someone my age to smoke."

"My grandfather had cancer. I don't want to go through what he did."

If someone offers you tobacco, here are some ways to say no. *Identify some other ways you can refuse tobacco.*

Photodisc/Getty Images

BREAKING THE TOBACCO HABIT

MAIN IDEA Many resources are available to people who want to be tobacco free.

Some of the damage done by smoking can never be reversed. However, quitting tobacco does prevent more damage to the body and will improve a person's overall health. The process can be difficult, but there are many resources available.

Some of the **damage** done by *smoking* **can never be reversed.**

The body goes through physical changes when a person no longer uses tobacco. Learning to live without tobacco takes time and a lot of willpower. Tobacco users often try a variety of methods to try to quit before they find one that works for them.

Some people may choose to stop by going cold turkey, which means *stopping all use of tobacco products immediately.* The cold turkey method can be difficult for people because they need help breaking the addiction to nicotine. They will experience withdrawal symptoms for up to six months. One source of help is nicotine replacement therapies (NRTs), which are *products that assist a person in breaking a tobacco habit.* NRTs include nicotine gums, lozenges, and patches worn on the skin.

Many organizations also help users quit. For example, tobacco users can find tips and support groups through the American Lung Association, the American Heart Association, or the American Cancer Society. Some schools now have programs to help teens who want to quit using tobacco. If you know someone who is trying to get rid of a tobacco habit, you can share the following information:

- List your reasons.
- Get support and encouragement from family or friends.
- Set small goals.
- Choose tobacco-free places to spend time.
- Change your tobacco-related habits.
- Be physically active.
- Keep trying.

In addition to the types of health organizations noted above, other good sources of information for people who want to quit include hospitals, web sites, and libraries. Doctors and nurses can also be helpful in helping people quit tobacco use. Medical professionals can offer advice about ways to deal with the symptoms of nicotine withdrawal. Doctors may also prescribe certain NRTs for those unable to quit on their own.

>>> **Reading Check**

DEFINE *What are nicotine replacement therapies?* ■

>>> **After You Read**

1. **DEFINE** Explain what the term *cold turkey* means. Use it in a complete sentence.
2. **IDENTIFY** List at least three benefits of being tobacco free.
3. **SUMMARIZE** Explain how people who wish to stop using tobacco can get help.

>>> **Thinking Critically**

4. **INFER** Why is it easier never to start using tobacco than it is to quit once you have started?
5. **APPLY** How would you influence a peer to make the healthful choice to quit smoking?

>>> **Applying Health Skills**

6. **GOAL SETTING** Make a plan to help someone quit using tobacco. Research what resources are available online. Include alternative activities the tobacco user can do when he or she experiences the urge to use tobacco.

⟳ Review

🔊 Audio

Hands-On HEALTH ACTIVITY

Inside *Your* Lungs

WHAT YOU WILL NEED
* 64 sugar cubes
* Cellophane tape
* One sheet of graph paper

WHAT YOU WILL DO

1 Use the sugar cubes to make a square that is 4 cubes long, 4 wide, and 4 deep.

2 Use tape to hold the sugar cubes together.

3 Use the graph paper to figure out how many paper squares can be covered by the large sugar rectangle.

4 Remove the tape and measure how many paper squares can be covered by a single cube. Remember to record all six sides.

5 Multiply this single cube measurement by 64.

WRAPPING IT UP
Which covers more graph paper squares: the large sugar rectangle or the 64 cubes? The cubes represent your alveoli. Just breathing in does not get oxygen to your body cells. It only gets it to your lungs. Alveoli pass oxygen to your blood. Dividing the lungs into many smaller sacs (alveoli) gets more oxygen to your blood faster. Warm-blooded animals like us need this trick. We need to get oxygen at a fast enough rate to perform all our activities.

You've learned that smoking affects your lungs. Do you remember what's inside your lungs? Each lung contains millions of little sacs called alveoli. When you inhale, oxygen and anything else you breathe makes its way into these 600 million little sacs. Blood vessels surround the alveoli. They pick up oxygen from the alveoli and carry it to your cells. Smoking makes the alveoli less able to handle the oxygen your body needs.

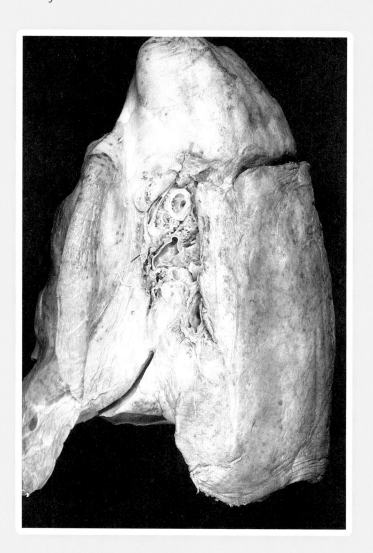

© McGraw-Hill Education/Dennis Strete

READING REVIEW

FOLDABLES **and Other Study Aids**

Take out the Foldable® that you created and any study organizers that you created.
Find a partner and quiz each other using these study aids.

LESSON 1 **Facts About Tobacco**

BIG IDEA The substances contained in tobacco products are very harmful to your health.

* Tobacco is a harmful and addictive substance in all its forms—cigarettes, cigars and pipes, and smokeless.
* Tobacco contains many chemicals that can harm your body, including nicotine, tar, and carbon monoxide.

LESSON 2 **Health Risks of Tobacco Use**

BIG IDEA Tobacco is a dangerous drug with serious health consequences.

* Health experts have been warning about the dangers of tobacco for many years.
* Tobacco use causes many serious health problems, such as emphysema, lung cancer, and heart disease.

LESSON 3 **Tobacco Addiction**

BIG IDEA Tobacco contains nicotine, which is an extremely powerful and addictive drug.

* Tobacco contains strong substances that make it difficult to stop using once a person has started.
* Tobacco use results in both physical and psychological dependence on nicotine.

LESSON 4 **Costs to Society**

BIG IDEA Tobacco has consequences in addition to the harm it causes tobacco users.

* Tobacco use is costly not only to smokers but also to nonsmokers.
* Added health care expenses and lost productivity are major costs to society as a result of tobacco use.
* Laws and education help protect nonsmokers and lower the cost of tobacco use to society.
* Efforts to counter the costs of tobacco use include tobacco taxes, smoke-free environments, labeling laws, and advertising limits.

LESSON 5 **Saying No to Tobacco Use**

BIG IDEA It is important to have strategies to resist the strong influences around tobacco use.

* Influences that can lead teens to try tobacco may come from peers, family, the media, and advertising.
* Knowing how to resist tobacco, including the reasons to say no, will help you stay tobacco free.
* Resources available to people who want to be tobacco free include certain medical products along with various organizations and support groups.

 Review

 Web Quest

ASSESSMENT

Reviewing Vocabulary *and* Main Ideas

> physical dependence
> withdrawal
> nicotine
> lungs
> smokeless tobacco
> carbon monoxide
> addiction
> cardiovascular
> emphysema

›› On a sheet of paper, write the numbers 1–9. After each number, write the term from the list that best completes each statement.

LESSON 1 Facts About Tobacco

1. _____ is an addictive drug found in tobacco leaves and in all tobacco products.

2. A poisonous, odorless gas found in tobacco smoke is _____.

3. _____ is ground tobacco that is chewed or inhaled through the nose.

LESSON 2 Health Risks of Tobacco Use

4. Tobacco smoke damages the air sacs in the lungs, which can cause _____.

5. Tobacco use affects the circulatory or _____ system, which includes the heart and blood vessels.

6. Cancer of the _____ is the leading cause of death among people who smoke.

LESSON 3 Tobacco Addiction

7. Anyone who stops using tobacco goes through nicotine _____.

8. Nicotine is a powerful drug that causes _____.

9. _____ on nicotine involves the body developing a chemical need for the drug.

›› On a sheet of paper, write the numbers 10–15. Write *True* or *False* for each statement below. If the statement is false, change the underlined word or phrase to make it true.

LESSON 4 Costs to Society

10. Whenever people smoke near you, you breathe their secondhand smoke, or environmental tobacco smoke (ETS).

11. Many state and local governments have tried to lower the cost to nonsmokers through laws that require smoking in all public spaces.

12. Tobacco products are required to carry clear labels or disclaimers saying that smoking is healthful.

LESSON 5 Saying No to Tobacco Use

13. Peer pressure can help you say no if you are pressured to try tobacco.

14. Some people may choose to stop using tobacco suddenly and completely, or by going cold turkey.

15. Ninety percent of adult smokers start smoking after age 18.

✔ eAssessment

>> Using complete sentences, answer the following questions on a sheet of paper.

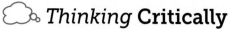 *Thinking* **Critically**

16. APPLY Nadine smells tobacco on her sister Shari's hair and clothes one afternoon. How could Nadine talk to her sister about the tobacco smell without accusing her of smoking?

17. SYNTHESIZE Explain how someone becomes addicted to tobacco. Include facts about addiction and a description of the path to addiction.

Write **About It**

18. PERSONAL WRITING Write a story about a teen who is faced with a difficult decision. The teen has to decide whether to join a club in which the members smoke and use other tobacco products. Use the six steps of the decision-making process to show how the teen makes a healthful choice.

 STANDARDIZED TEST PRACTICE

Math
Use the directions and chart below to answer the questions on the right.

In order to run the marathon, Melissa knows she needs to stay tobacco free. To qualify for the race, she needs to improve the time it takes her to run one mile. She tracks her time on a weekly basis. Refer to the chart below to see Melissa's progress.

Week	Time to run 1 mile (min)
1	16
2	11
3	9
4	8

1. By what percent did Melissa improve her time from Week 2 to Week 3?
   A. 15%
   B. 18%
   C. 20%
   D. 23%

2. During which time did Melissa show a 50 percent improvement?
   A. From Week 1 to Week 2
   B. From Week 1 to Week 4
   C. From Week 2 to Week 4
   D. From Week 3 to Week 4

Alcohol

LESSONS

PREMIUM ONLINE RESOURCES >

 Audio

 Videos

 Bilingual Glossary

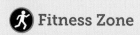 Fitness Zone

 Web Quest

 Review

346

Alcohol Use *and* Teens

BIG IDEA ▶ Most teens do not use alcohol, but several factors influence teens to try it.

Before You Read

QUICK WRITE Write a few sentences describing what you already know about alcohol use.

 Video

As You Read

FOLDABLES | Study Organizer

Make the Foldable® found in the FL pages in the back of the book to record the information presented in Lesson 1.

Vocabulary

› alcohol
› drug
› depressant
› inhibitions
› binge drinking
› minor

 Audio

 Bilingual Glossary

Developing Good Character

Respect Choosing to be alcohol free shows that you respect yourself. Create, sign, and date a pledge listing your reasons for choosing to avoid alcohol. Remember that this pledge is a promise to yourself to make a healthful choice. *Identify a way you could encourage your peers to also make healthful choices regarding alcohol.*

WHAT IS ALCOHOL?

MAIN IDEA ▶ Alcohol is a drug that affects the mind and body.

Have you seen spoiled food that has mold growing on it? This change is caused by a chemical reaction. **Alcohol** is *a drug created by a chemical reaction in some foods, especially fruits and grains.* The type of alcohol in beer, wine, and liquor is one of the most widely used and abused drugs in the United States.

Over time, using too much alcohol can damage body organs and cause disease. Alcohol also affects the brain and central nervous system, causing changes in behavior. Alcohol is considered a **drug,** or *a substance other than food that changes the structure or function of the body or mind.*

Alcohol acts as a **depressant,** or *a drug that slows down the body's functions and reactions, including heart and breathing rates.* These physical changes may make it difficult to think and act responsibly.

Using **too much** *alcohol* can **damage** body organs and cause **disease.**

Not all alcohol use is bad. For adults, a small amount of wine each day may help keep the cardiovascular system healthy. However, that amount is limited to one 5 oz. glass of wine per day for an adult female and two glasses for an adult male. For some people, though, even small amounts of alcohol can affect how they feel and behave.

Alcohol use can cause people to lose their **inhibitions,** meaning they have no *conscious or unconscious restraint on their behaviors or actions.* Some people may act in ways that are not typical for that person. Some people become relaxed and friendly. Others become depressed and angry. Under the influence of alcohol, people often say and do things they will later regret.

Aldo Murillo/Getty Images

WHY DO SOME TEENS USE ALCOHOL?

MAIN IDEA Teens face many influences that encourage them to try alcohol.

Studies show that most teens do not use alcohol. So why do some teens try it, even when they know alcohol is harmful to their health and also illegal? Curiosity is one reason a teen may try alcohol. Another reason is that they think it will make them more popular. Some teens think alcohol use makes them feel relaxed or more grown up. Others use it to feel some relief from emotions that they have not yet learned how to handle.

Alcohol *in the* Media

Television commercials, movies, Internet sites, and online ads often make using alcohol seem fun and exciting. You have probably seen an ad for some type of alcoholic drink. The people who appear in the ads look young and attractive. Companies that make and sell alcohol do this on purpose. They don't want people to see or think about the negative effects of their products.

Media images may lead some teens to feel that drinking alcohol is okay. Some teens may also think that using alcohol will add more fun and excitement to their lives, like the people in the ads.

> ### >>> Reading Check
>
> **CHARACTERIZE** *What are some typical reasons that a teen may try alcohol?*

Peer Pressure

"I want to be cool, too." That's a thought many teens often have before they try alcohol, even though they may not really want to. Negative peer pressure is another major reason why some teens use alcohol. Some may choose to use alcohol to try to fit in or to not be embarrassed in front of their friends.

However, having even one drink can be harmful to your health. Simply trying alcohol can lead to risky situations. For example, sometimes teens may dare one another to consume a lot of alcohol very quickly. *Having several drinks in a short period of time,* or binge drinking, is very dangerous and can even cause death. It is not always easy to say no, but negative peer pressure is not a good reason to choose alcohol.

Even one drink can be harmful to your health.

Media ads are designed to make products look fun and exciting. *What elements of alcohol ads might encourage a teen to try alcohol?*

Huntstock/Getty Images

REASONS NOT TO DRINK

MAIN IDEA The negative effects of alcohol use pose even greater risks for teens.

The USDA recommends that adults who choose to use alcohol consume it only in moderation. This is defined as no more than one drink per day for an adult female and no more than two for an adult male. However, some people should never consume alcohol. According to the CDC, those who should not use alcohol at all include people who are:

- **Minors,** or *people under the age of adult rights and responsibilities.* For alcohol, a minor is anyone younger than age 21.
- Pregnant or women who are trying to become pregnant.

- Taking medications that can be harmful when they are mixed with alcohol.
- Recovering from alcoholism or unable to control the amount they drink.
- Facing any type of medical condition that can be made worse by alcohol use.
- Driving, planning to drive, or engaging in any other activity that requires skill, coordination, and alertness.

The use of alcohol can have harmful effects for anyone, but teens are especially at risk. During the teen years, your body is still growing and developing.

When teens consume alcohol, their bodies do not grow and develop properly. Alcohol can seriously harm the brain's ability to learn and remember.

Teens also face many emotional changes. They might think that alcohol can help them better deal with their emotions. However, alcohol can make a teen's social and emotional life more difficult. Many people who use alcohol feel bad about themselves. They may have a hard time making friends and relating to others. Alcohol use can disrupt sleep and create more stress than it may seem to relieve.

Health SKILLS ACTIVITY

Stress Management

Dealing with *Emotions*

Dealing with difficult emotions is part of life for a teen. Rather than using alcohol, teens can use the following strategies to deal with emotions in healthful ways.

* Get enough sleep. Being well-rested can give you the energy you need to deal with difficult feelings and stress.

* Take some deep breaths. This can help you relax.

* Stay active. Physical activity can help you focus your energy and lower your stress level.

* Talk to someone you trust and respect about what you're feeling.

 On Your Own Work with a group to create a brochure or blog post for your fellow classmates that describes positive ways to deal with difficult emotions. Be sure to point out the negative effects of using alcohol to deal with difficult emotions.

Teens who consume alcohol also risk trouble with the law. It is illegal for a minor to use alcohol. Buying or drinking alcohol can lead to an arrest, a fine, or time in a youth detention center. A person of any age who is convicted of driving while intoxicated (DWI), or driving under the influence (DUI), risks losing his or her license. If that person causes an accident and someone is injured, the driver may face more serious consequences.

Teens who use *alcohol* also risk **legal trouble.**

Choosing not to use alcohol is a healthful decision. It shows that you understand how risky alcohol use can be. As you have learned, some teens may believe that using alcohol will help them fit in with their peers. In reality, most teens do not use alcohol. By choosing not to drink, you will already be fitting in with most people your age. Many teens realize the negative effects that alcohol can have on all sides of their health triangle and are saying no to alcohol use.

>>> Reading Check

DEFINE *What is a minor? How does this term relate to alcohol?* ■

>>> After You Read

1. **VOCABULARY** Define alcohol. Use it in a sentence.
2. **EXPLAIN** What are two reasons teens often give for using alcohol?
3. **STATE** What are two reasons not to drink alcohol?

>>> Thinking Critically

4. **APPLY** What do you think is the most important reason for a teen not to use alcohol?
5. **PREDICT** How can using alcohol affect a teen's development?

>>> Applying Health Skills

6. **ACCESSING INFORMATION** Some teens may believe myths about alcohol. With classmates, research several of these myths. Use your findings to create a poster showing the truth about these concepts.

🔄 Review

🔊 Audio

Many teens are making the decision to stay alcohol free. *Explain how abstaining from alcohol can improve your health.*

©Creatas/PunchStock

Effects *of* Alcohol Use

BIG IDEA Alcohol use has far-reaching effects to the body, other people, and personal relationships.

Before You Read

QUICK WRITE Write a few sentences describing what you already know about the harmful effects of alcohol.

 Video

As You Read

STUDY ORGANIZER Make the study organizer found in the FL pages in the back of the book to record the information presented in Lesson 2.

Vocabulary

› intoxicated
› blood alcohol concentration (BAC)
› alcohol poisoning
› ulcer
› fatty liver
› cirrhosis
› reaction time
› fetal alcohol syndrome (FAS)

 Audio

 Bilingual Glossary

Developing Good Character

Being a Responsible Friend One way of showing you are responsible is by looking out for the well-being of others. Don't let a friend get in a car with a driver who has been drinking. If a friend is using alcohol, urge that person to get help. Don't hesitate to talk to an adult if your friend is unwilling to seek help. This is not breaking your friend's trust. It is taking the first step in getting your friend the help that he or she needs. *Cite another example of showing responsibility in dealing with substance abuse.*

HOW ALCOHOL AFFECTS THE BODY

MAIN IDEA Alcohol has many short- and long-term effects on your body.

Alcohol begins to affect body systems soon after it is consumed. It is quickly absorbed by the bloodstream. Alcohol affects the brain and central nervous system as soon as 30 seconds after it is consumed. Alcohol does not affect everyone in the same way, however. Some people can consume more than others before they become intoxicated.

Intoxicated means *physically and mentally impaired by the use of alcohol.* In other words, it means "having consumed more alcohol than the body can tolerate." This is also known as "being drunk." However, the amount of alcohol someone consumes is only one factor in understanding the effects of alcohol use.

How Alcohol's Effects Vary

Different people react to alcohol use in different ways. One of the biggest factors is the user's blood alcohol concentration (BAC). This is *the amount of alcohol in the blood.* An alcohol user's BAC is expressed as a percentage.

Alcohol begins to **affect** body systems *soon* after it is consumed.

A blood alcohol concentration of 0.02 percent will cause most people to feel light-headed. A BAC of 0.08 percent interferes with a person's ability to drive a car safely. Police officers use this percentage to determine whether a person is legally intoxicated. A BAC of 0.40 percent can lead to coma and death. A number of other factors can also influence how alcohol affects an individual. A person's size and gender make a difference. So does how much alcohol someone consumes and how fast.

SHORT-TERM EFFECTS

MAIN IDEA The use of alcohol has an immediate effect on many parts of the body.

Alcohol has both short-term and long-term effects on the body. The use of alcohol not only causes immediate risks, but it can also cause serious health problems over time. Alcohol can affect the brain, stomach, liver, and kidneys right away. Using a lot of alcohol over time can cause serious damage to these organs.

The use of *alcohol* can cause **serious** health problems.

ALCOHOL AND THE BRAIN

Alcohol is absorbed into the bloodstream and reaches the brain very quickly. As a result, the brain and nervous system slow down immediately. Even one drink can make it difficult to think clearly. Alcohol blocks messages trying to get to the brain. After more drinks, it becomes harder to concentrate and remember. Also, it becomes difficult to speak clearly or walk in a straight line. People under the influence of alcohol may also feel dizzy, have blurred vision, and lose their balance. A person under the influence of alcohol is also more likely to engage in other risk behaviors such as driving while intoxicated, tobacco and other drug use, sexual activity, and acts of violence.

ALCOHOL AND THE STOMACH

In the stomach, alcohol increases the flow of acid used for digestion. Some people become sick to their stomach. Most of the alcohol passes into the small intestine. Some, however, is absorbed into the bloodstream and causes the blood vessels to expand. From the bloodstream, alcohol passes into the liver.

Beer 12 oz. **Wine** 5 oz. **Liquor** 1.5 oz.

Each of the drinks shown contains the same amount of alcohol. *How much beer would a person need to drink to consume the same amount of alcohol as in two glasses of wine?*

ALCOHOL AND THE HEART

Alcohol affects the way the heart pumps blood through the body. Because it makes the blood vessels wider, the blood comes closer to the surface of the skin. This makes a person who is consuming alcohol feel warm, but his or her body temperature is actually dropping. Alcohol also slows down a person's heart rate.

ALCOHOL AND THE LIVER AND KIDNEYS

Short-term use of alcohol also affects the liver and kidneys. The liver acts like a filter, taking alcohol from the bloodstream and removing it from the body. However, the liver can filter only about half an ounce of alcohol from the bloodstream each hour. Any additional alcohol that is consumed stays in the bloodstream and affects the body. In addition, alcohol causes the kidneys to produce more urine. Extra urine production can lead to dehydration, or the loss of important body fluids. When people consume too much alcohol, they often feel more thirsty than usual the next day.

ALCOHOL POISONING

If someone consumes a lot of alcohol very quickly, it can lead to alcohol poisoning. This is *a dangerous condition that results when a person drinks excessive amounts of alcohol over a short time period.* Binge drinking is a common cause of alcohol poisoning. A person who has too much too quickly may vomit, become unconscious, or have trouble breathing. Alcohol poisoning can result in death.

>>> **Reading Check**

IDENTIFY *Which parts of the body are affected by alcohol?*

LONG-TERM EFFECTS

MAIN IDEA Alcohol use affects all areas of a person's life.

Consuming alcohol regularly can lead to a number of serious health problems. Alcohol use can damage major organs and make existing health problems worse. It can also lead to learning and memory problems.

BRAIN Alcohol affects the parts of the brain which control memory and problem solving. Alcohol also destroys brain cells. Because brain cells do not grow back, this can be serious enough to limit everyday functions. Alcohol can also block messages sent to the brain. When this happens, people can have a hard time seeing, hearing, or moving.

HEART Heavy drinking makes the heart weak and enlarged, which leads to high blood pressure. The risk of congestive heart failure and stroke also increases with excessive alcohol use.

STOMACH Alcohol causes your body to create more acid. Stomach acid usually helps with digestion. However, the extra acid created by alcohol consumption can eventually cause sores called ulcers to develop in the stomach lining. An ulcer is *an open sore in the stomach lining.* Drinking alcohol also makes the valve between your stomach and esophagus weak. This valve usually works to keeps acid in the stomach. When it is weakened by alcohol use, acid comes up and causes heartburn.

*Alcohol use can **damage** major **organs** and make existing health problems even worse.*

LIVER Consuming alcohol regularly over a long period of time puts a serious strain on the liver. Fatty liver is *a condition in which fats build up in the liver and cannot be broken down.* Heavy drinkers are particularly at risk of developing cirrhosis, or *the scarring and destruction of liver tissue.* This condition can be deadly. Cirrhosis creates scar tissue that prevents blood from flowing normally through the liver. If the liver is not working correctly, it cannot filter out wastes or remove other poisons from the blood. These poisons can eventually reach the brain and cause more damage.

>>> **Reading Check**

ANALYZE *Which of the effects of alcohol use do you feel would have the most serious impact on your health? Explain your answer.*

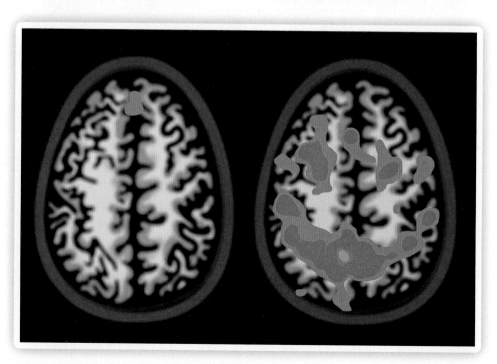

Alcohol affects the activity of the brain, as this CAT scan shows. *Name a possible effect of alcohol on brain activity.*

Courtesy of Dr. Susan Tapert

OTHER DANGERS OF ALCOHOL USE

MAIN IDEA Alcohol use can affect thoughts and behavior.

Because alcohol affects the brain, it also affects thoughts and behaviors. As a result, a person who consumes alcohol can cause arguments, physical fights, and vehicle accidents. The person may engage in risky behavior, such as using illegal drugs or engaging in sexual activity.

Alcohol *and* Driving

A person who uses alcohol experiences loss of coordination, concentration, and visual awareness. He or she also has slowed reaction time, or *the ability of the body to respond quickly and appropriately to situations.* Driving while intoxicated is extremely dangerous for the driver, his or her passengers, and others on the road. It is very important for your safety not to ride in a vehicle with a driver who has been using alcohol. If a person has been drinking, do your best to avoid letting them drive. You can always call someone else to come pick you up.

Alcohol *and* Behavior

Using alcohol can also damage your mental/emotional and social health. Teens who use alcohol are more likely to do poorly in school. A teen who uses alcohol may start to lose interest in his or her favorite activities. In addition, that person may risk losing friends as a result of his or her alcohol use.

Someone under the influence of alcohol might also engage in other risk behaviors such as tobacco use or sexual activity. Alcohol use can lead a person to make unhealthful decisions.

It is very **important** for your safety *not to ride* in a vehicle with a **driver** who has been **using alcohol.**

Alcohol *and* Pregnancy

If a pregnant woman consumes alcohol, it passes through her bloodstream to her baby. This can lead to what is known as fetal alcohol syndrome (FAS), or *alcohol-related birth defects that include both physical and mental problems.* A baby born with fetal alcohol syndrome can have low birth weight and a smaller-than-normal brain. FAS can also cause serious heart and kidney problems. As they grow older, babies who were born with FAS may also develop speech problems and have learning disabilities.

> ### Reading Check
> **INFER** *Why does slowed reaction time make driving while intoxicated so dangerous?* ■

LESSON 2

REVIEW

After You Read

1. **DESCRIBE** What kinds of long-term damage can alcohol use cause?
2. **RECALL** Describe how alcohol affects the mind.
3. **VOCABULARY** Define blood alcohol content. Use the term in a sentence.

Thinking Critically

4. **APPLY** You are at a park with friends. When it is time to leave, a friend's brother offers you a ride. You smell alcohol on his breath. What should you do, and why?
5. **ANALYZE** Why is a person under the influence of alcohol more likely to engage in other high-risk behaviors?

Applying Health Skills

6. **ACCESSING INFORMATION** Some teens may believe myths about alcohol. With classmates, research several of these myths. Use your findings to create a poster showing the truth about these concepts.

⟳ Review
🔊 Audio

Alcoholism *and* Alcohol Abuse

BIG IDEA Alcohol is a highly addictive drug that can lead to disease and damage relationships.

>>> **Before You Read**

QUICK WRITE Make a list of the reasons you can think of to avoid alcohol use.

 Video

>>> **As You Read**

STUDY ORGANIZER Make the study organizer found in the FL pages in the back of the book to record the information presented in Lesson 3.

>>> **Vocabulary**

› alcoholism
› malnutrition
› alcohol abuse
› substance abuse

🔊 Audio

🔤 Bilingual Glossary

ALCOHOL'S ADDICTIVE POWER

MAIN IDEA Alcohol is a powerful drug that can cause addiction.

One of the greatest dangers with alcohol use is that it is habit-forming. As with other drugs, using alcohol regularly can lead to addiction. Teens age 15 and younger are four times more likely to become addicted than older people. A person who uses alcohol in large amounts is even more likely to become addicted. An addiction to any drug can change a person's life. It takes the focus off healthful goals. It can also damage relationships with family and friends.

Alcohol use is a serious health issue. Estimates show that at least 17 million people in the U.S. have an alcohol problem. Alcohol addiction has a negative impact on that person's entire life. Alcohol addiction affects all three sides of a user's health triangle—physical, mental/emotional, and social.

> One of the **greatest dangers** with *alcohol use* is that it is **habit-forming.**

How can someone tell if a person has an alcohol problem? A person who is addicted to alcohol frequently uses it alone. He or she often uses alcohol to the point of becoming intoxicated. Using this drug typically becomes more important than anything else in a person's life. Someone with an alcohol problem may ignore friends or lose them entirely. That person often neglects his or her family. Many people who are addicted to alcohol do poorly in school or at work. They may lose their jobs or get in trouble with the law. Most people with an alcohol problem need to use it every day in order to function. Some may even forget to eat regularly and stop taking care of themselves.

🏃 **Fitness Zone**

Stress-relieving activities Some people use alcohol to relieve stress. I think of other ways to deal with my stress. I can write a list of things I like to do that help me get rid of stress and stay healthy: run, play basketball, read a book, have a healthful snack with my mom, or ride my bike.

McGraw-Hill Companies, Inc. Amy Mendelson, photographer

THE DISEASE OF ALCOHOLISM

MAIN IDEA The disease of alcoholism results from addiction and has physical, mental/emotional, and social consequences.

People who are addicted to alcohol suffer from alcoholism, or *a disease in which a person has a physical and psychological need for alcohol.* They are called *alcoholics.* Alcoholics typically experience some or all of the following symptoms of alcoholism:

- **Craving** is a strong feeling of need to consume alcohol. This is likely related to alcohol's effects on the brain.
- **Loss of control** means the user is unable to limit his or her alcohol consumption.
- **Tolerance** is when your body needs more and more of a drug to get the same effect. If someone is an alcoholic, he or she will need to consume more and more alcohol in order to feel intoxicated.
- **Physical dependence** can lead to painful symptoms. If an alcoholic stops using alcohol, he or she may experience sweating, shaking, or anxiety. Making excuses to drink is another common symptom. An alcoholic may be unable to limit how much he or she consumes at one time. Alcoholism may also cause a person to become irritable or violent. This can result in injury or abuse. Alcoholics may hurt themselves or others.

> ### ⟩⟩⟩ Reading Check
>
> **IDENTIFY** *What makes alcohol such a dangerous drug?*

Making excuses to **drink** is another common **symptom** of *alcoholism.*

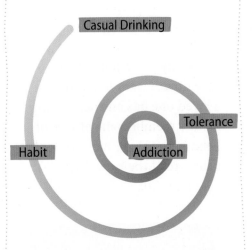

The spiral of addiction begins with casual alcohol use. *Draw a conclusion about where the spiral ends.*

Stages *of* Alcoholism

Alcoholism is a disease that develops over time. The cycle of addiction is sometimes represented by a downward spiral. The spiral in the figure above shows how addiction to alcohol starts and progresses along the way. Alcohol addiction typically starts with casual drinking. After a person starts using alcohol casually, it can then become a habit. Once alcohol use is a habit, the body builds up a tolerance, which spirals toward addiction.

Each of the three stages of alcoholism may be either long or short. How long each stage goes on depends on the individual and on how old he or she is when the alcohol use begins. All alcoholics do not go through each stage in the same way.

- **Stage One—Abuse.** The user may have short-term memory loss and blackouts. He or she may also begin to lie or make excuses for drinking. It is also common for a person who abuses alcohol to begin saying or doing hurtful things to friends and family members.
- **Stage Two—Dependence.** The alcoholic loses control and cannot stop drinking. The person's body begins to depend on the drug. The user can become aggressive, avoid family and friends, or have physical problems. The user typically tries to hide his or her alcohol problem but is unable to function well at home, school, or work.
- **Stage Three—Addiction.** The person may be intoxicated for long periods of time. The liver may already be damaged. Less alcohol may be needed to cause intoxication. Common at this stage are strange fears, hallucinations, and malnutrition. This is *a condition in which the body does not get the nutrients it needs to grow and function properly.*

Decision Making

Helping a *Friend*

Mariah and Jenelle have been best friends for a long time. Mariah recently told Jenelle that her mom drinks alcohol nearly every day and sometimes becomes violent. Mariah also said that sometimes she gets very scared. Jenelle wonders what she should do to help Mariah.

What Would You Do? Apply the six steps of the decision-making process to Mariah's situation.

1. State the situation.
2. List the options.
3. Weigh the possible outcomes.
4. Consider your values.
5. Make a decision and act on it.
6. Evaluate the decision.

With a partner, role-play what Jenelle would say to Mariah and how Mariah might respond.

How Alcoholism Affects Families

Alcoholism is a problem that affects more people than just the alcoholic. It can be a painful experience for family members as well. Children of alcoholics sometimes blame themselves, thinking they did something to drive a parent to alcohol. This is not the case. A child is never to blame for a parent's alcoholism.

Denial is also a problem for family and friends. Often, they do not want to admit that a loved one has an addiction. Family members may focus on helping the alcoholic and not take care of their own needs.

Using alcohol can lead to unhealthy relationships. *Describe some ways that addiction can cause problems in relationships.*

If the alcoholic is abusive, this can have a negative effect. Friends may try to help by making an alcoholic feel comfortable with his or her behavior. This only encourages the addiction. It can create an unhealthy pattern and keep the alcoholic from getting the help he or she needs.

How Alcoholism Affects Society

Teen alcohol use costs the U.S. more than $50 billion a year. The total cost of alcohol-related problems is estimated at $223.5 billion a year. That figure is higher than the total for smoking and other drug-related issues. The greatest impact is on health care, law enforcement, and the workplace. Doctors and nurses have to take care of people with alcohol problems. Police and the courts must deal with people who break alcohol laws. A business can lose money when an employee who uses alcohol does not work hard on the job.

>>> **Reading Check**

DETERMINE *Where does the cycle of addiction usually begin?*

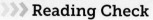

©PhotoAlto/PunchStock

Avoiding alcohol makes it easier to focus in school. *Explain how alcohol abuse can make it hard to concentrate.*

>>> **After You Read**

1. **VOCABULARY** Define alcohol abuse. Use the term in a complete sentence.
2. **EXPLAIN** What is the difference between alcoholism and alcohol abuse?
3. **DESCRIBE** Briefly describe the spiral of addiction.

>>> **Thinking Critically**

4. **ANALYZE** Layla has started drinking alcohol. Now she is forgetting things and lying to others about her alcohol use. She also has been in more arguments with her friends and family. What stage of alcoholism is Layla likely experiencing?
5. **EVALUATE** Briefly explain why alcohol use is even more dangerous for teens.

>>> **Applying Health Skills**

6. **DECISION MAKING** A friend has been irritable and moody lately. He tells you that he really needs alcohol and asks you to help him get some. He thinks only alcohol will make him feel better. Use the decision-making steps to make a responsible choice.

⟳ Review

🔊 Audio

ALCOHOL ABUSE

MAIN IDEA Alcohol abuse is different than alcoholism.

Although the terms *alcoholism* and *alcohol abuse* are sometimes used in the same way, there is a difference. Alcohol abuse means *using alcohol in ways that are unhealthy, illegal, or both.* People who abuse alcohol are not physically dependent on the drug. Their bodies are not in extreme need for the drug.

Alcohol abuse is different than **alcoholism.**

Alcohol abuse is a type of substance abuse, which involves *using illegal or harmful drugs, including any use of alcohol while under the legal drinking age.*

Alcohol abuse has four main symptoms. These include:

- Failing to complete major work tasks or ignoring responsibilities at home or school.
- Drinking in situations that are dangerous. For example, driving when intoxicated or riding with someone who has been drinking can result in an accident and serious injury.
- Having ongoing financial or legal problems.
- Continuing to drink even after the person has a problem with a friend or family member.

>>> **Reading Check**

IDENTIFY *Name a behavior that may occur when someone abuses alcohol.* ∎

Getting Help *for* Alcohol Abuse

BIG IDEA Many resources are available to help alcoholics, alcohol abusers, and their families.

>>> **Before You Read**

QUICK WRITE Name a person you might turn to when you need help with a problem. Write a brief description of this person's qualities, such as being a good listener

 Video

>>> **As You Read**

STUDY ORGANIZER Make the study organizer found in the FL pages in the back of the book to record the information presented in Lesson 4.

>>> **Vocabulary**

> enablers
> intervention
> recovery
> detoxification

 Audio

(ABC) Bilingual Glossary

HELP FOR ALCOHOL ABUSE

MAIN IDEA Family, friends, and organizations can all help someone with an alcohol problem.

People who are struggling with alcohol need help. However, many cannot admit they have a problem. Sometimes alcoholics surround themselves with enablers. These are *persons who create an atmosphere in which the alcoholic can comfortably continue his or her unacceptable behavior.* Enabling may allow an alcoholic to continue drinking without facing any negative consequences. A way to overcome this situation is to hold an intervention, or *a gathering where family and friends get a problem drinker to agree to seek help.* During an intervention, an alcoholic's family and friends can discuss their concerns about the alcohol abuse. They can try to convince the abuser to stop using alcohol. An intervention may include commitments by friends and family to no longer enable the alcoholic's behavior.

Before holding an intervention, family members and friends may choose to meet with a substance abuse counselor. A counselor can often make arrangements for the alcoholic to get treatment. Friends and family members can then use the intervention to encourage the person with the alcohol problem to seek help.

Ways *to* Seek Help

Groups such as Alcoholics Anonymous (AA) can help people who are addicted to alcohol. Similar groups help friends and families of alcoholics. Support groups work to help people break their patterns of addiction. They allow people to talk with others who are facing the same problem. You can search online to learn more about these groups.

People who are *struggling* with **alcohol** need **help.**

Developing Good Character
Citizenship Students Against Destructive Decisions (SADD) helps people understand the harmful effects of alcohol on teens. Do some research on SADD. Find out how you and your classmates can get involved with this organization. Discuss your findings with your class.

Stockbyte/SuperStock

The Road *to* Recovery

Before an alcoholic can get better, he or she must decide never to drink again. When that happens, **recovery,** or *the process of learning to live an alcohol-free life,* can begin. As we have learned, addiction is very powerful. Recovery is usually long and difficult. It involves several steps that each person must follow.

- **Admission.** The person must first admit that he or she has an addiction and ask for help.
- **Detoxification,** or *the physical process of freeing the body of an addictive substance.* Detoxification marks only the beginning of breaking the physical addiction. An alcoholic may go through withdrawal when he or she suddenly stops using alcohol.
- **Counseling.** Alcoholics need outside help from counselors and support groups to recover. As we have learned, there are local organizations that can help provide a support group.

- **Resolution.** The alcoholic commits to accepting responsibility for his or her actions. After recovery, people who have had alcohol problems are called recovering alcoholics. A recovering alcoholic is someone who has an addiction to alcohol but chooses to live without alcohol.

Even after a person goes through recovery, he or she must always fight addiction. Recovery is never final. Recovering alcoholics risk a relapse if they drink again. That is why it is important they decide to always make wise choices related to alcohol.

⟫ Reading Check

EXPLAIN *What makes recovery from alcoholism so difficult?*

A recovering alcoholic must always fight the addiction. *Give examples of some healthful choices that would help someone remain alcohol free.*

Health SKILLS ACTIVITY

Communication Skills

When *Communication* Counts Most

You may be able to help a friend or family member who has an alcohol problem. Here's how.

✻ **HAVE AN HONEST TALK.** Find a time when the person is sober. Discuss your concerns, and talk about the effects alcohol can have on a person's health.

✻ **ENCOURAGE THE PERSON TO SEEK HELP.** After expressing your concerns, explain why he or she needs help and support.

✻ **OFFER INFORMATION.** Provide details about where the person can go for help. Make sure the person understands what help is available and how to get it.

With A Group

Role-play a conversation in which you use the skills outlined above. Think about specific words that would express your concern and encourage the person to get help.

HELP FOR FAMILIES

MAIN IDEA Support groups can help families of alcoholics.

Families and friends of alcoholics can get help too. For example, Al-Anon teaches about the effects of alcoholism. This group also helps people learn strategies for dealing with an alcoholic.

Alateen helps teens who have parents who abuse alcohol. Organizations like these may offer group sessions or direct families to counseling and mental health resources. Al-Anon and Alateen also help educate the public.

>>> Reading Check

IDENTIFY *Name one example of a support group for families of alcohol abusers.*

This support group was formed to help families of people who suffer from alcoholism. *What are some support groups for alcoholism in your area?*

STAYING ALCOHOL FREE

MAIN IDEA Choosing not to use alcohol is the best way to avoid its dangers.

As you have learned, the use of alcohol has serious physical, mental/emotional, and social consequences. Choosing to be alcohol free is the best way not to experience these dangers. You can try to avoid situations where people are drinking alcohol. If someone pressures you to drink alcohol, use refusal skills.

Practice the S.T.O.P. strategy:
- **S**ay no in a firm voice.
- **T**ell why not.
- **O**ffer another idea.
- **P**romptly leave.

If you choose friends who are also alcohol free, you will have a support system. When you are around people who make healthful choices, it is easier for you to make healthful choices.

Positive peer pressure can make it more likely that you and your friends will choose activities that do not involve alcohol.

Choosing **not to use alcohol** is the best way to *avoid its dangers.*

Colleen Cahill/VIBE/Alamy

Benefits *of* Staying Alcohol Free

Staying alcohol free is a choice to lead a healthy lifestyle. When you choose not to use alcohol, you are showing respect for yourself and your body. You are choosing to remain in control of who you are.

Staying **alcohol free** is a *choice* to lead a **healthy** *lifestyle.*

Another benefit is being able to focus on your future. An alcohol-free lifestyle allows you to care for your family and friends. Better relationships are a benefit of choosing not to use alcohol.

Staying alcohol free keeps you focused to achieve your life goals. *What are some activities you can do to help focus on your goals?*

Healthy Alternatives

When someone offers you alcohol, use your refusal skills as a healthful alternative. Refer to the S.T.O.P. strategy on the previous page. If you are offered alcohol, say no and explain why you have chosen not to drink. Offer a suggestion of an alcohol-free activity. If those steps do not work, promptly leave the area.

Finding another way to think or act will also help you avoid alcohol use. Instead of using alcohol, find a healthful way to spend your time. Join a club or sports group at school. Volunteer at a local organization, such as a food bank or animal shelter. Volunteering can give you a sense of purpose and can make you feel good about yourself. Another idea is to start a hobby or business with your friends. Alcohol use will never help you reach your goals, but positive activities such as these can help.

>>> **Reading Check**

DESCRIBE *What are benefits of choosing an alcohol-free lifestyle?* ■

>>> **After You Read**

1. **EXPLAIN** Describe how a person can get help for an alcohol problem.
2. **VOCABULARY** Define intervention. Use the term in a complete sentence.
3. **IDENTIFY** What is the best way to avoid problems with alcohol?

>>> **Thinking Critically**

4. **HYPOTHESIZE** How might you be affected if one of your close friends or family members developed an alcohol problem? Where could you find help? Explain your answer.
5. **ANALYZE** How can healthy alternatives prevent alcohol use?

>>> **Applying Health Skills**

6. **GOAL SETTING** Think about personal goals you have, such as going to college or the kind of job you would like to have some day. Write one or two of these on a sheet of paper. Leave space under each one. Use that space to explain how alcohol use could prevent you from reaching your goals.

 Review

 Audio

©BananaStock/PunchStock

Hands-On HEALTH ACTIVITY

Refusing *to Get into a* Car *with a* Driver Who *Has* Been Drinking

WHAT YOU WILL NEED

* 1 index card per student
* Colored pencils or markers

WHAT YOU WILL DO

1 Working with a small group, brainstorm a list of refusal statements a teen can use to avoid riding in a car with a driver who has been drinking.

2 Write a skit that has dialogue showing successful use of refusal skills. Be sure that every group member has a part.

3 Act out your skit for the class.

WRAPPING IT UP

As a class, discuss the dialogue used in each of the skits. Decide which skit presented the most effective refusal statements. Then, on your own, take your index card and write "Don't Ride with a Drunk Driver" on the card. Then, write at least two statements you can use to refuse such a ride. Use markers or colored pencils to make the card creative and colorful. Display your cards in the classroom.

You have learned that you should not drink and drive. You should also avoid getting into a car with someone who has been drinking. If someone who has been drinking alcohol invites you to ride in a car with him or her, you should know how to refuse that invitation. The following are some suggestions that can help you avoid an unsafe situation.

* Make a decision to never ride with someone who has been drinking, and stick to it.
* Do not make arrangements to go places with a driver who you know will likely drink at an event you are going to.
* Find other ways to get a ride home if you are with a driver who has been drinking.
* Use direct statements: "I am not riding with you. You have been drinking. Don't drive. I'll find us another ride."

Brand X Pictures

READING REVIEW

FOLDABLES and Other Study Aids

Take out the Foldable® that you created and any study organizers that you created. Find a partner and quiz each other using these study aids.

LESSON 1 Alcohol Use and Teens

BIG IDEA Most teens do not use alcohol, but several factors influence teens to try it.

* Alcohol is a drug created by a chemical reaction in some foods, especially fruits and grains.
* Alcohol acts as a depressant by slowing down the functions and reactions of both the mind and body.
* Studies show that most teens do not use alcohol, but they do face a number of influences that encourage them to try it.
* Influences on teen alcohol use include peer pressure and images in the media.
* The negative effects of alcohol use pose even greater risks for teens than for adults.

LESSON 2 Effects of Alcohol Use

BIG IDEA Alcohol use has far-reaching effects to the body, other people, and personal relationships.

* Alcohol use has many short-term and long-term effects on a person's body.
* The use of alcohol has an immediate effect on many organs in a person's body, including the brain, stomach, heart, liver, and kidneys.
* A major factor in how a person reacts to alcohol use is blood alcohol concentration (BAC), or the amount of alcohol in the user's blood.
* Alcohol use affects all areas of a person's life.
* The use of alcohol can have negative effects on a person's thoughts and behavior.

LESSON 3 Alcoholism and Alcohol Abuse

BIG IDEA Alcohol is a highly addictive drug that can lead to disease and damage relationships.

* One of the greatest dangers of alcohol use is that it is habit-forming and can lead to addiction.
* Alcohol abuse is a serious health issue—an estimated 17 million people in the U.S. are either addicted to alcohol or have an alcohol problem.
* The disease of alcoholism results from a spiral of addiction and has physical, mental/emotional, and social consequences.
* Alcoholism is different from alcohol abuse, in which a person is not physically dependent on alcohol, but alcohol abuse is still a form of substance abuse.
* Alcoholism also affects families, friends, and society.

LESSON 4 Getting Help for Alcohol Abuse

BIG IDEA Many resources are available to help alcoholics, alcohol abusers, and their families.

* Family, friends, and various organizations can all help someone who has an alcohol problem.
* An intervention is a gathering where family and friends attempt to get a problem drinker to agree to seek help.
* Support groups can help alcoholics recover from their addiction and help families and friends of alcoholics learn to deal with the problem.
* Choosing not to use alcohol is the best way to avoid the dangers it poses to all aspects of a person's health.

 Review

 Web Quest

ASSESSMENT

Reviewing Vocabulary *and* Main Ideas

> inhibitions
> alcohol

> blood alcohol
>   concentration (BAC)

> cirrhosis
> binge drinking

> intoxicated

>> On a sheet of paper, write the numbers 1–6. After each number, write the term from the list that best completes each statement.

LESSON 1 Alcohol Use and Teens

1. The type of _____ found in beverages such as beer, wine, and liquor is one of the most widely used and abused drugs in the U.S.

2. Consuming a lot of alcohol very quickly, or _____, is very dangerous and can cause death.

3. Drinking can cause people to lose their _____, or act in ways that are not typical for that person.

LESSON 2 Effects of Alcohol Use

4. A person becomes _____ if he or she drinks more alcohol than his or her body can tolerate.

5. A person's size and gender, along with how much he or she drinks and how fast, can affect that person's _____.

6. Heavy drinkers are at risk of developing _____, a potentially deadly condition in which normal liver cells turn into scar tissue.

>> On a sheet of paper, write the numbers 7–12. Write *True* or *False* for each statement below. If the statement is false, change the underlined word or phrase to make it true.

LESSON 3 Alcoholism and Alcohol Abuse

7. Using alcohol regularly can lead to <u>addiction</u>.

8. The cycle of addiction to alcohol includes abuse, <u>malnutrition</u>, and addiction.

9. Alcohol abuse is a type of <u>substance abuse</u>.

LESSON 4 Getting Help for Alcohol Abuse

10. During an <u>enabler</u>, family and friends can discuss their concerns and try to convince an abuser to stop using alcohol.

11. An alcoholic may go through <u>resolution</u> when he or she suddenly stops using alcohol.

12. A <u>recovering alcoholic</u> is someone who has an addiction to alcohol but chooses to live without it.

 eAssessment

>> Using complete sentences, answer the following questions on a sheet of paper.

Thinking Critically

13. EVALUATE How can alcohol use as a teen cause health problems later in life?

14. ANALYZE Why is an alcoholic always said to be recovering rather than cured?

15. APPLY Imagine you are planning a birthday party for a friend. What are some fun activities you could choose that do not include the use of alcohol?

16. EVALUATE Explain how avoiding alcohol can have a positive effect. Be certain to include physical, mental/emotional, and social effects.

Write About It

17. EXPOSITORY WRITING Write a short advertisement encouraging teens to be alcohol free. Be sure to include ways to say no to negative peer influences and stay alcohol free.

18. OPINION What reasons might teens use to persuade others to use alcohol? What are some refusal responses to these statements?

Ⓐ Ⓑ Ⓒ Ⓓ STANDARDIZED TEST PRACTICE

Reading

Anish made the following concept map to organize his ideas for a paper. Review his concept map and then answer questions 1–3.

1. Under which subtopic should details about how alcohol affects the brain be placed?
   A. Drunk driving
   B. Affects relationships
   C. Harm body and mind
   D. Affects decision making

2. Which detail below supports the subtopic "Underage drinking illegal"?
   A. Binge drinking may lead to death.
   B. Alcohol is found in three forms: beer, liquor, and wine.
   C. It is illegal for anyone under the age of 21 to use alcohol.
   D. Alcoholism can be treated.

3. Based on this writing plan, what type of paper is Anish planning to write?
   A. a persuasive essay to convince teens not to drink alcohol
   B. a narrative that describes a personal experience with alcohol
   C. an explanatory paper that discusses the physical effects of alcohol
   D. none of the above

HOW SUBSTANCES AFFECT THE BODY

Respiratory System

The organs that supply your blood with oxygen

 CANCER

 SLOWS THE RESPIRATORY SYSTEM

Nearly **450,000 people** in the U.S. **die** each year from **smoking** or exposure to secondhand smoke.

Digestive System

The organs that break down foods into substances your cells can use

 DEHYDRATION

LIVER DISEASE

NAUSEA

 STOMACH ULCERS

 HEARTBURN

 INTOXICATION

Skeletal + Muscular Systems

Your body is supported by your skeleton. Your muscles move your body and control your organs.

 POOR BALANCE

 OSTEOPOROSIS

Nervous System

Your body's message and control center

 MEMORY LOSS

 STROKE

 DIZZINESS

 SLOWS BRAIN AND NERVOUS SYSTEM FUNCTIONS

 SEIZURES

DESTROYS BRAIN CELLS

ADDICTION

 BLURRED VISION

 VISION AND HEARING PROBLEMS

 IRRITABILITY

 SLEEPLESSNESS

Circulatory System

This system carries oxygen and nutrients to the cells, and waste products away from the cells.

 HIGH BLOOD PRESSURE

 SLOWS HEART RATE

 HEART FAILURE

"Sudden sniffing death" **can occur if you use an inhalant just one time.**

■ **TOBACCO**
■ **ALCOHOL**
■ **DRUGS**

WHAT'S IN TOBACCO?

If you use tobacco, here's what ends up inside you.

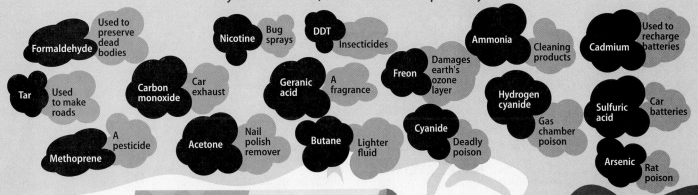

Formaldehyde — Used to preserve dead bodies

Tar — Used to make roads

Methoprene — A pesticide

Nicotine

Bug sprays

Carbon monoxide

Car exhaust

Acetone — Nail polish remover

DDT — Insecticides

Geranic acid — A fragrance

Butane — Lighter fluid

Freon — Damages earth's ozone layer

Cyanide — Deadly poison

Ammonia — Cleaning products

Hydrogen cyanide — Gas chamber poison

Cadmium — Used to recharge batteries

Sulfuric acid — Car batteries

Arsenic — Rat poison

55% of *high school* students have *never,* **EVER**, tried *cigarette smoking.*

Medicine misuse sends 700,000 people to the hospital in the United States each year.

ALCOHOL

Alcohol use affects EVERY aspect of your life, including your relationships, your school work, and your social activities.

Alcohol affects EVERY organ of your body.

SUPPORT GROUPS offer a place to:

 Share experiences with one another.

 Discuss difficulties.

 Learn effective ways to cope with problems.

 Encourage one another.

Alcohol **is a factor in nearly one-third of all traffic deaths in the United States.**

MEDICINES

Drugs that PROTECT your health can HARM your health if taken incorrectly.

DRUG MISUSE:

Misusing a drug means that you:

Avoid following label instructions.

Use a drug not prescribed for you.

Use for longer than advised.

Take more than the prescribed dose.

Let someone else use a drug prescribed for you.

TEENS + ILLEGAL DRUGS

 NARCOTICS
Less than **3%** have ever injected an illegal drug into their body.

MARIJUANA
Less than **10%** have tried marijuana.

STIMULANTS
Less than **4%** have ever tried methamphetamines.

Drug users are at higher risk for: HIV, all other STDs, hepatitis + tuberculosis.

Drugs

LESSONS

Spreading the word about the dangers of drug use helps others avoid drugs. *What are some ways to spread the word about the dangers of drug use?*

Drug Use and Abuse

BIG IDEA Using drugs affects your body, mind, emotions, and social life and can lead to consequences with the law.

WHAT IS A DRUG?

MAIN IDEA Some drugs can help heal the body, but drugs can also be harmful to your health.

You have likely heard the word *drug* many times before. A **drug** is *a substance other than food that changes the structure or function of the body or mind.* Some drugs are medicines, which may be able to save your life or the life of someone you love. Other drugs are illegal and dangerous.

All medicines are drugs, but not all drugs are medicines. Medicines prevent or cure illnesses or treat their symptoms. Some are available as over-the-counter (OTC) medicines, meaning you can find them on the shelf at a grocery store or pharmacy. Some OTC medicines can only be purchased by an adult, but most of them are available to anyone. Other medicines can only be legally obtained with a prescription, or written permission from a doctor.

All *medicines* are **drugs,** but not all *drugs* are **medicines.**

All types of drugs, including medicines, can be misused or abused. It is important to be careful when using any drug, including medicine prescribed by your doctor. Whether you use an OTC medicine or a prescription drug, always closely read and follow the directions on the label. With a prescription medicine, your doctor will include instructions for how much to take, when to take it, and for how long. Some drugs are illegal because they are harmful to your health. However, even medicines that are legal can be dangerous if they are not used correctly. Medicines are only effective when they are used properly.

>>> **Reading Check**

IDENTIFY *What are two categories of legal drugs?*

>>> **Before You Read**

QUICK WRITE List two legal drugs or types of medicine. Then list two drugs you know are illegal or harmful.

 Video

>>> **As You Read**

FOLDABLES Study Organizer

Make the Foldable® found in the FL pages in the back of the book to record the information presented in Lesson 1.

>>> **Vocabulary**

› drug
› drug misuse
› drug abuse

🔊 Audio

🔤 Bilingual Glossary

Developing Good Character

Citizenship Part of being a good citizen is helping to protect the health of others. The Partnership at Drugfree.org or the National Institute on Drug Abuse (NIDA) advocate for the health of others. *Conduct research on organizations like these to find out how you can get involved. Present your findings to the class. Encourage your peers to get involved too.*

DRUG MISUSE AND ABUSE

MAIN IDEA) Any drug can be harmful to your health if abused or misused.

As you have learned, illegal drugs as well as drugs that are medicines can be harmful if they are not used correctly. Many drugs are illegal for a good reason: they can be extremely dangerous or addictive. However, **drug misuse**, *taking or using medicine in a way that is not intended,* is also dangerous and can even lead to abusing drugs. A person who does any of the following is misusing drugs:

- Using the drug without following instructions on the label
- Using a drug not prescribed for you
- Allowing someone else to use a drug prescribed for you
- Taking more of the drug than the doctor prescribed
- Using the drug longer than advised by your doctor
Drug abuse is another danger to your health. **Drug abuse** is *intentionally using drugs in a way that is unhealthful or illegal.* When someone uses an illegal drug, such as marijuana or heroin, he or she is abusing drugs.

When someone uses legal drugs for nonmedical purposes, that person is also abusing drugs. Even prescription medicines can be dangerous if used improperly. You put your health at risk if you take prescription medicine that was not prescribed for you.

Misusing or abusing any drug can damage your body and lead to allergic reactions, illness, or even death. Drug abuse interferes with brain function, affecting your mental/emotional health. Some drugs make it difficult to concentrate, or they may cause depression or anxiety. Your social health is also affected. Teens who abuse drugs may withdraw from family and friends and lose interest in school or other activities.

> *Misusing* or **abusing** a drug **can damage** your body and lead to *allergic reactions,* **illness,** or even *death.*

Digital Vision/Getty Images

Knowing the facts about illegal drugs can help you avoid using drugs. *List resources for valid information about illegal drugs?*

Physical Consequences

Physical effects of drug use include sleeplessness, memory loss, irritability, nausea, heart failure, seizures, or stroke. If a drug user drives while taking drugs, that person is putting others at risk too. Many drug users develop an addiction. The symptoms of addiction can include:

TOLERANCE develops when a person uses a drug regularly. The user needs more and more of the drug to get the same effect.

CRAVING is a primary symptom of addiction. A person will feel a strong need, desire, or urge to use drugs and will feel anxious if he or she cannot use them.

LOSS OF CONTROL causes a person to take more drugs than he or she meant to take. Drug use may also happen at an unplanned time or place.

PHYSICAL DEPENDENCE makes it very difficult to quit using a drug. A person's body develops an actual physical need for drugs in order to function.

Drug use can reduce your motor skills, or the ability to move your muscles in normal ways. Simple tasks such as writing, speaking, or walking can be affected. This type of damage can often be permanent. Drug use can also have negative effects on teen growth and development.

Mental *and* Emotional Consequences

Drug abuse weakens a person's ability to think and learn, even though the person may not realize it at the time. Some drugs kill brain cells. The brain damage that results can interfere with the user's ability to think.

Using drugs keeps you from learning to handle difficult emotions in healthful ways. Drug users often experience depression, anxiety, and confusion. Drug use also often leads to poor decisions and bad judgment. It can cause a person to engage in other risk behaviors.

Social Consequences

Drug abuse can change someone's personality, cause mood swings, or even lead to violence. Drug users often have low self-esteem and difficulty dealing with others, even those closest to them. A person addicted to drugs will start to think only of his or her need for the drug.

Teens who abuse drugs can also lose their friends. Some may end their friendships or lie to friends in order to cover up their addictions. After a while, obtaining and using the drug becomes more important than maintaining relationships with friends.

Teens who use drugs often miss school or do not learn well because they cannot pay attention. Teens may not be able to participate in school activities if they are caught using drugs. As a result, they lose the opportunity to learn new skills or have interesting new experiences. Teens who abuse drugs often hurt their chances of reaching their long-term goals, such as going to college or having a career.

In females, drug use can negatively affect:

- height
- weight
- onset of first menstrual cycle
- regularity of periods
- breast development
- function of ovaries
- pregnancy
- the health of unborn babies

In males, drug use can negatively affect:

- height
- weight
- male hormone levels
- testicle size and function
- muscle mass and development
- the age at which the voice gets lower
- the age at which body and facial hair increases

The use of illegal drugs can have serious health consequences during the teen years. *Name two reasons why drug use is especially harmful to teens.*

>>> **Reading Check**

DESCRIBE *How can drug use affect a person's emotional health?*

DRUG USE AND THE LAW

MAIN IDEA Using drugs can lead to problems with the law.

The legal consequences of drug use are one of the many negative effects of this harmful behavior. Most drug use is illegal and dangerous. Federal and state laws say that harmful and addictive drugs may not be used or sold. When someone is caught using illegal drugs, the legal consequences are very serious. Teens may be arrested for possession of drugs. Teens can spend time in a detention center or be sentenced to probation where they must regularly check in with a court officer. Often they and their parents may have to pay fines. Teens can also get a criminal record.

A strong connection exists between drug use and crime. Someone who uses drugs may steal the drugs or steal money to buy the drugs. Stealing can lead to acts of violence, which increase the chances of being caught and sent to jail.

A **strong connection** exists between *drug use* and *crime*.

The legal consequences of drug use can also impact a teen's social health. A criminal record can last forever. Such a background can restrict access to certain jobs. A criminal record can also affect relationships with family and friends.

> ### >>> Reading Check
>
> **EXPLAIN** *How can drug abuse lead to crime?* ■

Selling or just having illegal drugs can result in serious legal consequences. *What is one legal consequence of using illegal drugs?*

REVIEW

>>> After You Read

1. **VOCABULARY** Define *drug misuse.* Give at least one example in a complete sentence.
2. **LIST** Name three physical effects drugs can have on a person's body.
3. **DESCRIBE** How could using illegal drugs negatively affect your social health?

>>> Thinking Critically

4. **EVALUATE** What is the difference between using drugs as medicine and abusing drugs?
5. **ANALYZE** Why is a person who uses drugs more likely to be involved in a crime?

>>> Applying Health Skills

6. **ACCESS INFORMATION** Use online or library resources to research drug misuse and abuse in the United States. Create a pamphlet or electronic presentation to educate others about the dangers.

⟳ Review

🔊 Audio

Brand X Pictures/Getty Images

LESSON 1 Drug Use and Abuse **375**

Types *of* Drugs *and* Their Effects

BIG IDEA All types of illegal drugs have both short- and long-term effects.

Before You Read

QUICK WRITE List two illegal drugs you know about. Then briefly describe how these two drugs are harmful to a person's health.

 Video

As You Read

STUDY ORGANIZER Make the study organizer found in the FL pages in the back of the book to record the information presented in Lesson 2.

Vocabulary

> marijuana
> stimulant
> depressant
> hallucinogen
> narcotics
> club drugs
> inhalant
> anabolic steroids

 Audio

 Bilingual Glossary

MARIJUANA

MAIN IDEA Marijuana is an illegal drug that affects the body.

Marijuana comes from *dried leaves and flowers of the hemp plant, called cannabis sativa.* It is an illegal drug that is usually smoked. You may have heard the different terms that people use when referring to marijuana, such as *pot* or *weed.*

Marijuana has a strong effect on the brain of a person who uses the drug. A chemical in marijuana changes the way the brain processes what a person sees, feels, hears, and perceives. Marijuana use can cause a variety of reactions in people. Some users may feel a pleasant sensation, but others do not react well to this drug. It is important to know that marijuana use harms the body in many ways. It has both short-term and long-term effects on body systems.

Short-Term Effects

Marijuana use increases reaction time and impairs coordination and makes it harder to make healthful decisions. It also increases heart rate and appetite. High doses of marijuana can cause anxiety and panic attacks.

Marijuana use *harms* the **body** in **many** *ways.*

Long-Term Effects

Marijuana users risk developing lung disease or cancer. The drug has many of the same harmful chemicals as tobacco smoke. In fact, the marijuana plant contains more than 400 chemicals. As with many drugs, marijuana use over time can cause depression, personality changes, or trouble at school or work. It can also affect relationships with friends and family.

> ## Reading Check
>
> **EXPLAIN** *How does marijuana affect the user?*

What Teens Want to Know

Teen Brain Development Studies by the National Institute of Mental Health show that the teen brain is very different from the brain of an adult. The part of the brain that helps you foresee control impulses and foresee consequences of your actions is not fully developed in teens. This is a factor that makes it especially dangerous for teens to try drugs.

Royalty-Free/Corbis

STIMULANTS, DEPRESSANTS, AND CLUB DRUGS

MAIN IDEA Stimulants, depressants, and club drugs have many negative effects on the body.

A **stimulant** (STIM·yuh·luhnt) is *a drug that speeds up the body's functions.* Stimulants raise the heart rate, blood pressure, and metabolism. A **depressant** (di•PRE•suhnt) is *a drug that slows down the body's functions and reactions, including heart and breathing rates.* Stimulants and depressants affect the body in opposite ways. **Club drugs** are *illegal drugs that are found mostly in nightclubs or at all-night dance parties called raves.* These are often used in social settings and are very dangerous.

Stimulants

Stimulants cause the heart to beat faster. Blood pressure and metabolism also rise. Someone who uses a stimulant will often move or speak more quickly than normal. That person may also feel excited or even anxious.

Illegal stimulants include cocaine, crack, and methamphetamine (meth). Some stimulants are legal and not necessarily harmful, such as caffeine found in coffee, tea, soda, and chocolate. Doctors sometimes prescribe stimulants to their patients for certain problems. However, stimulant abuse can be dangerous. See the figure below to view the harmful effects of different types of stimulants.

Depressants

In contrast to stimulants, depressants slow down a person's motor skills and coordination. They can affect someone mentally and emotionally by giving a false sense of well-being through feelings of reduced anxiety or relaxation. However, when a depressant wears off, the user may experience extreme mood swings or depression.

Most depressants come in tablet or capsule form. Depressants are legal when prescribed by a doctor to treat certain conditions. For example, doctors sometimes prescribe tranquilizers or barbiturates to treat people who suffer from anxiety or sleep disorders. Alcohol is also a depressant. It is illegal for people under the age of 21 to purchase alcohol. Misuse and abuse of depressants, including alcohol, can lead to coma or even death. The risk is even higher when a person combines alcohol with a depressant drug.

The biggest risk associated with stimulant abuse is damage to your heart, sometimes causing heart attacks or death. *What are some other harmful effects of stimulant abuse?*

Substance	Other Names	Forms	Methods of Use	Harmful Effects
Amphetamine	Crystal, ice, glass, crank, speed, uppers	Pills, powder, chunky crystals	Swallowed, snorted up the nose, smoked, injected	Uneven heartbeat, rise in blood pressure, physical collapse, stroke, heart attack, and death
Methamphetamine	Meth, crank, speed, ice	Pills, powder, crystals	Swallowed, snorted up the nose, smoked, injected	Memory loss, damage to heart and nervous system, seizures, and death
Cocaine	Coke, dust, snow, flake, blow, girl	White powder	Snorted up the nose, injected	Damage to nose lining and liver; heart attack, seizures, stroke, and death
Crack	Crack, freebase rocks, rock	Off-white rocks or chunks	Smoked, injected	Damage to lungs if smoked, seizures, heart attack, and death

Club Drugs

Club drugs are often used to make people feel more relaxed in a social setting. They are often made in home laboratories. Club drugs may be mixed with other drugs or harmful chemicals. **ECSTASY** increases the heart rate and body temperature, which can damage a person's organs. A person using Ecstasy may experience tingly skin or clenched jaws. He or she can also feel anxious and paranoid.

ROHYPNOL makes a person's blood pressure drop. The user feels dizzy and very sleepy. This drug also causes blackouts and memory loss. Rohypnol typically comes in pill form, although it can be crushed into a powder. It is also a drug that the user may not know that he or she has been given. If added to a drink, for example, it can make a person unconscious. As a result, this drug is unfortunately used to commit the crime of date rape.

KETAMINE, an anesthetic used in medical procedures, can be deadly if abused. It causes hallucinations, and people who use it often experience memory loss. An overdose of ketamine can cause a person to stop breathing.

>>> **Reading Check**
COMPARE AND CONTRAST *What is the main difference between the way stimulants and depressants affect the body?*

HALLUCINOGENS

MAIN IDEA Hallucinogens are dangerous drugs that mainly affect the user's mind.

Hallucinogens (huh·LOO·suhn·uh·jenz) are *drugs that distort moods, thoughts, and senses.* These drugs interfere with thought processes and the ability to communicate. Hallucinogen users may become disoriented or confused. Strange behavior is common because the user can no longer tell what is real and what is not.

LSD (acid) and PCP (angel dust) are common hallucinogens. LSD is one of the strongest. It may come in tablet form or on absorbent paper. Someone who uses LSD may not know who or where they are. Harmful behaviors are common with LSD use. Users may even have terrifying flashbacks weeks or months after using the drug. The effects of PCP are similar to those of LSD.

>>> **Reading Check**
DETERMINE *Name three harmful effects of LSD.*

Hallucinogens can cause many harmful effects, including death. *What are two examples of hallucinogens?*

Substance	Other Names	Forms	Methods of Use	Harmful Effects
PCP	Angel dust, supergrass, killer weed, rocket fuel	White powder, liquid	Applied to leafy materials and smoked	Loss of coordination; increased heart rate, blood pressure, and body temperature; convulsions; heart and lung failure; broken blood vessels; bizarre or violent behavior; temporary psychosis; false feeling of having superpowers
LSD	Acid, blotter, microdot, white lightning	Tablets; squares soaked on paper	Eaten or licked	Increased blood pressure, heart rate, and body temperature; chills, nausea, tremors and sleeplessness; unpredictable behavior; flashbacks; false feeling of having superpowers

NARCOTICS

MAIN IDEA Not all narcotics are illegal, but all are extremely addictive.

Narcotics (nar·KAH·tics) are *drugs that get rid of pain and dull the senses.* These are highly addictive drugs. Historically, narcotics have been made from opium—a plant liquid that can numb the body. When used under a doctor's supervision, narcotics such as morphine and codeine are effective in treating extreme pain. However, laws control how all narcotics are sold and used because they are so addictive.

Heroin

Heroin is an illegal narcotic made from morphine. It is often inhaled or injected and sometimes smoked. Injecting the drug gives users a pleasant feeling. However, this feeling quickly wears off, and the user experiences symptoms of withdrawal.

These symptoms may include nausea, cramps, and vomiting. People who use heroin risk unconsciousness and death. Since it is usually injected, heroin users also risk HIV or hepatitis infection from shared needles. Because it is highly addictive, users commonly experience tolerance and dependence.

Oxycodone

Oxycodone is available legally through a doctor's prescription. When used as directed, it can control severe pain in patients with cancer and other diseases. However, oxycodone is also frequently abused. The long-term effect of using the drug is increased tolerance, which leads to physical addiction.

>>> **Reading Check**

IDENTIFY *Name two types of narcotics that may be used as medicine when prescribed by a health care professional.*

Think of all the ways you can have fun without using drugs. *What are some drug-free activities that you and your friends enjoy?*

Health Effects of Narcotic Drug Abuse

Can cause drowsiness, constipation, and depressed breathing.

Taking a large single dose could cause severe respiratory depression or death.

Can cause death if taken with certain medications or alcohol.

Can lead to physical dependence and tolerance. The body becomes used to the substance and higher doses are needed to feel the same initial effect.

Narcotics are highly addictive, often causing uncontrollable drug use in spite of negative consequences.

Withdrawal symptoms occur if use is reduced abruptly. Symptoms can include restlessness, muscle and bone pain, insomnia, diarrhea, vomiting, cold flashes with goose bumps, and involuntary leg movements.

Withdrawal from narcotics usually requires detoxification in a hospital. Although withdrawal is often a painful experience, it is not life-threatening.

When narcotics are abused, there is a risk of addiction and other health consequences. *What are some of these consequences?*

iStock/SuperStock

INHALANTS

MAIN IDEA Using inhalants can cause brain damage and even death.

Inhalants (in·HAY·luhntz) are *the vapors of chemicals that are sniffed or inhaled to get a "high."* Most come from toxic household products. Toxic inhalants include hairspray, lighter fluid, air freshener, cleaning products, markers or pens, and paint. Inhalant use can cause nausea, dizziness, confusion, and loss of motor skills.

Inhalant use sends poisons straight to the brain. Inhalants cause permanent damage and affect a person's ability to walk, talk, or think. Inhalant abuse, even if it is only the first time, can also kill the user instantly. Experimenting with inhalants can cause death from choking, suffocation, or heart attack.

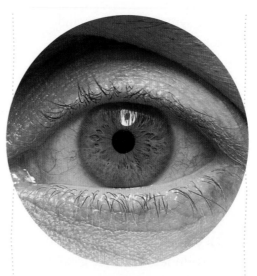

One symptom of inhalant abuse is red or runny eyes. *What other parts of the body can inhalant abuse damage?*

Warning Signs *of* Inhalant Abuse

A person who uses inhalants often shows symptoms of the abuse. Some common symptoms are listed below. If you notice these symptoms in someone you know, speak to a teacher or trusted adult about your concerns.

- Eyes that are red or runny
- Sores or spots near the mouth
- Breath that smells strange or like chemicals
- Holding a marker or pen near the nose

>>> **Reading Check**

GIVE EXAMPLES *What are two symptoms of inhalant abuse?*

Health SKILLS ACTIVITY

Accessing Information

Marijuana *Myths*

Marijuana is the most commonly used illegal drug. Because its effects are not as dramatic as some other illegal drugs, it is often mistakenly believed to be a harmless drug. Here are some common myths about marijuana use:

Myth: "Marijuana is not addictive. Users can stop whenever they want."

Myth: "Smoking marijuana is safer than using other drugs."

Myth: "Everyone smokes marijuana."

On Your Own

Using this lesson and other reliable online or print resources, find information that shows how these myths may not be true. Develop a fact sheet listing each myth and the truths behind each myth.

Radius/SuperStock

Fitness Zone

Improving Performance Some professional athletes use steroids to get stronger or better. I know that the best and healthiest way to get stronger is to practice. After school each day and over the weekend, I try to practice soccer drills and ball handling. I also try to stay active and get exercise at least a few days each week to get stronger.

STEROIDS

MAIN IDEA Steroid use can cause serious health problems.

Some drugs mimic the behavior of chemicals made by the body. For example, anabolic steroids (a·nuh·BAH·lik STAIR·oydz) are *substances that cause muscle tissue to develop at an abnormally high rate.* Doctors sometimes legally prescribe steroids to treat growth problems, lung diseases, and skin conditions.

> Any **nonmedical** use of *steroids* is **illegal.**

Some athletes use steroids to try to increase their body weight, strength, or endurance. However, steroid use causes a variety of serious health issues.

Effects include uncontrolled anger, shrunken testicles, severe acne, high blood pressure, and infertility. Females can develop a deeper voice, excess facial hair, and a masculine-looking body. Steroid use can also cause damage to other body systems.

Any nonmedical use of steroids is illegal. Athletes who use steroids can be dismissed from a team or an event. Illegal steroid users may also face fines and jail time. Athletes who are caught using steroids are often suspended or banned from their sport. Many have had their reputations damaged as a result.

>>> **Reading Check**

RECALL *How are anabolic steroids similar to chemicals in your body?* ■

LESSON 2

REVIEW

>>> **After You Read**

1. **VOCABULARY** Define depressant. Use the term in a complete sentence.
2. **EXPLAIN** What is the biggest long-term risk for marijuana users?
3. **SUMMARIZE** Name three types of drugs that are legal only when prescribed by a doctor.

>>> **Thinking Critically**

4. **APPLY** Suppose a friend told you inhalants were safe because they are items found in your own house. How would you respond? Is this valid health information?
5. **ANALYZE** What are some of the high-risk behaviors that could result from using hallucinogens or club drugs?

>>> **Applying Health Skills**

6. **ADVOCACY** Write a script for a public service announcement for radio or television, explaining the short- and long-term effects of narcotics use.

Ⓒ Review

🔊 Audio

The best way to improve your athletic performance is to practice. *How could using drugs stand in the way of reaching your goals?*

Staying Drug Free

BIG IDEA Many reasons and resources exist to help teens and their families stay drug free.

Developing Good Character

Citizenship You can demonstrate good citizenship by encouraging others to stay drug free. Find out about programs in your school or community that educate teens on the dangers of drug use and tell your classmates about them. Identify what methods they use to reach teens and find out how students can get involved. *What programs would you be interested in participating in? Why?*

WHY DO SOME TEENS USE DRUGS?

MAIN IDEA Responding to peer pressure, the media, and personal problems can influence teens to try drugs.

Teens use drugs for many different reasons. For example, the media often show people enjoying alcohol or even drugs, particularly in TV shows or movies. These images can make some teens feel they will be like the happy, attractive people they see if they begin to drink alcohol or use drugs too.

What these images may not show are the problems addiction can cause. Often it appears that using drugs or alcohol is harmless or without consequence. However, drug and alcohol use can have serious effects on all sides of your health triangle. The media does sometimes offer helpful information about drugs and alcohol. Television reports and magazine articles can provide accurate health information about the effects of drugs on the brain and body, for example.

Drug and **alcohol** use can have *serious* effects on all sides of your **health triangle.**

Additionally, teens might see adults in their lives using alcohol. Friends might pressure a teen to try drugs or alcohol. Peer pressure can have a strong influence even if it is negative. It can be difficult to say no to friends whose approval you would like.

Finally, some teens may try drugs to cope with emotional problems. To escape or forget, they may turn to substance abuse, or *using illegal or harmful drugs, including any use of alcohol while under the legal drinking age.* Substance abuse includes drug misuse and drug abuse.

>> **Reading Check**

EXPLAIN *Describe the influence that peers can have on a teen's decision to use alcohol or drugs.*

Sue Smith/age fotostock

WAYS TO STAY DRUG FREE

MAIN IDEA Staying drug free has many benefits.

The choice to avoid illegal drugs and the improper use of legal drugs may be the most healthful decision you can make. Being substance free shows self-control. It means you have taken charge of your life and your health. It is important to make decisions that promote both a healthy body and mind. Staying drug free can be difficult, but it has many benefits.

Reasons to Avoid Drugs

Making wise choices about drugs will have a positive effect on your physical, mental/emotional, and social health. When you avoid drugs, you will have a greater chance of reaching your goals.

These are a few activities you can enjoy without using alcohol or other drugs. *Can you think of some other positive alternatives to substance abuse?*

By deciding not to use drugs, you can enjoy these benefits:

- **Physical health** You show that you care for yourself and your health. You will not suffer the physical consequences of drug abuse on the body.
- **Control over your actions** You are better able to stay in control and act more responsibly.
- **Obeying the law** You show respect for the law and are a good citizen.
- **Protecting your future** You are able to set goals and work toward them. You are able to concentrate better and will do better in school.
- **Healthier relationships** You are able to enjoy other interests with family and friends.
- **Self-respect** You want to avoid harm to your body and mind. You can be confident about your decisions. You know you have made healthful choices.

The *choice* to **avoid illegal drugs** is the most *important* and *healthful* **decision** you could make.

Alternatives to Drug Use

If someone offers you drugs or alcohol, what would you do? You could suggest an alternative (ahl·TER·nuh·tihv), or *another way of thinking or acting.* Offering positive alternatives can help relieve some of the pressure you may be feeling. It also gives you the chance to be a positive influence on your peers. Possible alternatives include attending drug-free events and improving your talents and skills.

Have fun at drug-free and alcohol-free events. Avoid environments where alcohol and other drugs are present. Use positive peer pressure to help others avoid these environments.

Improve your talents or skills. Choose an activity you like, and practice it until you become an expert. Become a great skateboarder, a computer whiz, or the best artist at school.

Be part of a group. Join a sports team, a club, or a community group.

Start your own business. Make yourself available for babysitting, yard work, or other jobs. Let friends and neighbors know.

Fancy/Alamy

Refusing *Drugs*

Tyler has noticed that his best friend, Jonathan, has become a little distant. Jonathan recently changed schools, so Tyler no longer sees him every day. Tyler knows that Jonathan has made some new friends at his new school. When they get together, Tyler notices that Jonathan doesn't seem interested and has trouble following the conversation.

One day, Tyler meets up with Jonathan and some of his new friends. Jonathan's new friends suggest that everyone go back to Jonathan's house to smoke marijuana. Tyler wants to keep Jonathan as a friend, but he doesn't want to use drugs. What should Tyler do?

What Would You Do?

Apply the S.T.O.P. formula to Tyler's situation.

- **S**ay no in a firm voice.
- **T**ell why not.
- **O**ffer other ideas.
- **P**romptly leave.

Explain how Tyler might use this strategy to refuse the harmful suggestion made by Jonathan's friends.

Saying No *to* Drugs

Developing skills to refuse drugs is very important. The best way to avoid being pressured to use illegal substances is to use refusal skills. Refusal skills can help you say no to other unhealthy behaviors as well. You can resist negative peer pressure without feeling guilty or uncomfortable. Saying no in a clear and confident way lets others know you respect yourself and your health. If you feel pressure to experiment with drugs, remember the S.T.O.P. strategy: **S**ay no in a firm voice; **T**ell why not; **O**ffer alternative ideas or activities; and if all else fails, **P**romptly leave.

This strategy is helpful when you are faced with a difficult situation. However, you can also take steps to help you avoid even having to use your refusal skills.

You can choose to make friends with people who have also chosen to avoid drugs. Friends who are committed to being drug free will support your decision and help you avoid situations where drugs or alcohol may be present. You can also look for healthful ways to deal with whatever issues you may face. If you feel lonely or depressed, or if you need help solving a personal problem, talk to a parent or another adult you trust. Making wise choices about how you spend your time, who your friends are, and how you deal with your feelings can all have a positive effect on your ability to be drug free.

>>> **Reading Check**

IDENTIFY *What are two positive alternatives to drug use?*

Friends can be an important influence during the teen years. *How can friends help you stay drug free?*

©Purestock/PunchStock

HELP FOR DRUG USERS AND THEIR FAMILIES

MAIN IDEA › Resources are available to help drug users and their families face substance abuse.

Stopping drug abuse after it has started is much harder than resisting drugs in the first place. Some effects of drug abuse are permanent. However, drug addiction is treatable. Many resources are available to help drug users overcome the pattern of addiction.

Drug Treatment Options

Drug addiction is a disease. Treatment requires changes in behavior. People who are addicted to drugs must first admit that they have a problem. Then they need to seek help to recover. In some cases, drug rehabilitation is needed. This is *a process in which a person relearns how to live without an abused drug.* Sometimes a drug abuser may enter a treatment facility.

Support groups allow people to work together to help stay drug free. Often, recovering addicts find strength from talking with other people who are working toward the same goal.

Regular meetings encourage these positive relationships. Support groups for addiction include Narcotics Anonymous and Cocaine Anonymous.

Some people find the support and help they need to stay drug free through counseling. Counseling provides an opportunity to openly share thoughts and feelings with a trained expert. It can help addicts deal with their psychological dependency on drugs. Counseling may involve only the addict or the person's entire family.

Help *for* Families

When someone is addicted to drugs, that person's family also needs help. One of the many resources for families is Nar-Anon. Like Al-Anon, Nar-Anon helps family members learn how to deal with the problems caused by drug addiction.

>>> **Reading Check**

RECALL *What are some drug treatment options?* ■

LESSON 3

REVIEW

>>> **After You Read**

1. **VOCABULARY** Define *drug rehabilitation.* Use it in a complete sentence.
2. **LIST** What are two reasons why some teens might choose to use drugs?
3. **DESCRIBE** What are some of the ways support groups help people become drug free?

>>> **Thinking Critically**

4. **EVALUATE** How can suggesting a positive alternative to alcohol or drug use help you stay substance free? Explain your answer.
5. **APPLY** What do you think is the most important reason for you to stay drug free? Explain your reasoning in a short paragraph.

>>> **Applying Health Skills**

6. **REFUSAL SKILLS** Think about ways to say no to harmful behaviors. Team up with a classmate. Role-play a situation where you use these strategies to say no to illegal drugs.

🔄 Review

🔊 Audio

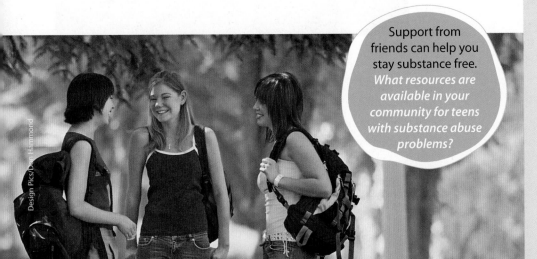

Support from friends can help you stay substance free. *What resources are available in your community for teens with substance abuse problems?*

Design Pics/Don Hammond

Memory Obstacles

Many drugs make it difficult to process information. In this activity, loud music and disruptive talk mimics the effects of some drugs. How will these distractions interfere with your ability to recall information?

WHAT YOU WILL NEED

* Paper for each member of the group
* Pencil for each member of the group
* 2 posters with 25 pictures of everyday items on each
* A source of loud music

WHAT YOU WILL DO

1 Using pictures from magazines, one person in the group will make 2 posters with 25 everyday items on each one.

2 The leader will hold up the first poster and give the group 30 seconds to look at it. Each student should try to remember as many items as possible.

3 The leader then puts the poster down, and the students write down as many of the items as they can recall.

4 Now loud music is turned on. The leader also asks two people on opposite sides of the room to have a loud conversation with each other.

5 Then the leader holds up the second poster and repeats step 3.

WRAPPING IT UP

How many people in the group had more trouble remembering items on the second poster than on the first? Why? Compare what just happened in your group to what can happen to your brain when under the influence of a drug.

Getty Images/SW Productions

READING REVIEW

FOLDABLES and Other Study Aids

Take out the Foldable® that you created and any study organizers that you created. Find a partner and quiz each other using these study aids.

LESSON 1 **Drug Use and Abuse**

BIG IDEA Using drugs affects your body, mind, emotions, and social life and can lead to consequences with the law.

* Drugs can be medicines that help the body, but many drugs are harmful to your health.
* Any drug can be harmful to your health if abused or misused.
* Drug misuse, or taking medicine in a way that is not intended, is dangerous and can lead to drug abuse.
* Drug abuse is intentionally using drugs in a way that is unhealthful or illegal.
* Using drugs can result in serious legal problems, including arrest, fines, jail time, and a criminal record.

LESSON 2 **Types of Drugs and Their Effects**

BIG IDEA All types of illegal drugs have both short- and long-term effects.

* Marijuana is an illegal drug that affects the body.
* Stimulants come in several forms and have effects on the user.
* Common stimulants include caffeine, which is legal, along with illegal drugs such as cocaine.
* Depressants slow down the body and mind and affect a person's motor skills and coordination.
* Many depressants are legal when prescribed by a doctor, but they are often misused or abused.

* Drugs that are considered to be hallucinogens, such as LSD, PCP, and Ecstasy, affect the user's mind and have dangerous effects.
* Club drugs are often used in social settings and can result in dangerous consequences.
* Not all narcotics are illegal, but these types of drugs are all extremely addictive.
* Using inhalants, even for the first time, can cause brain damage and even death.
* Anabolic steroid use can cause many serious health problems, including anger issues, shrunken testicles in males, masculine characteristics in females, severe acne, heart disease, certain types of cancer, and damage to other body systems.

LESSON 3 **Staying Drug Free**

BIG IDEA Many reasons and resources exist to help teens and their families stay drug free.

* Responding to peer pressure, the media, and personal problems can influence teens to try drugs.
* Practicing refusal skills can help you stay drug free.
* Staying drug free can benefit all sides of your health triangle—physical, mental/emotional, and social.
* Resources are available to help drug users and their families deal with the problems of substance abuse and break the pattern of addiction.

 Review

 Web Quest

ASSESSMENT

Reviewing Vocabulary *and* Main Ideas

> › depressant
> › drug
> › hallucinogens
> › drug misuse
> › marijuana
> › inhalants
> › drug abuse
> › stimulant

>> On a sheet of paper, write the numbers 1–8. After each number, write the term from the list that best completes each statement.

LESSON 1 Drug Use and Abuse

1. _____ involves taking or using medicine in a way that is not intended.

2. When someone takes an illegal drug, such as marijuana or heroin, that person is engaging in _____.

3. A _____ is a substance other than food that changes the structure or function of the body or mind.

LESSON 2 Types of Drugs and Their Effects

4. _____ are drugs that distort the moods, thoughts, and senses of the user.

5. _____ include any substance whose fumes are sniffed and inhaled to produce mind-altering sensations.

6. The most commonly used illegal drug is _____.

7. A drug that raises the heart rate, blood pressure, and metabolism is called a _____.

8. A _____ slows down a person's motor skills and coordination and can affect a user mentally and emotionally.

>> On a sheet of paper, write the numbers 9–14. Write *True* or *False* for each statement below. If the statement is false, change the underlined word or phrase to make it true.

LESSON 3 Staying Drug Free

9. Another way of thinking or acting is called an <u>alternative</u>.

10. The use of illegal or harmful drugs is called <u>rehabilitation</u>.

11. <u>Support groups</u> allow people to work together to help stay drug free.

12. <u>Few</u> resources are available to help drug users and their families deal with substance abuse.

13. The <u>S.T.O.P. strategy</u> involves saying no in a firm voice, telling why not, offering positive alternative ideas or activities, and promptly leaving if all else fails.

14. A process in which an addicted person learns how to live without drugs is called <u>refusal skills</u>.

✔ eAssessment

>> Using complete sentences, answer the following questions on a sheet of paper.

🗨️ *Thinking* **Critically**

15. HYPOTHESIZE What are three reasons someone might begin experimenting with drugs?

16. HYPOTHESIZE Why might someone ignore the risks of drug use?

🔌 *Write* **About It**

17. NARRATIVE WRITING Write a story about a teen athlete who is considering using steroids.

18. EXPOSITORY WRITING Write a paragraph describing how a teen can use positive peer pressure to influence others to avoid using drugs.

Ⓐ Ⓑ Ⓒ Ⓓ STANDARDIZED TEST PRACTICE

Reading
Read the two paragraphs and then answer the questions.

Heroin is a highly addictive drug that relieves pain and dulls the senses. Heroin users usually inject the drug into their body. The drug gives users a sense of euphoria, or a feeling of intense joy. This feeling is short-lived, however. Soon after, the users suffer from withdrawal, which can be extremely painful.

Because the symptoms of heroin withdrawal are so painful, users must go through a detoxification process. Most need professional help to quit. The process may include doses of legal drugs that relieve the withdrawal symptoms. One of the drugs used in heroin detoxification is called methadone. It is a synthetic drug not found in nature. It delays the feelings and cravings that users experience during withdrawal.

1. In the first paragraph, euphoria means
  A. a feeling of pain.
  B. a feeling of fullness.
  C. a feeling of intense joy.
  D. None of the above

2. From the information in the second paragraph, the reader can conclude that heroin
  A. is a safe drug.
  B. can be used with moderation.
  C. can be given up easily by users.
  D. is a dangerous and addictive drug.

3. According to the second paragraph, many heroin users must go through a detoxification process in order to quit because
  A. withdrawal symptoms are so painful.
  B. it is required by the law.
  C. it stops cravings for other drugs, as well.
  D. None of the above

Using Medicines Wisely

PREMIUM ONLINE RESOURCES

 Audio

 Videos

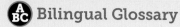 Bilingual Glossary

 Fitness Zone

 Web Quest

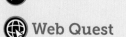 Review

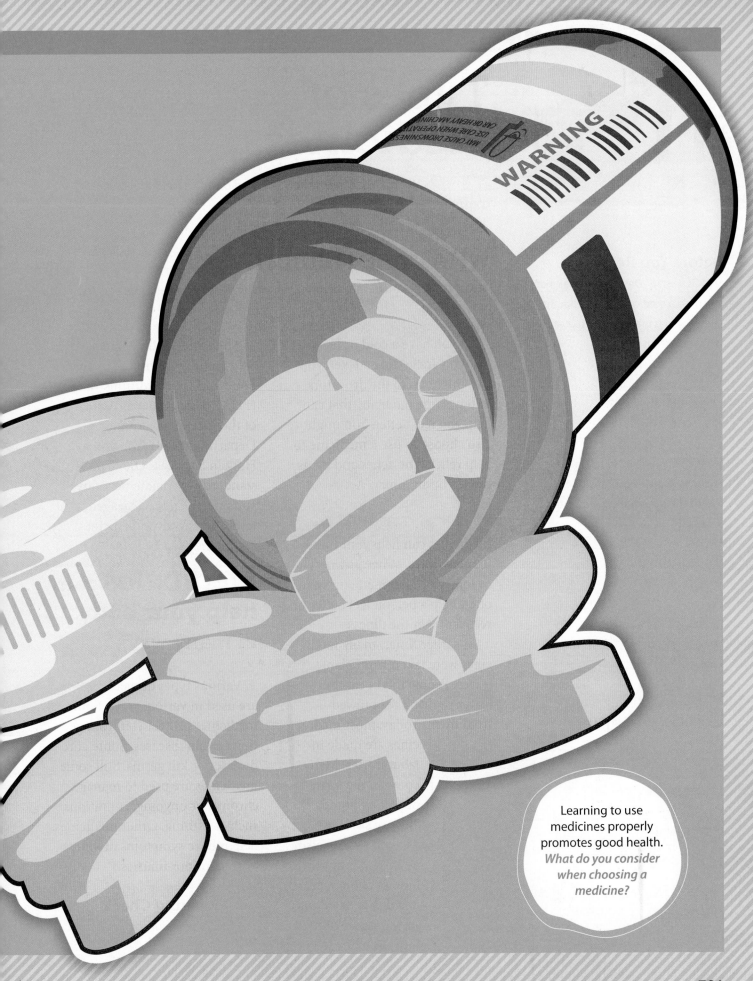

Learning to use medicines properly promotes good health. *What do you consider when choosing a medicine?*

Types *of* Medicines

BIG IDEA Using medicines wisely is a sign of good personal and consumer health.

Before You Read

QUICK WRITE Describe a time when you used a medicine. State what the medicine was intended to treat and how you used it.

 Video

As You Read

FOLDABLES | Study Organizer

Make the Foldable® found in the FL pages in the back of the book to record the information presented in Lesson 1.

Vocabulary

› medicine
› vaccine
› antibiotics
› over-the-counter (OTC) medicine
› prescription medicine

 Audio

 Bilingual Glossary

Myth vs. Fact

Myth: Only babies and toddlers need vaccines.
Fact: Several important vaccines are required for those between the ages of 10 and 18. Most preteens and teens will need to receive a vaccine or booster shot for diphtheria and tetanus, chicken pox, hepatitis B, measles, mumps, and rubella, and meningococcus.

WHAT ARE MEDICINES?

MAIN IDEA A medicine is a drug that can prevent or cure an illness or ease its symptoms.

If you have a cold or feel as if you are getting a fever, you might go to the drugstore to buy some medicine that can help you feel better. You might also choose to use a medicine to help relieve the aches and pains caused by an injury. A medicine is *a drug that prevents or cures an illness or eases its symptoms.* Medicines can help your body in many ways. When sickness or injury occurs, medicines can often help a person feel better or recover from the illness.

In earlier times, many medicines were taken from plant leaves. People might eat the leaves or drink tea brewed from them. Today, most medicines are made in laboratories. Many come in the form of pills or liquids and are swallowed. Medicines may also be injected into the bloodstream using needles, inhaled into the lungs, or rubbed into the skin.

Medicines in the United States are carefully controlled by the Food and Drug Administration (FDA). This agency is part of the federal government's Department of Health and Human Services. The FDA sets standards for medicine safety and effectiveness. The agency tests and approves medicines before they can be sold.

Medicines can **help your body** in many ways.

Various types of medicines are used in various ways. Some medicines protect you from getting certain diseases. Some cure diseases or kill germs. Still some medicines are used to manage chronic, or ongoing, conditions such as asthma. Other medicines help relieve symptoms of illness or treat minor injuries.

> **Reading Check**
>
> **DEFINE** *What are medicines?*

George Doyle/Getty Images

THE PURPOSE OF MEDICINES

MAIN IDEA Different medicines serve different purposes in the body.

Have you ever had to get a shot during a visit to a doctor or nurse? Sometimes, a doctor may give you a shot to help fight an illness. On other occasions, you may get a shot to help prevent a disease. A vaccine is one example of a type of medicine that is used to keep you from getting a disease. A medical professional may also give people certain medicines to help them recover from a disease or ease the symptoms of an illness. Scientists who focus on finding cures have developed medicines that help a sick person's body fight disease or infection. Others types of medicines can be used to relieve pain, soreness, or swelling.

Have you ever had to get a shot during a visit to a doctor or nurse?

Vaccines are medicines that prevent disease. *What are some vaccines that you may have received?*

Type of Medicine	Some Examples	Disease or Problem
Vaccines	• MMR vaccine • Varicella vaccine • HPV vaccine • Pertussis vaccine	• Measles, mumps, and rubella • Chicken pox • Human Papillomavirus • Whooping Cough
Antibiotics	• Penicillin • Cephalosporin • Tetracycline • Macrolides	• Strep throat, pneumonia, STDs • Meningitis, skin rash • Urinary tract infections, Rocky Mountain spotted fever • Given to patients allergic to penicillin
Pain Relievers	• Aspirin, acetaminophen, ibuprofen, codeine	• General pain relief

Various types of medicines are used to treat or prevent various illnesses. *What types of illnesses are antibiotics used to treat?*

Preventing Disease

A **vaccine** (vak·SEEN) is *a preparation of dead or weakened pathogens that is introduced into the body to cause an immune response.* Vaccines protect against diseases that can spread from person to person. When you are vaccinated, your body can make substances called antibodies that will attack or kill off the germs that cause a disease. Some vaccines provide protection for many years. Others, such as the flu vaccine, protect you for only about a year.

Fighting Germs

Antibiotics (an·tih·by·AH·tiks) are *medicines that reduce or kill harmful bacteria in the body.* An antibiotic may be prescribed to treat an infection. However, improper use can make bacteria resistant. That means certain antibiotics may no longer stop an infection. An antibiotic cannot help the body fight an illness such as the common cold. Other drugs are used to fight viruses and fungi.

Taking Medicine

Katie runs track. Because of all the training she does, Katie often has sore muscles. She takes an over-the-counter pain reliever for her sore muscles, but lately it hasn't been helping. What should Katie do?

Remember the six steps of the decision-making process:

1. State the situation.
2. List the options.
3. Weigh the possible outcomes.
4. Consider your values.
5. Make a decision and act on it.
6. Evaluate the decision.

Use these steps to determine what you would do if you were in Katie's situation.

Relieving Pain

Have you ever had sore muscles, a toothache, or a headache? If you took medicine to feel better, it was most likely a pain reliever. Pain relievers block pain signals sent through the nervous system. Many pain relievers are available as **over-the-counter (OTC) medicine,** or *a medicine that you can buy without a doctor's permission.* These include aspirin, ibuprofen, and acetaminophen. Aspirin can also reduce swelling. Some painkillers, such as codeine, treat more serious pain and are available only with a prescription. A **prescription** (prih·SKRIP·shuhn) **medicine** is *a medicine that can be obtained legally only with a doctor's written permission.* Prescription medicines require a written order because they typically carry more risks.

> Different medicines do different jobs. *Why are medicines regulated by the government?*

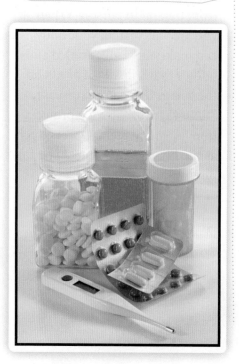

Managing Disease

Some medicines help people manage chronic diseases or conditions, such as allergies, asthma, diabetes, or mental illnesses such as anxiety and depression. People with diabetes take insulin to help control their blood sugar. People with allergies can take antihistamines to treat swelling and other allergy symptoms.

Often, medicine is taken by swallowing. However, medicine can be given in a number of different ways. The various methods of delivering medicine to the body include:

- **Swallowing, or ingestion.** A pill, tablet, capsule, or liquid moves through the stomach into the bloodstream and then through the body. Most pain relievers are taken this way.

- **Injection, or shot.** Injected medicines begin to work more quickly because they directly enter the bloodstream. These are administered by a needle that pierces the skin.

- **Inhalation.** Medicine can be inhaled, or breathed in, as a mist or fine powder. People with asthma may use an inhaler. Cold or sinus medication can also be inhaled through the nose.

- **Topical application.** Creams and ointments can be rubbed directly onto the skin. Patches containing medicine may also be applied to the skin.

⟫⟫⟫ Reading Check

EXPLAIN *Name three different kinds of medicine, and tell what each does.*

Tetra Images/Getty Images

PRESCRIPTION AND NON-PRESCRIPTION MEDICINES

MAIN IDEA Some medicines require a doctor's permission, while others are available without a prescription.

As you have learned, prescription medicine can be obtained only with a written order from a doctor. Written permission is required because prescription medicines carry more risks. You do not need a doctor's permission to buy an over-the-counter medicine. However, both prescription and over-the-counter medicines must be used very carefully.

If you are prescribed medicine, your doctor will write out instructions that explain how much to take, when to take it, and for how long. Prescriptions must be filled by a pharmacist who is trained to prepare and distribute medicines. Your doctor's instructions will appear on the prescription medicine label.

The FDA requires medicine labels to have specific information. Always read the label closely before taking any medicine, and follow the directions.

You can find over-the-counter (OTC) medicines in groceries and drugstores. They are considered safe to use without a doctor's permission. However, always be careful when you use OTC medicines. Follow the directions, because even OTC medicines can be harmful if not used correctly. If you have any questions about a medicine, ask a doctor or pharmacist.

>>> **Reading Check**

CONTRAST *What is the main difference between prescription and over-the-counter (OTC) medicines?* ■

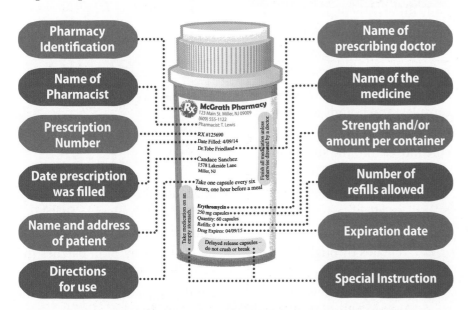

Pharmacy Identification

Name of Pharmacist

Prescription Number

Date prescription was filled

Name and address of patient

Directions for use

Name of prescribing doctor

Name of the medicine

Strength and/or amount per container

Number of refills allowed

Expiration date

Special Instruction

McGrath Pharmacy
123 Main St. Miller, NJ 09009
(609) 555-1122
Pharmacist: T. Lewis

RX #125690
Date Filled: 4/09/14
Dr. Tobe Friedland

Candace Sanchez
1578 Lakeside Lane
Miller, NJ

Take one capsule every six hours, one hour before a meal

Erythromycin
250 mg capsules
Quantity: 60 capsules
Refills: 0
Drug Expires: 04/09/15

Delayed release capsules – do not crush or break

Take medication on an empty stomach.

Finish all medication unless otherwise directed by a doctor.

Your doctor or pharmacist can help you understand a prescription medicine label. *What types of information can you find on a prescription medicine label?*

>>> **After You Read**

1. **VOCABULARY** Define the term *antibiotics*. Use it in an original sentence.
2. **NAME** What is the type of medicine that prevents a disease from developing?
3. **LIST** What are the four main purposes of medicines?

>>> **Thinking Critically**

4. **ANALYZE** A friend of yours on the football team wants an energy burst before a game. He wants to take a handful of vitamins. When you express concern, he says, "They're over-the-counter vitamins." Respond to this comment.

>>> **Applying Health Skills**

5. **APPLY** Lanie's doctor gave her a six-day prescription of an antibiotic for her sore throat. After only three days, all of her symptoms are gone. Should Lanie continue taking the antibiotic? Explain why or why not.

⟳ Review

 Audio

How Medicines Affect Your Body

BIG IDEA Medicines can contribute to good health when used properly.

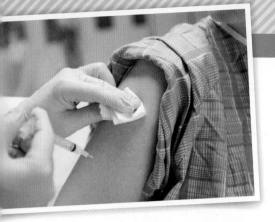

>>> **Before You Read**

QUICK WRITE Write about a time when you had to take medicine and how it affected you.

 Video

>>> **As You Read**

STUDY ORGANIZER Make the study organizer found in the FL pages in the back of the book to record the information presented in Lesson 2.

>>> **Vocabulary**

› side effect
› tolerance

 Audio

Ⓐ Ⓑ Bilingual Glossary

🏃 Fitness Zone

When I get regular exercise and eat healthfully, I don't get sick very often. As a result, I don't need to take medicine very often and risk possible bad side effects. Physical activity helps me stay healthy. *How can a physical fitness routine contribute to your good health?*

HOW MEDICINES ENTER THE BODY

MAIN IDEA A medicine may be swallowed, injected, inhaled, or applied to the skin.

Different medicines are used in different ways. For example, a mild sunburn or an itchy mosquito bite could be treated with a cream or lotion. Another form of over-the-counter medicine could help relieve pain from a headache or a pulled muscle. A person with a serious illness may need prescription pills or injections. Medicines affect your body differently depending on how they are administered.

Swallowing, or ingestion, is the most common way to take medicines. Pills, tablets, capsules, and liquids are taken orally, or by mouth. The medicine moves into the stomach and small intestine. From the digestive system, the medicine passes into the bloodstream and circulates throughout the body. Cold medicines and pain relievers are often delivered this way.

Injection, or a shot, is another way that medicines can enter the body.

A needle injects the medicine directly into the bloodstream. Injected medicines begin to work more quickly than other types.

Inhalation is yet another delivery system. Medicine can be inhaled as a mist or fine powder. People with asthma often use inhalers. You can breathe in, or inhale, cold or sinus medication through your nostrils.

Medicines affect your body *differently* **depending** on how they are **administered.**

Medicines are also given topically, or applied to the skin. You can apply creams and ointments this way. Skin patches that release medicine over time are another type of topical medicine.

>>> **Reading Check**

IDENTIFY *What is the most common way medicines are taken?*

PROBLEMS WITH MEDICINES

MAIN IDEA Medicines affect different people in different ways.

Because every person's body is unique, medicines affect people in different ways. Combining medicines may also affect the way they work. Some medicines do not interact well with others and can cause harmful reactions. Some people are allergic to certain medicines and cannot take them at all. Your age, weight, and overall health can determine how a medicine affects you.

Side Effects

Medicines can help you, but if you do not use them properly, they can also hurt you. A side effect is *a reaction to a medicine other than the one intended.* Sometimes, side effects are simply unpleasant. For example, a medicine might make you feel sick to your stomach, sleepy, dizzy, or cause a headache. A more serious side effect is an allergic reaction, which requires immediate medical attention.

You can avoid most side effects by following the instructions from your doctor and pharmacist. Always inform your doctor of any allergies you have to medicines. Read the label on any kind of medicine you take.

The label on any kind of medicine explains when and how to use it as well as how much to use. Some medicines should not be taken together. In addition, some activities can be dangerous if you take certain medicines. For example, labels on medicines that cause sleepiness warn against operating machinery or driving. Ask your doctor or pharmacist if you do not understand a medicine label.

Drug Interactions

Taking two or more medicines at once can be dangerous. They may cause unexpected drug interactions. For example, one drug may become more effective or less effective, or the combination may produce different, more dangerous side effects that neither drug would have caused when taken by itself. Taking medicines with certain foods can also cause interaction problems. Always let your doctor and pharmacist know what other medicines you are taking before starting a new medicine.

> You can buy over-the-counter (OTC) medicines without a prescription. *Why is it important to carefully read the label on any medicine?*

Decision Making

Scheduling a *Dosage*

Jason is on summer vacation. He gets sick, and his doctor prescribes medicine for him. The medication makes Jason feel sleepy and light-headed after he takes it. The label on the bottle warns patients not to use heavy machinery within four hours of taking the medicine. Jason is supposed to take it once a day, but he earns money by mowing lawns. Now Jason is worried about using the mower while he's on the medication.

Remember the six steps of the decision-making process:

1. State the situation.
2. List the options.
3. Weigh the possible outcomes.
4. Consider your values.
5. Make a decision and act on it.
6. Evaluate the decision.

What Would You Do? Apply the six steps above to help Jason make a healthful choice. What would you suggest that Jason do?

Thinkstock/Jupiterimages

IMPORTANT
USE EXACTLY AS DIRECTED
DO NOT DISCONTINUE OR SKIP DOSES
UNLESS DIRECTED BY YOUR DOCTOR

Tolerance

When someone uses a particular medicine for a long period of time, the person's body may develop a tolerance. **Tolerance** is *the body's need for larger and larger amounts of a drug to produce the same effect.* The concept of tolerance can apply to medicine as well as alcohol and other drugs. In some cases, tolerance may cause a medicine to lose its effectiveness over time.

The *more* an antibiotic is used, the **less effective** it becomes.

A person may go through withdrawal when he or she stops using the medicine. These symptoms will gradually ease over time. If you experience withdrawal after using a medicine, talk to your doctor. You may need to be prescribed a different medication. Symptoms of medicine withdrawal can include:

- Nervousness.
- Insomnia.
- Severe headaches.
- Vomiting.
- Chills.
- Cramps.

Antibiotic Resistance

The more an antibiotic is used, the less effective it becomes. This is especially true when an antibiotic is overused. Why? Antibiotics kill harmful illness-causing bacteria in the body. With frequent exposure, however, bacteria often build up a resistance to antibiotics. Bacteria adapt to, or overcome, the medicine. Bacteria can also develop a resistance when antibiotics are not taken as prescribed. For example, if the prescription label says to take the medicine for 14 days, and you stop after 7 days, the bacteria may still be in your body and could make you sick again. Medicines should always be used wisely, and they should be used only as directed.

>>> **Reading Check**

IDENTIFY *Name two risks of using medicines.* ■

The MRSA bacteria can infect a simple scrape and move to the lungs. Most antibiotics will not cure it. *Why might these bacteria be so resistant to antibiotics?*

REVIEW

>>> **After You Read**

1. **VOCABULARY** Define the term *side effect*. Use it in an original sentence.
2. **DESCRIBE** What factors determine a medicine's effect on the body?
3. **LIST** Name two ways medicines are taken into the body.

>>> **Thinking Critically**

4. **ANALYZE** Why might a doctor prescribe different medicines for two people with the same illness?
5. **HYPOTHESIZE** Milla's doctor has prescribed a medicine to treat a case of poison ivy. She also regularly takes medicine because she has trouble concentrating. Should Milla tell her doctor what medicine she is already taking? Why or why not?

>>> **Applying Health Skills**

6. **ACCESSING INFORMATION** Go online to research a popular drug that you have seen advertisements for. What condition does the drug treat? What are its side effects?

⟳ Review

◀)) Audio

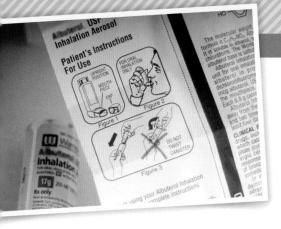

Using Medicines Correctly

BIG IDEA Using medicines improperly can cause harmful side effects.

 Before You Read

QUICK WRITE Write a short paragraph about what you think might happen if a painkiller were used incorrectly.

▶ Video

 As You Read

STUDY ORGANIZER Make the study organizer found in the FL pages in the back of the book to record the information presented in Lesson 3.

>>> **Vocabulary**

› medicine misuse
› medicine abuse

🔊 Audio

🔤 Bilingual Glossary

Developing Good Character

Responsibility You can show responsibility by sharing information about the dangers of drug misuse with your family. Urge parents or guardians to throw away any medicines that have expired. *What are some other steps you can take to demonstrate responsibility at home?*

IMPROPER USE OF MEDICINES

MAIN IDEA Misusing medicines can be as harmful as using illegal drugs.

Medicines are types of drugs. Medicines are intended to be helpful, not harmful. They can prevent and cure diseases, fight germs, and relieve pain. If they are not used properly, however, medicines can be as harmful as illegal drugs. They can cause addiction, injury, and even death. If you use medicines improperly as a teen, it can result in serious health problems not only now but also later in life.

Both prescription medicines and over-the-counter medicines carry labels warning of possible side effects. Using too much medicine or using it too often can cause serious damage to body systems. For example, using too much of a medicine could cause liver or kidney failure. Certain medicines can also be very dangerous to unborn babies, newborns, or young children. A woman who is pregnant or plans to become pregnant should talk to a doctor before she takes any medicine.

Medicine Misuse

Medicines have many benefits when they are used correctly. Research studies show that most teens—96 percent—do use medicines correctly. *Taking medicine in a way that is not intended* is known as medicine misuse.

If not used properly, *medicines* can be as **harmful** as *illegal drugs.*

Taking more medicine than a doctor instructs is one example of misusing medicines. Another example is failure to follow the directions on the label. Medicine misuse can be dangerous. It may only prevent you from getting the full benefits of a medicine. At worst, however, medicine misuse can seriously harm your health. This is why medicines need to be taken with great care.

>>> **Reading Check**

RECALL *How can medicine misuse affect a person's health?*

Medicine Abuse

Medicine abuse is *intentionally using medicines in ways that are unhealthful and illegal.* Medicine abuse is a form of drug abuse. Some teens believe that prescription and OTC medicines are safer than illegal drugs. However, medicines are safe only if used properly. People may abuse medicines for several reasons:

- **To lose weight** A healthy diet and exercise are the safest way to maintain a healthy weight.
- **To stay awake** Getting plenty of sleep and learning strategies to manage your time wisely will help you study effectively.

- **To get "high"** A dangerous trend is the practice of having "pill parties," in which party-goers mix whatever OTC and prescription medicines may be available. Using medicine that is not prescribed for you is both illegal and dangerous and could even cause death.

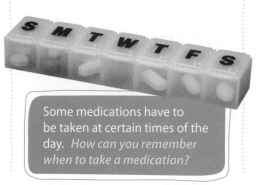

Some medications have to be taken at certain times of the day. *How can you remember when to take a medication?*

Using a medicine prescribed to someone else is also medicine abuse. Medicines are prescribed for a specific person to treat a specific illness. Using someone else's medicine, even if you think you have the same illness, is both illegal and unsafe.

A danger of both medicine abuse and medicine misuse is the risk of overdose. An overdose is when someone takes too much medicine at once. Misusing medicines can also lead to addiction. The best way to make sure you are using a medicine safely is to follow the instructions on the label.

Health SKILLS ACTIVITY

Practicing Healthful Behaviors

Handling *Medicine* Safely in Your *Home*

What do you know about medicine safety in the home? Follow these guidelines to store, use, and dispose of medicine safely.

* Know what medicines are in your home and what they are used to treat.

* Store medicines in a cool, dry place.

* Keep medicines safely sealed in childproof containers, and keep them out of the reach of children.

* Never share prescription medicines. They could cause serious harm to someone else.

* Do not use OTC medicines for more than ten days at a time without first checking with your doctor.

* Before taking two or more medicines at the same time, get your doctor's approval. Combining medicines can cause harmful side effects.

* Do not use medicines that have passed their expiration date.

* To safely dispose of outdated or unused liquid or pills, flush them down the toilet.

On Your Own

Create a "Medicine Safety Checklist" that you can use at home. Review the completed checklist with your family. Post the list in an appropriate spot in your home.

Royalty-Free/Corbis

HOW TO USE MEDICINES SAFELY

MAIN IDEA Medicines are helpful only when they are used properly.

When using medicines, it is important to keep them out of reach of young children. *Why is it important to be responsible when using medicines?*

Medicines can be helpful, but they can also cause serious harm. This is why they should be taken with great care. To avoid misuse, follow these guidelines:

- Follow the instructions on the label.
- Take the correct dosage for the recommended length of time. If you experience side effects from a prescription drug, contact your doctor before you stop using the medicine.
- Do not take medicines after their expiration date.
- Do not give prescription medicines to someone else. Do not use a medicine prescribed for an earlier illness without asking your doctor.

Never give **prescription** *medicines* to **someone** else.

- Contact your doctor if you do not understand the label instructions, if you experience any unusual or unpleasant side effects, or if you accidentally take too much medicine.
- Store all medicines safely—in a cool, dry place, in their original containers, and out of reach of children.

>>> **Reading Check**

GIVE EXAMPLES *What are three ways to use medicines safely?* ■

REVIEW

>>> **After You Read**

1. **VOCABULARY** Define the term *medicine misuse*. Use it in an original sentence.
2. **DESCRIBE** What are three ways you can avoid medicine abuse?
3. **COMPARE AND CONTRAST** How does medicine abuse differ from medicine misuse?

>>> **Thinking Critically**

4. **DISCUSS** How can taking medicines not prescribed for you, or mixing medicines, harm your health?
5. **EVALUATE** Tasha's friend has offered her some of her prescription medication. Tasha asks you whether she should take it. What advice would you give her and why?

>>> **Applying Health Skills**

6. **APPLY** Create a script for a commercial or PSA that explains how people can use their health care providers, pharmacists, and medicine labels to ensure that they are using their medicines properly.

🔄 Review

🔊 Audio

©Comstock/Alamy

Hands-On HEALTH ACTIVITY

Making Smart Choices about Medicines

WHAT YOU WILL NEED

* Three index cards per student
* One pencil or pen per student
* Internet or library access

WHAT YOU WILL DO

1 In teams of three, use each of your index cards to write one medicine you might find at a pharmacy. Each group should compile a total of nine different medications.

2 Research the health benefits and risks associated with each of the medicines you have listed. On the back of each card, list two benefits and two risks for each medication.

3 When you are finished, place all the completed cards face down in the middle of the table. Your teacher will then name a category, such as how the medicine enters the body, whether it is a prescription or OTC medicine, or which one carries the most risk.

4 For each round, your team will turn the cards over and sort them into the proper category. The winner is the team that completes the most rounds successfully.

A medicine can be either helpful or harmful to the body, depending on whether it is used properly and how it affects your body. Some medicines can protect you from certain diseases or are used to manage specific conditions. Others cure diseases, kill germs, relieve symptoms, or treat minor injuries. This activity will help you better understand the health risks and benefits of using different medications.

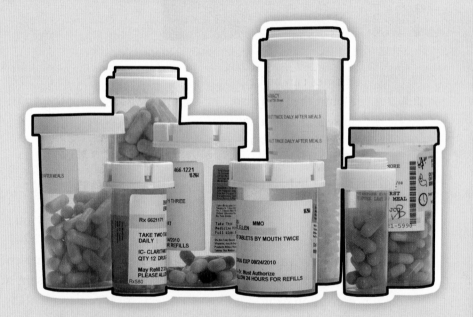

WRAPPING IT UP

Create a "How to Use Medicines Wisely" checklist that explains the steps to use when making a decision about using a medicine. Keep in mind the purpose of the medication, the way a medication affects your body, the health benefits and risks of a medication, and how the medication is used.

Jeffrey Coolidge/Getty Images

READING REVIEW

FOLDABLES and Other Study Aids

Take out the Foldable® that you created and any study organizers that you created. Find a partner and quiz each other using these study aids.

LESSON 1 Types of Medicines

BIG IDEA Using medicines wisely is a sign of good personal and consumer health.

* A medicine is a drug that can prevent or cure an illness or ease its symptoms.
* Different medicines are used in various ways and serve different purposes in the body.
* Medicines in the United States are carefully controlled by the Food and Drug Administration (FDA), which sets standards for medicine safety and effectiveness.
* Some medicines require written permission from a doctor (prescription medicines).
* Other medicines are available without a prescription (over-the-counter [OTC] medicines).

LESSON 2 How Medicines Affect Your Body

BIG IDEA Medicines can contribute to good health when used properly.

* Different types of medicines can be used or taken in different ways.
* A medicine may be swallowed, injected, inhaled, or applied to the skin.
* Medicines affect different people in different ways and can be harmful if not used properly.

* Even proper use of medicines can result in sometimes dangerous side effects.
* The body can build up a tolerance to certain medicines over time, and frequent exposure to antibiotics may cause bacteria to build up a resistance.
* Always use medicines only as directed, and always read and follow all the instructions on the label.

LESSON 3 Using Medicines Correctly

BIG IDEA Using medicines improperly can cause harmful side effects.

* Misusing or abusing medicines can be as harmful as using illegal drugs.
* Taking medicine in a way that is not intended is medicine misuse.
* Medicine abuse is intentionally using medicines in ways that are unhealthful and illegal.
* Medicines are helpful only when they are used properly.
* Using medicines properly involves reading and following label instructions and taking the correct dosage for the recommended length of time.
* Always store medicines safely—in a cool, dry, secure place that is out of the reach of children.

 Review

 Web Quest

ASSESSMENT

Reviewing Vocabulary *and* Main Ideas

> over-the-counter (OTC) medicine
> side effects
> resistance
> prescription medicine
> tolerance
> medicine
> vaccine
> antibiotic

≫ On a sheet of paper, write the numbers 1–8. After each number, write the term from the list that best completes each statement.

LESSON 1 Types of Medicines

1. A(n) _____ is a medicine that can be sold only with a written order from a doctor.

2. A medicine that prevents a disease from developing is called a(n) _____.

3. A(n) _____ is a type of medicine that is used to treat a bacterial infection.

4. A drug that prevents or cures a disease or illness or eases its symptoms is called _____.

5. You can buy a(n) _____ without a written order from a doctor.

LESSON 2 How Medicines Affect Your Body

6. Drowsiness and nausea are examples of _____ that you could have from taking medicine.

7. Frequent exposure to antibiotics can cause bacteria to adapt to, or build up a _____ to, these drugs.

8. If you use a particular medicine for a long period of time, your body may develop a _____ for it.

≫ On a sheet of paper, write the numbers 9–14. Write *True* or *False* for each statement below. If the statement is false, change the underlined word or phrase to make it true.

LESSON 3 Using Medicines Correctly

9. Taking more of a medicine than a doctor instructs is an example of <u>medicine misuse</u>.

10. An important part of storing medicines safely is to keep them out of the reach of <u>adults</u>.

11. One way to make sure you are using a medicine safely is to <u>ignore</u> the instructions on the label.

12. <u>Only prescription medicines</u> carry labels warning of possible side effects.

13. A danger of both medicine abuse and medicine misuse is the risk of <u>drug overdose</u>.

14. It is <u>always</u> okay to share prescription medicines.

 eAssessment

Thinking Critically

15. PREDICT How could medicine misuse or abuse affect each side of your health triangle?

16. INTERPRET You have gotten sick, so your doctor has prescribed you some medicine. After taking the medicine, however, you break out in a rash. What should you do?

17. ANALYZE What can you do to help make sure you use a prescription medicine properly?

Write About It

18. OPINION Write a blog post about the benefits and dangers of medicines. Explain ways that medicines can contribute to good health if used properly and ways they can be harmful if they are misused or abused.

19. DESCRIPTIVE WRITING Imagine that you have a medicine cabinet at home that is full of many different types of medicine. Explain how you would go about organizing the medicine cabinet to help make it safer for your family.

ⒶⒷⒸⒹ STANDARDIZED TEST PRACTICE

Writing

Read the prompts below. On a separate sheet of paper, write an essay that addresses each prompt. Use information from the chapter to support your writing. Refer to the tips in the right column.

1. Carl experiences back pain and visits his doctor, who prescribes a prescription painkiller. His friend Matt pulls a muscle playing basketball. Matt asks Carl if he can share his medicine. Explain why Carl should not share his prescription medicine with his friend Matt.

2. Imagine that you are a pharmacist working in a drugstore. A customer is confused by the label on the over-the-counter medicine. Write a dialogue in which you explain the different parts of a medicine label to ensure that they are using the medication properly.

Whenever you begin writing, first determine your task, audience, and purpose. Ask yourself the following questions: What is my topic? What are the guidelines of my essay? Is my purpose to explain a topic or persuade my audience? Who will be reading my writing? Understanding your task, audience, and purpose will help you accomplish your writing goal.

Unit 9

preventing disease

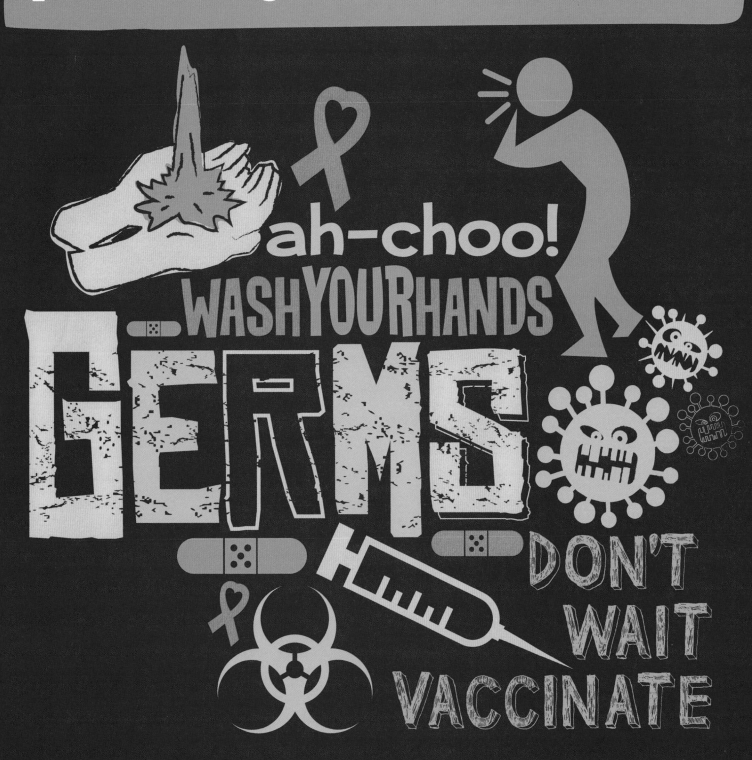

Communicable Diseases

LESSONS

PREMIUM ONLINE RESOURCES

 Audio

 Videos

 Bilingual Glossary

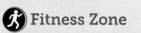 Fitness Zone

 Web Quest

 Review

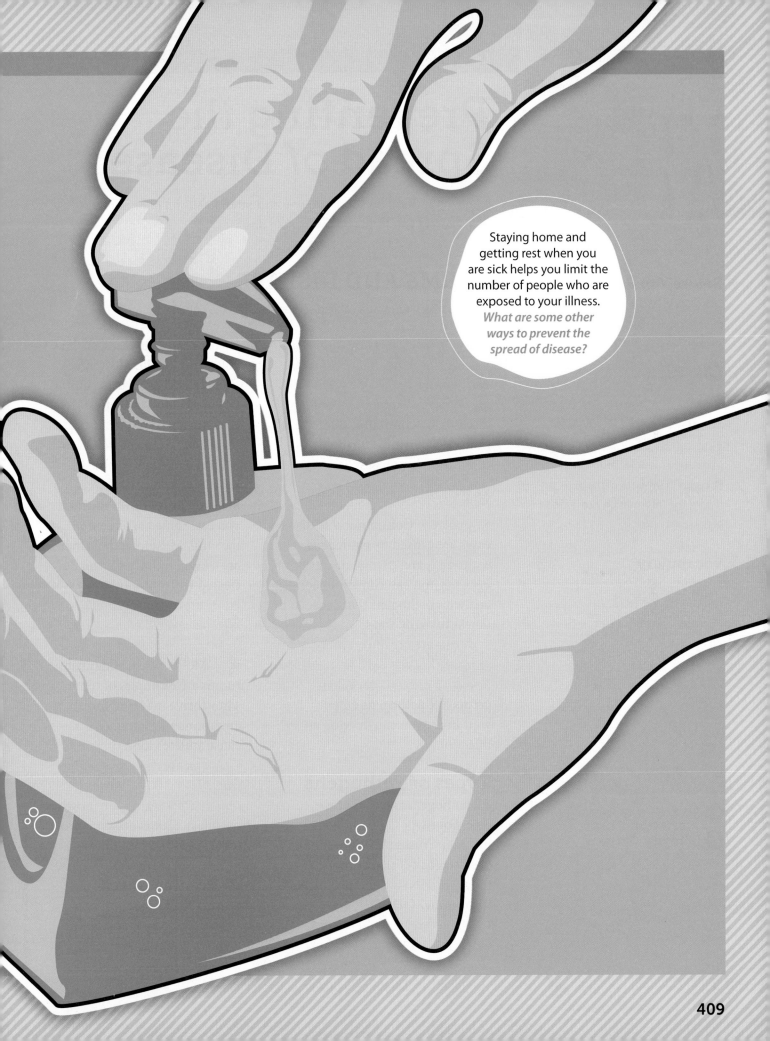

Staying home and getting rest when you are sick helps you limit the number of people who are exposed to your illness. *What are some other ways to prevent the spread of disease?*

Preventing *the* Spread *of* Disease

BIG IDEA Good personal hygiene and other healthful behaviors can help protect you from communicable diseases.

Before You Read

QUICK WRITE How do you think people catch colds? Explain your answer.

▶ Video

As You Read

Make the Foldable® found in the FL pages in the back of the book to record the information presented in Lesson 1.

Vocabulary

› disease
› communicable disease
› germs
› pathogens
› viruses
› bacteria
› fungi
› protozoa
› vector
› hygiene
› contagious

 Audio

🔤 Bilingual Glossary

GERMS AND DISEASE

MAIN IDEA Communicable diseases are caused by germs.

A disease is *any condition that interferes with the proper functioning of the body and mind.* Some diseases are communicable (kuh•MYOO•nih•kuh•buhl). A **communicable disease** is *a disease the can be spread to a person from another person, an animal, or an object.* Communicable diseases are caused by pathogens. **Pathogens** are *germs that cause diseases.* You can come into contact with pathogens that cause illness in many ways.

Germs are *organisms that are so small they can only be seen through a microscope.* When germs like pathogens enter your body, you can develop an infection.

Types *of* Pathogens

All pathogens can cause disease. Four types of pathogens include: viruses (VY•ruh•suhz), bacteria (bak•TIR•ee•uh), fungi (FUHN•jy), and protozoa (proh•tuh•ZOH•uh).

- **Viruses** are *the smallest and simplest pathogens.*

Viruses cause diseases, such as colds and the flu. Most infections caused by viruses cannot be cured with antibiotics.

- **Bacteria** are *simple one-celled organisms.* Some bacteria are helpful. For example, the bacteria in your digestive tract help you break down the food you eat. Some bacteria are harmful. Most infections that are caused by bacteria can be treated with antibiotics.
- **Fungi** are *organisms that are more complex than bacteria but cannot make their own food.* Molds, yeast, and mushrooms are examples of fungi. They thrive in warm, moist environments. Fungi can cause athlete's foot.
- **Protozoa** are *one-celled organisms that are more complex than bacteria.* They are mostly harmless. However, some protozoa, cause serious illnesses.

Reading Check

DEFINE *What is a virus?*

Jani Bryson/the Agency Collection/Getty Images

How Pathogens Spread

To understand how to avoid becoming infected with a communicable disease, it's important to understand how pathogens are spread. Most pathogens are spread in the following ways.

DIRECT CONTACT One of the most common ways pathogens are spread is by touching another person. For example, if someone with a cold sneezes into her hand and then you shake hands; you could get a cold.

Some pathogens are spread through contact with infected blood. People can come into contact with infected blood by using a needle that someone else has used. Dirty needles used for tattooing or piercing , and drug injection can spread pathogens. It is also possible for the blood of an infected person to infect someone else if the blood comes into contact with broken skin. Some pathogens are spread through sexual contact.

INDIRECT CONTACT If someone sneezes or coughs, pathogens are spread through the air. Use tissues to cover your nose when you sneeze and your mouth when you cough. Pathogens are also spread when people share certain items. Drinking glasses and eating utensils should never be shared. Items such as toothbrushes and razors should not be shared.

CONTACT WITH ANIMALS OR INSECTS Animals and insects can spread pathogens. *An organism, such as an insect, that transmits pathogens* is called a <u>vector</u>.

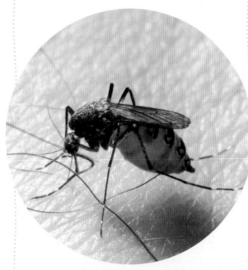

The West Nile virus is spread by infected mosquitoes. *Describe some ways you can protect yourself against mosquitoes.*

If an animal is sick with rabies, for example, that animal can spread rabies if it bites someone or another animal. Insects spread many diseases. For example, infected deer ticks can spread Lyme disease. Infected mosquitoes can spread malaria or West Nile virus.

Using insect repellent when you go outdoors and staying away from still or standing water can help protect you from insect bites.

CONTAMINATED FOOD AND WATER Pathogens can be spread through contaminated food or water. If food is undercooked or improperly stored, pathogens can cause illness. Illnesses caused by contaminated foods are called foodborne illnesses. To prevent foodborne illnesses, make sure meat is fully cooked. Undercooked meat may contain bacteria that will make you ill. Properly store foods that can spoil, such as dairy products and meat. Wash all fruits and vegetables. Meat, poultry, fish, and eggs need to be handled carefully. Wash knives and surfaces that meat, poultry, fish, and eggs have touched.

>>> **Reading Check**

LIST *How are pathogens spread?*

Communicable diseases are all caused by pathogens. *According to the chart, what common diseases do fungi cause?*

Pathogens	Diseases
Viruses	Colds, chicken pox, influenza, measles, mononucleosis, mumps, hepatitis, herpes, HPV, HIV, yellow fever, polio, rabies, viral pneumonia
Bacteria	Pinkeye, whooping cough, strep throat, tuberculosis, Lyme disease, most foodborne illnesses, diphtheria, bacterial pneumonia, cholera, gonorrhea
Fungi	Athlete's foot, ringworm
Protozoa	Dysentery, malaria, trichomoniasis

Renaud Visage/Digital Vision/Getty Images

STOPPING THE SPREAD OF PATHOGENS

MAIN IDEA Practicing certain healthy behaviors can stop the spread of pathogens.

Although you can't completely stop the spread of pathogens, you can protect yourself and others from the spread of pathogens. Good personal hygiene, or *cleanliness,* is one of the best ways to help stop the spread of pathogens. You can also help stop their spread by eating nutritious foods and getting enough sleep. Engaging in regular, physical activity will also help your body fight pathogens. Keep your environment, the space around you, clean. This will also keep down the number of pathogens. Stopping the spread of pathogens will help to protect your health and the health of others.

Protecting Yourself

In addition to the tips already mentioned, follow these guidelines to help keep yourself from getting sick.

- Stay away from people who are sick with a communicable disease. Be especially careful if they are still contagious, or *able to spread to others by direct or indirect contact.*
- Wash your hands thoroughly before preparing or eating food. Wash after using the bathroom, playing with pets, visiting a sick person, and touching trash or garbage.
- Do not share hygiene items.

Washing your hands is one way to stop the spread of pathogens. *Name two other ways to help stop the spread of pathogens.*

- Keep your hands and fingers away from your mouth, nose, and eyes. Don't bite your fingernails.

Ingram Publishing

Health SKILLS ACTIVITY

Practicing Healthful Behaviors

Keep Your *Hands Clean!*

One of the best ways of stopping pathogens from entering your body or spreading to others is to wash your hands frequently. Practice washing your hands thoroughly using the following steps:

1. Use warm water to wet your hands, and then apply soap.
2. Vigorously rub your hands together, scrubbing your whole hand for 30 seconds or more.

3. Rinse your hands thoroughly. Use a paper towel to turn off the water in public restrooms.
4. Use a clean towel or paper towel to dry your hands. In a public restroom, use a paper towel to open the door when you leave.

With A Group Create a poster encouraging students to wash their hands regularly. Include the handwashing steps listed above. With permission from school administrators, hang your poster in a school restroom or hallway.

- Handle and prepare food safely. This is especially important for meats, fish, poultry, and eggs. Meats, fish, poultry, and eggs should be cooked thoroughly.
- Wash vegetables and fruits before eating.
- Wash counters thoroughly with paper towels or a clean sponge or cloth. Use warm, soapy water. If you use a sponge or cloth, clean them thoroughly and frequently.
- Keep your environment clean. Empty the trash often. Keep the trash cans clean. Clean up after your pets.
  These lifestyle practices will also help keep you healthy.
- Eat a balanced diet.
- Bathe or shower regularly using soap. Be sure to wash your hair using shampoo.

Keeping yourself and your environment clean is one way to stay safe and healthy. *Explain how emptying out the trash often helps keep you healthy.*

John Howard/Lifesize/Getty Images

- Avoid tobacco products, alcohol, and other drugs.
- Get 8-9 hours of sleep.
- Rest when you are sick.
- Check with parents or guardians to make sure your immunizations are up to date.
- Learn healthy ways to manage your stress.
- Get regular physical checkups.

Protecting Others

Help protect the people you come into contact with. Think ahead and follow these healthful behaviors.

- If you are sick, stay home. Get medical help if you need it. If you become ill at school, tell your teacher or school nurse. Getting medical help can keep the illness from getting worse or spreading.
- When you sneeze or cough, cover your mouth with a tissue. Only use the tissue once. If you don't have a tissue, sneeze or cough into the crook of your elbow. Wash your hands immediately after you sneeze or cough.
- If you take medication, follow the directions exactly. If you stop taking the medicine before you are supposed to, you might become sick again.

>>> **Reading Check**

DEFINE *What is hygiene?* ■

>>> **After You Read**

1. **DEFINE** What is a *communicable disease*?
2. **IDENTIFY** What are four ways pathogens can be spread?
3. **DESCRIBE** How can staying home when you are sick help keep others healthy?

>>> **Thinking Critically**

4. **ANALYZE** How can hand washing help keep a community free from communicable diseases?
5. **APPLY** Imagine that you wake up with a sore throat and headache. Your team is playing in the soccer finals today and you're the starting goalie. What should you do?

>>> **Applying Health Skills**

6. **ADVOCACY** Create a brochure that explains to students how they can help keep themselves and others safe from the spread of pathogens. List at least five things students can do to keep themselves safe and five things students can do to keep others safe from the spread of pathogens.

 Review

 Audio

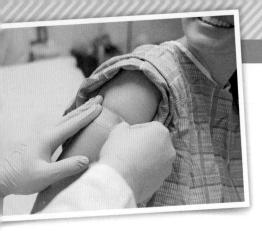

Defense Against Infection

BIG IDEA ▷ Your immune system protects you from infection.

Before You Read

QUICK WRITE Think about the last time you missed school because of an illness. Describe the illness and the steps you took to get better.

 Video

As You Read

STUDY ORGANIZER Make the study organizer found in the FL pages in the back of the book to record the information presented in Lesson 2.

Vocabulary

› infection
› immune system
› inflammation
› lymphatic system
› lymphocytes
› antigens
› antibodies
› immunity
› vaccine

 Audio

 Bilingual Glossary

Myth or Fact

Myth: Fevers are dangerous, especially when people have infections.

Fact: Fevers are part of the body's defense against infection. Low fevers are not dangerous.

KEEPING PATHOGENS OUT

MAIN IDEA ▷ Your body has five major barriers that defend against pathogens.

Pathogens are everywhere. They are in the air you breathe, the water you drink, and on objects you touch. Most viruses, bacteria, and other pathogens never make you sick. Your body has natural barriers between you and pathogens. The five major barriers are the skin, tears, saliva, mucous (MYOO-kuhs) membranes, and stomach acid. The skin is your body's largest organ. It acts like a wall to protect the inner organs. Another barrier is formed by body fluids such as tears and saliva. They contain chemicals that kill certain germs. The mucous membranes protect the insides of your mouth, throat, nose, and eyes. They are coated with a sticky fluid that destroys germs. Some germs, however, will be able to get past these barriers. Stomach acid kills germs that make it past the saliva and mucous membrane barriers in your mouth.

The *five major barriers* between you and pathogens are **skin, tears, saliva, mucous membranes,** and **stomach acid.**

Even with all of these barriers, pathogens sometimes find their way into your body. If pathogens enter the body, you might have an infection. Infection is *the result of pathogens or germs invading the body, multiplying, and harming some of your body's cells.* Infection can trigger a fever in the body. The increase in body temperature caused by the fever makes it difficult for pathogens to live.

Reading Check

NAME *What five barriers protect your body from infection?*

M. Constantini/PhotoAlto

NONSPECIFIC IMMUNE RESPONSE

MAIN IDEA Your immune system responds when pathogens get past the five major barriers of protection.

Sometimes your body's barriers can't keep out all of the pathogens. For example, when you get a splinter, it breaks the skin. The pathogens that are on the splinter enter your system through the broken skin. When this happens, your immune system responds to the invaders. The immune system is *a combination of body defenses made up of the cells, tissues, and organs that fight pathogens in the body.* The response is called a nonspecific immune response. It's called nonspecific because your body reacts the same no matter what foreign matter enters the body.

The splinter or other foreign object will cause the skin around it to become red and sore and swollen. The nonspecific immune system has responded with inflammation. Inflammation is *the body's response to injury or disease, resulting in a condition of swelling, pain, heat, and redness.* When the splinter enters your system, the brain tells white blood cells to rush to the area. White blood cells destroy the pathogens. When your body has inflammation or other infection, it produces a protein that stimulates the immune system. This stops pathogens from multiplying.

If pathogens do multiply, your body temperature may rise. This will cause a fever. A higher body temperature makes it difficult for pathogens to multiply. The fever also signals the body to produce more white blood cells, which attack and destroy the pathogens.

>>> **Reading Check**

EXPLAIN *What is your body's first response to invading pathogens?*

Barriers help keep pathogens out of your body. *What are the mucous membranes?*

Tears cover and protect the eye from dust and pathogens. As they flow, tears carry foreign material away from the eye. Tears contain chemicals that kill pathogens.

Mucous membranes are the soft skin that lines the nose, mouth, eyes, and other body openings. They are coated in a sticky material called mucus (MYOO•kuhs) that traps pathogens. When you cough, sneeze, or clear your throat, the pathogens trapped in the mucus are expelled.

Saliva washes germs away from your teeth. It contains chemicals that kill pathogens trying to enter through your mouth.

Skin provides a tough, outer protective surface that keeps pathogens from entering your blood. If you get a cut, burn, or scrape, pathogens can get past this barrier.

Stomach acid is a gastric juice produced by the lining of your stomach. It kills many of the pathogens that make it past the saliva and mucous membranes of your mouth.

Ron Levine/Digital Vision/Getty Images

SPECIFIC IMMUNE RESPONSE

MAIN IDEA The specific immune response attacks if pathogens get past the nonspecific immune response.

I t is possible that some pathogens will get past the body's nonspecific immune response. When this happens, the body responds with a specific immune response. This is the immune system's second response. Each specific response is customized to attack a particular pathogen and its poisons, or toxins.

It's possible that some **pathogens** will get past the body's **nonspecific immune response.**

Your immune system recognizes pathogens that have invaded before. The particular response cells that battled those pathogens stay in your body to help you fight a disease if you are infected a second time. If the same types of pathogens attack again, those response cells attack and fight the pathogens very quickly. The second response is faster than the first response.

Many different kinds of cells in your immune system work together to fight invading pathogens. *Describe the purpose of memory B and T cells.*

The Lymphatic System

The lymphatic system is *a secondary circulatory system that helps the body fight pathogens and maintains its fluid balance.* It contains fluid called lymph. The lymphatic system contains lymphocytes (LIM•fuh•sytes). Lymphocytes are *special white blood cells in the blood and lymphatic system.*

The three main types of lymphocytes include: B cells, T cells, and NK cells. B cells and T cells work together to defeat invading pathogens. T cells identify the invaders and alert the B cells. The B cells attack and kill the invading pathogen. Macrophages are also found in the lymph. Macrophages destroy foreign substances in the body. Then they remove the foreign material from the lymph.

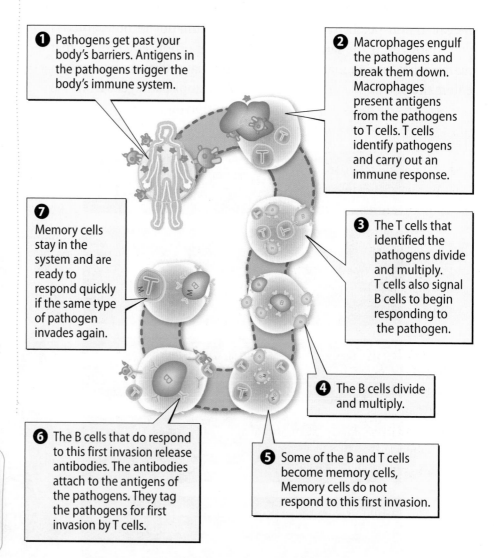

❶ Pathogens get past your body's barriers. Antigens in the pathogens trigger the body's immune system.

❷ Macrophages engulf the pathogens and break them down. Macrophages present antigens from the pathogens to T cells. T cells identify pathogens and carry out an immune response.

❸ The T cells that identified the pathogens divide and multiply. T cells also signal B cells to begin responding to the pathogen.

❹ The B cells divide and multiply.

❺ Some of the B and T cells become memory cells, Memory cells do not respond to this first invasion.

❻ The B cells that do respond to this first invasion release antibodies. The antibodies attach to the antigens of the pathogens. They tag the pathogens for first invasion by T cells.

❼ Memory cells stay in the system and are ready to respond quickly if the same type of pathogen invades again.

Antigens *and* Antibodies

The three types of lymphocytes—T cells, B cells, and NK cells—react to antigens. **Antigens** are *substances that send the immune system into action.* For example, if bacteria enter your system, lymphocytes will recognize substances on the surface of the bacteria as antigens. Also, blood that is a different type than yours will be recognized as an antigen.

Your body reacts to antigens by making more B cells and T cells. B cells make antibodies. **Antibodies** are *specific proteins that attach to antigens, keeping them from harming the body.* Antibodies fight a particular type of antigen. Using the example from above, special antibodies will attack the antigens on the bacteria. Different antibodies will attack the antigens on the blood. Some of the new B cells and T cells don't react to the antigens the first time. They wait to react if the same kind of pathogen enters the body again. These are known as memory B cells and memory T cells.

T cells stimulate the production of B cells or attack pathogens directly. There are two main types of T cells: helper cells and killer cells. Helper cells activate the production of B cells. Killer cells attach to invading pathogens and destroy them.

Health SKILLS ACTIVITY

Practicing Healthful Behaviors

Keeping a *Healthy Immune System*

Good health habits can keep your immune system healthy. A healthy immune system means your body is better able to fight off infection. Commit to these habits to keep your immune system in top condition!

* Follow a healthful eating plan. Eat plenty of vitamin-rich fruits, vegetables, and whole grains.

* Participate in regular physical activity.

* Learn to manage stress in healthy ways.

* Get plenty of sleep. Rest strengthens your body's defenses.

* Avoid tobacco, alcohol, and other drugs.

* Drink plenty of water.

* Bathe and shower regularly.

* Get regular checkups from a healthcare professional.

Using print or online resources, research other good health habits that can help you look and feel your best.

IMMUNITY

MAIN IDEA ▷ Your body has different types of immunity.

You have several types of **immunity**, or *the ability to resist the pathogens that cause a particular disease.* You have natural immunities. For example, some germs that affect animals do not affect you. When your pet gets sick, you won't get sick from your pet. The reverse is also true.

Your body's five barriers also provide immunity. When they are penetrated by pathogens, your immune system attacks the invading pathogens.

Before birth, a mother's antibodies pass to her infant's body. After birth, the mother's milk provides additional antibodies that help the infant's immune system fight pathogens.

Your body builds immunity when you get certain diseases. Memory cells remember the antigens the next time they invade your body. Then they produce antibodies to fight the antigens.

Immunity can also be developed by introducing a vaccine into the body. A **vaccine** is *a preparation of dead or weakened pathogens that is introduced into the body to cause an immune response.* Vaccines help the immune system make antibodies for certain diseases. It is important to keep vaccinations up to date to help keep you healthy. It also helps keep those around you healthy.

>>> **Reading Check**

EXPLAIN *How does your immune system react to a vaccination?* ■

This table lists common vaccines and the ages at which each vaccine is given. *Identify at what ages the vaccine for polio is given.*

Vaccine	Recommended Ages for Vaccine
Hepatitis B	Birth, 1–2 months, 6–18 months
DTaP: diphtheria, tetanus, pertussis (whooping cough)	2, 4, 6 and 15–18 months, 4–6 years, 11–12 years
HiB (*H. influenzae* type b)	2, 4, 6 and 12–15 months
IPV: Polio	2, 4, and 6–18 months, 4–6 years
MMR: measles, mumps, rubella	12–15 months; 4–6 years
Varicella: chicken pox	12–15 months; 4–6 years
Hepatitis A	12–23 months
Human Papillomavirus	(3 doses)11–12 years

Sources: http://www.cdc.gov/vaccines/parents/downloads/parent-ver-sch-0-6yrs.pdf
http://www.cdc.gov/vaccines/who/teens/downloads/parent-version-schedule-7-18yrs.pdf

REVIEW

>>> **After You Read**

1. **DEFINE** What is an *antigen*?
2. **IDENTIFY** What is the *lymphatic system*?
3. **DESCRIBE** How does natural immunity help a newborn baby?

>>> **Thinking Critically**

4. **ANALYZE** How can keeping your immunizations up to date help keep others healthy?
5. **APPLY** Your mom is making soup. As you walk through the kitchen, you decide to taste it. You take a spoon out of the drawer and use it to taste the soup. It's so good you want another bite. Should you use the same spoon? Why or why not?

>>> **Applying Health Skills**

6. **PRACTICING HEALTHFUL BEHAVIORS** Imagine that you have a cold. Explain steps you can take to keep from spreading the pathogens to others.

⟳ Review

 Audio

Communicable Diseases

BIG IDEA Common communicable diseases include colds, the flu, strep throat, pneumonia, mononucleosis, hepatitis, and tuberculosis.

Before You Read

QUICK WRITE Write down the names of three common diseases. What are the symptoms of each?

 ▶ Video

As You Read

STUDY ORGANIZER Make the study organizer found in the FL pages in the back of the book to record the information presented in Lesson 3.

Vocabulary

› influenza
› contagious period
› mononucleosis
› hepatitis
› hepatitis B
› tuberculosis
› pneumonia
› strep throat

 Audio

 Bilingual Glossary

THE COMMON COLD

MAIN IDEA Everyone will experience a cold on occasion.

The cold is the most common communicable disease. Cold symptoms include runny nose, headache, sore throat, coughing, sneezing, and mild fever. You might be wondering, if colds are so common, why they can't come up with a vaccine to guard against colds. Well, there are hundreds of different viruses that are responsible for the common cold. Because there are so many, scientists can't develop vaccines for all of them.

So, what can you do if you have a cold? There is no cure for the common cold. The first thing you should do, though, is rest in bed. During the first 24 hours your cold is contagious.

It is best if you stay at home and don't infect others. You have probably seen ads stating that when you have a cold, you should get plenty of rest. That advice is true. Rest will help your body recover from the cold.

Your parents or guardians might also give you over-the-counter medicines to help relieve your cold symptoms. You should also drink plenty of fluids when you have a cold. If your cold symptoms get worse, or your sore throat lasts for several days, you should see a doctor.

The *cold* is the most common **communicable disease.**

Reading Check

EXPLAIN *Why should you stay home for 24 hours after your cold symptoms first appear?*

What Teens Want to Know

Is it harmful to take antibiotics when you don't really need them? Antibiotics are useful for bacterial illness, not viral illness like colds, coughs, or influenza. Using antibiotics when they are not needed can contribute to the development of antibiotic-resistant infections. *What problems might occur due to a person's overuse of antibiotics?*

THE FLU

MAIN IDEA ❯ The flu is a common communicable disease.

Your joints and muscles ache. You have fever, chills, fatigue, and a headache. You have influenza, or the flu. Influenza (in•floo•EN•zuh), is *a highly communicable viral disease characterized by fever, chills, fatigue, headache, muscle aches and respiratory symptoms.* Flu symptoms usually affect you more quickly and more seriously than cold symptoms do. Like a cold, though, flu can be spread through both direct and indirect contact. Resting and drinking lots of fluids can help you recover faster from the flu. Some types of flu are serious and require a doctor's care.

Flu viruses are not the same as the viruses that cause colds. Every year, certain strains of the flu virus are different than

the year before. Scientists meet every year to determine which strains will spread fastest during the next flu season. Then they develop vaccines for those strains of flu.

❯❯❯ **Reading Check**

COMPARE *How are the flu and the common cold similar?*

Some people should get flu shots before each flu season including people 65 and older and anyone with a weakened immune system. *Explain why the flu vaccine changes every year.*

OTHER COMMUNICABLE DISEASES

MAIN IDEA ❯ All communicable diseases have a contagious period.

Communicable diseases have a contagious period. The contagious period is *the length of time that a particular disease can be spread from person to person.* Chicken pox, measles, and mumps all have specific contagious periods.

- Chicken pox is contagious for about a week before symptoms appear. Typical symptoms for chicken pox include an itchy, bumpy rash, fever, and aching muscles. The itchy bumps will blister and then dry up. When they are dry, chicken pox are no longer contagious. A vaccine for chicken pox has decreased its occurrence.

- Measles has symptoms that include a rash, fever, and head and body aches. Measles are contagious a few days before symptoms appear. The contagious period lasts about five days after the symptoms appear. Measles is a dangerous disease. Vaccines are available in the United States and some other countries. Because of the vaccine, measles is less common than it once was.

- Mumps causes a fever, headache, and swollen salivary glands. Mumps are contagious a week before symptoms appear and for about nine days after.

Most of the children in the United States are vaccinated against mumps. Fewer people get the disease now.

Communicable diseases have a *contagious* period.

❯❯❯ **Reading Check**

EXPLAIN *How have the vaccinations for chicken pox, measles, and mumps affected people?*

Mononucleosis

Mononucleosis (MAH•noh•nook•klee•OH•sis), or mono, is *a viral disease characterized by a severe sore throat and swelling of the lymph glands in the neck and around the throat area.* Mono most commonly infects teens and young adults. Known as the "kissing disease," it is spread through contact with the saliva of an infected person. It can also be spread by sharing contaminated drinking glasses and eating utensils.

Symptoms include a sore throat and swollen lymph glands in the neck and throat. Fatigue, loss of appetite, fever, and headache are also symptoms. Rest is the best treatment for mono.

Hepatitis

Hepatitis (hep•uh•TY•tis) is *a viral disease characterized by an inflammation of the liver and yellowing of the skin and the whites of the eyes.*

Other symptoms include fatigue, weakness, loss of appetite, fever, headaches, and sore throat. Hepatitis A, B, and C are three different virus types.

Hepatitis A is commonly found in areas that have poor sanitation. It spreads when infected human wastes contaminate the food and water sources. People can also become infected if they have open wounds that are exposed to contaminated water.

Hepatitis B, *a disease caused by the hepatitis B virus that affects the liver,* and hepatitis C can be more dangerous. These strains can cause permanent liver damage. They are usually spread through contact with infected blood or other body fluids. They can spread when drug users share needles and through sexual contact. Vaccinations for available for hepatitis A and B. Medications can help treat hepatitis C.

Tuberculosis

Tuberculosis (too•ber•kyuh•LOH•sis), or TB, is *a bacterial disease that usually affects the lungs.* Symptoms of tuberculosis include cough, fatigue, night sweats, fever, and weight loss. It is spread through the air when an infected person coughs or sneezes.

A person can have tuberculosis and not even know it. He or she may not show any symptoms. They may not be sick, but they can still spread the disease. Doctors and other health care providers often test people to be sure they do not carry TB. Tuberculosis can be treated with medications.

This figure shows information for several communicable diseases. *Identify which diseases have similar symptoms.*

Disease	Symptoms	Contagious Period	Vaccine
Chicken pox	Itchy rash, fever, muscle aches	One to five days before symptoms appear to when spots crust over	Yes
Pneumonia	High fever, chest pain, cough	Varies	For some types
Rubella	Swollen lymph nodes, rash, fever	Seven days before rash starts to five days after	Yes
Measles	Fever, runny nose, cough, rash	Three to four days before rash starts to four days after	Yes
Mumps	Fever, headache, swollen areas in neck and under jaw	Seven days before symptoms to nine days after	Yes
Whooping cough	Fever, runny nose, dry cough (with a whooping sound)	From inflammation of mucous membranes to four weeks after	Yes
Tuberculosis	Fever, fatigue, weight loss, coughing blood	Varies	Yes

Pneumonia

Pneumonia is *a serious inflammation of the lungs.* It is usually caused by bacteria or viruses. Symptoms include fever, cough, chills, and nausea. Pneumonia can also cause vomiting, chest pains, and difficulty breathing. Pneumonia is spread through direct or indirect contact with an infected person. You can catch it from people whether or not they show symptoms of illness.

Pneumonia caused by bacteria can be treated with antibiotics. Pneumonia caused by viruses can be treated with antivirals. People with pneumonia need to rest and drink plenty of fluids. People who have other illnesses such as diabetes or HIV/AIDS are at greater risk to catch pneumonia. Pneumonia can be prevented with vaccines. Also, practicing good hygiene can reduce your risk for pneumonia.

Eating healthy foods can help your body fight off communicable diseases. *Describe other choices you can make that will help you stay healthy.*

Strep Throat

Most sore throats are caused by a virus. **Strep throat** is *a sore throat caused by streptococcal bacteria.* Strep throat can be treated with antibiotics. It is spread through direct or indirect contact. When an infected person breathes or coughs, they release droplets into the air. If you happen to breathe in some of those droplets, you may get strep throat.

Symptoms include a red and painful throat, fever, and swollen lymph nodes in your neck, Other symptoms include headache, nausea, and vomiting. If you have these symptoms, get medical help. If left untreated, it can lead to more serious illnesses.

❯❯❯ Reading Check

COMPARE *How are the symptoms for pneumonia and strep throat similar?* ■

REVIEW

❯❯❯ After You Read

1. **DEFINE** What is the *contagious period* of a disease?
2. **EXPLAIN** How might a person become infected with hepatitis A?
3. **DESCRIBE** Why is it important to get treatment for diseases such as strep throat?

❯❯❯ Thinking Critically

4. **ANALYZE** Japan is a small country with lots of people. If someone is ill and must go out, she or he will wear a surgical mask. Why do you think they do this?
5. **EVALUATE** You read on the Internet that a scientist has come up with the cure for the common cold. How can you know if his claim is valid?

❯❯❯ Applying Health Skills

6. **PRACTICING HEALTHFUL BEHAVIORS** Create a poster that provides students with tips on how to stay healthy and how to keep others healthy. Ask for permission to post your finished product in a school hallway.

🄯 Review

🔊 Audio

Photodisc/Getty Images

Sexually Transmitted Diseases

BIG IDEA ▷ Sexually transmitted diseases are infections spread through sexual activity.

Before You Read

QUICK WRITE Write a paragraph about why it is important for teens to avoid sexual activity.

 Video

As You Read

STUDY ORGANIZER Make the study organizer found in the FL pages in the back of the book to record the information presented in Lesson 4.

Vocabulary

> sexually transmitted diseases
> chlamydia
> abstinence
> genital herpes
> genital warts
> trichomoniasis
> gonorrhea
> syphilis

 Audio

 Bilingual Glossary

Myth or Fact

Myth: Birth control pills will protect a female against sexually transmitted diseases.
Fact: Birth control pills do not protect against any kind of sexually transmitted disease.

MedicalRF.com

WHAT ARE SEXUALLY TRANSMITTED DISEASES?

MAIN IDEA ▷ Sexually transmitted diseases are a growing problem in the United States.

Sexually transmitted diseases (STDs) are *infections that are spread from person to person through sexual contact.* STDs are also called sexually transmitted infections (STIs). The pathogens that cause STDs are transferred from person to person through sexual contact. In other words, a person who engages in sexual activity with someone who has an STD can be infected with the disease.

STDs are a major health problem in the United States. The government agency Centers for Disease Control and Prevention (CDC) tracks health issues in the United States. The CDC estimates that there are 19 million new sexually transmitted infections each year. This costs the health care system around $17 billion every year. The bigger cost, though, is the long-term health effects of those infected with STDs.

STDs are a major **health problem** in the United States.

The CDC estimates that young people make up about one-fourth of the sexually active population in the United States. Unfortunately, those same young people represent about half of the new STDs reported each year.

Some STDs can be treated, but others cannot. In some cases, an STD cannot be cured. When this occurs, the risk for other diseases, such as cancer, may increase. The good news is that STDs are preventable. They can be prevented by avoiding high-risk activities like sexual activity or sharing drug needles.

>>> **Reading Check**

DEFINE *What is a sexually transmitted disease?*

COMMON STDS

MAIN IDEA Sexually transmitted diseases include a wide range of diseases.

STDs are passed from one partner to another through sexual activity. All STDs affect both males and females. STDs can cause serious health problems if they are not treated. If you or someone you know suspect you have an STD, seek medical attention. Some common STDs are described here.

Chlamydia

Chlamydia (kluh•MI•dee•uh) *is a bacterial STD that may affect the reproductive organs, urethra, and anus.* It is caused by bacteria. Chlamydia is transmitted from person to person through sexual contact. The greater the number of sexual partners a person has, the greater the risk for the infection. Chlamydia can also be passed from an infected mother to her baby during childbirth.

It is often called a "silent" disease. This is because most infected people don't show any symptoms. A person can have it and not know about it. During the early stages of infection, both males and females can have a genital discharge and pain when urinating. If symptoms of early infection do occur, they usually appear within one to three weeks of sexual contact with an infected person.

Untreated chlamydia can cause other infections in the body and infertility. Some females with untreated chlamydia will develop pelvic inflammatory disease, or PID.

A person who suspects that he or she is infected with an STD must see a doctor.

STDs can be prevented by saying no to high-risk behaviors, such as sexual activity.

Someone who has an STD may not have visible symptoms, or may have symptoms that come and go. However, such a person may be contagious even when there are no symptoms.

Not all STDs are curable, and some are even fatal.

Vaccines are not available for most STDs.

STDs can make a person sterile or infertile.

When you know the facts about STDs, you have the power to avoid them. *What is the best way to avoid getting STDs?*

PID is a painful infection of the uterus, fallopian tubes, and the ovaries. Untreated PID can cause permanent damage to these structures. It can cause severe pelvic pain and infertility. Chlamydia can be treated with antibiotics. However, reinfection is common if sexual partners have not been treated.

Abstinence, *the conscious, active choice not to participate in high-risk behaviors,* is the only way to prevent chlamydia.

Genital Herpes

Genital herpes (HER•peez) is *a viral STD that produces painful blisters on the genital area.* It is transmitted by skin-to-skin contact that can occur without having sexual contact. Herpes often does not cause obvious symptoms for many years. Sometimes, though, there are outbreaks of painful sores and blisters. Even when the blisters go away, the virus is still in the body and can be passed on to others.

If you or someone you know **suspect** you have an *STD,* **seek medical attention.**

Other symptoms include pain and burning in the lower genital region and genital discharge. Although, there is no known cure for gential herpes, medications can reduce the frequency of outbreaks.

>>> Reading Check

EXPLAIN *Why is it important to seek medical help for sexually transmitted diseases?*

Genital Warts

Genital warts are *growths or bumps in the genital area caused by certain types of the human papillomavirus (HPV)*. HPV is the most common sexually transmitted infection in the United States. About 6 million people become infected each year.

There are more than 40 types of HPV. They infect the genital areas of both males and females. HPV is passed through genital contact. Sexual intercourse does not have to occur for the virus to be transmitted. It can be passed to another person whether or not symptoms are present.

HPV is another "silent" disease. Most people don't develop symptoms or health problems. In 90 percent of cases, the immune system clears the body of the virus in about two years. In some cases the body does not clear the infections. Then genital warts may appear. Warts may appear as a small bump or several bumps. On rare occasions, warts appear in the throat. The warts can be treated, but there is no cure for the HPV infection itself. Cervical cancer and cancers of other reproductive organs can also be caused by HPV.

A vaccine has been developed that can protect against the most common types of HPV. When taken by females ages 11 through 26, the vaccines can protect against genital warts and most cervical cancers.

When males ages 9 through 26 take the vaccines, they may be protected against genital warts and anal cancers.

Condoms may also lower the risk of transmitting HPV. They may also lower the risk of genital warts and cervical cancer. However, condoms do not cover all of the areas that can be infected. So, condoms do not fully protect against HPV.

> *HPV* is the most common **sexually transmitted infection** in the United States.

Trichomoniasis

Trichomoniasis (TREE•koh•moh•NI•ah•sis) or "trich" is *an STD caused by the protozoan Trichomonas vaginalis*. It is a very common STD. It is considered to be the most common curable STD.

Most people who have trichomoniasis don't know they have it. It is also a silent disease. However, some people do have symptoms. Those symptoms include vaginal discharge and discomfort during urination. Irritation or itching in the genital area can also be symptoms. If left untreated, trichomoniasis can last for months or years. Having trichomoniasis can increase the chance of getting other STDs. Trichomoniasis can be treated and cured with prescription antibiotics.

Pubic Lice

Pubic lice or "crabs" are insects that that infect a person's genital area. Symptoms include itching around the genitals and crawling insects that are visible to the naked eye. Pubic lice are highly contagious. They can be treated effectively with medicated shampoo or prescription lotion.

Gonorrhea

Gonorrhea (gahn•uh•REE•uh) is *a bacterial STD that affects the mucous membranes of the body, particularly in the genital area*. Most often it affects the genital area. It is a very common disease.

Some males with gonorrhea do not have any symptoms. Those who do, generally have a thick yellowish discharge from the genitals. They will also have a burning sensation when urinating. Most females with gonorrhea don't have symptoms. If a female does have symptoms, they are usually mild. The symptoms are often mistaken for a vaginal or bladder infection.

If not treated, gonorrhea can spread to other parts of the body such as the joints and heart. These conditions can be life-threatening. Females who are not treated can become infertile. Gonorrhea can be treated and cured with antibiotics.

>>> **Reading Check**

EXPLAIN *Why is HPV known as a silent disease?*

Syphilis

Syphilis (SIH•fuh•luhs) is *a bacterial STD that can affect many parts of the body.* Syphilis is passed from one person to another by contact with a syphilis sore. Syphilis cannot be spread through contact with objects such as toilet seats, doorknobs, bathtubs, shared clothing, or swimming pools.

The *symptoms* of **syphilis** change as the disease **progresses.**

The symptoms of syphilis change as the disease progresses. In the first stage painless sores appear at the place of infection. If left untreated, the disease progresses into the second stage. In this stage, a severe body rash occurs. There may also be fever, swollen lymph glands, sore throat, patchy hair loss, headaches, weight loss, muscle aches, and fatigue. During the later stages of syphilis, the infection moves throughout the body. It may damage the brain, nerves, eyes, heart, blood, vessels, liver, bones, and joints.

If diagnosed and treated in the early stages, syphilis can be cured with antibiotics.

>>> Reading Check

IDENTIFY *Name two STDs that are caused by viruses.*

Health SKILLS ACTIVITY · Accessing Information

Finding Information *About STDs*

How can you find valid information about diseases or medical conditions? First, talk to your parents, guardians, or family doctor about any medical concerns you have. You can also learn more about a disease by doing your own research. Medical journals and scientific publications have strict rules about the information they publish. They are valid sources of information. Finding valid health information on the Internet is a little trickier. Be sure that the source of any information you find online is an expert on the subject you are reading about. Check who owns or operates the Web site that you are using. Is the owner or operator a university, hospital, or government office? Find out who wrote the information for the site. Check out the author just as you would for a print article.

When you check the reliability of a source, you might ask your health teacher, school nurse, family doctor, or other trusted adult about the source. Any media source that sells products should be considered with caution. Look for web sites that end in .gov or .edu. A good place to start is the information found at the Centers for Disease Control and Prevention (CDC) and the National Institutes of Health (NIH).

Research one of the STDs from this lesson. Find out more about the symptoms, effects, and any treatments for the disease. Prepare a short report on your findings.

PREVENTING STDS

MAIN IDEA › Abstinence is the only 100 percent effective way to prevent STDs.

You can't usually tell by looking that someone has an STD. The only 100 percent effective way to avoid STDs is to abstain from sexual activity.

Often, the media send the message that sexual activity is exciting. However, the message doesn't tell you about the possibility of an unplanned pregnancy or getting an STD. It doesn't tell you about social and emotional problems that can result from sexual activity. You need to be aware of the possible consequences of sexual activity.

It is normal to have sexual feelings as a teen. However, having these feelings does not mean you must act on them. Talk to a trusted adult about these feelings. Talking can help you better understand the feelings. Find out what your family values are and what is expected of you. Choose friends who share your values and who will support you in your decisions.

Avoid being alone on a date. By participating in group activities, you can avoid the pressures of sexual activity. If you do go on a one-on-one date, communicate your limits to your date before you go out. Practice how to respond to a date who tries to pressure you into sexual activity. For example, if your date says, "If you really care about me, you would have sex with me," you can say "If you really care for me, you would respect my decision."

››› Reading Check

APPLY *What might you say if a date wants to engage in sexual activity?* ■

Group activities can help you avoid situations where you may feel pressure to engage in sexual activity. *What activities do you and your friends like to do together?*

REVIEW

››› After You Read

1. **DEFINE** What is an *STD*?
2. **EXPLAIN** What are the consequences of syphilis if left untreated?
3. **APPLY** What is the best way to avoid becoming infected with STDs?

››› Thinking Critically

4. **APPLY** A teen thinks he or she has an STD. Why is it important for the teen to seek medical help?
5. **EVALUATE** Why might drinking alcohol increase your risk of getting an STD?

››› Applying Health Skills

6. **COMMUNICATION SKILLS** Using the S.T.O.P. strategy can help you refuse to participate in unhealthy activities. The steps to the S.T.O.P. strategy are: **S**ay no in a firm voice. **T**ell why not. **O**ffer other ideas. **P**romptly leave. Use these steps to develop a list of responses to pressure to engage in sexual activity.

⟳ Review

◉ Audio

Helping people impacted by HIV and AIDS build healthier lives.

HIV/AIDS

BIG IDEA HIV causes AIDS, which is a deadly disease that interferes with the body's immune system.

 Before You Read

QUICK WRITE Write a response to this question: What would happen if the cells that control your immune responses began to be destroyed?

▶ **Video**

As You Read

STUDY ORGANIZER Make the study organizer found in the FL pages in the back of the book to record the information presented in Lesson 5.

Vocabulary

> HIV (human immunodeficiency virus)
> AIDS (acquired immunodeficiency syndrome)
> opportunistic infection
> carrier

🔊 **Audio**

🔤 **Bilingual Glossary**

WHAT ARE HIV AND AIDS?

MAIN IDEA HIV causes AIDS, a serious disease that attacks the body's immune system.

HIV (human immunodeficiency virus) is *the virus that causes AIDS.* HIV attacks lymphocytes called T cells. When it attacks a T cell, it replaces the cell's genetic information with its own genetic information. Then it begins to multiply. As more T cells are taken over, the immune system weakens. Eventually, the T cell count drops so low that the immune system can no longer protect the body. When this happens, AIDS develops.

Drug therapy can help delay the onset of AIDS. Scientists continue to work on improved medical treatments for HIV and AIDS. Practicing abstinence from sexual activity and other high-risk behaviors is the only way to prevent HIV/AIDS.

AIDS (acquired immune deficiency) is *a disease that interferes with the body's ability to fight infection.* AIDS continues to weaken the immune system. When AIDS is present in the body, opportunistic infections occur.

AIDS continues to **weaken** the *immune system.*

An opportunistic infection is a *disease that attacks a person with a weakened immune system and rarely occurs in a healthy person.* For example, one opportunistic infection that is a danger for people with AIDS is pneumonia. Many AIDS patients develop a type of pneumonia that can cause death.

Reading Check

DESCRIBE *How does HIV develop into AIDS?*

Developing Good Character

Caring Practicing abstinence from sexual activity until marriage and avoiding illegal drug use, especially the use of injectable drugs, will help protect you from HIV infection. *How does avoiding these risk behaviors show that you care about yourself and others?*

How HIV Spreads

You cannot get HIV through casual contact with a person who has the virus. HIV is only transmitted through infected blood or body fluids. Body fluids that can transmit HIV include fluid from the vagina, breast milk, and semen. Semen is the fluid that carries sperm. There are several ways these fluids spread from one person to another.

- **Having any form of sexual contact with an infected person.** The most common way HIV spreads is through sexual activity. When people engage in sexual activity, semen and vaginal fluids are transferred from one person to the other. HIV can be transferred in these fluids. People can have the virus and not know it. Even so, they can still infect other people. Just one incident of sexual activity with an infected partner can spread the virus. People who have more than one sexual partner are at greater risk.

- **Using a contaminated needle.** One drop of blood left on a needle can contain enough HIV to infect someone. Never inject yourself with illegal drugs. Contaminated needles used for tattooing and piercing can also transmit the virus. People who use needles to take medication for diabetes or other illnesses should do so under the care of a doctor.

- **Other ways HIV is spread.** HIV can be spread from a pregnant female with HIV to her developing baby.

It can be transmitted to the child during birth or during breast feeding. Certain drugs can reduce the rate of transmission of HIV to the unborn child. In the years before it was known that HIV caused AIDS, some people were infected during blood transfusions through a contaminated blood supply. However, since 1985, all blood is screened for HIV. The blood supply in the United States is considered to be safe.

> HIV cripples the immune system by killing the T cells that control immune responses. *What kinds of diseases eventually harm people who have AIDS?*

How HIV is NOT Spread

HIV *cannot* be spread by:

- Being bitten by a mosquito
- Touching the tears or sweat of an infected person
- Shaking hands with or hugging someone with HIV
- Swimming with an infected person
- Sharing utensils with an infected person
- Donating blood
- Using the same shower or toilet as an infected person
- Sharing sports equipment with an infected person

>>> **Reading Check**

IDENTIFY *What are five ways HIV cannot be spread?*

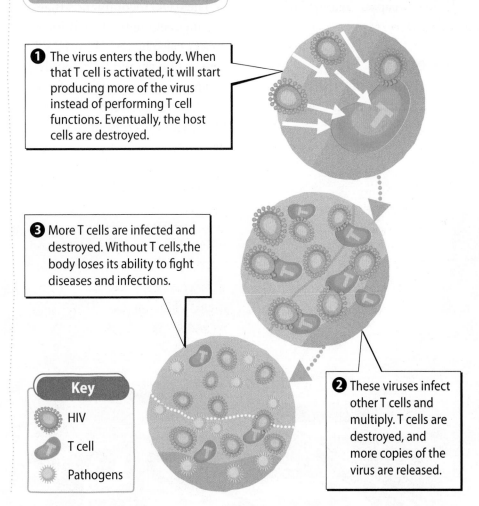

❶ The virus enters the body. When that T cell is activated, it will start producing more of the virus instead of performing T cell functions. Eventually, the host cells are destroyed.

❸ More T cells are infected and destroyed. Without T cells, the body loses its ability to fight diseases and infections.

❷ These viruses infect other T cells and multiply. T cells are destroyed, and more copies of the virus are released.

Key
- HIV
- T cell
- Pathogens

FIGHTING AIDS

MAIN IDEA AIDS is a worldwide problem.

The World Health Organization (WHO) estimated that in a recent year 34 million people worldwide were infected with HIV. In that same year 1.8 million people died from AIDS. Recent estimates by the Centers for Disease Control and Prevention (CDC) state that 12 million people in the United States are infected with HIV. Young people ages 13-29 account for 39 percent of the new HIV infections in recent years. The CDC also believes that 20 percent of the people with HIV don't know they have it. Knowledge and abstinence are the best weapons against HIV and AIDS.

Detecting HIV

Earlier, you learned that a person can be a carrier of HIV, without having AIDS.

A person can be a **carrier** of *HIV,* **without** having *AIDS.*

A carrier is *a person who appears healthy but is infected with HIV and can pass it to others.* Laboratory tests are the only way of knowing whether or not a person has HIV. If the lab test does not show antibodies for HIV present in the blood, the person should be retested in three months. A recently infected person may not have had time to develop antibodies.

The CDC recommends that anyone seeking treatment for an STD also be tested for HIV. The testing should be carefully explained to the patient beforehand. Testing should only take place with the consent of the patient.

Treating HIV *and* AIDS

There is no cure for HIV or AIDS. However, scientists continue to work to improve medical treatments for people with HIV and AIDS. Drugs are available to slow down the progress of the HIV infection. Many of the drugs are very expensive and some have serious side effects. There is evidence that some of the drugs are losing their effectiveness against the virus. As HIV is exposed to new drugs, it is changing in ways that make the drugs ineffective.

Fighting HIV infection is costly and difficult. So, scientists and educators work hard to keep people from getting HIV in the first place.

> ### ›› Reading Check
>
> **EXPLAIN** *What treatment is available for people with HIV?*

The AIDS quilt is a memorial to people who have died of AIDS. Each square represents one person who died from the disease. *What can you learn from seeing the AIDS quilt?*

©Hisham Ibrahim/Corbis

Millions of cases of STDs occur each year in the United States. *What can be done to stop the spread of these diseases?*

>>> After You Read

1. **LIST** What are three ways HIV is spread?
2. **IDENTIFY** What are four ways you cannot become infected with HIV?
3. **DEFINE** Provide definitions for *HIV* and *AIDS.*

>>> Thinking Critically

4. **ANALYZE** What is the relationship between HIV and AIDS?
5. **APPLY** Why is it important to see a health care provider if someone thinks they may have an STD?

>>> Applying Health Skills

6. **ADVOCACY** Create a brochure or pamphlet about HIV and AIDS. Include the basic facts about the diseases and explain how to avoid getting the disease. Include ways it can be transmitted and ways it cannot be transmitted. Provide copies for your school health office.

⟳ Review

◀)) Audio

Stopping *the* Spread *of* HIV

A vaccine for HIV is in the works, but it will likely be many years before it can be used by the general population. Because there are several different forms of HIV, a vaccine created for one form might not work on another form. One vaccine that works on all forms of HIV is possible, but likely won't be available for many years. The best way to prevent HIV is to avoid high-risk behaviors. There are three main ways to avoid HIV and AIDS:

- **Practice abstinence.**
  Avoiding sexual activities until marriage is one type of abstinence. People who participate in sexual activity expose themselves to risks. They may not know whether their partner has an STD. They may not know whether their partner has HIV or AIDS. The more sexual partners a person has, the greater their risk of getting an STD.

- **Avoid drugs and alcohol.**
  Using drugs and alcohol can make a person lose their ability to make good decisions. When their judgment is impaired, they are more likely to engage in risky behaviors. The risky behaviors might include sexual activity.

- **Avoid sharing needles.**
  Shared needles can carry enough blood to inject HIV into your blood stream. This includes needles used for injecting drugs or for tattoos or piercing.

  The only sure way to avoid getting HIV is to avoid contact with sources of the virus. Abstaining from sexual activity until marriage and not injecting drugs or sharing needles are ways to avoid contact with the virus. ∎

©Lars Niki

Hands-On HEALTH ACTIVITY

Healthy Habits

How healthy are your habits? This activity will help you find out.

WHAT YOU WILL NEED

* pencil and paper

WHAT YOU WILL DO

Write *yes* or *no* for each statement.

1. I avoid sharing eating utensils or drinking glasses with others.

2. I avoid drinking water taken directly from streams or lakes.

3. I cover my nose and mouth when I cough or sneeze.

4. I make sure leftover food is properly stored. I wash my hands before preparing or serving food and after using the bathroom.

5. When I am sick, I get medical care and I avoid others during the contagious period.

6. I avoid sharing personal hygiene items with others. I avoid contact with people who have a cold or other communicable diseases.

7. I have received all the recommended vaccinations.

WRAPPING IT UP

Give yourself 1 point for each yes. A score of 8–10 is very good. A score of 6–7 is good. A score of 4–5 is fair. If you score below 4, you need to work on improving your health behaviors.

READING REVIEW

FOLDABLES and Other Study Aids

Take out the Foldable® that you created and any study organizers that you created.
Find a partner and quiz each other using these study aids.

LESSON 1 Preventing the Spread of Disease

BIG IDEA Good personal hygiene and other healthful behaviors can help protect you from communicable diseases.

* Communicable diseases are caused by germs.
* Practicing certain healthy behaviors can stop the spread of pathogens.
* Pathogens are spread through direct and indrect contact.

LESSON 2 Defense Against Infection

BIG IDEA Your immune system protects you from infection.

* Your body has five major barriers that defend against pathogens.
* Your immune system responds when pathogens get past the five major barriers of protection.
* The specific immune response attacks quickly if pathogens get past the nonspecific immune response.
* Your body has different types of immunity.

LESSON 3 Communicable Diseases

BIG IDEA Common communicable diseases include colds, the flu, strep throat, pneumonia, mononucleosis, hepatitis, and tuberculosis.

* Everyone will experience a cold on occasion.
* Influenza, or the flu, is a common communicable disease.
* All communicable diseases have a contagious period.

LESSON 4 Sexually Transmitted Diseases

BIG IDEA Sexually transmitted diseases are infections spread through sexual activity.

* Sexually transmitted diseases are a growing problem in the United States.
* Sexually transmitted diseases include a wide range of diseases.
* Some sexually transmitted diseases can be cured, but some cannot be cured.
* Abstinence is the only 100 percent effective way to prevent STDs.

LESSON 5 HIV/AIDS

BIG IDEA HIV causes AIDS, which is a deadly disease that interferes with the body's immune system.

* HIV causes AIDS, a serious disease that attacks the body's immune system.
* HIV is spread through the exchange of bodily fluids.
* HIV is not spread by mosquito bites, by being exposed to someone with HIV, sharing utensils, or donating blood.
* AIDS is a worldwide problem.

 Review

 Web Quest

ASSESSMENT

Reviewing Vocabulary *and* Main Ideas

> immune system   > viruses   > influenza   > tuberculosis
> infection   > vaccine   > contagious period   > inflammation

» On a sheet of paper, write the numbers 1–7. After each number, write the term from the list that best completes each statement.

LESSON 1 Preventing the Spread of Disease

1. A(n) _____ is a condition that occurs when pathogens enter the body, multiply, and cause harm.

2. _____ are the smallest kinds of pathogens.

LESSON 2 The Body's Defenses Against Infection

3. _____ is the body's response to injury or disease, resulting in a condition of swelling, pain, heat, and redness.

4. The _____ is a combination of body defenses made up of the cells, tissues, and organs that fight pathogens in the body.

LESSON 3 Communicable Diseases

5. _____ is a communicable disease characterized by fever, chills, fatigue, headache, muscle aches, and respiratory symptoms.

6. The _____ is the length of time that a particular disease can be spread from person to person.

7. _____ is a bacterial disease that usually affects the lungs.

» On a sheet of paper, write the numbers 8–14. Write *True* or *False* for each statement below. If the statement is false, change the underlined word or phrase to make it true.

LESSON 4 Sexually Transmitted Diseases

8. Sexually transmitted diseases are infections that spread from person to person through <u>casual</u> contact.

9. Syphilis is a bacterial STD that can affect <u>many parts of the body</u>.

10. Genital herpes <u>can</u> be transmitted when symptoms are not present.

11. Chlamydia can cause <u>pelvic inflammatory disease.</u>

LESSON 5 HIV/AIDS

12. HIV is the virus that causes <u>AIDS</u>.

13. You <u>can</u> become infected with HIV by shaking hands with an infected person.

14. Scientists have developed powerful new drugs to <u>cure</u> HIV infections.

✔ eAssessment

>> Using complete sentences, answer the following questions on a sheet of paper.

☁ *Thinking* **Critically**

15. **EXPLAIN** Why do people who have AIDS actually die from other diseases?

16. **INTERPRET** Sometimes after you get a vaccination, years later you have to get more of the same vaccine. Why do you think you might need this "booster shot"?

✎ *Write* **About It**

17. **PERSONAL WRITING** Write a journal entry describing what factors can influence a teen's attitude toward sexual activity. Are these influences positive or negative?

ⒶⒷ STANDARDIZED TEST PRACTICE
ⒸⒹ

Reading
Read the passage to answer the questions.

People living during the Middle Ages witnessed one of the most horrific disasters that forever changed Western civilization: the bubonic plague. The bubonic plague swept across the known world and killed many people in a short period of time. The pathogen for the plague is a bacterium called *Yersinia pestis*. The bacteria lived inside fleas. The fleas lived on rats and infected the animals by biting them. Uninfected fleas that bit infected rats could also become infected. Fleas moved from rat to rat and spread the plague quickly. When a rat got the plague, it died. As the disease swept through the rat populations in cities, many rats died off. As rats became scarce, more and more infected fleas began living on and biting humans. People became hosts for the plague. From the years 1347 to 1350, the plague killed one-third of the population of Europe.

1. What is the main point of the passage?
   A. to explain how bacteria kill fleas
   B. to explain how the population of Europe became so low
   C. to explain how to prevent the plague from killing people
   D. to explain how the plague killed so many people so quickly

2. Fleas began living on and biting humans because
   A. fleas did not like to live on rats.
   B. rats were not good hosts for the plague.
   C. humans could live with the plaque for a long period of time.
   D. rats died off and the rat population became scarce.

PREVENTING THE SPREAD OF DISEASE

WHAT'S A COMMUNICABLE DISEASE?

A communicable disease can be spread to a person from another person, an animal, or an object. Communicable diseases are caused by pathogens.

How they spread

DIRECT CONTACT **INDIRECT CONTACT** **CONTACT WITH ANIMALS OR INSECTS** **CONTAMINATED FOOD AND WATER**

TYPES OF DISEASES AND ILLNESSES

The Common Cold

Runny nose, headache, sore throat, coughing, sneezing, and mild fever

No cure because cold is caused by wide range of viruses

The Flu

Fever, chills, fatigue, headache, muscle aches and respiratory symptoms

Vaccines available during each flu season

Chicken Pox

Itchy, bumpy rash, fever, and aching muscles

Vaccine

Measles

Rash, fever, and head and body aches

Vaccine

Mumps

Fever, headache, and swollen salivary glands

Vaccine

Mononucleosis

Sore throat, swollen lymph glands in neck and throat, fatigue, loss of appetite, fever, headache

Rest; once person is fully recovered, they can't get it again

Strep throat

Red and painful throat, fever, swollen lymph nodes in neck, headache, nausea, vomiting. Can be very serious if left untreated.

Antibiotics

STOPPING THE SPREAD OF PATHOGENS

Washing hands

Practice washing your hands thoroughly using the following steps:

1 Use warm water to wet your hands, and then apply soap.

2 Vigorously rub your hands together, scrubbing your whole hand for 30 seconds or more.

Singing the ABCs song = **30** seconds.

3 Rinse your hands thoroughly. Use a paper towel to turn off the water in public restrooms.

4 Use a clean towel or paper towel to dry your hands. In a public restroom, use a paper towel to open the door when you leave.

Other ways to stop pathogens

KEEP YOUR DISTANCE from people who are sick with a communicable disease.

DO NOT SHARE eating or drinking utensils or toothbrushes or other personal hygiene items.

HANDLE AND PREPARE food safely. Meats, fish, poultry, and eggs should be cooked thoroughly.

IF YOU ARE SICK, stay home and away from others.

KEEP YOUR ENVIRONMENT clean. Empty the trash often. Keep the trash cans clean. Clean up after your pets.

WASH COUNTERS with paper towels or a clean sponge or cloth. Use warm, soapy water. If you use a cloth, clean it frequently.

WHEN YOU SNEEZE or cough, cover your mouth with a tissue. Only use the tissue once. If you don't have a tissue, sneeze or cough into the crook of your elbow. Wash your hands immediately after you sneeze or cough.

IF YOU HAVE A PRESCRIPTION to take medication, follow the directions exactly. Take all of the medicine you are supposed to take.

NONCOMMUNICABLE DISEASES

Risk factors for noncommunicable diseases, include heredity, lifestyle choices, and environmental factors.

CANCER

Cancer occurs when abnormal cells multiply out of control.

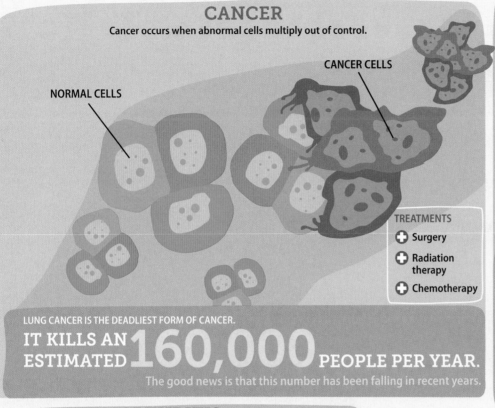

CANCER CELLS

NORMAL CELLS

TREATMENTS
- Surgery
- Radiation therapy
- Chemotherapy

LUNG CANCER IS THE DEADLIEST FORM OF CANCER.

IT KILLS AN ESTIMATED 160,000 PEOPLE PER YEAR.

The good news is that this number has been falling in recent years.

HEART DISEASE

Heart disease, or cardiovascular disease, reduces the function of the heart and blood vessels.

HEART ATTACK | STROKE | HIGH BLOOD PRESSURE

TREATMENTS
- Angioplasty
- Bypass surgery
- Medication
- Heart transplants
- Pacemakers

More than

half a million people per year

die from this condition.

Avoid smoking, get enough exercise, and eat a healthful diet. These behaviors can help keep your heart healthy your whole life.

ARTHRITIS

Arthritis is a disease of the joints marked by pain and swelling in body joints.

About **294,000** children under the age of 18 are affected by pediatric arthritis

Types of juvenile arthritis

JUVENILE RHEUMATOID ARTHRITIS
Typically affects five or more joints, most commonly affects knees, wrists and ankles

JUVENILE IDIOPATHIC ARTHRITIS
Typically affects four or fewer joints, usually: knees, ankles or wrists

SYSTEMIC ONSET JUVENILE RHEUMATOID ARTHRITIS
Affects the small joints of the hands, wrists, knees and ankles

Treatments
- Low-impact physical activity balanced with rest
- Healthy weight
- Joint braces
- Medication
- Surgery
- Massage
- Joint replacement

DIABETES

Diabetes is a disease that prevents the body from converting food into energy.

TYPES

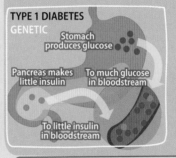

TYPE 1 DIABETES
GENETIC
Stomach produces glucose
Pancreas makes little insulin
To much glucose in bloodstream
To little insulin in bloodstream

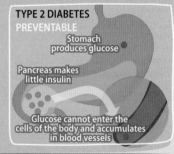

TYPE 2 DIABETES
PREVENTABLE
Stomach produces glucose
Pancreas makes little insulin
Glucose cannot enter the cells of the body and accumulates in blood vessels

TREATMENTS
- Diabetes can be managed through insulin treatment, medication, and lifestyle changes.

ASTHMA

In the United States, about 24 million people are reported to have asthma. More than 6 million asthma sufferers are people under the age of 18.

ASTHMA ATTACK TRIGGERS ALLERGENS SUCH AS

 MOLD | DUST | POLLEN | PETS

 PHYSICAL ACTIVITY | POLLUTANTS | RAPID BREATHING

TREATMENTS
- Asthma can be managed through medication, and reducing stress and exposure to allergens.

ALLERGIES

Between 40 million and 50 million Americans have some type of allergy.

COMMON ALLERGENS

 POLLEN | NUTS | BEE STINGS

TREATMENTS
- Medication
- Avoidence of allergens

 PET HAIR | SHELLFISH

Noncommunicable Diseases

 **PREMIUM ONLINE RESOURCES** 〉

 Audio

 Videos

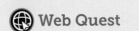 Bilingual Glossary

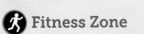 Fitness Zone

 Web Quest

 Review

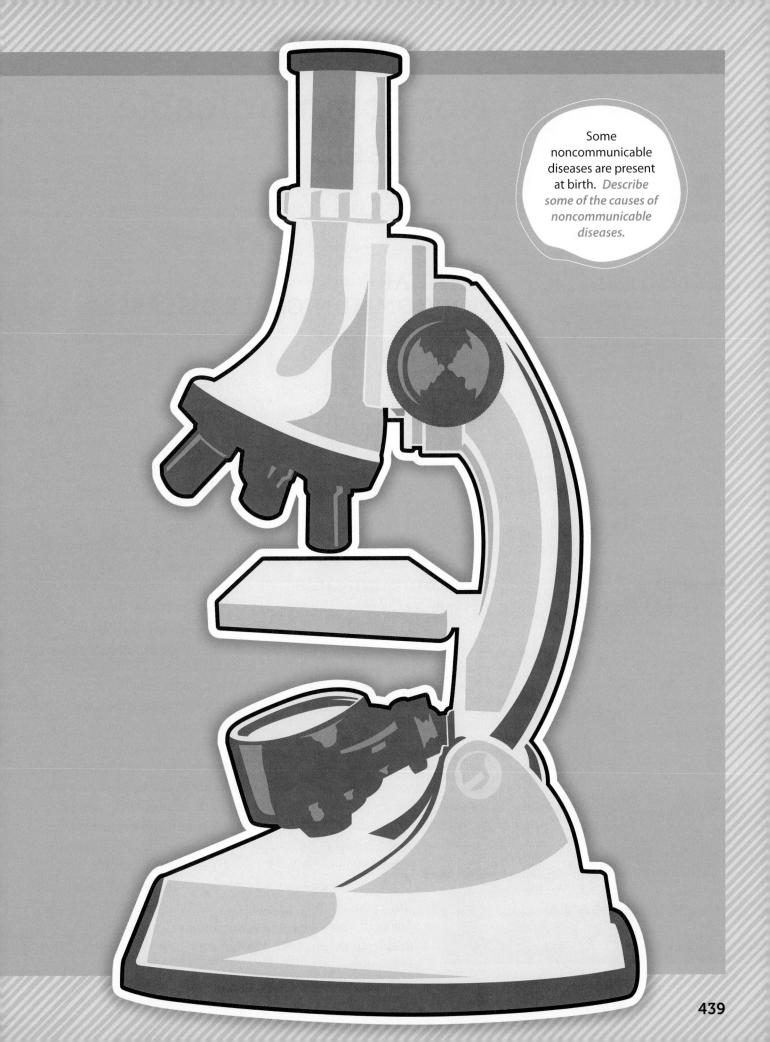

Some noncommunicable diseases are present at birth. *Describe some of the causes of noncommunicable diseases.*

Noncommunicable Diseases

BIG IDEA Noncommunicable diseases can result from heredity or lifestyle choices, or may have an unknown cause.

WHAT ARE NONCOMMUNICABLE DISEASES?

MAIN IDEA Noncommunicable diseases are diseases that cannot be spread from person to person.

Hannah is at her friend Danielle's house. Her eyes get red and itchy. She starts to sneeze. Why? Hannah is allergic to Danielle's dog. An allergy is a type of a **noncommunicable disease,** or *a disease that cannot be spread from one person to another.*

Noncommunicable *diseases* cannot be spread from **one person** to another.

Some noncommunicable diseases are **chronic,** or *present continuously on and off over a long period of time.* **Chronic diseases** are *diseases that are present continuously or off and on over a long time.* Asthma is one type of chronic disease. Other noncommunicable diseases are considered degenerative diseases.

Degenerative diseases *cause a further breakdown in body cells, tissues, and organs as they progress.* Multiple sclerosis (MS) is one type of degenerative disease. MS damages nerve cells. This causes nerve signals to slow down or stop. Over time, a person will notice an increased loss of nerve function.

Other diseases that are considered degenerative diseases include heart disease, osteoporosis, and arthritis. Many times, teens think these diseases occur mostly to older people. It's true that older people are the ones who suffer more frequently from these diseases. However, in some cases, such as arthritis and osteoporosis, taking good care of your body now will help prevent those diseases in the future.

Reading Check

COMPARE AND CONTRAST *What are the similarities and differences between chronic diseases and degenerative diseases?*

MIXA/Alamy

WHAT CAUSES NONCOMMUNICABLE DISEASES?

MAIN IDEA > Risk factors for noncommunicable diseases include heredity, lifestyle choices, and environmental factors.

Noncommunicable diseases can have many causes. Some may be present at birth. Others may be caused by lifestyle choices or a person's environment. Some noncommunicable diseases have no known cause.

Diseases Present *at* Birth

Some babies are born with congenital disorders. **Congenital disorders** are *all disorders that are present when a baby is born.* Babies may be born with physical or mental disabilities. These may be caused by birth defects or genetic disorders. The causes of many birth defects are unknown. Some congenital diseases are caused by heredity. **Heredity** is *the passing of traits from parents to their biological children.*

Another noncommunicable disease a person can be born with is congenital heart disease. Many types of congenital heart disease exist. In some cases it is a defect in a heart valve or a big blood vessel leading out of the heart. The cause of congenital heart disease is often unknown. In some cases, a congenital heart defect may be inherited. In some cases, it may result from medications or infections the mother may have taken or experienced during her pregnancy.

Some birth defects are caused by the mother's choice of lifestyle. For example, a pregnant female who drinks alcohol may give birth to a child with fetal alcohol syndrome (FAS).

Symptoms of FAS include heart defects, poor coordination, and problems with speech, thinking, or social skills.

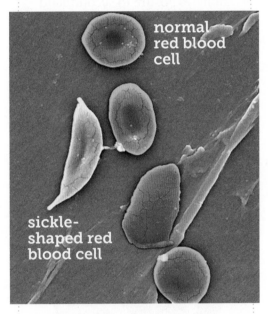

normal red blood cell

sickle-shaped red blood cell

The abnormal shape of red blood cells in sickle cell anemia affects their ability to carry oxygen. *Identify the symptoms a person with sickle cell anemia may have as a result.*

Lifestyle Choices *and* Disease

Certain risk factors increase a person's chance of developing a disease. Heredity, age, gender, and ethnic group are factors over which people have no control. People can, however, control their lifestyle choices.

When people make unhealthy lifestyle choices, they increase their risk of disease. For example, smoking tobacco can cause lung cancer. A lack of physical activity, being overweight, or eating foods high in fat can lead to heart disease. To decrease your risk of disease, practice the following healthy lifestyle behaviors:

- **Eat healthy foods.** Eat whole grains, fruits, and vegetables. Limit foods high in fat, sugar, or salt.
- **Stay physically active.** Teens should be physically active at least 60 minutes every day.
- **Maintain a healthy weight.** Keep your weight within the recommended range for your gender, age, height, and body frame.
- **Get enough sleep.** Teens need at least eight hours of sleep a night.
- **Manage stress.** Use time management and other strategies to reduce stress.
- **Avoid tobacco, alcohol, and other drugs.** These substances harm many parts of the body.

Environmental Factors *and* Disease

The environment affects your health. For example, air pollution can cause disease. Breathing polluted air can worsen respiratory problems. These can include asthma, emphysema, bronchitis, and even lung cancer.

You may have seen photos of cities covered by smog. Smog is a yellow-brown haze that forms when sunlight reacts with air pollution. Breathing smog can cause respiratory diseases in some people. When smog is heavy, people with respiratory diseases may need to limit their outdoor activities.

Carbon monoxide is a colorless, odorless gas, and is another harmful environmental substance. Fumes from car exhaust and some furnaces and fireplaces give off carbon monoxide. High levels of carbon monoxide gas can be dangerous. It can cause serious illness or even death.

High levels of **carbon monoxide** gas can be **dangerous.**

Exposure to tobacco smoke can lead to a number of health problems. Children exposed to tobacco smoke have an increased risk of asthma attacks. They also risk respiratory infections, and ear infections. In adults, exposure to secondhand smoke can cause heart disease and lung cancer.

>>> Reading Check

IDENTIFY *What are three causes of noncommunicable diseases?* ■

Smog is one environmental factor that can cause respiratory diseases. *Identify two other environmental factors that can cause disease.*

REVIEW

>>> After You Read

1. **VOCABULARY** Define *noncommunicable disease* and use it in a sentence.
2. **GIVE EXAMPLES** What are examples of two congenital disorders?
3. **IDENTIFY** What are three risk factors that can cause noncommunicable diseases?

>>> Thinking Critically

4. **EVALUATE** How can lifestyle choices affect a person's health?
5. **SUGGEST** How can communities lower the risk of diseases caused by the environment?

>>> Applying Health Skills

6. **ACCESSING INFORMATION** Research some common noncommunicable diseases. Choose one example and create a fact sheet describing what causes the disease and how it affects the body. Include information on how this disease is treated or managed. Share your findings with the class.

Review

Audio

©Ingram Publishing/SuperStock

Cancer

BIG IDEA Cancer occurs when abnormal cells multiply out of control.

Before You Read

QUICK WRITE Write two sentences that explain why people need to wear sunscreen when they go outdoors.

 Video

As You Read

STUDY ORGANIZER Make the study organizer found in the FL pages in the back of the book to record the information presented in Lesson 2.

Vocabulary

> cancer
> tumor
> benign
> malignant
> risk factors
> carcinogen
> biopsy
> radiation therapy
> chemotherapy
> remission
> recurrence

 Audio

🅰🅱 Bilingual Glossary

Myth vs. Fact

Myth: Indoor tanning beds and booths are not associated with skin cancer.
Fact: Indoor tanning has been linked to melanoma, the most deadliest type of skin cancer.

WHAT IS CANCER?

MAIN IDEA Cancer is characterized by the rapid and uncontrolled growth of abnormal cells.

Cancer is a noncommunicable disease that can affect people of any age. Cancer is *a disease that occurs when abnormal cells multiply out of control.* Any tissue in the body can become cancerous. A great deal of medical research is aimed at treating and finding cures for different cancers. Although many cancers can be treated successfully, cancer is the second leading cause of death in the United States. Only heart disease kills more Americans each year.

How does cancer develop? The adult human body contains more than 100 trillion cells. These cells constantly divide to make more cells so the body can grow and repair itself. Most of the body's cells are normal at any given time. However, even in healthy bodies, some cells can become abnormal. Your immune system usually destroys these cells.

However, some abnormal cells can survive and may even begin to divide. Some of these abnormal cells grow in a clump called a tumor (TOO·mer). A tumor is *a group of abnormal cells that forms a mass.*

Malignant **tumors** can multiply **out of control** and sometimes *spread* to other parts of the body.

Tumors are either benign (bi·NYN) or malignant (muh·LIG·n(schwa)nt). Benign tumors are *not cancerous* and do not spread. Malignant tumors are *cancerous.* Malignant tumors can multiply out of control and sometimes spread to other parts of the body.

Type of Cancer	Important Facts
Skin cancer	is the most common kind of cancer. Excessive exposure to direct sunlight is the major cause of skin cancer.
Breast cancer	most often occurs in females over 50. It can occur in younger females and occasionally in males.
Reproductive organ cancer	can occur in the testicles and prostate gland in males. It can occur in the ovaries, cervix, and uterus in females.
Lung cancer	is the leading cause of cancer deaths in the United States. Smoking is the biggest risk factor for both males and females.
Colon and rectal cancer	develop in the digestive tract. Early detection and better screening have greatly reduced the number of cases of colon and rectal cancers.
Leukemia	is a cancer of the white blood cells that starts in the bone marrow. An increase in cancerous white blood cells interferes with the healthy white blood cells' immune response.
Lymphoma	is a cancer that starts in the lymphatic system. It weakens the immune system and increases the risk of developing infections.

Types *of* Cancer

Almost any tissue in the body can become cancerous. Some types of cancer are more common than others. Skin cancer is the leading type of cancer. More than one million new cases of skin cancer are reported every year in the United States. Skin cancers make up about half of all the cancer cases reported. Most cases of skin cancer are highly curable if detected early and treated appropriately. Lung cancer kills an estimated 160,000 people per year. This number has been falling in recent years.

Causes *and* Risk Factors

Some types of cancers develop for unknown reasons. However, doctors have identified specific risk factors for several certain types of cancer.

Risk factors are *characteristics or behaviors that increase the likelihood of developing a medical disorder or disease.* Risk factors can include inherited traits, age, lifestyle choices, and environmental factors. For example, a high-fat, low-fiber diet may be a risk factor for developing colon and rectal cancers.

Almost any *tissue* in the body can become cancerous.

Some types of cancer have well-known causes. For example, breathing asbestos dust can cause lung cancer. Asbestos is a carcinogen (kar·SI·nuh·juhn), or *a substance that can cause cancer.* Other common carcinogens are the chemicals found in tobacco products. These are linked to lung and mouth cancers.

The table lists some of the most common types of cancer. *Explain why people with lymphoma usually struggle with other diseases too.*

Not all carcinogens are chemicals. For instance, ultraviolet light from the sun can cause skin cancer. Exposure to radiation in large doses, including X-rays, can also cause cancer. Infection from certain viruses has been linked to specific types of cancer. For example, a chronic infection of hepatitis B virus is known to cause liver cancer.

>>> **Reading Check**

IDENTIFY *What are two carcinogens and the cancers they cause?*

DIAGNOSING CANCER

MAIN IDEA Health care professionals use a variety of methods to detect and diagnose cancer.

Health care professionals use many methods to detect cancer. Some are very simple. For example, a doctor might spot a group of skin cells that don't look normal. He or she might feel a lump where the tissue should be soft.

Healthcare professionals also use more involved methods. They can use X-rays and other scanning equipment to look for unusual cell formations.

If tissue shows a suspicious lump or formation, it usually undergoes a biopsy. A **biopsy** is *the removal of a tissue sample from a person for examination.* The tissue from a biopsy goes to a lab for careful examination to determine whether the cells are cancerous.

Change in bowel or bladder habits

A sore that does not heal

Unusual bleeding or discharge

Thickening or lump in a breast or elsewhere

Indigestion or difficulty swallowing

Obvious change in a wart or mole

Nagging cough or hoarseness

Knowing the seven warning signs of cancer can help you detect cancer in its early stages. *Describe why early detection is helpful in treating cancer successfully.*

If they are, technicians will do other tests to learn more about the cancer. Together, a team of health care providers and the patient can decide on a plan for treatment.

One way to increase the chances of successful treatment is to detect cancer as early as possible. In order to detect cancer early, it is important to watch for seven warning signs. A person who shows a warning sign of cancer should be examined by a physician as soon as possible.

Healthcare professionals sometimes use X-rays to look for signs of cancer. *Explain how X-rays and other scanning tools help doctors detect more cases of cancer than they could without these tools.*

TREATING CANCER

MAIN IDEA Surgery, radiation therapy, and chemotherapy treat cancer.

The best way to treat cancer depends on many factors. These include the type of cancer, the stage of the disease, and the age and general health of the patient. The most common cancer treatments include surgery, radiation therapy, and chemotherapy. Cancer patients many times receive at least two of these treatments and sometimes all three.

Surgery removes cancer cells from the body. It is used to treat some cancers, including breast, lung, and colon. Surgery is most effective when the cancer is isolated in one part of the body and can be completely removed.

 Radiation therapy is *a treatment that uses X-rays or other forms of radiation to kill cancer cells.* This method of fighting cancer works best when cells are limited to a single area. It can also kill cells that may remain after surgery. More than half of all cancer patients are treated with radiation therapy. Doctors use chemotherapy to fight cancers that are found throughout the body. **Chemotherapy** is *the use of powerful medicines to destroy cancer cells.*

All cancer treatments have side effects. Side effects of radiation therapy and chemotherapy include nausea, fatigue, and temporary hair loss.

Side effects differ from person to person. They depend on the patient's age, the type of treatment, and the location of the cancer in the body. When cancer treatment is successful, the cancer is in remission. **Remission** is *a period during which cancer signs and symptoms disappear.* Cancer that is in remission can sometimes return. *The return of cancer after a remission* is called a **recurrence.**

>>> **Reading Check**

IDENTIFY *Which treatment is commonly used for a cancer that has spread?*

Health SKILLS ACTIVITY

Advocacy

Promote Ways to *Reduce Cancer Risk*

You can reduce the risk of getting certain cancers. Decisions you make now can help you reduce your risk of developing cancer later in life. Use reliable online and print sources to research how to help prevent certain types of cancer. Here is a head start:

* Lung cancer – stay tobacco free

* Skin cancer – limit exposure to UV rays from the sun

* Colon and rectal cancers – eat a diet rich in whole grains, fruits, and vegetables

With A Group

Write and illustrate a booklet or create a slideshow on ways to reduce cancer risks. Give your resources a catchy name, slogan, and logo. Share your booklet or slideshow with the class.

REDUCING THE RISK OF CANCER

MAIN IDEA Practicing healthful behaviors can reduce your risk of cancer.

Asymmetry
One side of a mole looks different from the other side.

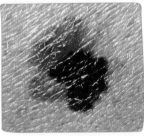

Border Irregularity
The edges are jagged or blurred.

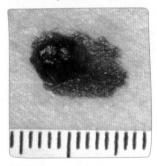

Color
The color is not uniform, or the same, throughout. If a mole is tan and brown, black, or red and white, have it checked.

Diameter
The diameter is greater than 6 millimeters, about the size of a pencil eraser. A growth that has expanded to this size over time should be checked.

You can protect yourself from some cancers. Below are some tips on how to reduce your cancer risk.

- **Avoid tobacco and alcohol.** Cigarette smoking and excessive alcohol use increase the risk of several types of cancer.
- **Eat well and exercise.** Eating well and staying fit can help you avoid some cancers.
- **Limit sun exposure.** Apply a sunscreen with an SPF of at least 15. Wear a hat that shades your neck and your ears. Avoid tanning beds.
- **Perform self-examinations.** Females should perform a breast self-exam once a month. Males should perform a testicular self-exam once a month. Check moles and see a healthcare provider if you notice changes.
- **Know the warning signs.** The American Cancer Society has identified seven possible signs of cancer. The first letter of each sign spells the word CAUTION.

Follow the American Cancer Society's ABCDs. Asymmetry, Border irregularity, Color, and Diameter. *Identify another way to reduce the risk of skin cancer.*

(t b)National Cancer Institute (NCI), (c)Skin Cancer Foundation

LESSON 2
REVIEW

After You Read

1. **DEFINE** Define *cancer*.
2. **IDENTIFY** Name three tools for diagnosing cancer.
3. **EXPLAIN** What are the differences among the three most common treatments for cancer?

Thinking Critically

4. **SYNTHESIZE** Moles get larger when the skin cells in the mole divide. Why do you think moles larger than 6 millimeters might be a warning sign of cancer?
5. **EVALUATE** Why is it important to diagnose cancer as early as possible?

Applying Health Skills

6. **ADVOCACY** During the summer, Zachary works as a lifeguard at the beach. He has noticed that a lot of teens don't wear sunscreen. He wants to create a pamphlet that talks about skin cancer and how it can be prevented. Create a similar pamphlet and hand it out to anyone who enjoys outdoor activities.

🔄 Review
🔊 Audio

Heart *and* Circulatory Problems

BIG IDEA Heart disease is any condition that reduces the strength or function of the heart.

Before You Read

QUICK WRITE Explain three ways to keep your heart healthy.

 Video

As You Read

STUDY ORGANIZER Make the study organizer found in the FL pages in the back of the book to record the information presented in Lesson 3.

Vocabulary

› arteriosclerosis
› atherosclerosis
› heart attack
› hypertension
› stroke
› angioplasty

 Audio

 Bilingual Glossary

WHAT IS HEART DISEASE?

MAIN IDEA Heart disease can be caused by a variety of factors.

Heart disease is any condition that reduces the strength or function of the heart and blood vessels. Heart disease is also known as cardiovascular disease. Common forms of heart disease include high blood pressure, high cholesterol, and atherosclerosis, or hardening of the arteries.

More than 27 million people in the United States have heart disease. More than half a million people per year die from this condition. Sometimes heart disease is due to heredity. However, most heart disease is related to making unhealthy lifestyle choices that increase risk.

More than *27 million* people in the United States have **heart disease.**

Tobacco use is one of the leading causes of heart disease. Other leading causes of heart disease include eating foods high in saturated fat and not getting enough regular physical activity.

Health experts say that the best way to reduce your risk of developing heart disease is to avoid tobacco use, eat healthy foods most of the time, and get regular exercise. They also say that if teens form these habits, they can avoid developing heart disease later in life.

> ## Reading Check
>
> **EXPLAIN** *What factors contribute to heart disease?*

Fitness Zone

Including Physical Activity in My Daily Life. Some days I walk for 30 minutes when I get home from school. After I finish my homework, I play basketball for 30 minutes with my friends. On other days, I jog for 30 minutes. Later, I ride my bicycle to and from my friend's house, which is about 15 minutes away. Getting at least 60 minutes of physical activity a day helps keep my heart and other body systems healthy.

Drew Myers/Corbis

Coronary Heart Disease

Body cells must have a constant supply of fresh oxygen to survive. When the arteries are damaged or blocked, the blood does not flow as well. Two disorders can result from problems with the blood flow through the arteries. Arteriosclerosis *is a group of disorders that cause a thickening and hardening of the arteries.* This means that less blood is able to flow through the arteries. Atherosclerosis is *a condition of arteriosclerosis that occurs when fatty substances build up on the inner lining of the arteries.* The buildup shrinks the space through which blood can travel.

Heart Attack

If the heart does not get enough oxygen, a heart attack is likely. A heart attack is *a serious condition that occurs when the blood supply to the heart slows or stops and the heart muscle is damaged.* If the blood to the heart is cut off for more than a few minutes, the heart muscle cells are damaged and die.

For males, symptoms of a heart attack include pain or pressure in the chest or pain in the upper body including the arms, left shoulder, jaw, back, or abdomen. Males may also be short of breath, experience dizziness and have cold skin. They may vomit, feel tightness in the chest, or pass out.

If the **heart** does not get enough oxygen, a *heart attack* is likely.

Females often have different symptoms of a heart attack than males. Females may experience symptoms including pain in the center of the chest, upper back, shoulder, arms, or jaw, and lightheadedness. Prior to having a heart attack, females also report feeling more tired than usual, feeling anxious, and having indigestion.

Eating nutritious foods can help you maintain a healthy heart. *List nutritious foods you eat every day.*

The heart's muscle tissue gets blood from the coronary arteries. When arteries become blocked, the heart is prevented from getting the blood it needs. *Describe what you think might happen if the heart does not get enough blood.*

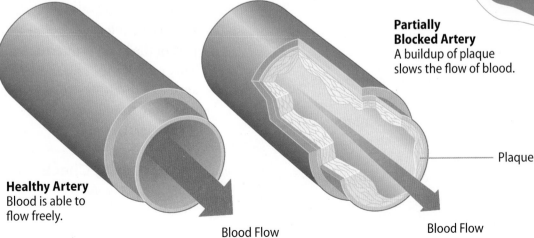

Healthy Artery
Blood is able to flow freely.

Blood Flow

Partially Blocked Artery
A buildup of plaque slows the flow of blood.

Plaque

Blood Flow

Heart-Healthy *Ads*

Teens are surrounded by food ads in magazines, television, radio, newspapers, billboards, and other media. With heart disease receiving so much attention in the media, food producers have begun advertising many foods as "heart healthy."

Find two advertisements in print, on the radio, on television, or online, that advertise heart-healthy foods. Study the ads. How do the ads promote the foods? What language do the ads include? How do you think the ad writers want you to feel about eating the food? Do the ads contain any facts? Are these facts reliable? How can you tell?

Create a video or print advertisement for a delicious, heart-healthy meal. Include creative lettering, magazine or newspaper pictures, and information about the food. Make your ad inviting so teens are persuaded to try your heart-healthy meal.

Other Cardiovascular Problems

One of the most common forms of heart disease is high blood pressure. **Hypertension** is *a condition in which the pressure of the blood on the walls of the blood vessels stays at a level that is higher than normal.* Hypertension can lead to a heart attack or a stroke. People with hypertension can manage the condition by following a healthful eating plan, exercising regularly, avoiding stress, and taking medicine if needed.

> One of the most common forms of *heart disease* is **high blood pressure.**

Like the heart, the brain needs oxygen and nutrients to function. It's possible for an artery to leak or develop a thick mass called a clot. When this happens, a stroke occurs. A **stroke** is *a serious condition that occurs when an artery of the brain breaks or becomes blocked.* During a stroke, blood flow to part of the brain is interrupted. That part of the brain is damaged as a result. A stroke patient may have trouble moving or speaking afterward.

>>> **Reading Check**

IDENTIFY *What are the two causes of a stroke?*

Purestock/SuperStock

TREATING HEART DISEASE

MAIN IDEA Heart disease treatments include angioplasty, medication, pacemakers, surgery, or transplants.

Healthcare advances have led to several treatments for heart disease. However, your goal should be to reduce your risk by practicing a healthy lifestyle. Some of the treatments for heart disease include:

- **Angioplasty** (AN·je·uh·plas·tee) is *a surgical procedure in which an instrument with a tiny balloon, drill bit, or laser attached is inserted into a blocked artery to clear a blockage.* Doctors inflate the tiny balloon. It pushes the blockage up and against the artery wall. Lasers or drill bits burn or cut away the blockage.
- **Medications** can break up blood clots that may block arteries. Sometimes aspirin is prescribed to prevent platelets from forming blood clots. Other medications may be prescribed to help lower cholesterol levels.

Healthcare providers can usually *treat* **heart disease.**

Medication is one way to treat heart disease. *Identify two healthful behaviors you can do now to help prevent heart disease later in life.*

- **Pacemakers** and implantable defibrillators are electronic devices placed inside the chest. A pacemaker sends electrical signals helping the heart beat regularly. An implantable defibrillator starts the heart beating when it has stopped.
- **Bypass surgery** creates new pathways for the blood. When a person has a blocked artery, surgeons take a blood vessel from another part of the body. The use it to make a new route for blood to flow.

Heart transplants completely replace a damaged heart with a new healthy heart. The new heart is from someone who has just died. Heart transplants are complex. They are done only when the heart is severely damaged and no other treatment will work.

>>> **Reading Check**

LIST *What are two ways to treat heart disease?*

Ingram Publishing

PREVENTING HEART DISEASE

MAIN IDEA Getting regular physical activity, eating healthful foods, and staying tobacco free can help prevent heart disease.

Symptoms of heart disease usually do not appear until adulthood. However, heart disease can begin developing in childhood. Making healthy choices today can lower your risk of developing heart disease when you are older. Strategies for keeping your heart healthy include:

- **Eat healthful foods.** Choose plenty of fresh fruits and vegetables, whole grains, and lean sources of protein.
- **Limit the amount of cholesterol, saturated fats, and trans fats that you eat.** Foods high in these are linked to cardiovascular disease.
- **Participate in regular physical activity.** Regular physical activity makes your heart stronger.
- **Maintain a healthy weight.** Your heart works best if your weight is within a healthy range. Talk to a healthcare provider about the range that is best for you.

Making healthy choices today can **lower your risk** of developing **heart disease.**

- **Manage stress.** Learning to relax will help you keep your blood pressure within a healthy range.
- **Stay tobacco free.** Chemicals in tobacco can cause heart disease, heart attacks, hypertension, and strokes. Staying tobacco free will help you avoid all of these problems.
- **Stay alcohol free.** Alcohol has been linked to high blood pressure and heart failure. Staying alcohol free helps you avoid these problems. ■

Regular physical activity strengthens the heart. It can help people with hypertension reduce their need for medication. *Name three types of physical activities that strengthen the heart.*

LESSON 3

REVIEW

⟫⟫ After You Read

1. **VOCABULARY** Define arteriosclerosis and atherosclerosis.
2. **EXPLAIN** What is angioplasty?
3. **DESCRIBE** How do the symptoms of heart attacks in females differ from those in males?

⟫⟫ Thinking Critically

4. **ANALYZE** Why is it a good idea to have your blood pressure checked regularly by a healthcare professional?
5. **SYNTHESIZE** Why do you think that medical professionals focus on preventing heart disease even though there are so many treatments for it?

⟫⟫ Applying Health Skills

6. **PRACTICING HEALTHFUL BEHAVIORS** Both of Madison's grandmothers died of heart disease. Madison wants to reduce her own risk of heart disease. Make a list of positive health behaviors Madison can do to reduce her risk of heart disease.

 Review

🔊 Audio

Exactostock/SuperStock

Diabetes *and* Arthritis

BIG IDEA Diabetes and arthritis can be treated medically and managed by making healthy lifestyle choices.

Before You Read

QUICK WRITE Name two facts you already know about diabetes.

▶ **Video**

As You Read

STUDY ORGANIZER Make the study organizer found in the FL pages in the back of the book to record the information presented in Lesson 4.

Vocabulary

› diabetes
› insulin
› type 1 diabetes
› type 2 diabetes
› arthritis
› osteoarthritis
› rheumatoid arthritis

🔊 **Audio**

🔤 **Bilingual Glossary**

Developing Good Character

Citizenship Part of being a good citizen is helping others when they need it. Volunteering is a great way to help people in your community. Research more about volunteer opportunities at organizations such as the American Diabetes Association and the Arthritis Foundation.

WHAT IS DIABETES?

MAIN IDEA Diabetes affects children and adults and can be managed with medical treatments and healthful behaviors.

Diabetes mellitus (dy·uh·BEE·teez MEH·luh·tuhs), or diabetes, is a noncommunicable disease that affects more than 25 million adults and children in the United States. **Diabetes** is *a disease that prevents the body from converting food into energy.* When you eat, your body breaks down the food to get the energy it contains. It turns food into a form of sugar called glucose. After your body digests food, glucose levels in the bloodstream rise. Glucose begins to enter cells with the help of a hormone called insulin (IN·suh·lin). **Insulin** is *a protein made in the pancreas that regulates the level of glucose in the blood.* Your cells use the glucose for energy. You need energy to help you get through school each day, to play sports, and to complete all of your other daily activities.

Diabetes affects more than **25 million** adults and children in the United States.

Some people who have diabetes do not have enough natural insulin. As a result, glucose cannot get into cells. Other people with diabetes make enough insulin, but the insulin does not do its job properly. When this occurs, the person feels the symptoms of diabetes. These symptoms include feeling tired, losing weight, excessive thirst, and increased urination. If left unmanaged, diabetes can cause other long-term health problems such as kidney disorders, blindness, and heart disease.

››› Reading Check

EXPLAIN *What is insulin?*

Healthful Eating Plan	A healthful eating plan can help keep blood glucose levels within a normal range.
Weight Management	Regular physical activity helps people with diabetes maintain a healthy weight.
Insulin	People with type 1 diabetes and some people with type 2 diabetes receive insulin through a syringe or a pump.
Medical Care	People with diabetes need to be under the care of a medical professional.

Types *of* Diabetes

Diabetes consists of two main types: type 1 and type 2. <u>Type 1 diabetes</u> is *a condition in which the immune system attacks insulin-producing cells in the pancreas.* When the cells that produce insulin are killed, the body cannot control how much glucose is in the bloodstream. Type 1 diabetes often starts in childhood. It may also begin in adulthood.

    <u>Type 2 diabetes</u> is *a condition in which the body cannot effectively use the insulin it produces.*

In the past, type 2 diabetes typically began in adulthood. Type 2 diabetes is closely linked to a lack of physical activity and being overweight. Type 2 diabetes is also becoming more common among children and teens.

Managing Diabetes

People with diabetes must learn how to manage the disease. People with type 1 diabetes usually need to have injections of insulin. They might receive insulin from a pump attached to their bodies.

People who have diabetes must manage the condition carefully. *Describe some of the dangers of untreated diabetes.*

People who have type 2 diabetes may also need insulin or other medications. Many of them, however, can control their disease by practicing healthy habits.

⟫⟫ Reading Check

IDENTIFY *What is the most common form of diabetes?*

WHAT IS ARTHRITIS?

MAIN IDEA ⟩ Arthritis is a disease marked by painful swelling and stiffness of the joints.

Arthritis (ar·THRY·tus) is *a disease of the joints marked by pain and swelling in body joints.* You may think of arthritis as a disease that strikes older adults, but children can develop it too. Two main types of arthritis include: osteoarthritis (ahs·tee·oh·ar·THRY·tus) and rheumatoid (ROO·muh·toyd) arthritis. Sometimes arthritis lasts a short period of time. Sometimes it is chronic. Chronic arthritis affects a person most days. When rheumatoid arthritis affects a young person, it's called juvenile rheumatoid arthritis (JRA).

You may think of *arthritis* as a disease that strikes older adults, but **children** can develop it too.

Realistic Reflections

Types of Arthritis

Osteoarthritis is the most common form of arthritis. Osteoarthritis is *a chronic disease that is characterized by the breakdown of the cartilage in joints.* The hard, slippery tissue in the joints between the bones is called cartilage. When cartilage in a joint wears down, the bones in the joints rub against each other. This rubbing causes pain, swelling, and stiffness. Risk factors include age, genetic factors, and being overweight.

Rheumatoid arthritis is *a chronic disease characterized by pain, inflammation, swelling, and stiffness of the joints.* It is usually more serious and disabling than osteoarthritis. People develop rheumatoid arthritis when their immune systems attack healthy joint tissue. These attacks damage joint tissue and cause painful swelling. Symptoms include soreness, joint stiffness and pain, aches, and fatigue.

Osteoarthritis is the most common form of arthritis.

Juvenile rheumatoid arthritis (JRA) is the most common form of arthritis in young people. JRA appears most often in young people between the ages of 6 months and 16 years. Early symptoms include swelling and pain in the joints. Joints may be red and warm to the touch. Sometimes the symptoms of JRA diminish after puberty.

Managing Arthritis

People with arthritis can develop a plan to reduce the symptoms. Many plans involve a combination of the following:

- **Physical activity and rest** People with arthritis suffer less if they balance rest with low-impact physical activity.
- **Healthy eating and maintaining a healthy weight** Maintaining a healthy weight reduces stress on arthritic joints in the knees and feet.
- **Joint protection** People can wear braces and splints to support arthritic joints.
- **Heat and cold treatments** Hot baths ease the pain of some kinds of arthritis. Cold treatments can help reduce the swelling.
- **Medication** Medicine can help slow the progress of some kinds of arthritis. Some medicines can also help ease arthritic joints.
- **Massage** A trained massage therapist can help some arthritis patients by gently massaging affected areas. This helps to relax the joints and increase blood flow.
- **Surgery and joint replacement** In some cases, surgeons can operate to repair a joint or correct its position. They may replace the damaged joint with an artificial one.

> **⟩⟩⟩ Reading Check**
>
> **EXPLAIN** *How do physical activity and rest help people who have arthritis manage the disease?* ■

LESSON 4

REVIEW

⟩⟩⟩ After You Read

1. **DEFINE** Define the terms *diabetes* and *arthritis*.
2. **DESCRIBE** What practices can people with type 2 diabetes use to manage their disease without medication?
3. **DESCRIBE** What happens to a person's joints when osteoarthritis develops?

⟩⟩⟩ Thinking Critically

4. **EVALUATE** How are type 1 diabetes and rheumatoid arthritis similar?
5. **SYNTHESIZE** Based on what you know about arthritis, how can you help someone with arthritis manage the disease?

⟩⟩⟩ Applying Health Skills

6. **PRACTICING HEALTHFUL BEHAVIORS** Participating in regular physical activity can help reduce your risk of developing type 2 diabetes. List ten physical activities you enjoy or might enjoy doing. Write a checkmark beside three activities you will participate in during the next month. Make a commitment to do each of these activities at least once a week.

ⓒ Review

🔊 Audio

Allergies *and* Asthma

> **BIG IDEA** Allergies and asthma are two kinds of noncommunicable diseases that can be managed with medicine and by avoiding allergens.

>> **Before You Read**

QUICK WRITE List three allergy symptoms. How can they be managed?

 Video

>> **As You Read**

STUDY ORGANIZER Make the study organizer found in the FL pages in the back of the book to record the information presented in lesson 5.

>> **Vocabulary**

> allergy
> allergens
> pollen
> histamines
> hives
> antihistamines
> asthma
> bronchodilator

 Audio

 Bilingual Glossary

Myth or Fact

Myth: People with asthma cannot participate in physical activity or sports.
Fact: Many people with asthma participate in a range of physical activities including jogging, hiking, swimming, and mountain biking. People with asthma are members of sports teams, such as basketball, swimming, and track and field. The key to participating in physical activity is to develop and follow a proper asthma management plan, which can help prevent asthma attacks.

WHAT ARE ALLERGIES?

> **MAIN IDEA** An allergy is an extreme sensitivity to a substance or allergen, such as pollen, insect bites or stings, or food.

Your immune system keeps you healthy as it helps your body fight off foreign substances. However, some people's immune systems react to seemingly harmless substances. These reactions are allergic responses. Between 40 million and 50 million Americans have some type of allergy. An **allergy** is *an extreme sensitivity to a substance.*

A person can be allergic to many different substances in the environment. **Allergens** are *a substances that cause allergic responses.* Allergens are very small. For example, people who are allergic to ragweed are allergic to the tiny pollen grains from the ragweed plant.

> Some people's *immune systems* react to fairly **harmless substances.**

Pollen is *a powdery substance released by the flowers of some plants.* It can be carried long distances in the air and easily inhaled. When inhaled, pollen causes an allergic reaction in a person who is allergic to it.

Many people have no allergies, and some people have minor allergies. Minor allergies may cause symptoms but are not severe enough to make the person with the allergy uncomfortable. When allergies are severe, however, a person's daily life can be affected.

>> **Reading Check**

EXPLAIN *What is an allergen?*

Jamie Grill/Getty Images

Allergic Reactions

When you are allergic to something, your immune system reacts quickly. In order to protect your body, your immune system produces antibodies to the allergen. Antibodies are a special kind of protein that locks onto cells. Antibodies cause certain cells in the body to release histamines (HIS·tuh·meenz). Histamines are *chemicals that the immune cells release to draw more blood and lymph to the area affected by the allergen.* When you are exposed to the same allergen again, the same antibody response will occur. You'll have an allergic reaction every time you come into contact with that allergen.

Allergic reactions can range from minor irritations to severe problems. For example, some people may have sneezing or a runny nose. Some people may get hives as part of an allergic reaction. Hives are *raised bumps on the skin that are very itchy.*

Some people are at risk for severe allergic reactions, including throat swelling or extreme difficulty breathing. These people may need to carry medicine called epinephrine (eh·pin·EFF·rihn). Epinephrine slows down or stops the allergic reaction by preventing the body from releasing histamines.

>>> **Reading Check**

LIST *What are three ways that the body responds to allergens?*

Managing Allergies

Although no cure exists for allergies, they can be managed in three basic ways.

- **Avoid the allergen.**
  If you have a food allergy, check the ingredient labels on food products. People with allergies to peanuts or other nuts need to be especially careful about what they eat.
- **Take medication.** People with allergies often take medicines called antihistamines, or *medicines that reduce the production of histamines.*
- **Get injections.** Injections contain a tiny amount of the allergen. They can gradually desensitize the immune system to the allergen.

>>> **Reading Check**

DESCRIBE *What are three ways to manage allergies?*

Common allergic reactions vary depending on the allergen. *Evaluate which of the allergic reactions listed here you think to be the most serious.*

Eyes can be red, watery, and itchy.

Nose can be runny and irritated. Sneezing is common.

Throat can become irritated and swollen. With severe swelling the throat can close shut.

Respiratory system can become irritated. May lead to coughing and difficulty breathing.

Digestive system can be upset. Cramping, stomach pains, and diarrhea are common.

Skin can become irritated and break out in a rash or hives.

RubberBall Productions

WHAT IS ASTHMA?

MAIN IDEA Asthma occurs when the airways in the lungs narrow, making it difficult to breathe.

Asthma is *a chronic inflammatory disorder of the airways that causes air passages to become narrow or blocked, making breathing difficult.* It is a noncommunicable disease that is a growing problem in many countries. In the United States, about 24 million people are reported to have asthma. More than 6 million of these asthma sufferers are under the age of 18.

Many substances and conditions can cause an asthma attack. Common triggers include:
- Allergens such as mold, dust, pollen, and pets.
- Physical activity.
- Air pollutants such as paint and gas fumes, cigarette smoke, industrial smoke, and smog.
- Respiratory infections, such as colds or the flu.
- Weather changes, especially dramatic changes.
- Rapid breathing which often happens under stress, when laughing, or when crying.

What are the symptoms of an asthma attack? A person may wheeze, cough, or feel short of breath during an attack. Symptoms can also include tightness or fullness in the chest.

>>> **Reading Check**

RECALL *Name three common asthma triggers.*

In the United States, about **24 million** people are reported to **have asthma.**

An asthma attack makes breathing more difficult. Airways become narrower and clogged with mucus. *Describe some symptoms of an asthma attack.*

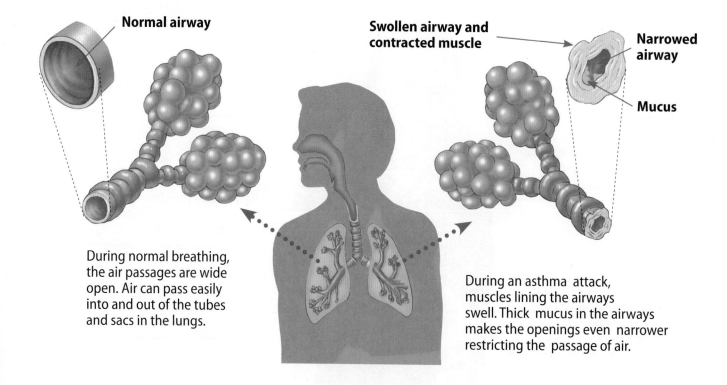

Normal airway

Swollen airway and contracted muscle

Narrowed airway

Mucus

During normal breathing, the air passages are wide open. Air can pass easily into and out of the tubes and sacs in the lungs.

During an asthma attack, muscles lining the airways swell. Thick mucus in the airways makes the openings even narrower restricting the passage of air.

Managing Asthma

Since there is no cure for asthma, people with the disease must learn to manage it and lead active lives. Managing asthma is often a team effort. Parents or trusted guardians, healthcare professionals, and friends can help people with asthma stay healthy. Strategies that people with asthma can use to help avoid asthma attacks include:

- **Monitoring the condition.** People with asthma must pay attention to the early signs of an attack. That way, they can act quickly if they sense an attack coming. They also can track their long-term lung capacity, or ability to take in air. An airflow meter measures lung capacity. When used regularly, an airflow meter helps people know when their airways are narrowing.
- **Managing the environment.** For example, if dust and mold trigger your asthma, reduce them in your environment. It helps to keep floors, bedding, and pets clean.
- **Managing stress.** Stress is a major cause of asthma attacks. Panicking during an attack can make it even worse.

Relaxing and staying calm can help those with asthma avoid attacks. Relaxing will help even during an attack.

- **Taking medication.** Two kinds of medicines can treat asthma: relievers and controllers. Relievers, such as bronchodilators (brahng·koh·DY·lay·turhz) help reduce symptoms during an asthma attack. A Bronchodilator is *a medication that relaxes the muscles around the air passages.* People usually use an inhaler to take a bronchodilator. Controller medicines are taken daily, and help prevent attacks by making airways less sensitive to asthma triggers.

⟩⟩⟩ Reading Check

DESCRIBE *How can a person with asthma manage his or her condition?* ■

Many people with asthma carry inhalers. These medicines ease breathing. *Describe other strategies people can use to avoid an asthma attack.*

©moodboard/Corbis

⟩⟩⟩ After You Read

1. **VOCABULARY** Define allergy and asthma.
2. **IDENTIFY** Name four common types of allergens.
3. **DESCRIBE** What are the symptoms of an asthma attack?

⟩⟩⟩ Thinking Critically

4. **HYPOTHESIZE** Why is it important for people with food allergies to be careful when eating out?
5. **SYNTHESIZE** If your friend is having an asthma attack, what are two ways you could help?

⟩⟩⟩ Applying Health Skills

6. **COMMUNICATION SKILLS** Samantha tried out for the basketball team and made it! She has asthma, however, and is too embarrassed to tell anyone about it. Write a short letter to Samantha and tell her why you think she should tell her coach and teammates that she has asthma. How can she manage her asthma during practice and games?

 Review

🔊 Audio

Hands-On HEALTH ACTIVITY

Determining Lung Capacity

WHAT YOU WILL NEED

* Round balloon for every student
* String
* Ruler
* Graph paper
* Strip of paper on the wall that is longer than the tallest student in class and marked in inches

WHAT YOU WILL DO

1 With a partner, stretch the balloon several times to loosen the rubber. Take a deep breath and then blow up the balloon as far as you can.

2 Your partner will wrap the string around the widest part of the balloon, then measure the length of string with a ruler. This measurement is the balloon's diameter. Record this number.

3 Switch places with your partner and repeat steps 2 and 3. Go to the measuring strip on the wall and determine your height in inches.

4 Make a graph on a sheet of paper. Mark the vertical (y) axis in inches or centimeters for balloon diameter. Mark the horizontal (x) axis in inches or centimeters for height. Plot the results from steps 2 and 3 on the graph.

In this activity, you will work with a partner to measure the air capacity of your lungs, or their ability to take in air. Then you will graph the results so that you can see how air capacity is related to body size.

WRAPPING IT UP

The average adult is able to exhale about 4.5 liters, or 8.4 pints, of air. A well-trained athlete may be able to exhale 6.5 liters, or 12 pints, of air. Did you find a relationship between height and lung capacity? How would smoking affect lung capacity? How do you think lung capacity changes during an asthma attack?

READING REVIEW

FOLDABLES and Other Study Aids

Take out the Foldable® that you created and any study organizers that you created.
Find a partner and quiz each other using these study aids.

LESSON 1 Causes of Noncommunicable Diseases

BIG IDEA Noncommunicable diseases can result from heredity or lifestyle choices, or may have an unknown cause.

* Noncommunicable diseases are diseases that cannot be spread from person to person.
* Risk factors for noncommunicable diseases include heredity, lifestyle choices, and environmental factors.

LESSON 2 Cancer

BIG IDEA Cancer occurs when abnormal cells multiply out of control.

* Cancer is characterized by the rapid and uncontrolled growth of abnormal cells.
* Health care professionals use a variety of methods to detect and diagnose cancer.
* Three common cancer treatments are surgery, radiation therapy, and chemotherapy.
* The risk of cancer can be reduced by practicing certain healthful behaviors.

LESSON 3 Heart and Circulatory Problems

BIG IDEA Heart disease is any condition that reduces the strength or function of the heart.

* Heart disease can be caused by a variety of factors.
* Heart disease treatments include angioplasty, medication, pacemakers, surgery, or transplants.
* Getting regular physical activity, eating healthful foods, and staying tobacco free. can help prevent heart disease.

LESSON 4 Diabetes and Arthritis

BIG IDEA Diabetes and arthritis can be treated medically and managed by making healthful lifestyle choices.

* Diabetes affects children and adults and can be managed with medical treatments and practicing healthful behaviors.
* Arthritis is a disease marked by painful swelling and stiffness of the joints.

LESSON 5 Allergies and Asthma

BIG IDEA Allergies and asthma are two kinds of noncommunicable diseases that can be managed with medicine and by avoiding allergens.

* An allergy is an extreme sensitivity to a substance or allergen, such as pollen, insect bites or stings, or food.
* Asthma occurs when the airways in the lungs narrow, making it difficult to breathe.

 Review

 Web Quest

ASSESSMENT

Reviewing Vocabulary *and* Main Ideas

> atherosclerosis
> congenital disorders

> chronic
> hypertension

> heart attack
> benign

> risk factors
> tumor

> noncommunicable disease

> degenerative diseases

> stroke

>> On a sheet of paper, write the numbers 1–8. After each number, write the term from the list that best completes each statement.

LESSON 1 Causes of Noncommunicable Diseases

1. A _____ disease is one that is present on and off over a period of time.

2. Diseases that cause further breakdown in body cells, tissues, and organs as they progress are called _____.

3. _____ are all disorders that are present when the baby is born.

4. A _____ is a disease that cannot be spread from person to person.

LESSON 2 Cancer

5. A _____ is a clump of abnormal cells.

6. A tumor that is not cancerous is _____.

7. A _____ is a substance that causes cancer.

8. _____ are characteristics or behaviors that increase the likelihood of developing a medical disorder or disease.

LESSON 3 Heart Disease and Circulatory Problems

9. A _____ can happen if the heart does not get enough oxygen-rich blood.

10. A condition in which a fatty substance builds up on the inner lining of the arteries is called _____.

11. When someone's blood pressure stays at a level that is higher than normal, that person has _____.

12. A _____ is a serious condition that occurs when an artery of the brain breaks or becomes blocked.

>> On a sheet of paper, write the numbers 9–15. Write *True* or *False* for each statement below. If the statement is false, change the underlined word or phrase to make it true.

LESSON 4 Diabetes and Arthritis

13. <u>Arthritis</u> is a disease that prevents the body from converting food into energy.

14. <u>Type 2</u> diabetes is the most common form of the disease.

15. <u>Being active</u> helps control diabetes.

LESSON 5 Allergies and Asthma

16. <u>Histamines</u> cause allergic reactions in people.

17. <u>Asthma</u> is a condition that makes breathing difficult.

18. <u>Antihistamines</u> are medications used to treat asthma.

eAssessment

 Using complete sentences, answer the following questions on a sheet of paper.

Thinking Critically

19. APPLY Why might playing the piano help people with arthritis in their hands?

20. HYPOTHESIZE Why do people with pollen allergies often have fewer symptoms in winter?

Write About It

21. EXPOSITORY WRITING Write an article for your school newspaper about the connection between weight and arthritis symptoms. Be sure that your article explains how a person's weight can affect the joints in the knees and feet. Explain the benefits of maintaining a healthy weight.

STANDARDIZED TEST PRACTICE

Math

Acute lymphoblastic leukemia (A.L.L.) is a serious disease that affects the white blood cells in the body. This table shows the survival rates of patients with A.L.L.

Use the table to answer the questions.

Survival Rates of A.L.L. Patients

	% of Patients Surviving at Least 5 Years
Adults	50%
Children	85%

1. Which is a *true* statement about the disease, according to the table?
   A. Fifty percent of adults don't survive.
   B. More adults survive than children.
   C. Ninety percent of children survive.
   D. Twenty percent of children don't survive.

2. By reading this chart, you can determine
   A. the survival rate for adults in ten years.
   B. that most children who have this disease survive for at least five years.
   C. the survival rate for children in ten years.
   D. that all adults who have this disease survive for at least five years.

Unit 10

safety + a healthy environment

Safety

LESSONS

PREMIUM ONLINE RESOURCES

 Audio

 Videos

 Bilingual Glossary

 Fitness Zone

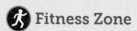 Web Quest

 Review

Performing activities safely lowers the risk of accidents. *Describe two ways you can avoid accidents.*

Building Safe Habits

BIG IDEA Being safety conscious means being aware that safety is important and acting safely.

Before You Read

QUICK WRITE List three examples of accidental injuries that people experience. After reading the lesson, check to see whether you were correct.

 Video

As You Read

FOLDABLES Study Organizer

Make the Foldable® found in the FL pages in the back of the book to record the information presented in Lesson 1.

Vocabulary

› accident
› accidental injuries
› safety conscious
› hazard
› accident chain

 Audio

 Bilingual Glossary

Developing Good Character

Responsibility When you put your belongings in their proper place, they are not in the way, so they're less likely to cause accidents. Putting away clothes and equipment also helps cut down on clutter. *Describe what else you can do to help prevent accidents.*

SAFETY FIRST

MAIN IDEA Accidents and accidental injuries can affect people of all ages.

You've probably heard warnings such as, "Buckle up!" and "Look both ways before you cross the street," for as long as you can remember. You might have helped teach these safe habits to a younger brother or sister. An **accident** is *any event that was not intended to happen.* Accidents do happen, but you can prevent many of them. When you stay safe and avoid accidents, you help yourself and those around you stay healthy.

You might not think serious accidents can happen to you. However, the Centers for Disease Control and Prevention (CDC) reports that accidental injuries are the leading cause of death in teens.

Accidental injuries are *injuries resulting from an accident.* Accidental injuries are the fifth leading cause of death in the United States across all age groups.

When you *stay safe* and *avoid accidents,* you **help yourself** and those around you *stay healthy.*

The leading causes of nonfatal accidental injuries include falls, being struck or cut by something, being bitten or stung, overexertion, and poisoning.

The first step in staying safe is being safety conscious. To be **safety-conscious** means *being aware that safety is important and careful to act in a safe manner.* It is easier to prevent injuries than to treat them. Think ahead. Know how to spot a **hazard** or *potential source of danger.*

Reading Check

EXPLAIN *What is the difference between an accident and an accidental injury?*

Fuse/Getty Images

HOW ACCIDENTS HAPPEN

MAIN IDEA For any accident to occur, three elements must be present.

Accidents happen when people stop being safety conscious. Think back to your last accident and the accident chain that led up to it. An accident chain is *a series of events that include a situation, an unsafe habit, and an unsafe action.* For any accident to occur, these three elements must be present.

Breaking *the* Accident Chain

Look at the links in Tony's accident chain below. Breaking just one link would have kept Tony from being injured.

> Many accidents can be avoided. *How could Tony avoid this accident?*

- **Change the Situation.** Tony could have gotten up earlier. He could have set his alarm for a reasonable time. He could have asked a family member to wake him if he overslept.
- **Change the Unsafe Habit.** Tony may have avoided the accident by storing his books on a bookshelf or placing them in his book bag.
- **Change the Unsafe Action.** Tony could have paid attention to where he was going. He could have slowed down and watched his step.

Being safety conscious might have kept Tony from tripping and falling. By changing the situation, the unsafe habit, or the unsafe action, Tony could have prevented his accident.

1 **The Situation**
Tony has overslept. He wakes up in a panic. The bus is coming in 15 minutes.

3 **The Unsafe Action**
Without looking where he is going, Tony runs to the bathroom to wash up.

5 **The Result**
When Tony falls, he sprains his wrist. He misses his bus and is in a lot of pain.

2 **The Unsafe Habit**
Tony didn't put his books away from the night before. He just left them on the floor.

4 **The Accident**
Tony trips over his books and falls down.

After You Read

1. **VOCABULARY** Define *accident*. Use the word in an original sentence.
2. **IDENTIFY** What are the five links in the accident chain?
3. **GIVE EXAMPLES** What are three ways to break the accident chain?

Thinking Critically

4. **APPLY** Benny has always had a bookshelf on the wall next to his bed. Now that he is taller, the bookshelf has become a problem. In fact, this year Benny has bumped his head on the shelf three times. What should he do to be safer?
5. **ANALYZE** Grant's friend dared him to walk across a narrow 12-foot-high fence. What should Grant do, and why?

Applying Health Skills

6. **DECISION MAKING** Tina wants to go bike riding with a friend, but she left her helmet in her dad's truck. What are Tina's options? Use decision making skills to help Tina make a safe decision.

 Review

 Audio

Safety *at* Home *and* School

BIG IDEA Following safety rules can keep you safe both at home and away from home.

⟫⟫ Before You Read

QUICK WRITE What is the first thing that you should do if you think the building you are in is on fire? Write a sentence or two describing the action you would take.

 Video

⟫⟫ As You Read

STUDY ORGANIZER Make the study organizer found in the FL pages in the back of the book to record the information presented in Lesson 2.

⟫⟫ Vocabulary

› flammable
› electrical overload
› smoke alarm
› fire extinguisher

 Audio

 Bilingual Glossary

Myth vs. Fact

Myth: Small fires are not a major concern and can be controlled.

Fact: A fire of any size needs immediate attention. Small fires can become uncontrollable fires very quickly. A fire that starts with material burning in a wastebasket in the room can spread to two rooms within four minutes. Smoke alarms and sprinkler systems play an important role in fire safety. In homes with smoke alarms, the risk of dying in a fire is reduced by 50 percent compared to homes without working smoke alarms. In homes with both smoke alarms and a sprinkler system, the risk of death is reduced by 82 percent.

SAFETY IN THE HOME

MAIN IDEA Your home may be filled with many potential safety hazards, such as stairs or appliances.

Jessica and her family are moving into a new home. As Jessica looks around her new room on the second floor, her father says, "We'll need to store an emergency ladder in your room." He then added, "This window is your emergency exit in case of fire."

Home is a place where everyone should feel safe and comfortable. Yet, homes can contain hazards. Stairways or spilled water can lead to falls, the most common type of home injury. Appliances can cause electrical shocks.

> *Home* is a place where everyone should **feel safe** and *comfortable.*

Sharp tools in the kitchen or garage, such as knives, saws, and screwdrivers can lead to cuts. Other injuries can result from poisonings, choking, drowning, and guns. Fires are the third leading cause of unintentional injury and death in the home. Overall, about 45 percent of deaths from accidental injuries occur in and around the home.

However, injuries from these hazards can be avoided. Following safety rules can reduce the risks of home hazards.

⟫⟫ Reading Check

EXPLAIN *Why is it important to follow safety rules at home?*

Fuse/Getty Images

Preventing Falls

Most falls in the home occur in the kitchen, the bathroom, or on the stairs. These safety rules can help you prevent falls.

KITCHEN SAFETY Clean up spills right away. Use a stepstool, not a chair, to get items that are out of reach. Avoid running on wet or waxed floors.

BATHROOM SAFETY Put a nonskid mat near the tub or shower. Use rugs that have a rubber backing to prevent the rug from slipping. Keep personal products in plastic bottles.

STAIRCASE SAFETY Keep staircases well lit and clear of all objects. Apply nonslip treads to slippery stairs. Make sure handrails are secure and stable. If small children live in the house, put gates at the top and bottom of the stairs.

Preventing Poisonings

Many common household products are poisonous. Items such as cleaning products, insecticides, medications, and vitamins can be harmful if not used or taken as directed. Accidental poisoning can occur by swallowing, absorbing through the skin, injection from a needle, or breathing poisonous fumes.

To help keep the people in your home safe from poisoning, make sure that labels on containers of household products are clearly marked. Store cleaning products, insecticides, and other potential poisons out of the reach of young children.

Never refer to a child's medicine or vitamins as candy. Be sure that all medicines are in bottles with childproof caps. Always use a product or medication as directed by the label.

Picking up toys that are left on steps can help prevent falls. *Describe other ways you can help prevent accidents in the home.*

Make sure that *labels* on containers of household products are **clearly marked.**

Gun Safety

In many states it is illegal for most teens to own a gun. If you find a gun, do not touch it. Call a parent, guardian, or other trusted adult immediately. If someone at school is carrying a gun or other weapon, tell an adult or school official.

The best way to prevent a gun accident in the home is to not have guns in the home. If a gun must be kept in the home, use the following safety precautions:

- Guns should have trigger locks and be stored unloaded in a locked cabinet.
- Ammunition should be stored in a separate locked cabinet.
- Any person who handles a gun should be trained in gun safety.
- All guns should be handled as if they are loaded.
- Avoid playing with a gun or pointing it at someone.

Electrical Safety

Improper use of electrical appliances or outlets can cause dangerous electrical shocks. To prevent electrical shocks, never use an electrical appliance around water or if you are wet. Unplug small appliances such as hair dryers when they are not in use. To unplug electrical appliances, gently pull the plug, not the cord. If a cord becomes frayed, don't use the appliance until it is repaired. Unplug any appliance that is not working properly. Avoid running cords under rugs. In homes with small children, cover unused outlets with plastic outlet protectors.

⟫⟫ Reading Check

LIST *What are two ways to prevent electrical shocks?*

Adam Gault/OJO Images/Getty Images

FIRE SAFETY

MAIN IDEA It is a healthy practice to have a fire safety plan, fire extinguisher, and working smoke alarms.

Fires happen in about 370,000 homes in the United States each year, killing approximately 2,800 people. The kitchen is where most fires start. A fire needs three elements to start—fuel, heat, and air. Sources of heat include cigarettes, matches, or electrical wires. If a heat source comes into contact with anything **flammable,** *substances that catch fire easily,* a fire can result. Household chemicals, rags, wood, or newspapers and magazines are all examples of flammable materials.

Other fires start from **electrical overload,** *a dangerous situation in which too much electric current flows along a single circuit.* Here are some safety guidelines to help you prevent home fires:

- Keep stoves and ovens clean. This will prevent food or grease from catching fire.
- Keep objects that can catch fire easily at least three feet away from portable heaters.
- Remind adults who smoke that they should never smoke in bed or on overstuffed furniture. Remind smokers to never toss a lit cigarette into the trash.

- Regularly inspect electrical wires, outlets, and appliances to make sure they are in proper working order. Never pull on the cord to unplug an appliance. Never run cords under rugs or carpets. If you notice worn or shredded cords, stop using the cord and tell an adult about it.

Fires happen in about **370,000 homes** in the United States each year.

- Discard old newspapers, oily rags, and other materials that burn easily.
- Use and store matches and lighters properly. Keep them out of the reach of young children. Don't leave candles burning unattended.

Careless cooking
Spattered grease and oil can cause kitchen fires. Unattended cooking pots can spill onto burners or in the oven.

Careless smoking
Cigarettes can start fires if people leave them unattended or fall asleep while they are still burning. Cigarettes can also start fires if people toss them into the trash when they are still burning.

Incorrect storage of flammable materials
Examples of flammable materials are paint, chemicals, oil, rags, and newspapers.

Damaged electrical systems or electrical overload
Fires can start due to too much current flowing through overloaded circuits. Shredded wires or torn cords can also lead to fires. Broken appliances can cause fires as well.

Gas leaks
Gas lines can leak and catch fire. Natural gas is odorless and colorless, so it has an additive that makes it smell. If you smell gas, first get out of the house, then call 911.

 Reading Check

RECALL *What are three safety guidelines to help prevent fires?*

Almost all people killed in fires are children and older adults. *Explain why children and older adults are the most likely to die in house fires.*

Ingram Publishing

Being Prepared *in* Case *of* Fire

The earlier you receive warning of a fire, the better your chances of getting out safely. Every level of a house should have smoke alarms. A smoke alarm is *a device that sounds an alarm when it senses smoke.* Smoke alarms are especially useful when you are sleeping. Install them as close to bedrooms as possible. Test smoke alarms every month. Replace batteries once a year.

Water will put out fires in which paper, wood, or cloth is burning. However, you should never use water to put out a fire that involves grease, oil, or electricity. That will actually make the fire worse. Instead, use a fire extinguisher, *a device that sprays chemicals that put out fires.* Every home should have a fire extinguisher. Read the directions, and make sure that you know how to use it properly.

Create a fire escape plan with your family. Choose a meeting point outside where everyone can gather in the event of a fire. Practice the escape plan with your family every six months.

If you are in a fire, you need to know what to do to escape safely. *Explain why you should get out first and then call 911.*

1. If possible, leave quickly. Get out of the building before calling 911 or the fire department.

2. Before opening a closed door, feel it to see if it is hot. If it is hot, do not open it. There may be flames just outside the door.

3. If you must exit through smoke, crawl along the floor. Smoke and hot air rise, so it is important to stay as low as possible. The air you breathe will be cleaner. The smoke will not be as likely to overcome you.

4. If you can't get out, stay in the room with the door closed. Roll up a blanket or towel and put it across the bottom of the door to keep out smoke. If there is a telephone in the room, call 911 or the fire department. If possible, open the window and yell for help.

5. If your clothing catches fire, stop, drop, and roll. Rolling on the ground will smother the flames. Never run; the rush of air will fan the flames.

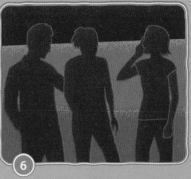

6. Once outside, go to the prearranged meeting point. Let everyone know that you are safe. Then someone should call 911 or the fire department. Never go back into a burning building.

SAFETY AT SCHOOL

MAIN IDEA Following rules at school helps you, other students, and teachers stay safe.

Your school probably has rules in place to keep students and teachers safe. Many accidental injuries at school can be avoided. Safety rules that protect the health and safety of students and teachers, may suggest that you:

Play by the rules. Rules are made to protect you and others. The cafeteria, classrooms, halls, gym, and auditorium may all have a separate set of important safety rules to follow.

Report weapons or unsafe activities. It's essential to follow rules prohibiting weapon possession at school. If you think that someone has brought a gun or other weapon to school, report the person immediately to a teacher or principal.

Wear necessary safety gear. Working in a science lab or playing sports are two places where appropriate gear will help keep you safe.

> ### ❯❯❯ Reading Check
> **RECALL** *What strategies can you and your peers use to stay safe at school?* ■

Some schools have security guards patrolling campus to help keep students safe. *Discuss what you can do to improve safety at school.*

❯❯❯ After You Read

1. **VOCABULARY** Define the terms *smoke alarm* and *fire extinguisher*.
2. **IDENTIFY** List three strategies for preventing poisoning.
3. **DESCRIBE** How can you be prepared for a fire that might happen in your home?

❯❯❯ Thinking Critically

4. **APPLY** In what ways is a cluttered room a hazard?
5. **ANALYZE** Why is it a bad idea to call medicine "candy" to get children to take it?

❯❯❯ Applying Health Skills

6. **REFUSAL SKILLS** Will wants to see the hunting rifle that Troy's dad just bought. The rifle is in a locked case, but Troy knows where the key is. How could Troy refuse Will's request?

🔄 Review

🔊 Audio

Realistic Reflections

Safety *on the* Road *and* Outdoors

BIG IDEA ⟩ Following safety rules can help prevent injury on the road and outdoors.

⟩⟩⟩ **Before You Read**

QUICK WRITE Write down three actions you take to stay safe when participating in outdoor activities.

▶ Video

⟩⟩⟩ **As You Read**

STUDY ORGANIZER Make the study organizer found in the FL pages in the back of the book to record the information presented in Lesson 3.

⟩⟩⟩ **Vocabulary**

› pedestrian
› defensive driving
› hypothermia

 Audio

 Bilingual Glossary

©Image Source/PunchStock

 **Fitness Zone**

Increasing My Fitness for Hiking To increase my fitness level for hiking, I can fill my backpack with items I would take hiking and camping, put on my hiking boots or shoes, and walk for one hour several times a week. I can go to the park or walk in my neighborhood. This will help me get used to walking with some weight in my backpack and break in my shoes, while increasing my fitness level.

STAYING SAFE ON THE ROAD

MAIN IDEA ⟩ Safety in vehicles includes wearing a safety belt and not distracting the driver.

In the United States, motor vehicle crashes are the leading cause of accidental deaths in people 1 to 24 years old. To be a safety-conscious passenger, wear a safety belt whenever you ride in a vehicle. Safety belts help keep you in your seat if your vehicle gets into a crash. Do not distract the driver of the vehicle. As many as 3,000 people per year are killed in accidents involving a distracted driver. Another way to stay safe is to never get in a car with a driver who has been drinking alcohol or using drugs. Call a trusted adult to pick you up instead.

Many cars have air bags, too. Air bags can help keep people in the front seats from colliding with the steering wheel and dashboard. However, the force of air bags can hurt small children.

The safest place for children to ride is in the back seat. Infants and small children should ride in an appropriate car seat or booster seat until they are large enough to use a safety belt.

Motor vehicle crashes are the **leading cause** of accidental deaths.

If you take the bus to school, don't bother the bus driver while he or she is driving. Don't get up while the bus is moving or put your arms out the window. When you get off the bus, make sure the bus driver and all drivers of the vehicles around the bus can see you clearly. Don't cross the street behind the bus. If you are in a bus during an emergency, cooperate with the driver so that you and everyone else on the bus remain safe.

Safety *on* Foot

Ever since you learned to walk, you have been a **pedestrian,** *a person who travels on foot.* Safety is important for pedestrians. Pay attention to what is happening around you. Follow these rules to become a safer pedestrian.

- Walk on the sidewalk if there is one. If there is no sidewalk, walk facing oncoming traffic, staying to the left side of the road.
- Cross streets only at crosswalks. Do not cross the street in the middle of the block.
- Look both ways several times before crossing, and keep looking and listening for oncoming cars.
- If you cross in front of a stopped vehicle, be sure the driver can see you. Make eye contact with him or her before stepping in front of the vehicle.
- Obey all traffic signals.
- During the day, wear bright clothing. If you walk at night, take a well-lit route. Wear light-colored or reflective clothing and carry a flashlight.
- Do not talk on a cell phone or wear headphones as you walk. Always be aware of your surroundings.

Safety *on* Wheels

Riding bicycles and using skates, in-line skates, skateboards, and scooters are activities many teens enjoy. One way to prevent injury while participating in these activities is to learn about the risks and then follow rules to avoid them.

Always wear an appropriate helmet and other safety gear. Make sure your clothing fits well and does not prohibit you from moving freely.

Bicycle riders need to obey the same traffic rules as drivers. *Describe what else bicycle riders should do to stay safe.*

Head injuries cause 70 to 80 percent of the deaths from bicycle accidents. Wearing a helmet every time you get on your bike can reduce your risk of head injury by 85 percent. It is also important for bicycle riders to ride with the flow of traffic and obey traffic signs and signals. Bicyclists should never weave through traffic. When riding with a friend, ride in single file, not side-by-side. Learn hand signals, and use them before turning or changing lanes.

Both bicyclists and drivers must practice **defensive driving,** which means *watching out for other people on the road and anticipating unsafe acts.*

Wearing a helmet every time you get **on your bike** can **reduce** your **risk** of *head injury* by *85%.*

Bicyclists should be visible to others by wearing bright, reflective clothes. Bicycles should have lights and reflectors. To reduce risk of injury while bicycling, do not ride at night or in bad weather.

Skates, in-line skates, skateboards, and scooters can be a fun, but only when they are used safely. Here are some guidelines for having fun while staying injury free.

- Wear protective gear, including a hard-shell helmet, wrist guards, gloves, elbow pads, and knee pads.
- Do not speed.
- Follow your community's rules on where you can ride your skateboard or scooter. Avoid skating or riding in parking lots, streets, and other areas with traffic.
- Practice a safe way to fall on a soft surface.
- Don't skate or ride a scooter after dark.
- Avoid riding or skating on wet, dirty, or uneven surfaces.

>>> **Reading Check**

GIVE EXAMPLES *Name three ways to stay safe on wheels.*

STAYING SAFE OUTDOORS

MAIN IDEA Safety precautions make outdoor activities more fun.

Outdoor activities are more fun when you "play it safe." To stay safe while enjoying outdoor activities, remember to:

Use good judgment. Plan ahead. Check the weather forecast. Make sure you have the proper safety gear for each activity and that what you are doing is safe. If you are unsure, ask a trusted adult.

Take a buddy or two. Be sure you are with at least one other person. Always tell your parent or guardian where you are going.

Warm up and cool down. To prevent injuries, stretch after your warm-up and cool down.

Stay aware. Learn the signs of weather emergencies. When necessary, move quickly to shelter for safety.

Know your limits. Be aware of your skills and abilities. For example, if you are a beginning swimmer, don't try to swim a long distance.

Protect your skin. Wear bug protection and sunscreen. It is important to wear sunscreen to protect your skin from the sun's damaging rays.

Planning your trip can make it safer and more fun. *Explain what else you can do to stay safe while participating in outdoor activities.*

Planning *for* Weather

When planning your activity, always check the weather. One major risk is an electrical storm. If you are outdoors during an electrical storm, try to find shelter in a building or car. Follow safety tips for hot weather and cold weather.

When planning your activity, **always check** the *weather.*

During hot weather, your body can overheat. If you feel dizzy, out of breath, or have a headache, take a break. Keep cool by drinking plenty of water. Rest in the shade when you can. Overworking your body in the heat can lead to two dangerous conditions: heat exhaustion and heatstroke. Signs of heat exhaustion can include cold, clammy skin, dizziness, or nausea.

Signs of heatstroke can include an increase in body temperature, difficulty breathing, and a loss of consciousness. Heatstroke can be deadly. If someone shows signs of heatstroke, call 911 and get medical help right away.

Cold weather can be dangerous if your body or parts of your body get too cold. When you are outside in cold weather, dress in layers. Wear a hat, warm footwear, and gloves or mittens. Anyone who starts to feel very cold or shiver should go inside and get warm. It is important to drink plenty of water and take a break when you feel tired. Also, wearing sunscreen is important in both warm and cold weather. While performing some winter activities, such as skiing, you are exposed to the sun. Wearing sunscreen and sunglasses will help to protect your skin from harmful UV rays.

©Fancy/Veer

Preventing *Drowning*

Use library and Internet resources to research drowning and how it can be prevented. Take notes on what you learn. With your teacher, invite a water safety instructor to visit your class. Have the instructor explain ways to avoid drowning in both warm and cold water.

On Your Own Create a video or poster that shows and describes the different drowning prevention techniques for warm and cold water.

Water Safety

Water activities can be a lot of fun. To avoid injury, you should learn and follow water safety rules. Know how to swim well. Good swimmers are less likely to panic in an emergency.

Follow these other tips to stay safe in the water.
- Swim only when a lifeguard or other trusted adult is present.
- Swim with a buddy.
- Monitor yourself. Don't swim if you are tired or cold, or if you have been out in the sun for too long.

- Look around your environment often. Watch for signs of storms. If you are swimming when a storm begins, get out of the water right away.
- Avoid swimming in water with strong currents.
- Dive only in areas that are marked as safe for diving. The American Red Cross suggests that water be at least nine feet deep for diving or jumping. Avoid diving into unfamiliar water or into above-ground pools that may be shallow.
- If you are responsible for children, watch them closely. Avoid letting them near the water unless there is a trained lifeguard on duty. Accidents can happen even in small wading pools.
- When boating or waterskiing, wear a life jacket at all times. If the water is cold, wear a wetsuit. This will protect you from developing **hypothermia** (hy·poh·THER·mee·uh), *a sudden and dangerous drop in body temperature.*

Swimming in an area where a lifeguard is present is one way to stay safe while in the water. *Identify what else you can do to stay safe while in the water.*

Brand X Pictures/Getty Images

Safety *on the* Trail

Preparation is the first step in a safe and enjoyable hike or camping trip. Follow the tips listed below to help you stay safe while hiking and camping.

Never camp or hike alone. Make sure family members know your route and your expected date and time of return. Carry a cell phone or long-range walkie-talkie.

Dress properly. Be aware of the weather and dress accordingly. Dress in layers and wear long pants to protect yourself against ticks. If you are hiking up a mountain, know that the weather may change as you change altitude. Wear sturdy footwear. Before you hike in any shoes or boots, break them in to avoid getting blisters.

Bring equipment and supplies. You should have a map of the area in which you will be hiking or camping. Learn how to read a compass. Take along a first aid kit, flashlight, and extra batteries. Be sure to bring an adequate supply of drinking water and food for your trip. Bring food that will not spoil.

Know the plants and animals. Learn to recognize the dangerous plants and animals in your area so that you can avoid them. For example, learn what poison ivy and poison oak look like. To avoid insect bites and stings, tuck your pant legs into your socks and apply insect repellent. Learn first aid to treat reactions to poisonous plants, insects, and snakebites.

Use fire responsibly. Learn the proper way to build a campfire. Light campfires only where permitted. Put out all campfires completely before you go to sleep or leave the campsite. To put out a campfire, soak it with water or cover it completely with sand or dirt that is free of debris.

Reading Check

IDENTIFY *What safety items should you bring on a hike or camping trip?* ■

Safe hiking and camping takes planning. *Explain why it is important to let people know where you are hiking and when you plan to return.*

Anne Ackermann/Getty Images

LESSON 3

REVIEW

After You Read

1. **DEFINE** What is a *pedestrian*?
2. **EXPLAIN** Why is it important to learn how to fall when skating or riding a bike?
3. **IDENTIFY** Name three ways to stay safe while in the water.

Thinking Critically

4. **ANALYZE** What are some safety factors that can reduce your risk of traffic injuries?
5. **EVALUATE** How does practicing the buddy system help keep you safe outdoors?

Applying Health Skills

6. **ADVOCACY** Create a poster that displays the dangerous plants in your area. Show plants such as poison ivy or poison oak as well as plants that are poisonous if eaten. Post a phone number to call if someone has eaten a poisonous plant.

Review

Audio

Personal Safety *and* Online Safety

BIG IDEA › You can protect yourself from violence by avoiding dangerous situations.

›› Before You Read

QUICK WRITE What steps do you take to keep yourself safe? Write a short paragraph about these strategies.

▶ Video

›› As You Read

STUDY ORGANIZER Make the study organizer found in the FL pages in the back of the book to record the information presented in Lesson 4.

›› Vocabulary

› precautions

◆)) Audio

🅐🅑🅒 Bilingual Glossary

Developing Good Character

Responsibility When you take steps to protect yourself when you are at home, on the street, or online, you are taking responsibility for your own safety. Taking responsibility for your personal safety is an important part of the health triangle. You can protect yourself from physical violence, you can feel confident about yourself, and you may feel less stress about situations that made you nervous before.

PERSONAL SAFETY

MAIN IDEA › You can reduce your risk of becoming a victim of violence by avoiding unsafe situations.

Did you know that teens are the victims of violence more than any other age group? Violence is physical force used to harm people or damage property. Teens are more likely than children to go out at night and less likely to protect their personal safety than adults.

If a situation **feels** *unsafe,* it probably **is.**

Personal safety refers to the steps you take to prevent yourself from becoming the victim of a crime. You can reduce your risk of becoming a victim of violence by avoiding unsafe situations. Be alert to your surroundings, and trust your instincts. If a situation feels unsafe, it probably is unsafe.

Staying Safe *at* Home

Staying safe at home involves observing safety rules such as the ones listed below.

- When you're home, keep your doors and windows locked. Only open the door for someone you know, or not at all if your parents tell you not to answer the door.
- When you answer the phone or use the Internet, avoid offering personal information. Never tell a stranger that you are home alone. If the person asks to speak to your parents. say that parents are busy and are not able to come to the phone at this time.
- When you come home, have your key ready before you reach the door.
- If someone comes to the door or window and you feel you are in danger, call 911.

›› Reading Check

DESCRIBE *How can you reduce the risk of becoming a victim of violence?*

Onoky/Getty Images

Staying Safe *on the* Street

Before going out, tell your family where you are going and how you will get there. Make sure they also know when you expect to return. Don't walk by yourself, if possible. After dark, walk in well-lit areas. Stay in familiar neighborhoods; avoid deserted streets and shortcuts. If you think someone is following you, go into a public place.

Some public places that offer safety include a store or a well-lit area with other people nearby.

Try not to look like an easy target. Stand tall and walk confidently. Never carry your wallet, purse, or backpack in a way that is easy for others to grab. If someone wants your money or possessions, give them up.

Avoid strangers. Never get into or go near a stranger's car or hitchhike. Do not enter a building with a stranger. Don't agree to run errands or do other tasks for strangers. If someone tries to grab you, scream and run away. Go to the nearest place with people. Ask them to call 911 or your parents.

>>> **Reading Check**

EXPLAIN *Why is it important to tell your family where you are going and when to expect you home?*

ONLINE SAFETY

MAIN IDEA Staying safe online is an important part of your overall safety.

Whether for school-work or fun, it's almost impossible to avoid using the Internet. However, you can avoid the dangers that might be lurking online by taking precautions, or *planned actions taken before an event to increase the chances of a safe outcome.* The Internet is like any tool. You've got to use it right to stay safe.

Protecting Your Data

Protecting your personal data helps keep people from stealing your usernames and passwords, which can lead to identity theft. Use the following tips to help protect your data while using the Internet.

Don't give out personal information. Some people may not be telling the truth about themselves. By not including personal information, you are taking the best action to avoid becoming the victim of a cyberbully as well.

The Internet is like any tool. You've got to **use it right** to *stay safe.*

Taking precautions when using the Internet helps protect your safety. *What are ways you can protect your personal data?*

Be wary of attachments or links in e-mails. Don't click on links that are sent in e-mails from strangers. If you open an attachment from a stranger, you might be inviting a virus into your computer. Run a virus scan before opening any attachments—even if you know the sender. He or she could be sending a virus without even knowing it.

Be careful about downloads. Download applications from reputable sites. Otherwise, the applications could come with viruses and spyware attached.

Keep passwords private. Never reveal them to people online.

Use filtering software. Search engines can filter information that's not meant for you. Use the "preferences" tab to set it up.

Get permission. Ask a trusted adult before heading online—and especially before filling out any forms on the Internet.

©Corbis/PunchStock

Avoiding Internet Predators

Internet predators use online communication to build up trust so they can try to get a potential victim to meet with them in person. To avoid falling victim to Internet predators, follow these guidelines for online safety.

Avoid sending photos to strangers online. In addition, be sure to get permission from a trusted adult before sending photos to friends.

Avoid responding to inappropriate messages. If anyone sends you a message that makes you feel uncomfortable for any reason, tell a parent or other trusted adult. You can also report the person to your internet provider.

Be cautious about meeting an online friend. If you decide to meet, do it in a public place with a trusted adult present.

Talk to a trusted adult. Feeling uncomfortable or scared about something that happened online? Tell a trusted adult about it immediately.

Although you can *hide* some **information,** it may not be **completely private.**

>>> **Reading Check**

LIST *What are three things you can do to protect yourself from online predators?*

Understanding Privacy Settings

Many social media networks and other websites have privacy settings. Privacy settings allow you to control who reads and comments on your personal pages and profiles. However, it is important to understand how each site's settings work.

Although you may be able to hide some information, such as your phone number, or address, it may not be completely hidden. For example, if a friend posts a comment on your page or personal website that includes your address or current location, another person can read the post and find out where you are or what even your address.

Most websites prompt you to read a privacy statement before you sign up. Always read the website's privacy statements. Review the privacy settings page whenever changes are made to the website. Remember to think about the information you post to a website or blog. Even if you delete the information or pictures, a copy will stay on the Internet. People with knowledge of the Internet can find the pictures or information even after it is deleted. It is easier not to post information than to try and remove it after it is posted.

>>> **Reading Check**

EXPLAIN *Why is it important to understand a website's privacy settings?* ■

>>> **After You Read**

1. **DEFINE** Define *personal safety.*
2. **DESCRIBE** What strategies should you use to keep yourself safe if you are walking home alone?
3. **IDENTIFY** If you were approached by a stranger on the Internet, how would you protect yourself?

>>> **Thinking Critically**

4. **APPLY** Imagine that you are home alone and a stranger comes to the door. He tells you that he needs help and asks if he can use your phone. What should you do to avoid risks to your safety?
5. **EVALUATE** Greg has been at a friend's house all day. Now, it is dark outside and he is uncomfortable walking home alone. What could Greg do?

>>> **Applying Health Skills**

6. **COMMUNICATION SKILLS** Erica has been chatting with a stranger online. At first the messages were harmless, but now she is uncomfortable with the conversations. Write a short dialogue that your friend could use to tell her parents about this issue.

 Review

Audio

Weather Safety *and* Natural Disasters

BIG IDEA Early warning signs give time to plan and stay safe during weather emergencies and natural disasters.

>>> Before You Read

QUICK WRITE What is one kind of weather emergency that is common in your area? List two things you could do to stay safe during that emergency.

 Video

>>> As You Read

STUDY ORGANIZER Make the study organizer found in the FL pages in the back of the book to record the information presented in Lesson 5.

>>> Vocabulary

› weather emergency
› tornado
› hurricane
› blizzard
› frostbite
› natural disaster
› earthquake
› aftershocks

 Audio

 Bilingual Glossary

Myth vs. Fact

Myth: The area under a highway overpass provides shelter from a tornado.

Fact: You should seek shelter in a building. If there are no buildings, lie down in a ditch and cover your head with your hands. If you cannot exit the vehicle, leave your seat belt on and put your head down below the windows. Cover your head and body with a blanket, if possible.

WHAT ARE WEATHER EMERGENCIES?

MAIN IDEA Weather emergencies are dangerous situations brought on by changes in the atmosphere.

Weather events make the news on a fairly regular basis. These events often happen with little warning. People cannot prevent them. A weather emergency is *a dangerous situation brought on by changes in the atmosphere.* Weather emergencies are natural events. Examples include thunderstorms, tornadoes, hurricanes, and blizzards.

Weather emergencies can impact a person's safety and health. As a result, the National Weather Service (NWS) works to track the progress of storms and sends out bulletins to the public. The bulletins keep people informed about possible weather emergencies. This helps keep people and communities safe.

Storm bulletins may involve watches or warnings. A storm watch indicates that a storm is likely to develop. A storm warning indicates that a severe storm has already developed and a weather emergency is happening. If your area is under a storm warning, turn on the television or radio. Follow the instructions of the NWS and local officials.

Weather emergencies can **impact** a person's *safety* and *health.*

Technology has helped scientists who watch the weather. Satellites gather data very quickly and feed it into powerful computers. Computers can also help predict the paths of storms. Television and the Internet can warn the public of danger very quickly. These early warnings give people more time to plan and stay safe.

Thunderstorms *and* Lightning

How can you protect yourself during thunderstorms? Whenever you see lightning or hear thunder, seek shelter and stay there. Do not use the telephone, unless it is a cordless or cell phone. Be prepared for a power loss. If you are outdoors, look for the nearest building. An alternative is an enclosed metal vehicle with the windows completely shut. If you are in an open field with no shelter nearby, lie down. Wait for the storm to pass. Avoid all metal objects including electric wires, fences, machinery, motors, and power tools. Unsafe places include underneath canopies, small picnic or rain shelters, or near trees.

Tornadoes

A tornado is *a whirling, funnel-shaped windstorm that drops from storm clouds to the ground.* If a tornado watch is issued for your area, listen to the radio for updates. Prepare to take shelter if you need to protect yourself. If a tornado warning is issued for your area, move to the shelter immediately.

You are safest underground in a cellar or basement. If you cannot go underground, take shelter in an inner hallway or any central, windowless room such as a bathroom or closet. If you are outdoors, lie in a ditch or flat on the ground. Stay away from trees, cars, and anything that could fall on you.

If you are in a room with furniture, stay under a heavy table. Lying in a bathtub under a cushion, mattress, or blanket may also offer good protection.

Hurricanes

A hurricane is *a strong windstorm with driving rain that forms over the sea.* Hurricanes occur in coastal regions.

Weather experts track the path of a hurricane by recording the coordinates of its location at regular intervals. *Describe how tracking a hurricane might help people who live in a coastal area.*

The National Weather Service tracks hurricanes and can estimate when and where a hurricane will hit land. This gives people time to plan ahead. Take the following steps to stay safe during a hurricane.

Before the hurricane arrives, secure your home. Board up windows. Close storm shutters before the winds start blowing.

Bring inside items such as furniture and bikes that wind could pick up and smash into houses.

Stay alert to TV or radio reports. Sometimes residents will be instructed to leave their homes, or evacuate, and head inland. It is necessary to follow these safety instructions. If no evacuation is ordered, stay indoors. Avoid standing near windows and doors.

Blizzards

Do you live in an area hit by snow in the winter? If you do, you may have experienced a blizzard, *a very heavy snowstorm with winds up to 45 miles per hour.* Protect yourself during blizzards and winter storms by following these precautions:

Stay indoors. The safest place during a blizzard is inside. Visibility is reduced during a blizzard, making it easy to get lost or disoriented if walking outside or even in a car.

Bundle up. A leading danger in a blizzard is hypothermia. Another health risk is frostbite, or *freezing of the skin.* Frostbite can cause severe injury to the skin and deeper tissues. If you must go out, wear layers of loose-fitting lightweight clothing under layers of outerwear that is both wind- and waterproof. Add a scarf, hat, gloves and boots.

>>> **Reading Check**

RECALL *What are three ways to stay safe during a blizzard?*

WHAT ARE NATURAL DISASTERS?

MAIN IDEA Natural disasters are dramatic events caused by Earth's processes.

Like weather emergencies, natural disasters can cause serious health and safety problems. A natural disaster is *an event caused by nature that results in widespread damage.* Natural disasters include floods and earthquakes. One way to stay safe during a natural disaster is to plan ahead. Keep some basic supplies on hand such as fresh water, a radio, a flashlight, batteries, blankets, canned food, a can opener, and a first-aid kit.

Floods

Floods can occur almost anywhere. Hurricanes and heavy rainfall can cause floods.

Flash floods, floods that occur with little or no warning, are the most dangerous of all. Flashflood waters rise very quickly and are surprisingly powerful. Two feet of moving water has enough force to sweep away cars. Water levels higher than two feet can carry away trucks and houses.

> *Two feet* of moving water has enough *force* to **sweep away** cars.

If the NWS issues a flood watch for your area, take your emergency kit, and go to the highest place in your home. Listen to a battery-powered radio for a flood warning.

If a flood warning is issued and you are told to evacuate, do so immediately. The following tips can help you survive a flood:

- Head for higher ground. If it is possible to travel safely, go to the home of a relative or neighbor who lives outside the warning area.
- Avoid walking, swimming, riding a bike, or driving a car through flooding water. You could be swept away, electrocuted by downed power lines, or drown.
- Drink only bottled water. Floodwater carries harmful pathogens that can make you sick. Flood water is easily polluted by trash and other waste.
- If you have evacuated the area, return home only after you are told it is safe for you to do so. Returning home before the authorities give the all-clear can put you at risk. It may also break the law.
- Make sure that everything that came in contact with the floodwater is cleaned and disinfected. Wear rubber or latex gloves during the cleanup. Throw out all food that may have become contaminated or if refrigeration was lost during the flood. Make sure the water supply is safe before drinking water.

>>> **Reading Check**

EXPLAIN *Why are flashfloods dangerous?*

Earthquakes

An **earthquake** is *a shifting of the earth's plates, resulting in a shaking of Earth's surface.* Earthquakes usually involve more than a single event. A large quake typically is followed by a series of aftershocks. **Aftershocks** are *smaller earthquakes as the earth readjusts after the main earthquake.* Collapsing walls and falling debris cause most injuries in an earthquake. To reduce your risk of injury from an earthquake, follow these precautions:

- **Stay indoors.** Crouch under heavy furniture. Stay away from objects that might fall, shatter, or cave in.
- **Get in the open if outdoors.** Avoid buildings, trees, power lines, streetlights, and overpasses.
- **Be careful afterward.** Stay out of damaged buildings. Damaged electrical and gas lines could be hazardous.

Scientists cannot predict when and where an earthquake will strike. However, they can measure how strong earthquakes are when they happen. The Richter scale rates the force of ground motion during an earthquake. An earthquake that measures 1 on this scale is slight. One that measures a 2 is 10 times stronger than 1. Likewise, one that measures a 3 is 10 times stronger than 2, and so on.

The most destructive earthquakes have a magnitude of 7 or more on the Richter scale. They are much less common. Scientists have never recorded an earthquake that measures more than 9 on the Richter scale.

During an earthquake, stay clear of falling objects. *Name two ways to stay safe during an earthquake.*

>>> **After You Read**

1. **DEFINE** Define *hurricane*, and use it in a sentence.
2. **DESCRIBE** Tell how to protect yourself during a flood.
3. **COMPARE** What is the difference between a weather emergency and a natural disaster?

>>> **Thinking Critically**

4. **APPLY** You are playing soccer in a field, and you see a flash of lightning from an approaching thunderstorm. What should you do?
5. **SYNTHESIZE** After a major earthquake, your friend wants you go with him to inspect a damaged building. How would you respond? Explain your answer.

>>> **Applying Health Skills**

6. **COMMUNICATION SKILLS** With a partner, choose a weather emergency from this lesson. Write and perform a skit to demonstrate strategies for staying safe during that event.

 Review

 Audio

Get under a sturdy piece of furniture. Cover your head with your arms or a pillow.

Stay away from windows, mirrors, and other objects that might shatter.

Stay away from trees, buildings, and power lines. They may fall.

Stay away from tall or heavy objects that could fall on you.

Find a clear, open area. Drop to the ground and protect your head with your arms. To do so, clasp your hands together at the back of your head, and bring your elbows together in front of your face.

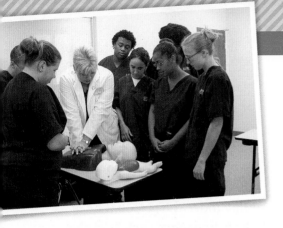

First Aid *and* Emergencies

BIG IDEA Knowing how to administer basic first aid can save a person's life in an emergency.

>>> **Before You Read**

QUICK WRITE Write a couple of sentences describing how you would treat a minor burn.

 Video

>>> **As You Read**

STUDY ORGANIZER Make the study organizer found in the FL pages in the back of the book to record the information presented in Lesson 6.

>>> **Vocabulary**

> first aid
> universal precautions
> abdominal thrust
> cardiopulmonary resuscitation (CPR)
> rescue breathing
> fracture
> dislocation

 Audio

 Bilingual Glossary

GIVING FIRST AID

MAIN IDEA First aid is the immediate care given to someone who becomes injured or ill until regular medical care can be provided.

Dania was taking a walk in her neighborhood when she saw someone lying on the ground. The man was wearing a bicycle helmet. A bicycle was lying nearby. Dania gently touched his shoulder and said, "Are you alright?" The man did not respond. Dania immediately pulled out her cell phone and dialed 911.

Some emergencies are minor. You cut your fingertip and it bleeds. A friend falls while skateboarding and injures his or her knee. These types of minor injuries should be cleaned with soap and warm water. They may also be wrapped or covered with a breathable bandage.

You can *prevent* further **injury** and may even *speed recovery* if you **know what to do** in an emergency.

Other emergencies can be life-threatening. Taking immediate action can mean the difference between life and death. **First aid** is *the immediate care given to someone who becomes injured or ill until regular medical care can be provided.*

Knowing basic first aid may help you deal with some emergencies while you wait for help to arrive. You can prevent further injury and may even speed recovery if you know what to do in an emergency. Knowing what *not* to do is equally important. Anyone who has received first aid should be taken to a medical provider as soon as possible.

Developing Good Character

Citizenship A good neighbor and citizen is prepared to report accidents, fires, serious illnesses, injuries, and crimes. Familiarize yourself with emergency phone numbers to call in your community. Make a list to keep handy by the telephone.

Steps *to* Take *in* an Emergency

How can you tell if an emergency is life-threatening? A person's life is considered in danger if he or she: (1) has stopped breathing, (2) has no heartbeat, (3) is bleeding severely, (4) is choking, (5) has swallowed poison, or (6) has been severely burned. These situations require immediately help. Call 911. Next, begin to treat the person. Proper training is needed to give first aid. In an emergency, the American Red Cross suggests the following strategy: Check-Call-Care.

Check the scene and the victim. Often something you see, hear, or smell will alert you to an emergency. Is someone calling out in trouble? Have you heard glass shattering? Do you smell smoke or anything unusual that makes your eyes sting or causes you to cough or have difficulty breathing? These sensations can signal a chemical spill or toxic gas release. Make sure the area is safe for you and the victim.

Move the person only if he or she is in danger of additional injury.

Call for help. Call 911 or the local EMS number. *EMS* stands for "emergency medical service." When making a call for help, stay calm. Describe the emergency to the operator and give a street address or describe the location by using landmarks. The operator will notify the police, fire, or emergency medical service departments. Stay on the phone until the operator tells you to hang up.

How can you tell if an *emergency* is **life-threatening?**

Care for the person until help arrives. After you have called for help, stay with the person until help arrives.

When someone is injured, giving first aid can help prevent further injury and possibly speed recovery. *What is first aid?*

Loosen any tight clothing on the person's body. Use a coat or blanket to keep the person warm or provide shade if the weather is warm. This will help maintain a normal body temperature. Avoid moving the person to prevent further pain or injury. Only move the person if he or she is in danger, such as in the path of traffic. Hands-Only™ Cardiopulmonary Resuscitation (CPR) may be necessary if the person is unconscious and unresponsive.

Universal Precautions

Viruses such as HIV, hepatitis B, and hepatitis C can be spread through contact with an infected person's blood. As a result, steps should be taken to minimize contact with another person's blood. To protect yourself when giving first aid, follow universal precautions, or *actions taken to prevent the spread of disease by treating all blood as if it were contaminated.* Wear protective gloves while treating someone. If possible, use a facemask or shield, when giving first aid for breathing emergencies. Cover any open wounds on your body with sterile dressings. Avoid touching any object that was in contact with the person's blood. Always wash hands thoroughly after giving first aid.

>>> **Reading Check**

EXPLAIN *What information should you give when calling 911 or another emergency number?*

For adults and children

1 Give 5 back blows with the heel of your hand.

2 Place the thumb of your fist against the person's abdomen, just above the navel. Grasp your fist with your other hand. Give quick, inward and upward thrusts until the person coughs up the object. If the person becomes unconscious, call 911 or the local emergency number. Begin CPR.

For infants

1 Hold the infant facedown on your forearm. Support the child's head and neck with your hand. Point the head downward so that it is lower than the chest. With the heel of your free hand, give the child five blows between the shoulder blades. If the child doesn't cough up the object, move on to chest thrusts (step 2).

2 Turn the infant over onto his or her back. Support the head with one hand. With two or three fingers, press into the middle of the child's breastbone—directly between and just below the nipples—five times. Repeat chest thrusts until the object comes out or the infant begins to breathe, cry, or cough. Make sure a health care professional checks the infant. If the infant becomes unconscious, call 911.

FIRST AID FOR CHOKING

MAIN IDEA Abdominal thrusts can help save someone who is choking.

> Follow these steps to help a person who is choking. *Why do you think chest thrusts are used to help a choking infant?*

If a person is clutching his or her throat, that is the universal sign for choking. Symptoms of choking include gasping or wheezing, a reddish-purple coloration, bulging eyes, and an inability to speak.

If a person can speak or cough, it is not a choking emergency. However, if the choking person makes no sound and cannot speak or cough, give first aid immediately.

If an adult or child is choking, give the person five blows to the back. Stand slightly behind the person who is choking.

Place one of your arms diagonally across the person's chest and lean him or her forward. Strike the person between the shoulder blades five times. If this does not dislodge the object, give five abdominal thrusts. An **abdominal thrust** is a *quick inward and upward pull into the diaphragm to force an obstruction out of the airway.*

If an infant is choking, hold the infant face down along your forearm, using your thigh for support. Give five back blows between the shoulder blades.

If this does not dislodge the object, turn the infant over and perform five chest thrusts with your fingers to force an object out of the airway.

If you are alone and choking, give yourself an abdominal thrust. Make a fist and position it slightly above your navel. With your other hand, grasp your fist and thrust inward and upward into your abdomen until the object dislodges. You can also lean over the back of a chair, or any firm object, pressing your abdomen into it.

RESCUE BREATHING AND CPR

> **MAIN IDEA** CPR is for older children and adults whose hearts have stopped beating.

All organs need oxygen-rich blood. If the heart stops beating, the flow of blood to the brain stops, too. When the brain stops functioning, breathing also stops. If you are confident that the victim is not breathing, it is necessary to begin **cardiopulmonary resuscitation** (CPR)—*a first-aid procedure to restore breathing and circulation.*

The American Heart Association (AHA) recommends two forms of CPR. A trained person will perform CPR that combines chest compressions with **rescue breathing**, *a first aid procedure where someone forces air into the lungs of a person who* *cannot breathe on his or her own.* A person who is untrained in giving CPR can perform *Hands-Only™ CPR.* This form of CPR focuses only on chest compressions. In an emergency, if no trained person is present, an untrained person should begin Hands-Only™ CPR before medical professionals arrive.

Hands-Only™ CPR *for* Adults

The first step is to call 911. Before performing CPR, tap the victim and shout, "Are you OK?" Check the victim for signs of normal breathing. Put your ear and cheek close to the victim's nose and mouth.

Listen and feel for exhaled air. Look to see if the chest is rising and falling. If the victim is unconscious and there is no response, begin Hands-Only™CPR.

The Steps of Hands-Only™ CPR for Adults, shows where to position your hands over the victim's chest and begin compressions. Try to give 100 chest compressions each minute, until the victim responds or until paramedics arrive.

> Follow these steps to help a person whose heart has stopped beating. *In what type of situation should you use CPR?*

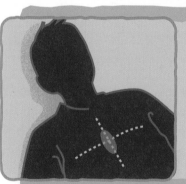

1. Use your fingers to find the end of the victim's sternum (breastbone), where the ribs come together.
2. Place two fingers over the end of the sternum.
3. Place the heel of your other hand against the sternum, directly above your fingers (on the side closest to the victim's face).
4. Place your other hand on top of the one you just put in position. Interlock the fingers of your hands and raise your fingers so they do not touch the person's chest.

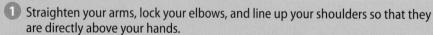
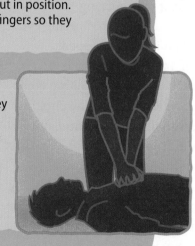

1. Straighten your arms, lock your elbows, and line up your shoulders so that they are directly above your hands.
2. Press downward firmly on the person's chest, forcing the breastbone down by 1.5 to 2 inches (3.8 to 5 cm).
3. Begin compressions at a steady pace. You can maintain a rhythm by counting. "One and two and three and…" Press down each time you say a number. Emergency medical experts recommend pressing down 100 times a minute.

FIRST AID FOR SEVERE BLEEDING

MAIN IDEA › To control bleeding, apply a cloth and direct pressure to the wound and elevate the wound, if possible.

When providing first aid to a person who is bleeding severely, follow universal precautions. Avoid touching the person's blood or wear gloves. Always wash your hands when you are finished. If the person has a wound that is bleeding severely or needs other medical help, call 911. Wash the wound with mild soap and water to remove dirt and debris. Follow these steps to control the bleeding:

- Raise the wounded body part above the level of the heart.
- Cover the wound with sterile gauze or a clean cloth. Apply steady pressure to the wound for five minutes, or until help arrives. Do not stop to check the wound.
- If blood soaks through the gauze, add another gauze pad on top of the first and continue to apply pressure.
- Once the bleeding slows, secure the pad in place with a bandage or strips of gauze. The pad should be snug.
- Stay with the victim until help arrives.

FIRST AID FOR BURNS

MAIN IDEA › Major burns require medical attention as soon as possible.

A **first-degree burn,** or *superficial burn,* is a burn in which only the outer layer of skin has burned. There may be pain and swelling. Flush the burned area with cold water for at least 20 minutes. Loosely wrap the burn in a clean, dry dressing.

A **second-degree burn,** or *partial-thickness burn,* is a moderately serious burn in which the burned area blisters.

There is usually severe pain and swelling. Flush the burned area with cold water for at least 20 minutes. Do not use ice. Elevate the burned area. Loosely wrap the cooled burn in a clean, dry dressing. Do not pop blisters or peel loose skin. If the burn is larger than 2 or 3 inches in diameter, or is on the hands, feet, face, groin, buttocks, or a major joint, get medical help immediately.

A **third-degree burn,** or *full-thickness burn,* is a very serious burn in which all the layers of skin are damaged. There may be little or no pain felt at this stage. They usually result from fire, electricity, or chemicals. Call 911 immediately. Do not remove burned clothing. Cover the area with a cool, clean, moist cloth. Only a medical professional should treat full-thickness burns.

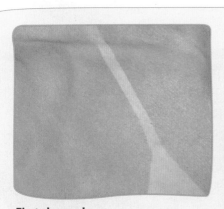

First-degree burn

Second-degree burn

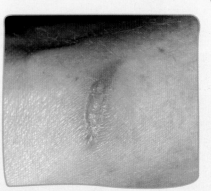

Third-degree burn

FIRST AID FOR OTHER EMERGENCIES

MAIN IDEA Animal bites, bruises and sprains, broken or dislocated bones and poisonings require different types of treatment.

Other common emergencies that you may encounter include insect and animal bites, bruises and sprains, broken or dislocated bones, and poisonings.

Insect *and* Animal Bites

Insect bites and stings can be painful but are not usually dangerous unless the person is allergic to the venom of the insect. If an allergic person has been stung, get medical help immediately. For all other bites and stings follow these steps:

- Remove the stinger by scraping it off with a firm, straight-edged object. Do not use tweezers.
- Wash the site thoroughly with mild soap and water.
- Apply ice (wrapped in a cloth) to the site for ten minutes to reduce pain and swelling. Continue to apply ice, alternating ten minutes on and off.

- To treat animal bites, wash the bite with soap and water. Apply pressure to stop any bleeding. Apply antibiotic ointment and a sterile dressing. For any bite that has broken the skin, contact your doctor.

If an allergic person has been *stung*, **get medical help** immediately.

Bruises *and* Sprains

A bruise forms when an impact breaks blood vessels below the surface of the skin. This allows blood to leak from the vessels into the tissues under the skin, leaving a black or bluish mark. A sprain is a condition in which the ligaments that hold the joints in position are stretched or torn. Symptoms of sprains include swelling and bruising.

While a doctor should evaluate serious sprains, minor sprains can be treated using the P.R.I.C.E. method:

- **Protect** the injured part by keeping it still. Moving it could cause further injury.
- **Rest** the affected joint for 24 to 48 hours.
- **Ice** the injured part to reduce swelling and pain. A cloth between the skin and ice bag will reduce discomfort. Be sure to remove the ice every 15–20 minutes so that it does not become too cold.
- **Compress** the injured area by wrapping it in an elastic bandage.
- **Elevate** the injured part above the level of the heart to reduce swelling.

Ice helps slow swelling after a sprain. *How can you reduce the risk of sprains during physical activity?*

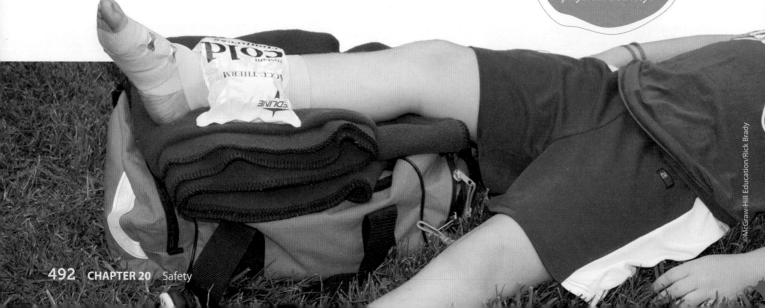

©McGraw-Hill Education/Rick Brady

Broken *or* Dislocated Bones

A fracture is *a break in a bone.* Fractures usually happen along the length of a bone. An open fracture is a complete break with one or both sides of the bone piercing the skin. A closed fracture does not break the skin and may be difficult to identify. Pain, swelling and a misshapen appearance are typical symptoms of a closed fracture. However, not all broken bones cause immediate pain. An X ray is the only method to confirm that a bone is broken.

Problems can also develop where bones meet at a joint. A dislocation is *a major injury that happens when a bone is forced from its normal position within a joint.* For example, if your upper arm bone is pulled out of your shoulder socket, it is dislocated. Moving a broken bone or dislocated joint could cause further injury. For both fractures and dislocations, call for help at once. While you wait for help, keep the victim still. Once a trained medical professional arrives, he or she can then immobilize the fracture or dislocation.

Poisoning

A poison is a substance that causes harm when swallowed, inhaled, absorbed by the skin, or injected into the body. Medicines and household products play a role in about half of all poisonings. All poisonings require immediate treatment.

In the event of a poisoning, call 911 or the nearest poison control center. Be ready to provide information about the victim and the suspected poison. The poison control center will advise you about how to proceed. Keep the person warm and breathing. Look for extra traces of poison around the victim's mouth. Remove these with a damp, clean cloth wrapped around your finger. Save the container of poison. and show it to the emergency medical care providers.

Medicines and household products play a role in about half of all *poisonings.*

If a poisonous chemical has made contact with someone's skin, remove all clothing that has touched the chemical and rinse the skin with water for 15 minutes. Wash with soap and water. Call the poison control center while the skin is being washed.

Some cases of poisoning are caused by contact with a poisonous plant. Poison ivy, poison oak, and poison sumac are three such plants. Most of these injuries can be treated at home using soap and water and over-the counter creams. For severe cases, see a doctor for treatment.

>>> **Reading Check**

LIST *Give two ways poisons can enter the body.* ■

LESSON 6

REVIEW

>>> **After You Read**

1. **VOCABULARY** Define *first aid.* Use the term in a sentence.
2. **RECALL** Name four universal precautions to take when administering first aid.
3. **LIST** Briefly give the steps in controlling severe bleeding.

>>> **Thinking Critically**

4. **INFER** Why can you infer that a person who cannot speak or cough is choking?
5. **APPLY** If you come upon an injured person on a hiking trail, should you try to move the person off the trail? Why or why not?

>>> **Applying Health Skills**

6. **STRESS MANAGEMENT** Emergency situations are often very stressful. With classmates, discuss strategies for reducing stress while dealing with a medical emergency.

 Review

 Audio

Hands-On HEALTH ACTIVITY

A Home Emergency Kit

WHAT YOU WILL NEED

* one piece of poster board
* marker
* paper
* pencil or pen

Weather emergencies and natural disasters are situations that no one can prevent. You can, however, be prepared. Creating a home emergency kit for your family can help keep you safe and secure until the emergency is over.

WHAT YOU WILL DO

1 Working in a small group, brainstorm all of the supplies you would include in a home emergency kit. Your kit should include enough items to last for three days.

2 Have one member of the group write all the items on the poster board.

3 Discuss why you feel certain items should or should not be included.

4 Make your list final. Then, compare your list with other groups in the class. Are your lists similar? How are they different?

WRAPPING IT UP

Write down the final list on a piece of paper. At home, discuss creating a home emergency kit with your family using the list you created.

©McGraw-Hill Education/Ken Karp

READING REVIEW

FOLDABLES and Other Study Aids

Take out the Foldable® that you created and any study organizers that you created. Find a partner and quiz each other using these study aids.

LESSON 1 Building Safe Habits

BIG IDEA Being safety conscious means being aware that safety is important and acting safely.

* Accidents and accidental injuries can affect people of all ages.
* For any accident to occur, three elements must be present.

LESSON 2 Safety at Home and School

BIG IDEA Following safety rules can keep you safe both at home and away from home.

* Your home may be filled with many potential safety hazards, such as stairs or appliances.
* It is a healthy practice to have a fire safety plan, fire extinguisher, and working smoke alarms.
* Following rules at school helps you, other students, and teachers stay safe.

LESSON 3 Safety on the Road and Outdoors

BIG IDEA Following safety rules can help prevent injury on the road and outdoors.

* Safety in vehicles includes wearing a safety belt and not distracting the driver.
* Safety precautions make outdoor activities more fun.

LESSON 4 Personal Safety and Online Safety

BIG IDEA You can protect yourself from violence by avoiding dangerous situations.

* You can reduce your risk of becoming a victim of violence by avoiding unsafe situations.
* Staying safe online is an important part of your overall safety.

LESSON 5 Weather Safety and Natural Disasters

BIG IDEA Weather emergencies include thunderstorms, tornadoes, hurricanes, and blizzards. Natural disasters include floods and earthquakes.

* Weather emergencies are dangerous situations brought on by changes in the atmosphere.
* Natural disasters are dramatic events caused by Earth's processes.

LESSON 6 First Aid and Emergencies

BIG IDEA Knowing how to administer basic first aid can save a person's life in an emergency.

* First aid is the immediate care given to someone who becomes injured or ill until regular medical care can be provided.
* Abdominal thrusts can save someone who is choking.
* Hands-Only™ CPR is for older children and adults who have stopped breathing.
* To control bleeding, apply a cloth and direct pressure to the wound and elevate the limb, if possible.
* Major burns require medical attention.
* Animal bites, bruises and sprains, broken or dislocated bones and poisonings require different treatments.

 Review

 Web Quest

ASSESSMENT

Reviewing Vocabulary *and* Main Ideas

> hazards
> flammable

> defensive driving
> accident

> fire extinguisher
> hypothermia

> accident chain
> electrical overload

>> On a sheet of paper, write the numbers 1–8. After each number, write the term from the list that best completes each sentence.

LESSON 1 Building Safe Habits

1. A(n) _____ is any event that was not intended to happen.

2. _____ are potential sources of danger.

3. A(n) _____ is a series of events that include a situation, an unsafe habit, and an unsafe action.

LESSON 2 Safety at Home and School

4. A(n) _____ can be used to put out a grease fire.

5. Materials that are _____ are able to catch fire easily.

6. _____ is a dangerous situation in which too much electric current flows along a single circuit.

LESSON 3 Safety on the Road and Outdoors

7. _____ involves watching out for other people on the road and anticipating unsafe acts.

8. When outside in cold weather, it is best to wear layers to reduce the risk of _____.

>> On a sheet of paper, write the numbers 9–16. Write *True* or *False* for each statement below. If the statement is false, change the underlined word or phrase to make it true.

LESSON 4 Personal Safety and Online Safety

9. Personal safety is the steps you take to prevent yourself from becoming the victim of a crime.

10. Precautions are actions taken after an event to prevent harm.

LESSON 5 Weather Safety and Natural Disasters

11. The center of a hurricane is called the eye.

12. In a tornado, you are safest underground.

LESSON 6 First Aid and Emergencies

13. A sprain is an invisible break in a bone.

14. A third-degree burn damages all layers of skin.

15. The abdominal thrusts maneuver is used to help a victim of shock.

16. Gasping is the universal sign for choking.

 eAssessment

>> Using complete sentences, answer the following questions on a sheet of paper.

Thinking **Critically**

17. EVALUATE How would you assess whether a person needed CPR?

18. ANALYZE How does an understanding of accident chains help prevent injuries?

Write **About It**

19. NARRATIVE WRITING Write a short story about a teen involved in a situation that leads to an accident. Describe the situation and the events that make up the accident chain. Then write an alternate ending describing how the teen used strategies to prevent the accident from happening.

STANDARDIZED TEST PRACTICE

Writing

Read the paragraph below. Write a short essay about other places that you think automatic external defibrillators should be placed to save people who are having heart attacks. Explain your choices.

CPR can save the lives of people having a heart attack. However, public access to automatic external defibrillators (AEDs) can save even more. AEDs are machines that detect an irregular or missing heartbeat. When the machine detects something is wrong, it sends an electric shock through the chest that restarts the heart. In the past, all hospitals had AEDs. Today, some cities are beginning to place them in areas where the public can easily access them.

City officials are also training people on how to use the AEDs. One city trained police to use AEDs, and every police car now carries one of these machines. In that city, the survival rate for people having heart attacks rose from 28 percent to 40 percent. Many airports and airplanes have AEDs that have already saved people's lives. The more AEDs that are available, the more lives they can save.

KEEPING SAFE—AT HOME AND ELSEWHERE

The first step in staying safe is being safety conscious.

FIRE

It is a healthy practice to have:

- A fire escape plan
- A fire extinguisher
- Working smoke alarms

ON THE WEB

- Don't give out personal information.
- Keep passwords private.
- Don't send photos to strangers.
- Don't respond to inappropriate messages.
- If you want to meet an online friend, do so in a public place with a trusted adult present.
- Remember to think before you post!

PREVENTING FALLS

Most falls in the home occur in the kitchen, the bathroom, or on the stairs.

Bathroom

- Put a nonskid mat near the tub or shower.
- Use rugs that have a rubber backing to prevent the rug from slipping.

Rubber

Staircase

- Keep staircases well lit and clear of objects.
- Apply nonslip treads to stairs.
- Make sure handrails are secure.

PREVENT POISONING

- Make sure medications are in containers with childproof caps.
- Avoid referring to children's medicine as candy.
- Store medications out of the reach of children.

ELECTRICITY

- Never use an electrical appliance around water.
- Unplug electrical appliances when they are not in use.
- Avoid running cords under rugs.
- Cover unused outlets with plastic protectors.

Kitchen

- Clean up spills right away.
- Use a stepstool to get items that are out of reach.

ON FOOT

- Walk on the sidewalk if there is one.
- If there is no sidewalk, walk facing oncoming traffic, staying to the left side of the road.
- Avoid talking on a cell phone or wearing headphones as you walk.

ON WHEELS

- Wear a helmet, wrist guards, elbow pads, and knee pads.
- Avoid riding on wet, dirty, or uneven surfaces.

KEEPING THE ENVIRONMENT SAFE

Many activities can pollute the environment and make it harmful for for all living things.

ACID RAIN

Sulfur dioxide and nitrogen oxides are produced when fossil fuels are burned. They mix with moisture in the air to form acid rain. Over time, acid rain can:
- harm plants and whole forests.
- contaminate water supplies.
- eat away at rock and stone.

AIR POLLUTION

The main cause of air pollution is the burning of fossil fuels. Energy from fossil fuels provides heat for homes, electricity to power factories and cities, and power to most motor vehicles. Fossils fuels are:
- oil
- coal
- natural gas

SMOG

Smog is a yellow-brown haze that forms when sunlight mixes with gases formed by burning fossil fuels.

DAMAGE TO THE OZONE LAYER

The ozone layer provides protection from the sun. Without it, people are more likely to develop skin cancer and eye damage. Chemicals that damage the ozone layer can be found in :
- aerosol cans
- refrigerators
- air conditioners
- automobile emissions

GLOBAL WARMING

The trapping of heat by carbon dioxide and other gases in the air is known as the greenhouse effect. This may be the cause of global warming, or a rise in Earth's temperatures. Global warming can affect weather patterns and ocean water levels.

How can you help?

- Walk or ride your bike.
- Use public transportation or carpool.
- Stay tobacco free.
- Plant trees and other plants.
- Don't burn trash, leaves, and brush.

WATER POLLUTION

Water is vital to all forms of life. However, wastes, chemicals, and other harmful substances pollute Earth's water. Water pollution is a widespread problem. Forty percent of all the nation's rivers, lakes, streams, and estuaries, or coastal waters, are too polluted to use for swimming, fishing, or drinking.

 Sewage

 Garbage

 Detergents

 Other household waste

 Chemicals

How can you help?

- Pick up pet waste from public areas to reduce toxic runoff.
- Use soaps, detergents, and cleaners that are biodegradable.
- Pick up any litter that is not hazardous.
- Dispose of chemicals properly and legally. Never pour them into a drain.
  - Take hazardous waste materials to the appropriate collection sites.

Green Schools + Environmental Health

 **PREMIUM ONLINE RESOURCES**

 Audio

 Videos

Bilingual Glossary

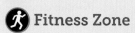 Fitness Zone

 Web Quest

Review

Picking up trash is one way to keep the environment clean. *How do you keep the environment clean in your neighborhood?*

Pollution *and* Health

BIG IDEA ▶ Pollution harms the environment and your health, and is often ugly.

Before You Read

QUICK WRITE Give three examples of pollution that affect your local community.

▶ Video

As You Read

FOLDABLES Study Organizer

Make the Foldable® found in the FL pages in the back of the book to record the information presented in Lesson 1.

Vocabulary

> environment
> pollute
> pollution
> fossil fuels
> acid rain
> ozone
> smog
> ozone layer
> greenhouse effect
> global warming
> sewage

 Audio

 Bilingual Glossary

Myth vs. Fact

Myth Carbon dioxide is the only gas that causes global warming.
Fact Gases such as methane and fluorinated gases also contribute to global warming. These gases are referred to as High Global Warming Potential gases.

YOUR ENVIRONMENT

MAIN IDEA ▶ Pollution is made up of dirty or harmful substances in the environment.

Vivian has been in the habit of turning on the television as soon as she gets home. She leaves it on, even when she isn't watching it. Lately, she has been wondering how much electricity she is wasting by doing this - and how much air pollution she might be causing. She decides to change her habit and turn off the television when she leaves the room.

Your **environment** is *all the living and nonliving things around you.* The environment includes forests, mountains, rivers, and oceans. It also includes your home, school, and community. The air you breathe, the water you drink, the plants and animals that live nearby, and the climate are all part of the environment. All living things are affected by the environment.

It's *important* for **each person** to *do* his or her part to **keep the environment clean.**

Sometimes people pollute the environment. To **pollute** means *to make unfit or harmful for living things.* The result is **pollution,** *dirty or harmful substances in the environment.* Pollution can harm your health. It affects everything in your environment. On days when air pollution is high, you may hear news reports suggesting that people with breathing problems limit the time they spend outside. Our health depends on a healthy environment. It's important for each person to do his or her part to keep the environment clean.

>>> **Reading Check**

IDENTIFY *What is pollution?*

AIR POLLUTION

> **MAIN IDEA** Burning fossil fuels such as coal, oil, and natural gas pollutes the air.

The leading cause of air pollution is the burning of fossil fuels. **Fossil** (FAH·suhl) **fuels** are *the oil, coal, and natural gas that are used to provide energy.* Burning fossil fuels releases toxic gases into the atmosphere that harm humans.

Acid Rain

Certain chemicals in gases, such as sulfur dioxides and nitrogen oxides, mix with moisture in the air to form **acid rain,** which is *rain that is more acidic than normal rain.* Over time, acid rain can harm plants and forests. It can also contaminate water supplies. Acid rain can even eat away at rock and stone.

Smog

Fossil fuels create other gases that are changed by heat and sunlight into **ozone,** *a gas made of three oxygen atoms.* In the upper atmosphere, ozone occurs naturally. It helps protect you from the sun's harmful rays. Closer to ground level, ozone mixes with other gases to form smog. **Smog** is *a yellow-brown haze that forms when sunlight reacts with air pollution.*

Ozone and smog can worsen existing health problems. For example, a person with bronchitis, asthma, or emphysema may have a very hard time breathing when smog is in the air. Many cities issue warnings on days when there is too much smog or ozone in the air.

On such days, people sensitive to smog or ozone should limit the time they spend outside.

You can **do your part** by using products that **won't cause** more *damage.*

Damage *to the* Ozone Layer

The **ozone layer** is *a shield above the earth's surface that protects living things from ultraviolet (UV) radiation.* In the 1970s, scientists discovered that the ozone layer was being damaged. This damage was caused by the chemicals used in aerosol cans, refrigerators, and air conditioners. It was also caused by emissions from automobiles. Without the ozone layer, people are more likely to develop skin cancer and eye damage.

Many countries are banning the use of the chemicals that damage the ozone layer. You can do your part by using products that will not cause more damage.

Global Warming

The trapping of heat by carbon dioxide and other gases in the air is known as the **greenhouse effect.** The greenhouse effect warms Earth which helps to support life. However, the release of carbon dioxide and other gases from burning fossil fuels increases the greenhouse effect. This may be the cause of **global warming,** or *a rise in Earth's temperatures.* Global warming can affect weather patterns as well as ocean water levels.

Increases in the greenhouse effect may cause global warming. *Describe what you can do to prevent the greenhouse effect.*

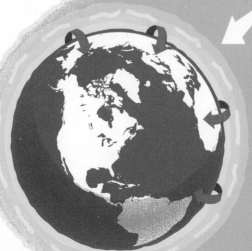

1. Light energy from the sun reaches the earth's lower atmosphere and is converted to heat.

2. A layer of carbon dioxide and other gases surrounding the earth traps the heat.

3. The surface of the earth and the lower atmosphere become warmer because of the trapped heat.

Pete Ryan/National Geographic Stock

WATER POLLUTION

MAIN IDEA Chemicals used on land are the primary source of water pollution.

Water is vital to all forms of life. However, wastes, chemicals, and other harmful substances pollute Earth's water. Water pollution is a widespread problem. Forty percent of all the nation's rivers, lakes, streams, and estuaries, or coastal areas, are too polluted to use for swimming, fishing, or drinking.

One type of pollution is sewage. **Sewage** is *human waste, garbage, detergents, and other household wastes washed down drains and toilets.* Sewage in the United States is treated. However, many countries do not properly treat water.

Chemicals used in industry also contribute to water pollution. Some enter the water from factories. Agriculture also contributes to water pollution. Oil spills from large tanker ships kill plants and animals and harm delicate habitats. Oil spills also run off into nearby lakes, rivers, and wetlands.

Water polluted with sewage can spread diseases. Eating shellfish from polluted water can cause hepatitis, a disease of the liver. Drinking water contaminated by metals such as lead or mercury can damage the liver, kidneys, and brain. It can also cause birth defects.

Pollution can affect your health as well as the health of every living thing around you. Before disposing of wastes, think about how they may affect the environment. What can you do to help protect the environment?

>>> Reading Check

DESCRIBE *How could water pollution affect your health?* ■

Oil spills drop 37 million gallons of oil into the oceans yearly. Another 363 million gallons of oil reach oceans from oil changes done at home. *How much more oil comes from oil changes than from oil spills?*

>>> After You Read

1. **DEFINE** Define *fossil fuels*.
2. **RECALL** Name two sources of air pollution and two sources of water pollution.
3. **IDENTIFY** How is smog formed?

>>> Thinking Critically

4. **EXPLAIN** How do fossil fuels contribute to global warming?
5. **ANALYZE** What is the difference between ozone in the upper atmosphere and ozone nearer to ground level?

>>> Applying Health Skills

6. **ACCESSING INFORMATION** Use reliable sources to research the dangers of exposure to lead in water and explain how to avoid this potentially harmful substance. Report your findings to the class.

Review

Audio

KEENPRESS/Getty Images

Preventing *and* Reducing Pollution

BIG IDEA You have the power to prevent and reduce pollution.

Before You Read

QUICK WRITE Make a list of actions you already take to reduce pollution.

▶ Video

As You Read

STUDY ORGANIZER Make the study organizer found in the FL pages in the back of the book to record the information presented in Lesson 2.

Vocabulary

› Environmental Protection Agency (EPA)
› Occupational Safety and Health Administration (OSHA)
› conservation
› biodegradable

 Audio

 Bilingual Glossary

🏃 Fitness Zone

Increasing My Fitness I can increase my fitness by walking, riding a bicycle, skateboarding, or inline skating instead of riding in a vehicle. All of these activities are examples of aerobic exercise, which help keep my heart and lungs healthy. At the same time, I am reducing air pollution by not using fossil fuels!

KEEPING THE ENVIRONMENT CLEAN

MAIN IDEA The Environmental Protection Agency is a government agency committed to protecting the environment.

Lately, Brian has noticed how much trash his family throws away. A lot of the trash in the garbage seems to be plastic bottles and aluminum cans. He wonders if convincing his family to recycle would help reduce some of the waste.

We can all do our part to help reduce pollution. When we work together as a community, we can do even more. Governments around the world are committed to reducing and preventing pollution. In the United States, the **Environmental Protection Agency (EPA)** is *the governmental agency that is committed to protecting the environment.* The **Occupational Safety and Health Administration (OSHA)** is *a branch of the U.S. Department of Labor that protects American workers.*

It is the responsbility of OSHA to make sure that work environments are safe and free of hazardous materials. If a workplace requires the use of these materials, OSHA will help the company to store the hazardous materials safely.

Local governments play a role in protecting the environment too. Many local governments maintain air and water quality. Some methods they use are waste management strategies and controlling auto emissions.

We can all *do our part* to help reduce pollution.

Waste management is the disposal of wastes in a way that protects the health of the environment and the people.

›› Reading Check

EXPLAIN *What is the Environmental Protection Agency (EPA)?*

REDUCING AIR POLLUTION

MAIN IDEA Using alternative transportation, staying tobacco free, and planting trees can all help reduce air pollution.

Any time you use an electrical appliance, ride in a car, or run a lawn mower, you are burning fossil fuels to produce energy. You are also contributing to air pollution. Here are some strategies to help reduce air pollution in your community.

- **Walk or ride your bike.** When you ride a bike or walk rather than ride in a vehicle, you save fuel and reduce pollution. You can also get the benefit from some physical activity.

- **Use public transportation or carpool.** Carpooling, or taking a bus, train, or subway, cuts down on the number of cars producing exhaust fumes.
- **Stay tobacco free.** Tobacco smoke is not only unhealthy for people who smoke, it pollutes the air.
- **Plant trees and other plants.** Plants convert remove carbon dioxide from the air during photosynthesis, helping reduce the amount of carbon dioxide in the atmosphere.

When you use less of a resource, such as fossil fuels, you are practicing conservation. **Conservation** is *the saving of resources.* You can conserve energy resources in your own home for example, by turning off lights when you leave a room. When you conserve energy, you are also using less fossil fuels.

>>> **Reading Check**

DESCRIBE *What can you do to promote cleaner air?*

Whenever you can, ride your bike rather than in a motor vehicle. It's better for the environment.
What health benefits do you get from riding a bike?

Thinkstock Images/Comstock Images/Getty Images

REDUCING WATER POLLUTION

MAIN IDEA ➤ Picking up after pets, picking up litter, using environment-friendly products, and disposing of chemicals properly can all reduce water pollution.

We all need clean drinking water. Clean water is important for all plants and animals. The industries and farms that use water to produce the foods and beverages we eat and drink need clean water. We also need clean water for water recreation activities. To help keep water clean, follow these tips.

- Pick up pet waste from public areas to reduce toxic runoff.
- Use soaps, detergents, and cleaners that are **biodegradable**—*broken down easily in the environment.*

- Pick up any litter that is not hazardous.
- Dispose of chemicals properly and legally. Never pour them into a drain.
- Take hazardous waste materials to the appropriate collection sites.

➤➤➤ Reading Check

LIST *What are two ways to reduce water pollution?* ■

The household products shown in this picture are all hazardous materials. *How do you safely dispose of hazardous wastes in your community?*

➤➤➤ After You Read

1. **DEFINE** What is the *Environmental Protection Agency*?
2. **GIVE EXAMPLES** What can you do to reduce air pollution?
3. **LIST** Name three ways you can help keep water clean.

➤➤➤ Thinking Critically

4. **INFER:** Why is it a good idea to turn off lights when you leave a room?
5. **ANALYZE** If conservation is a good idea, why do you think people might still need to be reminded to conserve resources?

➤➤➤ Applying Health Skills

6. **ADVOCACY** Write and illustrate a comic book that encourages teens to conserve electricity and water. In your comic book, be sure to explain why conservation of these resources is important.

🔄 Review

🔊 Audio

Protecting *the* Environment

BIG IDEA There are many things that you can do to protect the environment.

Before You Read

QUICK WRITE Write down three ways you can use less water.

 Video

As You Read

STUDY ORGANIZER Make the study organizer found in the FL pages in the back of the book to record the information presented in Lesson 3.

Vocabulary

› nonrenewable resources
› groundwater
› landfill
› hazardous wastes
› precycling
› recycle
› recycling

 Audio

 Bilingual Glossary

Developing Good Character

Citizenship Part of being a good citizen includes protecting the environment. You can do this as an individual by following the tips outlined in this chapter. You can participate in community clean ups and community rallies, and you can attend community meetings to discuss environmental issues in your area. You can also join a group that advocates for environmental protection.

PROTECTING NATURAL RESOURCES

MAIN IDEA Some resources, such as fossil fuels, are nonrenewable, meaning that they can be used only once.

Fossil fuels are natural materials known as nonrenewable resources. **Nonrenewable resources** are *substances that cannot be replaced once they are used.* Fossil fuels such as oil, natural gas, or coal are extracted from underground. Once they are used, they cannot be replaced.

> *Conservation* is a good way to **protect resources** such as water and trees.

Other resources are always being renewed. For example, the supply of freshwater is constantly being renewed through the water cycle. The water cycle refers to the natural movement of water through, around, and over the earth.

Even renewable resources, however, need to be protected. There is a limited amount of freshwater. Pollution makes freshwater more expensive because polluted water has to be cleaned before it is used. Trees are cut down to make paper and lumber. Removing too many trees upsets the balance of nature. By upsetting this balance, the lives of all living things are endangered. Conservation is a good way to protect resources such as water and trees.

Reading Check

EXPLAIN *What is a nonrenewable resource?*

Conserving Water Resources

As a teen, you may wonder how you can conserve water resources. How can one person conserve water? You can do several things to conserve water both at home and elsewhere.

INSIDE THE HOUSE

- Avoid letting water run unnecessarily. For example, turn off the faucet while brushing your teeth, or washing your face.
- Wash clothes in warm or cold water, which uses less energy than hot water.
- Run the washing machine or dishwasher only when you have a full load, and use the short cycle when appropriate.

How can *one person* conserve water?

- If you have an older toilet, place a 1 liter bottle filled with water inside your toilet tank. This will reduce the amount of water used for flushing. Another option is to replace an older toilet with a newer model that requires less water per flush.
- Fix leaky faucets to avoid the loss of water throughout the day. Leaky faucets in a home can waste up to 10,000 gallons of water a year. That is enough water to fill a swimming pool.
- Install water-saving showerheads or take shorter showers.

OUTSIDE THE HOUSE

- Turn the hose off when you are washing the car. Use the hose only for rinsing the car.
- Water lawns only when needed. Use soaker hoses for watering gardens. Avoid letting a sprinkler run while you are away from home.
- Garden with plants that conserve water.

This teen is conserving water by doing a full load of laundry. *If water is a renewable resource why do we have to conserve it?*

ThinkStock/age fotostock

DEALING WITH WASTES

Land pollution results from littering. It also occurs as a result of the careless disposal of household and industrial garbage. Many of the items we use in daily life are made of plastic and metal. When they are thrown away, these materials take a long time to break down, if they ever do. This affects not only the soil but also the air and groundwater, or *water that collects under the earth's surface.*

Types *of* Wastes

The average U.S. citizen produces about four pounds of trash, or solid waste, daily. The solid waste produced by households and businesses usually ends up in landfills. A landfill is *a huge specially designed pit where waste materials are dumped and buried.*

Landfills may have walls or linings so that water flowing through the landfill does not carry chemicals or other material into water supplies. In time, all landfills get filled up. When this happens, they are capped and sealed. A new landfill is made somewhere else.

Some wastes are hazardous to the health of all living things. Hazardous wastes are *Human-made liquid, solid, sludge or radioactive wastes that may endanger human health or the environment.* All hazardous wastes require special disposal. Hazardous wastes should be placed into a landfill. Some examples of hazardous wastes are dangerous industrial chemicals, asbestos, radioactive materials, as well as some medical wastes.

Hazardous substances from our homes include: motor oil, paint, insecticides, nail polish remover, antifreeze, bleach, and drain cleaner. Batteries, computers, and air conditioners also contain hazardous wastes.

Hazardous wastes are dangerous to the environment. To dispose of household hazardous waste, contact your local health department or environmental agency. They will explain how to get rid of it safely. Many communities have drop-off centers to collect household hazardous waste. Never put household hazardous wastes in the trash or pour chemicals down the drain.

>>> **Reading Check**

RECALL *What are two types of wastes?*

Even computers can harm the environment. *What other household products need to be disposed of as hazardous waste?*

Johner Royalty-Free/Getty Images

REDUCING, REUSING, AND RECYCLING

MAIN IDEA Precycling is reducing waste before it is used, and recycling conserves energy and natural resources.

Reducing wastes by _precycling—reducing waste before it occurs_—is one way to reduce the consumption of resources. Below are some basic guidelines for precycling:

- Buy products in packages made of glass, metal, or paper. It is possible to reuse or _recycle_ these materials, meaning _to change items in some way so that they can be used again._
- Carry purchases in your own reusable bags.
- Avoid using paper plates, plastic cups and utensils.
- Buy products in bulk to reduce packaging.

Reusing objects is another way to cut down on waste. For example, consider buying reusable food containers.

Reuse plastic grocery bags as trash bags. Donate unwanted clothes to charity rather than throwing the clothing out.

Recycling is _recovering and changing items so they can be used for other purposes._ Paper, aluminum, glass, plastics, and yard waste are the most commonly collected recycling materials. More and more people are becoming involved in recycling through drop-off centers and curbside programs.

As awareness of environmental health grows, people take a more active role in recycling. _What else can you infer from this graph?_

LESSON 3
REVIEW

>>> **After You Read**

1. **DEFINE** What is a _nonrenewable resource_?
2. **DESCRIBE** Name five common products that contain hazardous materials that contribute to pollution.
3. **IDENTIFY** Name the three Rs and tell how they are related to your health.

>>> **Thinking Critically**

4. **SYNTHESIZE** Explain why recycling and precycling are keys to a cleaner environment.
5. **ANALYZE** How does properly disposing of hazardous waste affect your environment as well as your personal health?

>>> **Applying Health Skills**

6. **GOAL SETTING** Working with family members, evaluate your current approach to the three Rs and set specific goals to improve your household record for reducing, reusing, and recycling.

🔄 Review

🔊 Audio

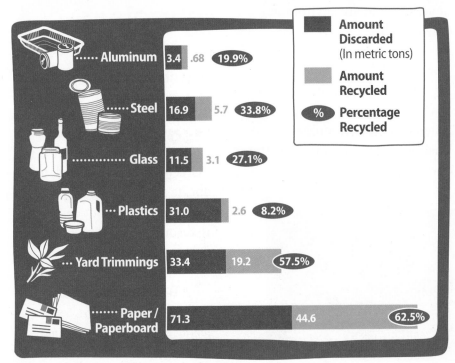

	Amount Discarded (In metric tons)	Amount Recycled	Percentage Recycled
Aluminum	3.4	.68	19.9%
Steel	16.9	5.7	33.8%
Glass	11.5	3.1	27.1%
Plastics	31.0	2.6	8.2%
Yard Trimmings	33.4	19.2	57.5%
Paper / Paperboard	71.3	44.6	62.5%

Source: U.S. Environmental Protection Agency, 2010.

Green Schools

BIG IDEA ▶ You can help your school become green.

Before You Read

QUICK WRITE What actions do you take at school to help protect the environment?

▶ **Video**

As You Read

STUDY ORGANIZER Make the study organizer found in the FL pages in the back of the book to record the information presented in Lesson 4.

Vocabulary

› green school
› pesticide

◀》 **Audio**

ⒶⒷ **Bilingual Glossary**

WHAT IS A GREEN SCHOOL?

MAIN IDEA ▶ Green schools are environmentally friendly and provide a healthy environment for students.

Robert noticed a lot of paper was being thrown away in his school's garbage bins. He wondered if it might be a good idea to start a school recycling program. He asked his friends, other students, his teachers, and the school principal to join his program to help recycle the school's waste.

You may have heard the term *green* used to refer to programs or actions that are environmentally friendly. A green school is *one that is environmentally friendly in several different ways.* Green schools have made changes to conserve energy, water, and other resources. Their goal is to provide a healthy environment for students.

Green schools eliminate toxins such as mold, certain cleaning chemicals, and pesticides. A pesticide is a *product used on crops to kill insects and other pests.* Green schools make sustainable food choices, such as having a school garden, and reducing solid wastes. Solid wastes include food and paper.

Green schools have made changes to conserve **energy, water,** and other **resources.**

Green schools also work to educate students, parents, and the community about conserving resources and protecting the environment.

> ⟩⟩⟩ **Reading Check**
>
> **EXPLAIN** *How can green schools provide a healthy environment for students?*

Myth vs. Fact

Myth: Changing our actions with regard to energy use at school doesn't really save energy.

Fact: When students and schools take steps to turn out lights in unoccupied rooms and turn off computers at night and on weekends, it can save energy and reduce energy bills by thousands of dollars per year.

Lisafx/age fotostock

BEING GREEN AT SCHOOL

MAIN IDEA You can protect the environment while at school.

Are you wondering what you can do to help protect the environment at school? The tips listed below may give you some ideas.

- **Don't litter.** Throw trash into trashcans and recycling bins. Pick up litter when you see it. Organize litter clean-up events in your community.
- **Conserve energy.** Walk, ride your bicycle, take the bus, or carpool. Turn off lights and other electronic equipment when they are not in use.
- **Conserve water.** Turn off the faucet while you wash your hands. Report leaking or dripping faucets or toilets to your teacher or principal.
- **Reduce solid waste.** Reuse paper by writing on both sides. Make double-sided copies or printouts. Pack a lunch from home in reusable containers and carry it in a reusable lunchbag.

Helping Your School Go Green

You can do many things to conserve resources. The following list contains steps you can take to help protect the environment.

- **Reduce exposure to chemicals.** Set up a program to properly dispose of hazardous wastes including batteries, fluorescent light bulbs, and electronic waste.
- **Start a recycling or composting program.** Recycle paper, plastic bottles, glass bottles, and aluminum cans. Build a compost pile so that food wastes from the kitchen can be composted.
- **Start a school garden.** Grow vegetables for your school's cafeteria. Have the kitchen incorporate food that you grow into lunches.
- **Start a group or club.** Create campaigns that explain how environmentally-friendly actions can be practiced.

> ### ⟩⟩⟩ Reading Check
> **DESCRIBE** *Identify two actions to help protect the environment.* ■

Photographer's Choice/Getty Images

Some schools provide bins for collecting different kinds of recyclable materials. *What ways does your school encourage recycling?*

⟩⟩⟩ After You Read

1. **VOCABULARY** Define *green school*.
2. **DESCRIBE** What actions can you take to protect the environment at school?
3. **IDENTIFY** List three actions you can take to help your school be more green.

⟩⟩⟩ Thinking Critically

4. **APPLY** Briefly describe how each of the following strategies help your school become more green and helps protect the environment: recycling paper and plastic products, planting a school garden.
5. **EVALUATE** Why is it important for schools to become actively involved in protecting the environment?

⟩⟩⟩ Applying Health Skills

6. **COMMUNICATION SKILLS** Choose one action your school could take to become more green. Write a presentation that could be given at a school board meeting to advocate for this change. Include facts and statistics to support your request.

 Review

🔊 Audio

Managing *the* Packaging

WHAT YOU WILL NEED

* small bag of potato chips
* wrapped slices of cheese
* video game (in original packaging)
* batteries (in original packaging)
* graph paper and pencil

WHAT YOU WILL DO

1 Work in small groups to create a graph.

2 Determine the unnecessary packaging of each of the four products. Rate them on a scale from 1 to 5, with 5 being the most unnecessary. Mark the product's rating on the graph's vertical (*y*) axis.

3 On a scale from 1 to 5, with 5 being the highest, rate the likelihood that a product could be recycled or reused. Mark the product's rating on the horizontal (*x*) axis.

4 Draw a horizontal line out from the product's packaging rating. Draw a vertical line up from the product's recycling rating. Where the lines meet, write the name of the product.

Packaging can be useful when it protects a product. Packaging can also be wasteful. Unnecessary packaging uses up the Earth's resources and can harm the environment when discarded.

©Comstock Images/Getty Images

WRAPPING IT UP

What did you consider before deciding where to place the product on the graph? Where on the graph do you find the objects that are the most environment-friendly? Where are the least environment-friendly objects?

READING REVIEW

FOLDABLES and Other Study Aids

Take out the Foldable® that you created and any study organizers that you created. Find a partner and quiz each other using these study aids.

LESSON 1 Pollution and Health

BIG IDEA Pollution harms the environment and your health, and is often ugly.

* Pollution is made up of dirty or harmful substances in the environment.
* Burning fossil fuels such as coal, oil, and natural gas pollutes the air.
* Chemicals used on land are the primary source of water pollution.

LESSON 2 Preventing and Reducing Pollution

BIG IDEA You have the power to prevent and reduce pollution.

* The Environmental Protection Agency is a government agency committed to protecting the environment.
* Using alternative transportation, staying tobacco free, and planting trees can all help reduce air pollution.
* Picking up after pets, picking up litter, using environment-friendly products, and disposing of chemicals properly can all reduce water pollution.

LESSON 3 Protecting the Environment

BIG IDEA There are many things that you can do to protect the environment.

* Some resources, such as fossil fuels, are nonrenewable, meaning that they can be used only once.
* Recycling and conservation have a positive impact on the environment.
* Precycling is reducing waste before it is used, and recycling conserves energy and natural resources.

LESSON 4 Green Schools

BIG IDEA You can help your school become green.

* Green schools are environmentally friendly and provide a healthy environment for students.
* You can protect the environment while at school.

 Review

 Web Quest

ASSESSMENT

Reviewing Vocabulary *and* Main Ideas

> pollution
> acid rain
> ozone

> sewage
> Environmental
   Protection Agency

> conservation
> biodegradable

> Occupational
   Safety and Health
   Administration

›› On a sheet of paper, write the numbers 1–8. After each number, write the term from the list that best completes each statement.

LESSON 1 Pollution and Health

1. _____ can carry pathogens that cause disease.

2. Up in the atmosphere, _____ protects people from UV rays; closer to earth, it is part of smog.

3. Dirty or harmful substances in the environment are _____.

4. _____ can ruin forests and even eat away at stone.

LESSON 2 Preventing and Reducing Pollution

5. Waste that is _____ can easily break down in the environment.

6. _____ is the saving of resources.

7. The _____ is a governmental agency that is committed to protecting the environment.

8. A branch of the U.S. Department of Labor that protects workers is called _____.

›› On a sheet of paper, write the numbers 9–14. Write *True* or *False* for each statement below. If the statement is false, change the underlined word or phrase to make it true.

LESSON 3 Protecting the Environment

9. Buying a food container that can be used many times to store food is an example of <u>recycling</u>.

10. Oil, natural gas, and coal are <u>renewable</u> resources.

11. Using fluorescent light bulbs can <u>save</u> electricity.

12. Conservation is the <u>wasting</u> of resources.

LESSON 4 Green Schools

13. A green school is environmentally <u>unfriendly</u>.

14. <u>Pesticides</u> are toxic chemicals used to kill insects.

✔ eAssessment

>> Using complete sentences, answer the following questions on a sheet of paper.

🗣 *Thinking* **Critically**

15. INFER What do you think is meant by the saying "We all live downstream"?

16. INTERPRET A volcanic eruption can send tons of smoke and ash into the air. Do volcanoes pollute? Explain your answer.

🎗 *Write* **About It**

17. OPINION Write a short essay explaining why you think some products have more packaging than necessary. Include ideas as to how reducing excess packaging could help the environment.

Ⓐ Ⓑ Ⓒ Ⓓ STANDARDIZED TEST PRACTICE

Reading
Read the passage below and then answer the questions that follow.

Think about how often you throw away plastic, such as milk cartons or soda bottles. Using and then throwing away plastic causes landfills to fill up quickly. In the past, objects made out of wood, paper, cotton, or wool would biodegrade fairly easily. These objects broke down naturally over time. However, plastic items do not break down. That means plastic wastes must be stored in a landfill. Recently, scientists have discovered a way to make a sturdy, durable plastic that biodegrades when buried in dirt. Scientists are hopeful that using biodegradable plastic will help reduce the amount of waste buried in landfills.

1. What is the main point of the passage?
   A. Life was better years ago.
   B. Wood and paper are biodegradable.
   C. Biodegradable plastic will reduce the waste in landfills.
   D. Throwing away more plastic will actually reduce the waste in landfills.

2. What does *durable* mean in this sentence?
   Recently, scientists have discovered a way to make a sturdy, durable plastic that biodegrades when buried in dirt.
   A. weak
   B. flexible
   C. tough
   D. large

Chapter 1 Foldables®

Make this Foldable® to help organize what you learn in Lesson 1.

1 Begin with a plain sheet of notebook paper. Fold the sheet of paper in half along the long axis.

2 Turn the paper, and fold it into thirds.

3 Unfold and cut the top layer along both fold lines. This makes three tabs. Draw two overlapping ovals.

4 Label the left tab *Verbal,* the middle tab *Communication,* and the right tab *Nonverbal.*

Write the definitions and examples of verbal and nonverbal communication under the appropriate tab. Under the middle tab, describe how both types of communication help to share feelings, thoughts, and information.

Chapter 2 Foldables®

Make this Foldable® to help organize what you learn in Lesson 1.

1 Begin with a plain sheet of notebook paper. Draw a straight line across the middle of the page to divide it in half.

2 Fold the top and bottom quarters of the page to meet the line in the middle.

3 Next, fold the page in half.

4 Cut along the creases you have made in the top and bottom flaps.

5 On the four flaps you have created, write *Thinking about Dating, Group Dating, Individual Dating,* and *Healthy Ways to Show Affection.*

List the main ideas from this lesson under each of the appropriate tabs.

Chapter 3 Foldables®

Make this Foldable® to record what you learn Lesson 1.

1 Begin with a plain sheet of notebook paper. Fold it in half so its long edges meet.

2 Fold the paper into six smaller sections.

3 Open your Foldable® and cut six tabs along one edge of the paper.

Define key terms and record facts about the six traits of good character.

Chapter 4 Foldables®

Make this Foldable® to record what you learn Lesson 1.

1 Begin with a sheet of notebook paper. Fold it along the long axis. Leave a ½" tab along the side.

2 Unfold and cut the top layer along the three fold lines. This makes four tabs.

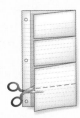

3 Turn the paper vertically and label the tabs bullying, cyberbullying, harassment, and strategies to stop bullying.

Define key terms and record facts about bullying, cyberbullying, harassment, and strategies to stop bullying.

Chapter 5 Foldables®

Make this Foldable® to help you organize the main ideas in Lesson 1.

1 Begin with a plain sheet of notebook paper. Fold the bottom edge up to form a pocket.

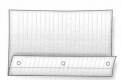

3 Open your Foldable® and glue the sides of each pocket.

2 Fold the paper into three sections.

On index cards or quarter sheets of notebook paper, take notes on personality, self-concept, and self-esteem.

Chapter 6 Foldables®

Make this Foldable® to help you record what you learn in Lesson 1.

1 Begin with a plain sheet of notebook paper. Fold the sheet of paper along the long axis. Leave a ½" tab along the side.

3 Unfold and cut the top layer along the three fold lines.

2 Turn the paper. Fold in half, then fold again.

4 Turn the paper vertically and label the tabs: anxiety disorders, mood disorders, personality disorders, and schizophrenia.

Write the definitions of each type of mental and emotional disorder. List characteristics of each mental and emotional disorder under the appropriate tab.

Chapter 7 Foldables®

Make this Foldable® to help you organize what you learn in Lesson 1.

1 Begin with a plain sheet of 11" x 17" paper. Fold the paper into thirds along the long axis, and then fold in half again lengthwise. This forms six rows.

2 Open the paper and refold the sheet into thirds along the short axis. This forms three columns.

3 Unfold and draw lines along the folds. List causes of conflict in the left column. Use the top column to list two places where conflict may occur.

As you read the lesson, fill in the chart with examples of behaviors that might cause conflicts.

Chapter 8 Foldables®

Make this Foldable® to help you organize what you learn in Lesson 1.

1 Begin with a plain sheet of 11" x 17" paper. Holding the paper lengthwise, fold down a 2-inch flap at the top of the sheet.

2 Next, fold the paper into thirds.

3 Unfold and write at the top of each of the three sections, *Types of Violence, Factors in Teen Violence,* and *Effects of Violence.*

As you read the lesson, fill in the chart under each of the three headings you have created with information you learn about each concept.

FOLDABLES®

Chapter 9 Foldables®

Make this Foldable® to record what you learn in Lesson 1.

1 Place the two sheets of paper 1 inch apart.

2 Fold up the bottom edges, stopping them 1 inch from the top edges. This makes all tabs the same size.

3 Crease the paper to hold the tabs in place.

4 Turn and label the tabs as shown.

Under the appropriate tab of your Foldable®, define terms and record information on nutrients and influences on food choices.

Chapter 10 Foldables®

Make this Foldable® to record what you learn in Lesson 1.

1 Fold the sheet of paper into thirds along the short axis.

2 Open and fold up the bottom edge to form a pocket. Glue the edges.

3 Label each pocket as shown. Place an index card or quarter sheet of notebook paper into each pocket.

Write down key points on each of these elements of fitness on index cards. Store the cards in the appropriate pocket of your Foldable®.

Chapter 11 Foldables®

Make this Foldable® to help you organize what you learn in Lesson 1.

1 Begin with a plain sheet of notebook paper. Fold the sheet of paper along the long axis, leaving a 2" tab along the side.

2 Turn the paper and fold it into thirds.

3 Unfold and cut the top layer along both fold lines. This makes three tabs.

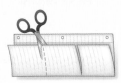

4 Label the long tab *Adolescence*, and label the three smaller tabs *Physical Changes*, *Mental/Emotional Changes*, and *Social Changes*.

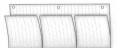

As you read the lesson, record what you learn about changes during adolescence.

Chapter 12 Foldables®

Make this Foldable® to help you organize what you learn in Lesson 1.

1 Begin with a plain sheet of notebook paper. Fold the sheet of paper along the long axis, leaving a 2" tab along the side.

2 Fold the paper in half, and then fold it again into fourths.

3 Unfold the paper.

4 Cut along the three fold lines on the front flap.

5 Label the long tab *Personal Care*, and label the four smaller tabs *Hygiene*, *Hair Care*, *Nail Care*, and *Consumer Skills*.

Under the appropriate tab, record what you learn about personal health care.

Chapter 13 Foldables®

Make this Foldable® to help you organize what you learn in Lesson 1.

1 Begin with a plain sheet of notebook paper. Fold the sheet of paper along the long axis, leaving a 1/2" tab along the side.

2 Fold the paper in half, and then fold it in half again.

3 Unfold the paper and cut the top layer along the three fold lines. This makes four tabs.

4 Turn the paper vertically, and label the four tabs *Skeletal System, Muscular System, Problems,* and *Care*.

Under the appropriate tab, record what you learn about your skeletal and muscular systems, problems with your bones and muscles, and caring for them.

Chapter 14 Foldables®

Make this Foldable® to help you organize what you learn in Lesson 1.

1 Begin with a plain sheet of notebook paper. Fold up the bottom edge to form a pocket. Glue the edges.

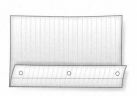

2 Fold the paper in half along the short axis to form a booklet.

3 Label the front of the booklet *Facts About Tobacco,* and label the inside pockets *What Is Tobacco?* and *Chemicals in Tobacco.* Place an index card or quarter sheet of notebook paper into each pocket.

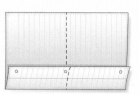

Take notes on index cards about the different types of tobacco products and the harmful chemicals they contain. Store your notes in the Foldable®.

Chapter 15 Foldables®

Make this Foldable® to help you organize what you learn in Lesson 1.

1 Begin with a plain sheet of 11″ x 17″ paper. Fold the short sides inward so that they meet in the middle.

2 Fold the top to the bottom.

3 Unfold and cut along the inside fold lines to form four tabs.

4 Label the tabs *What Is Alcohol, Why Do Some Teens Use Alcohol, Reasons Not to Drink,* and *Consequences of Alcohol Use.*

Under the appropriate tab, record what you learn about these four topics.

Chapter 16 Foldables®

Make this Foldable® to help you organize what you learn in Lesson 1.

1 Begin with a plain sheet of notebook paper. Fold the sheet along the long axis, leaving a ½″ tab along the side.

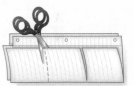

2 Fold the paper into thirds.

3 Unfold the paper and cut the shorter flap along both folds to create two tabs.

4 Label the long tab at the top *Drug Use and Abuse,* and label the three shorter tabs *Physical Consequences, Mental/Emotional Consequences,* and *Social Consequences.*

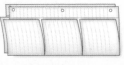

Under the tabs, record what you learn about the consequences of drug use and abuse.

Chapter 17 Foldables®

Make this Foldable® to help you organize what you learn in Lesson 1.

1 Begin with a plain sheet of notebook paper. Fold the sheet of paper along the long axis, leaving a 1/2" tab along the side.

2 Fold the paper in half, and then fold it in half again.

3 Unfold the paper and cut the top layer along the three fold lines. This makes four tabs.

4 Turn the paper vertically, and label the four tabs *What Are Medicines, The Purpose of Medicines, Prescription Medicines,* and *Non-Prescription Medicines.*

Record what you learn about the types and purposes of medicines.

Chapter 18 Foldables®

Make this Foldable® to help you organize what you learn in Lesson 1.

1 Begin with a plain sheet of notebook paper. Fold the sheet of paper along the long axis, leaving a 1/2" tab along the side.

2 Turn the paper. Fold in half, then fold in half again.

3 Unfold the paper and cut the top layer along the three fold lines. This makes four tabs.

4 Turn the paper vertically, and label the four tabs *Germs and Pathogens, How Pathogens Spread, Protecting Yourself,* and *Protecting Others.*

Under the appropriate tab, record what you learn about communicable diseases.

Chapter 19 Foldables®

Make this Foldable® to help you organize what you learn in Lesson 1.

1 Begin with a plain sheet 11" x 17" paper. Fold it into thirds along the short axis.

2 Open and fold the bottom edge up to form a pocket. Glue the edges.

3 Label each pocket *Causes, Effects,* and *Treatment.* Place an index card or quarter sheet of notebook paper into each pocket.

Record what you learn about the causes, effects, and treatments of noncommunicable diseases.

Chapter 20 Foldables®

Make this Foldable® to help you organize what you learn in Lesson 1 about safety.

1 Begin with three plain sheets of 8 ½ " x 11" paper. Place the sheets ½" apart.

2 Roll up the bottom edges, stopping them ½" from the top edges. This makes all tabs the same size.

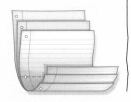

3 Crease the paper to hold the tabs in place and staple along the fold.

Record information from each lesson in the chapter about safety.

Chapter 21 Foldables®

Make this Foldable® to help you organize what you learn in Lesson 1.

1 Begin with a plain sheet 8 ½" x 11" paper from top to bottom, leaving a 2" tab at the bottom.

3 Unfold the paper once. Cut along the center fold line of the top layer only. This makes two tabs.

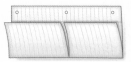

2 Fold in half from side to side.

Under the appropriate tab, take notes on the causes and effects of pollution.

Study Organizers

Use the following study organizers to record the information presented in the Lessons.

Four Section Chart

Key Word Cluster

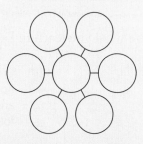

Two-Column Chart

Bulls-eye

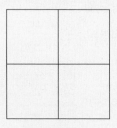

Three-Column chart

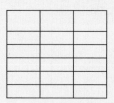

Flow Chart

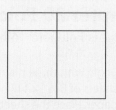

Index Cards

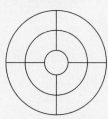

Outline

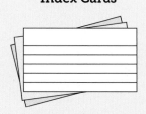

I.
  A.
    1.
    2.
  B.
II.

Venn Diagram

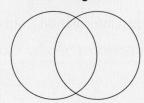

FLIP 4 FITNESS

Flip for Fitness is for everyone. Non-athletes who avoid joining organized sports can develop a personal fitness plan to stay in shape. Even athletes can use some of the tips to cross train for their favorite sport.

Planning a Routine

Flip for Fitness helps you plan a fitness routine that helps your body slowly adjust to activity. Over time, you will increase both the length of time you spend and the number of times that you are physically active each week. Teens should aim to get at least one hour of physical activity each day. These periods of physical activity can be divided into shorter segments, such as three 20 minute segments each day. Exercise includes any physical activity, such as completing a fitness plan, playing individual or group sports, or even helping clean at home. The key is to keep your body moving.

Before You Start Exercising

Every activity session should begin with a warm-up to prepare your body for exercise. Warm-ups raise your body temperature and get your muscles ready for physical activity. Easy warm-up activities include walking, marching, and jogging, as well as basic calisthenics or stretches. As you increase the time you spend doing a fitness activity, you should increase the time you spend warming up. Check the Sample Physical Fitness Plan in Teen Health in Connect Ed.

Alistair Berg/Digital Vision/Getty Images

Fitness Information *and* Resources

Fitness Apps *and* Other Resources

» **USDA's MyPlate** The MyPlate Super Tracker is a free online fitness and diet tracker. To review the tracker, go online to https://www.choosemyplate.gov and search for "Super Tracker".

» Additionally, organizations such as the **American Heart Association** and **KidsHealth** provide resources on developing walking programs. The online addresses are: http://startwalkingnow.org and http://kidshealth.org.

» Finally, smartphone and tablet users can download several nutrition and fitness tracking apps. Many are free of charge. Use the terms "fitness", "exercise", or "workout" when searching for apps.

Accessing Information

» The *Teen Health* online program includes resources to develop your own fitness plan. Check out the **Fitness Zone** resources in ConnectEd.

» The Centers for Disease Control and Prevention's, **Body & Mind (BAM)** web site also provide fitness information. The online address is: http://www.bam.gov. Search for "physical activity" or "activity cards."

Safety Tips

On the following pages, you'll find fitness activities for groups or individuals. Each activity includes information on what you'll need, how to start, and how to stay safe. Safety is the most important factor.

⚠ Always be aware of where you are and don't take any unnecessary chances.

⚠ Obey the rules of the road while riding your bicycle, avoid unsafe areas, and use the proper safety equipment when working out.

⚠ Finally, remember to drink water and to rest between exercise sessions.

5 Elements *of* Fitness

When developing a fitness plan, it's helpful to have a goal. Maybe your goal is to comfortably ride your bike to school each day or maybe you want to complete the Tour de France in the future. Regardless of the reasons why you develop a fitness plan, focusing on the five elements of fitness will help you achieve overall physical fitness. The five elements are:

1 Cardiovascular Endurance

The ability of the heart and lungs to function efficiently over time without getting tired. Familiar examples are jogging, walking, bike riding, and swimming.

2 Muscle Endurance

The ability of a muscle or a group of muscles to work non-stop without getting tired. Many activities that build cardiovascular endurance also build muscular endurance, such as jogging, walking, and bike riding.

3 Muscle Strength

The ability of the muscle to produce force during an activity. Activities that can help build muscle strength include push-ups, pull-ups, lifting weights, and running stairs.

4 Flexibility

The ability to move a body part freely, without pain. Improve your flexibility by stretching gently before and after exercise.

5 Body Composition

The amount of body fat a person has compared with the amount of lean mass, which is bone, muscle, and fluid. A healthy body is made up of more lean mass and less body fat. Body composition is a result of diet, exercise, and heredity.

Fitness Circuit

Are you looking for a quick workout that will develop endurance, strength, and flexibility? A Fitness Circuit may be just what you need. Many public parks have Fitness Circuits (sometimes called Par Courses) with exercise stations located throughout a park. You walk or run between stations as part of your workout. A fitness circuit can also be created in your backyard or even a basement.

What Will I Need?

» Access to a public park or a home-made Fitness Circuit course.

» Comfortable workout clothes that wick away perspiration.

» Athletic shoes.

» Stopwatch (optional).

» Jump rope, dumbbells, exercise bands, or check out the Fitness Zone Clipboard Energizer Activity Cards, Circuit Training for ideas.

How Do I Start?

» In the park, read the instructions at each exercise station and perform the exercises as shown. Use the correct form. Try to do as many repetitions as you can for 30 seconds.

» After you finish the exercise, walk or run to the next station and complete that exercise.

» Check your heart rate to see how intensely you exercised at the end of the Fitness Circuit.

» Every month or so, consider adding a new exercise.

How Can I Stay Safe?

» Be alert to your surroundings in a public park. It is best to have a friend with you. It's also more fun to exercise with a friend.

» At home, leave enough room between stations to allow you to move and exercise freely. Avoid clutter in your exercise area.

» Perform the exercises correctly and at your own pace.

Walking

Walking is more than just a way to get from one place to another. It's also a great physical activity. By walking for as little as 30 minutes each day you can reduce your risk of heart disease, manage your weight, and even reduce stress. Walking requires very little equipment and you can do it almost anywhere. More good news: Walking is also something you can do by yourself or with friends and family.

Colin Hawkins/Cultura/Getty Images

What Will I Need?

» Running or walking shoes. Many athletic shoe stores sell both.

» Loose comfortable clothes that wick away perspiration. Layering is also a good idea. Consider adding a hat, sunglasses, and sunscreen if needed.

» Stopwatch and water bottle unless there are water fountains on your route.

» A pedometer or GPS to track your distance.

How Do I Start?

» Five minutes of easy stretching.

» Walk upright with good posture. Do not exaggerate your stride or swing your arms across your body.

» Build your time and distance slowly. One mile or 20 minutes every other day may be enough for the first couple weeks. Eventually you will want to walk at least 30-60 minutes five days a week.

How Can I Stay Safe?

» Let your parents know where you will be walking and how long you will be gone.

» Avoid wearing headphones if by yourself or if walking on a road or street.

Running *or* Jogging

Running or jogging is one of the best all-around fitness activities. Running uses the large muscles of the legs thereby burning lots of calories and also gives your heart and lungs a good workout in a shorter amount of time. Running also helps get you into condition to play team sports like basketball, football, or soccer. More good news is that running can be done on your schedule although it's also fun to run with a friend or two.

What Will I Need?

» A good pair of running shoes. Ask your Physical Education teacher or an employee at a specialist running shop to help you choose the right pair.

» Socks made of cotton or another type of material that wicks away perspiration.

» Bright colored or reflective clothing and shoes.

» A stopwatch or watch with a second sweep to time your runs or track your distance.

» Optional equipment might include a jacket or other layer depending on the weather, sunscreen, and sunglasses.

How Do I Start?

Your ultimate goal is to run at least 20 to 30 minutes at least 3 days a week. Use the training schedule shown below. Start by walking and gradually increasing the amount of time you run during each exercise session. Starting slowly will help your muscles and tendons adjust to the increased work load. Try spacing the three runs over an entire week so that you have one day in-between runs to recover.

How Can I Stay Safe?

» Use the correct equipment for the sport you have chosen.

» Running on a track, treadmill, or in a park with level ground will help you avoid foot or ankle injuries.

» Avoid running on the road, especially at night.

» Avoid wearing headphones unless you are on a track, treadmill, or another safe place. Safety experts agree that headphones can distract you from being alert to your surroundings.

Here is a plan to get you started as a runner:

» Start each run with a brisk 3-5 minute walk to warm-up.

» Take some time to slowly stretch the muscles and areas of the body involved in running. Avoid "bouncing" when stretching or trying to force a muscle or tendon to stretch when you start to feel tightness.

» Begin slowly and gradually increase your distance and speed. A good plan for the first several weeks is to alternate walking with easy running. The running plan included in this section can give you some tips on how to train for a 5K run.

» Use the "talk test." Can you talk in complete sentences during your training runs? If not, you are running too fast.

Training Schedule

M W F	Split Schedule	Duration
Week 1	Brisk 5 min. walk Walk: 60 - 90 seconds Run: 60 seconds	Repeat for 20 min.
Week 2 and 3	Brisk 5 min. walk Walk: 60 seconds Run: 60 - 90 seconds	Repeat for 20 min.
Week 4	Brisk 5 min. walk Walk: 60 seconds Run: 3 - 5 minutes	Repeat for 20 min.
Week 5+	Brisk 5 min. walk Run: 20 to 30 minutes	

Preparing *for* Sports and Other Activities

Do you want to play a sport? If so, think about developing a fitness plan for that sport. Some of the questions to ask yourself are: Does the sport require anaerobic activity, like running and jumping hurdles? Does the sport require aerobic fitness, like cross-country running? Other sports, such as football and track require muscular strength. Sports like basketball require special skills like dribbling, passing, and shot making. A workout plan for that sport will help you get into shape before organized practice and competition begins.

What Will I Need?

Each sport has different equipment requirements. Talk to a coach or physical education teacher about how to get ready for your sport. You can also conduct online research to learn what type of equipment you will need, such as:

» Proper footwear and workout clothes for a specific sport.

» What facilities are available for training and practice, such as a running track, tennis court, football or soccer field, or other safe open area.

» Where you can access weights and others form of resistance training as part of your training.

How Do I Start?

Now that your research is done, you can create your fitness plan. Include the type of exercises you will do each training day.

» Include a warm-up in your plan.

» List the duration of time that you will work out.

» Plan to exercise 3–5 days a week doing at least one kind of exercise each day. Remember to include stretching before every workout.

How Can I Stay Safe?

» Get instruction on how to use free weights and machines

» Make sure you start every activity with a warm-up.

» Ease into your fitness plan gradually so you do not pull a muscle or do too much too soon.

» Practice good nutrition and drink plenty of water to stay hydrated.

Interval Training

Getting fit takes time. One method, interval training, can show improvement in two weeks or less. Interval training consists of a mix of activities. First you do a few minutes of intense exercise. Next, you do easier, less-intense activity that enables your body to recover. Interval training can improve your cardiovascular endurance. It also helps develop speed and quickness.

Intervals are typically done as part of a running program. Not everyone wants to be a runner though. Intervals can also be done riding a bicycle or while swimming. On a bicycle, alternate fast pedaling with easier riding. In a pool, swim two fast laps followed by slower, easier laps.

What Will I Need?

» A running track or other flat area with marked distances like a football or soccer field.

» If at a park, 5–8 cones or flags to mark off distances of 30 to 100 yards.

» A training partner to help you push yourself (optional).

How Do I Start?

» After warming up, alternate brisk walking (or easy jogging). On a football field or track, walk 30 yards, jog 30 yards, and then run at a fast pace for 30 yards. Rest for one minute and repeat this circuit several times. If at a park, use cones or flags to mark off similar distances.

» Accelerate gradually into the faster strides so you stay loose and feel in control of the pace.

» If possible, alternate running up stadium steps instead of fast running on a track. This will help your coordination as well as your speed. Running uphill in a park would have similar benefits.

How Can I Stay Safe?

» Interval training works the heart and lungs. For this reason, a workout using interval training should be done only once or twice a week with a day off between workouts.

» Check with your doctor first. If you have any medical condition like high blood pressure or asthma, ask your doctor if interval training is safe for you.

Glossary/Glosario

English

Español

Abdominal thrust Quick inward and upward pull into the diaphragm to force an obstruction out of the airway.

presión abdominal Presión rápida, hacia adentro y arriba sobre el diafragma, para desalojar un objeto que bloquea la vía respiratoria de una persona.

Abstinence (AB stuh nuhns) The conscious, active choice not to participate in high-risk behaviors.

abstinencia Opción activa y conciente de no participar en comportamientos de alto riesgo.

Abuse (uh BYOOS) The physical, emotional, or mental mistreatment of another person.

abuso Maltrato físico, emocional o mental de otra persona.

Accident Any event that was not intended to happen.

accidente Suceso que ocurre de manera no intencional.

Accident chain A series of events that include a situation, an unsafe habit, and an unsafe action.

accidente en cadena Serie de sucesos que incluye una situación, un hábito peligroso y un acto peligroso.

Accidental injuries Injuries resulting from an accident.

lesiones accidentales Lesiones que resultan de un accidente.

Accountability A willingness to answer for your actions and decisions.

responsabilidad Voluntad de responder de tus acciones y decisiones.

Acid rain Rain that is more acidic than normal rain.

lluvia ácida Lluvia que es más ácida de lo normal.

Acne (AK nee) Skin condition caused by active oil glands that clog hair follicles.

acné Condición de la piel causada por glándulas de aceite activas que obstruyen folículos de cabello.

Acquaintance Someone you see occasionally or know casually.

conocido Alguien a quien ves ocasionalmente o conoces casualmente.

Acquired Immunodeficiency Syndrome (AIDS) A deadly disease that interferes with the body's natural ability to fight infection.

síndrome de inmunodeficiencia adquirida (SIDA) Enfermedad mortal que interfiere con la habilidad natural del cuerpo de combatir infecciones.

Active listening Hearing, thinking about, and responding to another person's message.

audición activa Oír el mensaje de otra persona, pensar en el mensaje y responder.

Addiction A mental or physical need for a drug or other substance.

adicción Necesidad mental o física de una droga u otra substancia.

Addictive Capable of causing a user to develop intense cravings.

adictivo Capaz de ocasionar que el consumidor desarrolle una necesidad repentina intensa.

Adolescence (a duhl EH suhns) The stage of life between childhood and adulthood, usually beginning somewhere between the ages of 11 and 15.

adolescencia Periodo de vida entre la niñez y la adultez que empieza generalmente entre los 11 y los 15 años.

Adrenaline (uh DRE nuhl in) A hormone that increases the level of sugar in the blood, giving your body extra energy.

adrenalina Hormona que aumenta el nivel de azúcar en la sangre, y por lo tanto, proporciona energía adicional al cuerpo.

Advocacy Taking action in support of a cause.

promoción Actuar en apoyo de una causa.

Aerobic (ah ROH bik) exercise Rhythmic, moderate-to-vigorous activity that uses large amounts of oxygen and works the heart and lungs.

ejercicio aeróbico Actividad rítmica, moderada o fuerte que usa grandes cantidades de oxígeno y trabaja el corazón y los pulmones.

Affection Feelings of love for another person.

afecto Sentimiento de amor hacia otra persona.

Aftershocks Smaller earthquakes as the earth readjusts after the main earthquake.

réplica sísmica Temblores mas pequeños que ocurren mientras la tierra se reajusta después de un terremoto principal.

English

Aggressive Overly forceful, pushy, hostile, or otherwise attacking in approach.

Alcohol (AL kuh hawl) A drug created by a chemical reaction in some foods, especially fruits and grains.

Alcohol abuse Using alcohol in ways that are unhealthy, illegal, or both.

Alcohol poisoning A dangerous condition that results when a person drinks excessive amounts of alcohol over a short time period.

Alcoholism A disease in which a person has a physical and psychological need for alcohol.

Allergens (AL er juhnz) Substances that cause allergic responses.

Allergy Extreme sensitivity to a substance.

Alternative (ahl TER nuh tihv) Another way of thinking or acting.

Alveoli (al VEE oh lye) The tiny air sacs in the lungs.

Anabolic steroids (a nuh BAH lik STAIR oydz) Substances that cause muscle tissue to develop at an abnormally high rate.

Angioplasty A surgical procedure in which an instrument with a tiny balloon, drill bit, or laser attached is inserted into a blocked artery to clear a blockage.

Anorexia nervosa (a nuh REK see ah ner VOH sah) An eating disorder in which a person strongly fears gaining weight and starves herself or himself.

Antibiotics (an ti by AH tiks) Medicines that reduce or kill harmful bacteria in the body.

Antibodies Proteins that attach to antigens, keeping them from harming the body.

Antigens (AN ti genz) Substances that send the immune system into action.

Antihistamines Medicines that reduce the production of histamines.

Anxiety a state of uneasiness, usually associated with a future uncertainty.

Anxiety disorder Extreme fears of real or imaginary situations that get in the way of normal activities.

Appetite The psychological desire for food.

Arteries Blood vessels that carry blood away from the heart to various parts of the body.

Español

agresivo(a) Excesivamente forzoso, hostil o de otra manera, que ataca durante el acercamiento.

alcohol Droga producida por una reacción química en algunos alimentos, especialmente frutas y granos.

abuso de alcohol Uso de alcohol en formas que son no saludables, ilegales, o ambas.

intoxicación con alcohol Condición peligrosa que ocurre cuando una persona consume cantidades de alcohol excesivas en un corto periodo de tiempo.

alcoholismo Enfermedad que se caracteriza por la necesidad física y psicológica de consumir alcohol.

alérgenos Sustancias que causan reacciones alérgicas.

alérgia Sensibilidad extrema a una sustancia.

alternativa Otra forma de pensar o actuar.

alveolos Pequeños sacos de aire en los pulmones.

esteroides anabólicos Sustancias que causan que los tejidos musculares se desarrollen rápida y anormalmente.

angioplastia Proceso quirúrgico en el cual un instrumento con un globo pequeño, un pedacito de taladro, o láser es insertado en una arteria bloqueada para desbloquearla.

anorexia nerviosa Trastorno alimenticio en el cual una persona teme mucho subir de peso y se mata de hambre.

antibióticos Medicinas que disminuyen o matan bacterias dañinas en el cuerpo.

anticuerpos Proteínas que se pegan a los antígenos, impidiendo que éstos le hagan daño al cuerpo.

antígenos Sustancias que provocan el funcionamiento del sistema inmunológico.

antihistamínicos Medicinas que reducen la producción de histaminas.

ansiedad Estado de intranquilidad, usualmente asociado con una incertidumbre futura.

trastorno de ansiedad Temores extremos a situaciones reales o imaginarias que se interponen en el desarrollo de actividades normales.

apetito Deseo psicológico de alimentarse.

arterias Vasos sanguíneos que transportan sangre desde al corazón hacia otras partes del cuerpo.

Glossary/Glosario

English

Arteriosclerosis (ar TIR ee oh skluh ROH sis) A group of disorders that cause a thickening and hardening of the arteries.

Arthritis (ar THRY tus) A disease of the joints marked by painful swelling and stiffness.

Assault An attack on another person in order to hurt him or her.

Assertive response A response that declares your position strongly and confidently.

Assertive Willing to stand up for yourself in a firm but positive way.

Asthma Chronic inflammatory disorder of the airways that causes air passages to become narrow or blocked, making breathing difficult.

Astigmatism (uh STIG muh tiz uhm) An eye condition in which images appear wavy or blurry.

Atherosclerosis (a thuh roh skluh ROH sis) A condition of arteriosclerosis that occurs when fatty substances build up on the inner lining of the arteries.

Attitude (AT ih tood) A personal feeling or belief.

Español

arteriosclerosis Conjunto de trastornos que provoca el engrosamiento y endurecimiento de las arterias.

artritis Enfermedad de las articulaciones caracterizada por inflamación dolorosa y anquilosamiento.

asalto Ataque hacia otra persona con la intención de herirla.

reacción agresiva Reacción que establece tu posición con fuerza y confianza.

firme Dispuesto a defenderse de manera resuelta y positiva.

asma Trastorno crónico inflamatorio que causa que los pasajes de aire se hagan más pequeños o que se bloqueen, haciendo que la respiración se dificulte.

astigmatismo Afección del ojo que causa que las imágenes se vean distorsionadas y los objetos aparezcan ondulados o borrosos.

aterosclerosis Condición de arteriosclerosis que ocurre cuando sustancias grasosas se forman en el interior de las arterias.

actitud Sentimiento o creencia.

B

Bacteria (bak TIR ee uh) Simple one-celled organisms.

Battery The beating, hitting, or kicking of another person.

Benign (bi NYN) Not cancerous.

Binge drinking Having several drinks in a short period of time.

Binge eating An eating disorder in which a person repeatedly eats too much food at one time.

Biodegradable (by oh di GRAY duh buhl) Easily broken down in the environment.

Biological age Age determined by how well various body parts are working.

Biopsy The removal of a sample of tissue from a person for examination.

Blizzard A very heavy snowstorm with winds up to 45 miles per hour.

Blood alcohol concentration (BAC) The amount of alcohol in the blood.

Blood pressure The force of blood pushing against the walls of the blood vessels.

bacterias Organismos simples de una sola célula.

asalto Dar palizas, golpear o dar puntapiés a otra persona.

benigno No canceroso.

borrachera Ingerir varias bebidas en un periodo de tiempo corto.

trastorno de la alimentación compulsiva Trastorno en la alimentación por el cual una persona repetidamente come grandes cantidades de alimentos de una vez.

biodegradable Que se descompone fácilmente en el medio ambiente.

edad biológica Medida de la edad, determinada según el funcionamiento de varias partes del cuerpo.

biopsia Quitar una muestra de tejido de una persona para examinarlo.

ventisca Tormenta de nieve fuerte, con vientos que llegan a 45 millas por hora.

concentración de alcohol en la sangre Cantidad de alcohol en la sangre.

presión arterial Fuerza que ejerce la sangre contra las paredes de los vasos sanguíneos.

English

Body composition The proportions of fat, bone, muscle, and fluid that make up body weight.

Body image The way you see your body.

Body language Postures, gestures, and facial expressions.

Body Mass Index (BMI) A method for assessing your body size by taking your height and weight into account.

Body odor Smell from the body

Body system A group of organs that work together to carry out related tasks.

Brain The command center, or coordinator, of the nervous system.

Bronchi (BRAHNG ky) Two passageways that branch from the trachea, one to each lung.

Bronchodilator (brahng koh DY lay tur) A medication that relaxes the muscles around the air passages.

Bulimia (boo LEE mee ah) nervosa An eating disorder in which a person repeatedly eats large amounts of food and then purges.

Bullying A type of violence in which one person uses threats, taunts, or violence to intimidate another again and again.

Bullying behavior actions or words that are designed to hurt another person.

Calorie (KA luh ree) A unit of heat that measures the energy available in foods.

Cancer A disease that occurs when abnormal cells multiply out of control.

Capillaries Tiny blood vessels that carry blood to and from almost all body cells and connect arteries and veins.

Carbohydrates The starches and sugars found in foods.

Carbon monoxide (KAR buhn muh NAHK syd) A colorless, odorless, poisonous gas produced when tobacco burns.

Carcinogen (kar SIN un juhn) A substance that can cause cancer.

Cardiac muscle Muscle found in the walls of your heart.

Español

composición del cuerpo Proporción de grasa, hueso, músculo y líquidos que componen el peso del cuerpo.

imagen corporal Forma en la que ves tu cuerpo.

lenguaje corporal Posturas, gestos, y expresiones faciales.

indice de masa corporal Método que evalúa el tamaño indicado de tu cuerpo utilizando tu peso y tu estatura.

olor corporal Olor del cuerpo.

sistema del cuerpo Grupo de órganos que trabajan juntos para ejecutar funciones relacionadas.

cerebro Centro de mando, o el coordinador, del sistema nervioso.

bronquios Dos pasajes que se ramifican desde la tráquea hacia los dos pulmones.

broncodilatador Medicina que relaja los músculos alrededor de los bronquios.

bulimia nerviosa Trastorno en la alimentación por el cual una persona come grandes cantidades y después se induce el vómito.

intimidar Tipo de violencia en la cual una persona usa amenazas, burlas, o actos violentos para intimidar a otra persona una y otra vez.

El comportamiento de intimidación cualquier comportamiento que se dirige a otra persona que use los comentarios hirientes, amenazas o violencia.

caloría Unidad de calor que mide la energía que contienen los alimentos.

cáncer Enfermedad causada por células anormales cuyo crecimiento está fuera de control.

vasos capilares Péqueños vasos sanguíneos que transportan sangre desde y hacia casi todas las células del cuerpo y conectan arterias y venas.

hidratos de carbono Almidones y azúcares que proporcionan energía.

monóxido de carbono Gas incoloro, inodoro y tóxico que produce el tabaco al quemarse.

carcinógeno Sustancia en el medio ambiente que produce cáncer.

músculo cardiaco Músculo de las paredes del corazón.

Glossary/Glosario

English

Cardiopulmonary resuscitation (CPR) A first-aid procedure to restore breathing and circulation.

Cardiovascular (KAR dee oh VAS kyoo ler) system Organs and tissues that transport essential materials to body cells and remove their waste products.

Carrier A person who is infected with a virus and who can pass it on to others.

Cartilage (KAHR tuhl ij) A strong, flexible tissue that allows joints to move easily, cushions bones, and supports soft tissues.

Cell The basic unit of life.

Central nervous system (CNS) The brain and the spinal cord.

Character The way a person thinks, feels, and acts.

Chemotherapy The use of powerful medicines to destroy cancer cells.

Chlamydia (kluh MI dee uh) A bacterial STD that may affect the reproductive organs, urethra, and anus.

Chronic Present continuously on an off over a long period of time.

Chronic diseases Diseases that are present either continuously or off and on over a long time.

Chronological (krah nuh LAH ji kuhl) age Age measured in years.

Circulatory (SER kyuh luh tohr ee) system The group of organs and tissues that carry needed materials to cells and remove their waste products.

Cirrhosis (suh ROH suhs) The scarring and destruction of liver tissue.

Clinical social worker (CSW) A licensed, certified mental health professional with a master's degree in social work.

Clique A group of friends who hang out together and act in similar ways.

Club drugs Illegal drugs that are found mostly in nightclubs or at all-night dance parties called raves.

Cold turkey Stopping all use of tobacco products immediately.

Collaborate Work together.

Colon The large intestine.

Commitment A pledge or a promise.

Español

resucitación cardiopulmonar Procedimiento de primeros auxilios para restaurar la respiración y la circulación de la sangre.

sistema cardiovascular Órganos y tejidos que transportan materia esencial a las células del cuerpo y eliminan los desechos.

portador Persona que parece saludable pero esta infectada con el VIH y puede transmitirlo a otros.

cartílago Tejido fuerte y flexible que permite que las articulaciones se muevan fácilmente, amortigua huesos y sirve de soporte para tejidos suaves.

célula Unidad basica de la vida.

sistema nervioso central Cerebro y médula espinal.

carácter Manera en que piensas, sientes y actúas.

quimioterapia Uso de medicina poderosa para destruir células cancerosas.

clamidia Infección de transmisión sexual bacterial que puede afectar los órganos de reproducción, la uretra y el ano.

crónico Que está siempre presente o reaparece repetidamente durante un largo periodo de tiempo.

enfermedades crónicas Enfermedades que están siempre presentes o reaparecen repetidamente durante un largo periodo de tiempo.

edad cronológica Edad medida en años.

aparato circulatorio Grupo de órganos y tejidos que transportan materiales necesitados hacia células las y eliminam los desperdicios.

cirrosis Cicatrización y destrucción del tejido del hígado.

trabajador social clínico Profesional en salud mental licenciado y certificado en trabajo social.

camarilla Grupo de amigos que salen juntos y que se comportan de manera similar.

drogas de clubs Drogas ilegales que normalmente son utilizadas en discotecas y otras fiestas que duran toda la noche llamadas raves.

parar en seco Acto de parar inmediatamente el uso de productos que contienen tabaco inmediatamente.

colaborar Trabajar juntos.

colon Intestino grueso.

compromiso Promesa o voto.

English

Communicable (kuh MYOO nih kuh buhl) disease A disease that can be spread to a person from another person, an animal, or an object.

Communication The exchange of information through the use of words or actions.

Community service Volunteer programs whose goal is to improve the community and the life of its residents.

Comparison shopping Collecting information, comparing products, evaluating their benefits, and choosing products with the best value.

Compromise When both sides in a conflict agree to give up something to reach a solution that will satisfy everyone.

Conditioning Training to get into shape for physical activity or a sport.

Confidence Belief in your ability to do what you set out to do.

Conflict A disagreement between people with opposing viewpoints, interests, or needs.

Conflict resolution A life skill that involves solving a disagreement in a way that satisfies both sides.

Conflict-resolution skills The ability to end a disagreement or keep it from becoming a larger conflict.

Congenital disorders All disorders that are present when the baby is born.

Consequences The results of actions.

Conservation The saving of resources.

Constructive criticism Using a positive message to make a suggestion.

Consumer A person who buys products and services.

Contagious (kuhn TA juhs) Able to spread to others by direct or indirect contact

Contagious period Length of time that a particular disease can be spread from person to person.

Cool-down Gentle exercises that let the body adjust to ending a workout.

Coping strategies Ways of dealing with the sense of loss people feel when someone close to them dies.

Crisis hot line A toll-free telephone service where abuse victims can get help and information.

Cultural background The beliefs, customs, and traditions of a specific group of people.

Español

enfermedad contagiosa Enfermedad que se puede propagar de una persona a otra persona, un animal o un objeto.

comunicación Intercambio de información a través del uso de palabras y acciones.

servicio comunitario Programas voluntarios desarrollados con la meta de mejorar la comunidad y la vida de los residentes.

comparación de productos Recolectar información, comparar productos, evaluar sus beneficios, y escoger el producto que tiene mejor valor.

compromiso Cuando los dos lados de un conflicto concuerdan con dejar algo de lado para alcanzar una solución que satisfaga a todos.

acondicionamiento Entrenamiento para ponerse en forma para alguna actividad física o deporte.

confianza Creer en tu habilidad de hacer lo que te propones a hacer.

conflicto Desacuerdo entre dos personas con puntos de vista, intereses o necesidades opuestas.

resolución de un conflicto Habilidad que implica el hecho de resolver un desacuerdo satisfaciendo a los dos lados.

habilidades de solución de conflicto Habilidad de poder solucionar un desacuerdo o hacer que el desacuerdo no se convierta en algo más grande.

desórdenes congénitos Todos los desórdenes que se presentan cuando el bebé nace.

consecuencias Resultados de los actos.

conservación Protección de los recursos naturales.

La crítica constructiva Spanish Definition: Con un mensaje positive a hacer una sugerencia.

consumidor Persona que compra productos y servicios.

contagioso Capaz de propagarse a otros por contacto directo o indirecto.

periodo de contagio Periodo de tiempo en que se puede transmitir una enfermedad determinada de una persona a otra.

recuperación Ejercicios moderados que permiten que el cuerpo se ajuste al ir finalizando el plan de ejercicios.

estrategias Formas de tratar con el sentido de pérdida que las personas sienten cuando alguien cercano fallece.

linea de reporte de crisis Servicio telefónico sin pago en cual víctimas de abusos pueden recibir ayuda e información.

base cultural Creencias, costumbres y tradiciones de un grupo específico de personas.

Glossary/Glosario

English

Culture the collected beliefs, customs, and behaviors of a group

Cumulative (KYOO myuh luh tiv) risk When one risk factor adds to another to increase danger.

Cyberbullying the electronic posting of mean-spirited messages about a person often done anonymously.

Cycle of abuse Pattern of repeating abuse from one generation to the next.

D

Dandruff When too many dead skin cells flake off the outer layer of the scalp.

Dating violence When a person uses violence in a dating relationship to control his or her partner.

Decibel The unit for measuring the loudness of sound.

Decision making The process of making a choice or solving a problem.

Defensive driving Watching out for other people on the road and anticipating unsafe acts.

Degenerative diseases Diseases that cause further breakdown in body cells, tissues, and organs as they progress.

Dehydration The excessive loss of water from the body.

Depressant (di PRE suhnt) A drug that slows down the body's functions and reactions, including heart and breathing rates.

Dermatologist (DER muh TAHL uh jist) A physician who treats skin disorders.

Dermis (DER mis) The skin's inner layer.

Detoxification (dee tahk si fi KAY shuhn) The physical process of freeing the body of an addictive substance.

Developmental tasks Events that need to happen in order for you to continue growing toward becoming a healthy, mature adult.

Diabetes (dy uh BEE teez) A disease that prevents the body from converting food into energy.

Diaphragm (DY uh fram) A large, dome-shaped muscle below the lungs that expands and compresses the lungs, enabling breathing.

Digestion (di JES chuhn) The process by which the body breaks down food into smaller pieces that can be absorbed by the blood and sent to each cell in your body.

Español

cultura Colección de creencias, costumbres y comportamientos de un grupo.

riesgo acumulativo Cuando un factor riesgoso se suma a otro e incrementa el peligro.

El acoso cibernético la publicación electronica de la media mensajes animados sobre una persona hace a menudo anónimamente.

ciclo de abuso Patrón de repetición del abuso de una generación a la siguiente.

caspa Cuando demasiadas células de piel muertas se descaman de la capa exterior del cuero cabelludo.

relacion violenta Cuando una persona usa violencia en una relación amorosa para poder controlar a su pareja.

decibel Unidad que se usa para medir el volumen del sonido.

tomar decisiones Proceso de hacer una selección o de resolver un problema.

conducir de manera defensiva Estar atento a las otras personas en la carretera y anticipar acciones peligrosas.

enfermedades degenerativas Enfermedades que causan la destrucción progresiva de las células, tejidos y órganos del cuerpo a medida que avanzan.

deshidratación Pérdida excesiva de agua del cuerpo.

sedante Droga que disminuye las funciones y reacciones del cuerpo, incluso el ritmo cardiaco y la respiración.

dermatólogo Médico que trata trastornos de la piel.

dermis Capa interior de la piel.

desintoxicación Proceso físico de liberar al cuerpo de una sustancia adictiva.

tareas requeridas para el desarrollo Sucesos que deben ocurrir para que continúes desarrollándote hasta llegar a convertirte en un adulto saludable y maduro.

diabetes Enfermedad que impide que el cuerpo convierta los alimentos en energía.

diafragma Un músculo grande en forma de cúpula debajo de los pulmones que expande y comprime los pulmones, posibilitando la respiración.

digestión Proceso por el cual el cuerpo desintegra las comidas en pedazos más pequeños que pueden ser absorbidos por la sangre y enviados a cada célula en el cuerpo.

English

Digestive (dy JES tiv) system The group of organs that work together to break down foods into substances that your cells can use.

Disease (dih ZEEZ) Any condition that interferes with the proper functioning of the body or mind.

Dislocation A major injury that happens when a bone is forced from its normal position within a joint.

Domestic violence Physical abuse that occurs within a family.

Drug A substance other than food that changes the structure or function of the body or mind.

Drug abuse Intentionally using drugs in a way that is unhealthful or illegal.

Drug misuse Taking or using medicine in a way that is not intended.

Drug rehabilitation a process where the person relearns how to live without the abused drug

Español

aparato digestivo Conjunto de órganos que trabajan juntos para descomponer los alimentos en sustancias que tus células puedan usar.

enfermedad Toda afección que interfiere con el buen funcionamiento del cuerpo o de la mente.

dislocación Daño mayor que ocurre cuando un hueso es forzado fuera de lugar en la articulación.

violencia doméstica Abuso físico que ocurre dentro de una familia.

droga Sustancia no alimenticia que causa cambios en la estructura o el funcionamiento del cuerpo o la mente.

abuso de drogas Uso de drogas intencionalmente en una forma no saludable o ilegal.

mal empleo de drogas Tomar o usar medicina de una forma que no es la indicada.

rehabilitacion de las drogas Proceso por el cual una persona vuelve a aprender como vivir sin el abuso de una droga.

E

Earthquake A shifting of the earth's plates resulting in a shaking of the earth's surface.

Eating disorder Extreme eating behavior that can lead to serious illness or even death.

Electrical overload A dangerous situation in which too much electric current flows along a single circuit.

Embryo The developing organism from two weeks until the end of the eighth week of development.

Emotional needs Needs that affect a person's feelings and sense of well-being.

Emotions Feelings such as love, joy, or fear.

Empathy Identifying with and sharing another person's feelings.

Emphysema (em fuh SEE muh) A disease that results in the destruction of the alveoli in the lungs.

Enablers Persons who create an atmosphere in which the alcoholic can comfortably continue his or her unacceptable behavior.

Endocrine (EN duh krin) system The system of glands throughout the body that regulate body functions.

Endurance (en DUR uhnce) The ability to perform difficult physical activity without getting overly tired.

terremoto Cambio de las placas terrestres que hace que la superficie terrestre tiemble.

trastorno en la alimentación Conducta de alimentación extrema que puede causar enfermedades graves o la muerte.

sobrecarga eléctrica Situación peligrosa en que demasiada corriente eléctrica fluye a través de un solo circuito.

embrión Organismo en desarrollo desde las dos semanas hasta las ocho semanas de desarrollo.

necesidades emocionales Necesidades que afectan los sentimientos y el bienestar de una persona.

emociones Sentimientos como el amor, la alegría o el miedo.

empatía Identificar y compartir sentimientos de otra persona.

enfisema Enfermedad que resulta de la destrucción de los alvéolos en los pulmones.

habilitadores Personas que crean una atmósfera en la cual el alcohólico puede continuar su comportamiento inaceptable de una forma cómoda.

sistema endocrino Sistema de glándulas a través del cuerpo que regulan las funciones corporales.

resistencia Habilidad de desempeñar actividades físicas sin cansarse demasiado.

Glossary/Glosario

English

Environment (en VY ruhn muhnt) All the living and nonliving things around you.

Environmental Protection Agency (EPA) An agency of the U.S. government that is dedicated to protecting the environment.

Enzymes Proteins that affect the many body processes.

Epidermis The outermost layer of skin.

Escalate To become more serious.

Excretory (EKS kru to ree) system The group of organs that work together to remove wastes.

Exercise Planned physical activity done regularly to build or maintain one's fitness.

Family The basic unit of society and includes two or more people joined by blood, marriage, adoption, or a desire to support each other.

Family therapy Counseling that seeks to improve troubled family relationships.

Farsightedness a condition in which faraway objects appear clear while near objects look blurry

Fats Nutrients that promote normal growth, give you energy, and keep your skin healthy.

Fatty liver A condition in which fats build up in the liver and cannot be broken down.

Fertilization The joining of a male sperm cell and a female egg cell to form a fertilized egg.

Fetal (FEE tuhl) alcohol syndrome (FAS) A group of alcohol-related birth defects that include both physical and mental problems.

Fetus The developing organism from the end of the eighth week until birth.

Fiber A complex carbohydrate that the body cannot break down or use for energy.

Fight-or-flight response The body's way of responding to threats.

Fire extinguisher A device that sprays chemicals that put out fires.

First aid The immediate care given to someone who becomes injured or ill until regular medical care can be provided.

Español

medio Todas las cosas vivas y no vivas que te rodean.

Agencia de Protección Agencia del gobierno de Estados Unidos a cargo de la protección del medio ambiente.

enzimas Proteínas que afectan varios processos corporals.

epidermis Capa más externa de la piel.

intensificar Llegar a ser más grave.

sistema excretor El grupo de órganos que trabaja en conjunto para eliminar desperdicios del cuerpo.

ejercicio Actividad física planeada realizada regularmente para crear o mantener un buen estado físico.

familia Unidad básica de la sociedad que incluye dos o mas personas unidas por sangre, matrimonio, adopción o el deseo de ser soporte el uno del otro.

terapia familiar Asesoramiento cuyo propósito es mejorar relaciones problemáticas entre familiares.

hipermetropía Capacidad de ver claramente los objetos a la distancia, mientras los objetos cercanos se ven borrosos.

grasas Nutrientes que promueven crecimiento normal, dan energía, y mantienen la piel saludable.

higado adiposo Condición en la cual la grasa se forma en el hígado y no puede ser deshecha.

fertilización Unión de un espermatozoide masculino con una óvulo femenino para forma un óvulo fertilizado.

síndrome de alcoholismo fetal Conjunto de defectos de nacimiento causados por el alcohol que incluyen problemas físicos y mentales.

feto Organismo en desarrollo desde el final de la octava semana hasta el nacimiento.

fibra Carbohidratos complejos que el cuerpo no puede deshacer para usarlos como energía.

respuesta de lucha o huida Forma del cuerpo de responder a las amenazas.

extintor de fuego Aparato que rocia productos químicos que apagan fuegos.

primeros auxilios Cuidado inmediato que se da a una persona herida o enferma hasta que sea posible proporcionarle ayuda médica normal.

English

FITT principle A method for safely increasing aspects of your workout without injuring yourself.

Flammable Substances that catch fire easily.

Flexibility The ability to move joints fully and easily through a full range of motion.

Fluoride A chemical that helps prevent tooth decay.

Fossil (FAH suhl) fuels The oil, coal, and natural gas that are used to provide energy.

Fracture A break in a bone.

Fraud A calculated effort to trick or fool others.

Frequency The number of days you work out each week.

Friendship A relationship with someone you know, trust, and regard with affection.

Frostbite freezing of the skin

Fungi (FUHN jy) Organisms that are more complex than bacteria, but cannot make their own food.

G

Gallbladder A small, saclike organ that stores bile.

Gang A group of young people that comes together to take part in illegal activities.

Genital herpes (HER peez) A viral STD that produces painful blisters on the genital area.

Genital warts Growths or bumps in the genital area caused by certain types of the human papillomavirus (HPV).

Germs Organisms that are so small they can only be seen through a microscope.

Gingivitis (jin juh VY tis) A common disorder in which the gums are red and sore and bleed easily.

Gland A group of cells, or an organ, that secretes a chemical substance.

Global warming A rise in the earth's temperatures.

Goal setting The process of working toward something you want to accomplish.

Español

principio FITT Método mediante el cual es posible incrementar con seguridad aspectos del ejercicio sin hacerte daño a ti mismo.

inflamable Sustancias que se encienden fácilmente.

flexibilidad La habilidad de poder mover articulaciones completa y fácilmente a través de una extensión completa de movimiento.

fluoruro Un producto químico que ayuda a prevenir caries en los dientes.

combustible fósil Petróleo, carbón y gas natural que se usan para proporcionar energía.

fractura Rotura de un hueso.

fraude Esfuerzo calculado para engañar a otros.

frequencia Número de horas en las que haces ejercicios cada semana.

amistad Relación con una persona que conoces, en la que confías, y aprecias con afecto.

congelación Congelamiento de la piel.

fungi Organismos que son más complejos que bacterias, pero no pueden producir su propio alemento.

vesícula biliar Pequeño órgano en forma de bolsa que almacena bilis.

pandilla Grupo de jóvenes que se juntan para participar en actividades ilegales.

herpes genitales Infección de transmisión sexual, causada por un virus, que produce ampollas dolorosas en el área genital.

verrugas genitales Erupciones o protuberancias en el área genital causadas por ciertos tipos del virus papiloma de los seres humanos.

gérmenes Organismos tan diminutos que se ven sólo a través de un microscopio.

gingivitis Trastorno común que se caracteriza por el enrojecimiento y dolor de las encías que sangran con facilidad.

glándula Grupo de células, o un órgano, que secreta una sustancia química.

calentamiento del planeta Aumento en las temperaturas de la Tierra.

establecer metas Proceso de esforzarte para lograr algo que quieres.

Glossary/Glosario

English

Gonorrhea (gah nuh REE uh) A bacterial STD that affects the mucous membranes of the body, particularly in the genital area.

Greenhouse effect The trapping of heat by carbon dioxide and other gases in the air.

Green School A school that is environmentally-friendly in several different ways.

Grief The sorrow caused by loss of a loved one.

Grief reaction The process of dealing with strong feelings following any loss.

Groundwater Water that collects under the earth's surface.

Gynecologist A doctor who specializes in the female reproductive system.

Hallucinogens (huh LOO suhn uh jenz) Drugs that distort moods, thoughts, and senses.

Harassment (huh RAS muhnt) Ongoing conduct that offends another person by criticizing his or her race, color, religion, physical disability, or gender.

Hazard Potential source of danger.

Hazardous wastes Human-made liquid, solid, sludge or radioactive wastes that may endanger human health or the environment.

Health The combination of physical, mental/emotional, and social well-being.

Health care any services provided to individuals or communities that promote, maintain, or restore health

Health care system All the medical care available to a nation's people, the way they receive the care, and the way the care is paid for.

Health fraud The selling of products or services to prevent diseases or cure health problems which have not been scientifically proven safe or effective for such purposes.

Health insurance A plan in which a person pays a set fee to an insurance company in return for the company's agreement to pay some or all medical expenses when needed.

Health maintenance organization (HMO) A health insurance plan that contracts with selected physicians and specialists to provide medical services.

Health skills skills that help you become and stay healthy

Español

gonorrea Infección de transmisión sexual causada por bacterias que afecta las membranas mucosas del cuerpo, en particular en el área genital.

efecto invernadero Retención del calor por la presencia de dióxido de carbono y otros gases en el aire.

escuela verde Una escuela que es respetuoso con el medio ambiente de varias maneras diferentes.

pena Pesar provocado por la muerte de un ser querido.

reacción a la desgracia El proceso de tratar con algún sentimientos fuertes que se ocasionan por alguna perdida.

agua subterránea Agua acumulada debajo de la superficie de la tierra.

ginecólogo Médico que se especializa en el aparato reproductor femenino.

alucinógeno Droga que altera el estado de ánimo, los pensamientos y los sentidos.

acoso Conducta frecuente que ofende a otra persona con críticas sobre su raza, color, religión, incapacidad física, o sexo.

peligro Fuente posible de peligro.

desperdicios peligrosos Líquidos, sólidos, sedimentos, o desperdicios radiactivos producidos por humanos que ponen en peligro la salud humana o el medio ambiente.

salud Combinación de bienestar físico, mental/emocional y social.

cuidado medico Cualquier servicio proporcionado a individuos o comunidades que promueve, mantiene y les hace recobrar la salud.

sistema de cuidado de la salud Servicios médicos disponibles para a la gente de una nación y las formas en las cuales estos son pagados.

fraude médico Venta de productos o servicios para prevenir enfermedades o curar problemas de salud que no han sido aprobados científicamente o hechos efectivos para ese uso.

seguro médico Plan en el que una persona paga una cantidad fija a una compañía de seguros que acuerda cubrir parte o la totalidad de los gastos médicos.

organización para el mantenimiento de la salud Plan de seguro de salud que contrata a ciertos médicos y especialistas para dar servicios médicos.

habilidades de salud Habilidades que ayudan a ser y mantenerte saludable.

English

Heart The muscle that acts as the pump for the circulatory system

Heart and lung endurance A measure of how efficiently your heart and lungs work when you exercise and how quickly they return to normal when you stop.

Heart attack A serious condition that occurs when the blood supply to the heart slows or stops and the heart muscle is damaged.

Heat exhaustion An overheating of the body that can result from dehydration.

Hepatitis (hep uh TY tis) A viral disease characterized by an inflammation of the liver and yellowing of the skin and the whites of the eyes.

Hepatitis B A disease caused by the hepatitis B virus that affects the liver.

Heredity (huh RED I tee) The passing of traits from parents to their biological children.

Hernia An internal organ pushing against or through a surrounding cavity wall.

Histamines (HIS tuh meenz) The chemicals that the immune cells release to draw more blood and lymph to the area affected by the allergen.

HIV (human immunodeficiency virus) The virus that causes AIDS.

Hives Raised bumps on the skin that are very itchy.

Homicide A violent crime that results in the death of another person.

Hormones (HOR mohnz) Chemical substances produced in certain glands that help to regulate the way your body functions.

Hospice care Care provided to the terminally ill that focuses on comfort, not cure.

Hunger The physical need for food.

Hurricane A strong windstorm with driving rain that forms over the sea.

Hygiene (HY jeen) Cleanliness.

Hypertension A condition in which the pressure of the blood on the walls of the blood vessels stays at a level that is higher than normal.

Hypothermia (hy poh THER mee uh) A sudden and dangerous drop in body temperature.

Español

corazón Músculo que funciona como una bomba para el aparato circulatorio.

resistencia del corazón y los pulmones Medida de qué tan eficientemente tu corazón y tus pulmones cuando haces ejercicios y qué tan rápido regresan a lo normal cuando paras.

ataque cardiaco Afección seria que se presenta cuando el flujo de sangre al corazón disminuye o cesa, dañando el músculo cardiaco.

agotamiento por calor Recalentamiento del cuerpo que puede dar como resultado una deshidratación.

hepatitis Enfermedad viral caracterizada por inflamación del hígado, el torno amarillo de la piel y la parte blanca del ojo.

Hepatitis B Enfermedad causada por el virus de la hepatitis B que afecta el hígado.

herencia Transferencia de características de los padres biológicos a sus hijos.

hernia Órgano interno que está empujando contra o a través de una pared de cavidad.

histaminas Substancias químicas que las células inmunes suelta sueltan para atraer mas sangre y linfa hacia el area afectada por los alergenos.

VIH (virus de inmunodeficiencia humana) Virus que causa el SIDA.

urticaria Granos en la piel que pican mucho.

homicidio Crimen violento que resulta en la muerte de otra persona.

hormonas Sustancias químicas, producidas por ciertas glándulas que ayudan a regular las funciones del cuerpo.

asistencia para enfermos Asistencia para personas con enfermedades incurables que apunta a brindar comodidad, no a la cura.

hambre Necesidad física de alimentos.

huracán Tormenta de vientos y lluvia torrencial que se origina en alta mar.

higiene Limpieza.

hipertensión Afección en que la presión arterial de una persona se mantiene a niveles más altos de lo normal.

hipotermia Descenso rápido y peligroso de la temperatura del cuerpo.

Glossary/Glosario

English

Español

I

"I" message A statement that presents a situation from the speaker's personal viewpoint.

mensaje yo Declaración que presenta una situación desde el punto de vista personal del orador.

Immune (i MYOON) system A combination of body defenses made up of the cells tissues and organs that fight pathogens in the body.

sistema inmunológico Combinación de las defensas del cuerpo, compuesta de células, tejidos y órganos que combaten agentes patógenos.

Immunity The ability to resist the pathogens that cause a particular disease.

inmunidad Habilidad del cuerpo de resistir los agentes patógenos que causan una enfermedad en particular.

Infection A condition that happens when pathogens enter the body, multiply, and cause harm.

infección Afección que se produce cuando agentes patógenos invaden el cuerpo, se multiplican y dañan las células.

Inflammation The body's response to injury or disease, resulting in a condition of swelling, pain, heat and redness.

inflamación Reacción del cuerpo a lesiones o enfermedades que resulta en hinchazón, dolor, calor y enrojecimiento.

Influenza A highly communicable viral disease characterized by fever, chills, fatigue, headache, muscle aches and respiratory symptoms.

influenza Enfermedad viral muy contagiosa que se caracteriza por fiebres, escalofríos, dolores de cabeza, dolor muscular, y síntomas respiratorios.

Inhalants (in HAY luhntz) The vapors of chemicals that are sniffed or inhaled to get a "high."

inhalantes Los vapores de substancias químicas que son olidos o inhalados para drogarse.

Inhibition A conscious or unconscious restraint on his or her behaviors or actions.

inhibición Reprimir comportamientos o acciones consiente o inconscientemente.

Insulin (in suh lin) A protein made in the pancreas that regulates the level of glucose in the blood.

insulina Proteína hecha en el páncreas que regula el nivel de glucosa en la sangre.

Integrity Being true to your ethical values.

integridad Ser fiel a tus valores éticos.

Intensity How much energy you use when you work out.

intensidad En cuanto al estado físico, la cantidad de energía que usas cuando haces ejercicio.

Intervention A gathering in which family and friends get the problem drinker to agree to seek help.

intervención Reunión en la cual familia y amigos hacen que la persona con problemas alcohólicos busque ayuda.

Intimidation Purposely frightening another person through threatening words, looks, or body language.

intimidación Asustar a otra persona a propósito con palabras amenazantes, miradas o lenguaje corporal.

Intoxicated (in TAHK suh kay tuhd) Being drunk.

intoxicado(a) Estar borracho.

J

Joints The places where two or more bones meet.

articulación Lugar en donde se unen dos o más huesos.

K

Kidneys Organs that filter water and dissolved wastes from the blood and help maintain proper levels of water and salts in the body.

Riñones Órganos que filtran el agua y los desechos disueltos de la sangre y contribuyen a mantener los nivels adecuados de agua y sales en el cuerpo.

English

Español

L

Landfill Huge specially designed pit where waste materials are dumped and buried.

Larynx (LA ringks) The upper part of the respiratory system, which contains the vocal cords.

Lifestyle factors Behaviors and habits that help determine a person's level of health.

Ligament A type of connecting tissue that holds bones to other bones at the joint.

Limits Invisible boundaries that protect you.

Liver A digestive gland that secretes a substance called bile, which helps to digest fats.

Long-term goal A goal that you plan to reach over an extended period of time.

Loyal faithful

Lungs Two large organs that exchange oxygen and carbon dioxide.

Lymphatic system A secondary circulatory system that helps the body fight pathogens and maintains its fluid balance.

Lymphocytes (LIM fuh sytes) Special white blood cells in the lymphatic system.

terraplén sanitario Pozo enorme con diseño específico donde se arrojan y se entierran desechos.

laringe Parte superior del aparato respiratorio que contiene las cuerdas vocales.

factores del estilo de vida Conductas y hábitos que ayudan a determinar el nivel de salud de una persona.

ligamento Tipo de tejido conjuntivo que mantiene en su lugar los huesos en las articulaciones.

límites Barreras invisibles que te protegen.

hígado Glándula digestiva que secreta una sustancia llamada bilis, que ayuda a digerir las grasas.

meta a largo plazo Objetivo que planeas alcanzar en un largo periodo de tiempo.

leal Fiel.

pulmones Dos órganos grandes que intercambian oxígeno y monóxido de carbono.

sistema linfático Aparato circulatorio secundario que le ayuda al cuerpo a defenderse de agentes patógenos y a mantener el equilibrio de los líquidos.

linfocitos Glóbulos blancos especiales en la linfa.

M

Mainstream smoke The smoke that is inhaled and then exhaled by a smoker.

Major depression A very serious mood disorder in which people lose interest in life and can no longer find enjoyment in anything.

Malignant (muh LIG nuht) Cancerous.

Malnutrition A condition in which the body doesn't get the nutrients it needs to grow and function properly.

Managed care A health insurance plan that saves money by encouraging patients and providers to select less costly forms of care.

Managed care plans Health insurance plans that emphasize preventative medicine and work to control the cost and maintain the quality of health care.

Marijuana Dried leaves and flowers of the hemp plant, called cannabis sativa.

humo directo Humo que el fumador aspira y exhala.

depresión mayor Desorden del estado deanimo en el cual las personas pierden interés en su vida y no encuentran placer en nada.

maligno Canceroso.

desnutrición Cual es una afección en la que el cuerpo no recibe los nutrientes que necesita para crecer y funcionar de forma adecuada.

cuidado controlado Plan de seguro médico que ahorra dinero al limitar la selección de doctores de las personas.

planes de medicina administrada Planes de seguro médico que enfatizan medicina preventiva y el trabajo para controlar el costo y el mantenimiento del cuidado de salud de calidad.

mariguana Hojas y flores secas de la planta de cáñamo, llamada cannabis sativa.

Glossary/Glosario

English

Mediation Resolving conflicts by using another person or persons to help reach a solution that is acceptable to both sides.

Media Various methods for communicating information.

Media literacy The ability to understand the goals of advertising and the media.

Medicine A drug that prevents or cures an illness or eases its symptoms.

Medicine abuse Intentionally using medicines in ways that are unhealthful and illegal.

Medicine misuse Taking medicine in a way that is not intended.

Melanin Substance that gives skin its color.

Menstruation (men stroo AY shuhn) The flow from the body of blood, tissues, and fluids that result from the breakdown of the lining of the uterus.

Mental and emotional disorders Illnesses that affect a person's thoughts, feelings, and behavior.

Mental and emotional health The ability to handle the stresses and changes of everyday life in a reasonable way.

Metabolism The process by which the body gets energy from food.

Mind-body connection How your emotions affect your physical and overall health and how your overall health affects your emotions.

Minerals (MIN uh ruhls) Substances the body uses to form healthy bones and teeth, keep blood healthy, and keep the heart and other organs working properly.

Minor A person under the age of adult rights and responsibilities.

Mixed message A situation in which your words say one thing but your body language says another.

Mob mentality Acting or behaving in a certain and often negative manner because others are doing it.

Mononucleosis (MAH noh nook klee OH sis) A viral disease characterized by a severe sore throat and swelling of the lymph glands in the neck and around the throat area.

Mood disorder A mental and emotional problem in which a person undergoes mood swings that seem extreme, inappropriate, or last a long time.

Mood swings Frequent changes in emotional state.

Español

mediación Resolución de conflictos por medio de otra persona que ayuda a llegar a una solución aceptable para ambas partes.

medios de difusión Diversos métodos de comunicar información.

conocimiento de los medios de comunicación Habilidad de entender las metas de la publicidad en los medios publicitarios.

medicina Droga que previene o cura enfermedades o que alivia sus síntomas.

medicina abuso Intencionalmente el uso de medicamentos en formas que son insalubres e ilegales.

el abuso de medicamentos Tomar un medicamento de una manera que no es la intención.

melanina Sustancia que le da el color a la piel.

menstruación Fluido de sangre, tejidos y fluidos del cuerpo que resulta del desprendimiento del forro del útero.

trastornos mentales y emocionales Enfermedades que afectan los pensamientos, sentimientos y comportamiento de una persona.

salud mental y emocional Habilidad de hacerle frente de manera razonable al estrés y a los cambios de la vida diaria.

metabolismo Proceso mediante el cual el cuerpo obtiene energía de los alimentos.

conexión de la mente con el cuerpo Forma en la cual tus emociones afectan tu salud física y general, y como tu salud general afecta tus emociones.

minerales Sustancias que el cuerpo utiliza para formar huesos y dientes saludables, mantener la sangre saludable y mantener el corazón y otros órganos funcionando como deben.

menor Persona que es menor de en que se tienen la edad de derechos y responsabilidades de adultos.

mensaje contradictorio Situación en que tus palabras expresan algo pero tu lenguaje corporal lo contradice.

mentalidad de movimiento Actuar o comportarse de cierta manera, normalmente maneras negativas, sólo porque otros lo están haciendo.

mononucleosis Una enfermedad viral que se caracteriza por dolores de garganta severos y hinchazón de las glándulas linfas en el cuello y alrededor del area de la garganta.

trastorno de humor Problema emocional y mental en el cual la persona tiene cambios de humor que parecen extremos inapropiados, o que duran mucho tiempo.

cambios de humor Cambios frecuentes en el estado emocional.

English

Muscle endurance The ability of a muscle to repeatedly use force over a period of time.

Muscle strength The most weight you can lift or the most force you can exert at one time.

Muscular system Tissues that move parts of the body and control the organs.

MyPlate A visual reminder to help consumers make healthier food choices.

Narcotics (nar KAH tics) Drugs that get rid of pain and dull the senses.

Natural disaster an event caused by nature that result in widespread damage, destruction, and loss

Nearsightedness objects that are close appear clear while those far away look blurry

Neglect Failure to provide for the basic physical and emotional needs of a dependent.

Negotiation (neh GOH shee AY shuhn) The process of talking directly to the other person to resolve a conflict.

Nervous system The body's message and control center.

Neurons (NOO rahnz) Cells that make up the nervous system.

Neutrality A promise not to take sides.

Nicotine (NIH kuh teen) An addictive, or habit-forming, drug found in tobacco.

Nicotine replacement therapies (NRT) Products that assist a person in breaking a tobacco habit.

Noncommunicable disease A disease that cannot be spread from person to person.

Nonrenewable resources Substances that cannot be replaced once they are used.

Nurture Fulfill physical, mental, emotional, and social needs.

Nutrient dense Foods having a high amount of nutrients relative to the number of calories.

Nutrients (NOO tree ents) Substances in foods that your body needs to grow, have energy, and stay healthy.

Nutrition (noo TRIH shuhn) The process of taking in food and using it for energy, growth, and good health.

Español

resistencia muscular Habilidad de un músculo de usar fuerza repetitivamente en un periodo de tiempo.

fuerza muscular Medida del peso máximo que puedes cargar o la fuerza máxima que puedes emplear a un mismo tiempo.

sistema muscular Tejidos que mueven partes del cuerpo y controlan los órganos.

MiPlato Un recuerdo visual para ayudar a los consumidores elegir alimentos más saludables.

narcótico Droga que alivia el dolor y entorpece los sentidos.

desastre natural Evento causado por la naturaleza que resulta en daños extensos, destrucción y pérdida.

miopía Capacidad de ver claramente los objetos cercanos, mientras los objetos lejanos se ven borrosos.

abandono Fallas en el proceso de proveer las necesidades físicas y emocionales de una persona considerada como dependiente.

negociación Proceso de hablar directamente con otra persona para solucionar un conflicto.

sistema nervioso Centro de mensajes y control del cuerpo.

neuronas Células que forman el sistema nervioso.

neutralidad Promesa de no tomar partido durante un conflicto entre otros.

nicotina Droga adictiva, que forma hábito, encontrada en el tabaco.

terapias para remplazar la nicotina Productos que ayudan a las personas a romper el hábito del tabaco.

enfermedad no contagiosa Enfermedad que no se puede transmitir de una persona a otra.

recursos no renovables Sustancias que no se pueden reemplazar después de usarse.

criar Satisfacer necesidades físicas, mentales, emocionales y sociales.

denso en nutrientes Tener una alta cantidad de sustancias nutritivas en comparación con la cantidad de calorías.

nutrientes Sustancias en los alimentos que el cuerpo necesita para desarrollarse, tener energía y mantenerse saludable.

nutrición Proceso de ingerir alimentos y usarlos para la energía, el desarrollo y el mantenimiento de la buena salud.

Glossary/Glosario

English

Español

Occupational Safety and Health Administration (OSHA) A branch of the U.S. Department of Labor that protects American workers.

Administración de Salud y Seguridad Ocupacional Rama del Ministerio del Trabajo que protege la seguridad de los trabajadores estadounidenses.

Ophthalmologist (ahf thahl MAH luh jist) A physician who specializes in the structure, functions, and diseases of the eye.

oftalmólogo Médico que se especializa en la estructura, funciones y enfermedades del ojo.

Opportunistic infection A disease that attacks a person with a weakened immune system and rarely occurs in a healthy person.

infección oportunista Infección que ocurre raramente en personas saludables.

Optometrist (ahp TAH muh trist) A health care professional who is trained to examine the eyes for vision problems and to prescribe corrective lenses.

optómetra Profesional de la salud que está preparado para examinar la vista y recetar lentes correctivas.

Organ A body part made up of different tissues joined to perform a particular function.

órgano Parte del cuerpo que comprende distintos tejidos unidos para cumplir una función particular.

Orthodontist (or thuh DAHN tist) A dentist who prevents or corrects problems with the alignment or spacing of teeth.

ortodoncista Dentista que previene o corrige problemas del alineamiento o del espacio entre los dientes.

Osteoarthritis (ahs tee oh ahr THRY tuhs) A chronic disease that is common in older adults and results from a breakdown in cartilage in the joints.

osteoartritis Enfermedad crónica, común en los ancianos, que es el resultado de la degeneración del cartílago de las articulaciones.

Ovaries The female endocrine glands that release mature eggs and produce the hormones estrogen and progesterone.

ovarios Glándulas endocrinas femeninas que sueltan huevos maduros y producen hormonas como estrógeno y progesterona.

Over-the-counter (OTC) medicine A medicine that you can buy without a doctor's permission.

medicina sin receta Medicina que se puede comprar sin receta de un médico.

Overweight More than the appropriate weight for gender, height, age, body frame, and growth pattern.

sobrepeso Más del peso apropiado de acuerdo al sexo, estatura, edad, estructura corporal y ritmo de crecimiento.

Ovulation The process by which the ovaries release mature eggs, usually one each menstrual cycle.

ovulación Proceso en el cual los ovarios sueltan huevos maduros, normalmente uno en cada ciclo menstrual.

Ozone (OH zohn) A gas made of three oxygen atoms.

ozono Gas hecho de tres átomos de oxígeno.

Ozone layer A shield above the earth's surface that protects living things from ultraviolet (UV) radiation.

capa de ozono Capa protectora sobre la superficie de la Tierra que protege a los seres vivos de la radiación ultravioleta.

P

Pancreas (PAN kree uhs) A gland that helps the small intestine by producing pancreatic juice, a blend of enzymes that breaks down proteins, carbohydrates, and fats.

páncreas Glándula que le ayuda al intestino delgado, a través de la producción de jugo pancreático, que está formado por una mezcla de varias enzimas que descomponen las proteínas, hidratos de carbono y grasas.

Passive A tendency to give up, give in, or back down without standing up for rights and needs.

pasivo Tendencia a renuncia, dejar de lado, o hacerse atrás sin reclamar derechos o necesidades.

Passive smoker A nonsmoker who breathes in secondhand smoke.

fumador pasivo Persona que no fuma pero que inhala humo secundario.

Pathogens Germs that cause diseases.

patógenos Gérmenes que causan enfermedades.

Pedestrian A person who travels on foot.

peatón Persona que se traslada a pie.

English

Peers People close to you in age who are a lot like you.

Peer mediation (mee dee AY shuhn) a process in which a specially trained student listens to both sides of an argument to help the people reach a solution

Peer pressure The influence that your peer group has on you.

Peripheral nervous system (PNS) The nerves that connect the central nervous system to all parts of the body.

Personality A combination of your feelings, likes, dislikes, attitudes, abilities, and habits.

Personality disorder A variety of psychological conditions that affect a person's ability to get along with others.

Pesticide Product used on crops to kill insects and other pests.

Phobia Intense and exaggerated fear of a specific situation or object.

Physical abuse the use of physical force

Physical activity Any movement that makes your body use extra energy.

Physical dependence An addiction in which the body develops a chemical need for a drug.

Physical fitness The ability to handle the physical demands of everyday life without becoming overly tired.

Plaque (PLAK) A thin, sticky film that builds up on teeth and leads to tooth decay.

Pneumonia A serious inflammation of the lungs.

Point of sale promotion Advertising campaigns in which a product is promoted at a store's checkout counter.

Pollen A powdery substance released by the flowers of some plants.

Pollute (puh LOOT) to make unfit or harmful for living things

Pollution Dirty or harmful substances in the environment.

Precautions Planned actions taken before an event to increase the chances of a safe outcome.

Precycling Reducing waste before it occurs.

Prejudice (PREH juh dis) A negative and unjustly formed opinion.

Español

compañeros Personas de tu grupo de edad que se parecen a ti de muchas maneras.

mediación de compañeros Proceso en el cual un estudiante especialmente capacitado escucha los dos lados de un argumento para ayudar a las personas a llegar a un acuerdo.

presión de pares Influencia que tu grupo de compañeros tiene sobre ti.

sistema nervioso periférico Nervios que unen el sistema nervioso central con todas partes del cuerpo.

personalidad Combinación de tus sentimientos, gustos, disgustos, actitudes, habilidades y hábitos.

desorden de personalidad Condición sicológica que afecta la habilidad de una persona de actuar normalmente con otras.

pesticida Producto que se usa en las cosechas para matar insectos y otras plagas.

fobia Miedo exagerado hacia una situación o objeto específico.

abuso físico Uso de fuerza física.

actividad física Todo movimiento que cause que el cuerpo use energía.

dependencia física Adicción por la cual el cuerpo llega a tener una necesidad química de una droga.

buen estado físico Capacidad de cumplir con las exigencias físicas de la vida diaria sin cansarse demasiado.

placa bacteriana Película delgada y pegajosa que se acumula en los dientes y contribuye a las caries dentales.

neumonía Inflamación seria de los pulmones.

punto de promoción de venta Campañas de publicidad en las cuales un producto puede adquirirse en la caja.

polen Sustancia en forma de polvo, que despiden las flores de ciertas plantas.

contaminar Hacer lo impropio o dañoso para las cosas vivientes.

contaminación Sustancias sucias o dañinas en el medio ambiente.

precauciones Acciones planeadas que son tomadas antes de un evento para incrementar la seguridad.

preciclaje Proceso de reducir los desechos antes de que se produzcan.

prejuicio Opinión formada negativa e injustamente.

Glossary/Glosario

English

Preschooler A child between the ages of three and five.

Prescription (prih SKRIP shuhn) medicine A medicine that can be obtained legally only with a doctor's written permission.

Prevention Taking steps to avoid something.

Preventive care Steps taken to keep disease or injury from happening or getting worse.

PRICE formula Protect, rest, ice, compress, and elevate.

Primary care provider Health care professional who provides checkups and general care.

Product placement A paid arrangement a company has made to show its products in media such as television or film.

Proteins (PROH teenz) The nutrient group used to build and repair cells.

Protozoa (proh tuh ZOH uh) One-celled organisms that are more complex than bacteria.

Psychiatrist (sy KY uh trist) A medical doctor who treats mental health problems.

Psychological dependence A person's belief that he or she needs a drug to feel good or function normally.

Psychologist (sy KAH luh jist) A mental health professional who is trained and licensed by the state to counsel.

Puberty (PYOO bur tee) The time when you develop physical characteristics of adults of your own gender.

R

Radiation therapy A treatment that uses X rays or other forms of radiation to kill cancer cells.

Rape Forced sexual intercourse.

Reaction time The ability of the body to respond quickly and appropriately to situations.

Recovery The process of learning to live an alcohol-free life.

Recurrence The return of cancer after a remission.

Recycle To change items in some way so that they can be used again.

Recycling recovering and changing items so they can be used for other purposes

Español

niño preescolar Niño entre las edades de tres y cinco años.

medicinas bajo receta Medicina que puede ser obtenida legalmente solo con el permiso escrito de un doctor.

prevención Tomar pasos para evitar algo.

cuidado preventivo Medidas que se toman para evitar que ocurran enfermedades o daños o que empeoren.

PRICE fórmula Protege, descansa, hiela, comprime, y eleva.

profesional médico principal Profesional de la salud que proporciona exámenes médicos y cuidado general.

colocación de un producto Acuerdo pagado hecho por una compañía para mostrar su producto en los medios de publicidad.

proteínas Nutrientes que se usan para reparar las células y tejidos del cuerpo.

protozoarios Organismos unicelulares que tienen una estructura más compleja que las bacterias.

psiquiatra Médico que trata trastornos de la salud mental.

dependencia psicológica Cuando una persona cree que necesita una droga para sentirse bien o para trabajar normalmente.

psicólogo Profesional de la salud mental que está licenciado por el estado para hacer terapias.

pubertad Etapa de la vida en la cual una persona comienza a desarrollar ciertas características físicas propias de los adultos del mismo sexo.

terapia de radiación Tratamiento que usa rayos X u otra forma de radiación para matar células cancerosas.

violación Relaciones sexuales forzadas.

tiempo de reacción Habilidad del cuerpo de responder rápida y apropiadamente a diferentes situaciones.

recuperación Proceso de aprender a vivir una vida libre de alcohol.

reaparición Regreso del cáncer después de una remisión.

reciclar Cambiar un objeto de alguna manera para que se pueda volver a usar.

reciclaje Recuperar y cambiar un objeto para usarlo con otro propósito.

English

Refusal skills Strategies that help you say no effectively.

Relapse A return to the use of a drug after attempting to stop.

Reliable Trustworthy and dependable.

Remission A period during which cancer signs and symptoms disappear.

Reproduction The process by which living organisms produce others of their own kind.

Reproductive (ree pruh DUHK tiv) system The body organs and structures that make it possible to produce children.

Rescue breathing A first-aid procedure where someone forces air into the lungs of a person who cannot breathe on his or her own.

Resilience The ability to recover from problems or loss.

Respiratory system The organs that supply your blood with oxygen.

Resting heart rate The number of times your heart beats per minute when you are relaxing.

Revenge Punishment, injury, or insult to the person seen as the cause of the strong emotion.

Rheumatoid (ROO muh toyd) arthritis A chronic disease characterized by pain, inflammation, swelling, and stiffness of the joints.

Risk The chance that something harmful may happen to your health and wellness.

Risk behavior An action or behavior that might cause injury or harm to you or others.

Risk factors Characteristics or behaviors that increase the likelihood of developing a medical disorder or disease.

Role model A person who inspires you to think or act a certain way.

Role A part you play when you interact with another person.

S

Safety conscious Being aware that safety is important and careful to act in a safe manner.

Saliva (suh LY vuh) A digestive juice produced by the salivary glands in your mouth.

Español

habilidades de rechazo Estrategias que ayudan a decir no efectivamente.

recaída Regresar al uso de la droga después de haber intentado parar.

confiable Confiable y seguro.

remisión Periodo durante el cual desaparecen las señales y síntomas del cáncer.

reproducción Proceso mediante el cual los organismos vivos producen otros de su especie.

aparato reproductor Órganos y estructuras corporales que hacen posible la producción de niños.

respiración de rescate Procedimiento de primeros auxilios en el que una persona llena de aire los pulmones de una persona que no está respirando.

capacidad de recuperación Habilidad de recuperarse de un problema o perdida.

aparato respiratorio Órganos que suministran el oxígeno en la sangre.

ritmo cardiaco de descanso Número de veces el corazón late por minuto cuando estás relajado.

venganza Castigo, daño, o insulto hacia una persona por causa de una emoción fuerte.

artritis reumatoide Enfermedad crónica caracterizada por dolor, inflamación, hinchazón y anquilosamiento de las articulaciones.

riesgo Posibilidad de que algo dañino pueda ocurrir en tu salud y bienestar.

conducta arriesgada Acto o conducta que puede causarte daño o perjudicarte a ti o a otros.

factor de riesgo Característica o conducta que aumenta la posibilidad de llegar a tener un problema médico o enfermedad.

modelo ejemplo Persona que inspira a otras a que se comporten o piensen de cierta manera.

papel Parte que tú desempeñas cuando actúas con otra persona.

consciente de la seguridad Que se da cuenta de la importancia de la seguridad y actúa con cuidado.

saliva Líquido digestivo producido por las glándulas salivales de la boca.

Glossary/Glosario

English

Schizophrenia (skit zoh FREE nee uh) A severe mental disorder in which a person loses contact with reality.

Secondhand smoke Air that has been contaminated by tobacco smoke.

Self-concept The way you view yourself overall.

Self-esteem How you feel about yourself.

Semen (SEE muhn) A mixture of sperm and fluids that protect sperm and carry them through the tubes of the male reproductive system.

Sewage Human waste, garbage, detergents, and other household wastes washed down drains and toilets.

Sexual abuse Sexual contact that is forced upon another person.

Sexual harassment Uninvited and unwelcome sexual conduct directed at another person.

Sexually transmitted diseases (STDs) Infections that are spread from person to person through sexual contact.

Short-term goal A goal that you can achieve in a short length of time.

Side effect A reaction to a medicine other than the one intended.

Sidestream smoke Smoke that comes from the burning end of a cigarette, pipe, or cigar.

Skeletal muscle Muscle attached to bones that enables you to move your body.

Skeletal system The framework of bones and other tissues that supports the body.

Small intestine A coiled tube from 20 to 23 feet long, in which about 90 percent of digestion takes place.

Smog Yellow-brown haze that forms when sunlight reacts with air pollution.

Smoke alarm A device that sounds an alarm when it senses smoke.

Smokeless tobacco Ground tobacco that is chewed or inhaled through the nose.

Smooth muscle Type of muscle found in organs and in blood vessels and glands.

Snuff finely ground tobacco that is inhaled or held in the mouth or cheeks

Social age Age measured by your lifestyle and the connections you have with others.

Español

esquizofrenia Trastorno mental grave por el cual una persona pierde contacto con la realidad.

humo secundario Aire que está contaminado por el humo del tabaco.

autoconcepto Manera en que te ves a ti mismo.

autoestima Como te sientes sobre ti mismo.

semen Mezcla de esperma y fluidos que protegen a los espermatozoides y los transportan a través de los tubos del sistema reproductivo masculino.

aguas cloacales Basura, detergentes y otros desechos caseros que se llevan las tuberías de desagüe.

abuso sexual Contacto sexual forzado por una persona.

acoso sexual Conducta sexual no solicitada y fuera de lugar dirigida a otra persona.

enfermedades de transmisión sexual (ETS) Enfermedades que se propagan de una persona a otra, a través del contacto sexual.

meta a corto plazo Meta que uno puede alcanzar dentro de un breve periodo de tiempo.

efecto colateral Reacción inesperada de una medicina.

humo indirecto Humo que procede de un cigarrillo, pipa o cigarro encendido.

músculo del sistema osteoarticular Músculo ligado a huesos que te permiten mover el cuerpo.

sistema osteoarticular Armazón de huesos y otros tejidos que sostiene el cuerpo.

intestino delgado Tubo enrollado que mide entre veinte y veintitrés pies, en el cual ocurre el noventa por ciento de la digestión.

smog Neblina de color amarillo-café que se forma cuando la luz solar reacciona con la contaminación del aire.

alarma contra incendios Aparato que hace sonar una alarma cuando detecta humo.

rapé Tabaco molido que es masticado o inhalado a través de la nariz.

músculo liso Tipo de músculo que se encuentra en los órganos, los vasos sanguíneos y las glándulas.

rapé Tabaco molido finamente que es inhalado o mantenido en la boca o las mejillas.

edad social Edad calculada de acuerdo con tu estilo de vida y las conexiones que tienes con los demás.

English

Specialist (SPEH shuh list) Health care professional trained to treat a special category of patients or specific health problems.

Sperm Male reproductive cells.

Spinal cord A long bundle of neurons that sends messages to and from the brain and all parts of the body.

Sports gear sports clothing and safety equipment

Sprain An injury to the ligament connecting bones at a joint.

Stimulant (STIM yuh luhnt) A drug that speeds up the body's functions.

Strength The ability of your muscles to use force.

Strep throat Sore throat caused by streptococcal bacteria.

Stress fracture A small fracture caused by repeated strain on a bone.

Stress The body's response to real or imagined dangers or other life events.

Stressor Anything that causes stress.

Stroke A serious condition that occurs when an artery of the brain breaks or becomes blocked.

Subcutaneous (suhb kyoo TAY nee uhs) layer A layer of fat under the skin.

Substance abuse using illegal or harmful drugs, including any use of alcohol while under the legal drinking age

Suicide The act of killing oneself on purpose.

Sympathetic (simp uh THET ik) Aware of how you may be feeling at a given moment.

Syphilis (SIH fuh luhs) A bacterial STD that can affect many parts of the body.

Español

especialista Profesional del cuidado de la salud que está capacitado para tratar una categoría especial de pacientes o un problema de salud específico.

espermatozoides Células reproductoras masculinas.

médula espinal Largo conjunto de neuronas que transmiten mensajes entre el cerebro y todas las otras partes del cuerpo.

accesorios deportivos Ropa para deportes y equipo de seguridad.

torcedura Daño al ligamento que conecta huesos con articulaciones.

estimulante Droga que acelera las funciones del cuerpo.

fuerza Capacidad que tienen los músculos para ejercer una fuerza.

amigdalitis estreptocócica Dolor de garganta causado por bacterias estreptocócicas.

fractura leve Fractura pequeña causada por una torcedura repetitiva del mismo hueso.

estrés Reacción del cuerpo hacia peligros reales o imaginarios u otros eventos en la vida.

factor estresante Todo lo que provoca el estrés.

apoplejía Afección seria que se produce cuando una arteria en el cerebro se rompe o se obstruye.

capa subcutánea Capa de grasa debajo de la piel.

abuso de sustancias Consumo de drogas ilegales o nocivas, incluso el consumo del alcohol en cualquiera de sus formas antes de la edad legal para beber.

suicidio Matarse intencionalmente.

solidario Estar consciente que como te puedes estar sintiendo en un momento indicado.

sífilis Infección de transmisión sexual, causada por una bacteria, que puede afectar muchas partes del cuerpo.

T

Tar A thick, dark liquid that forms when tobacco burns.

Target audience A group of people for which a product is intended.

alquitrán Líquido espeso y oscuro que forma el tabaco al quemarse.

publico objetivo Grupo de gente al cual es dirigido un producto específico.

Glossary/Glosario

English

Target heart rate The number of heartbeats per minute that you should aim for during moderate-to-vigorous aerobic activity to help your circulatory system the most.

Tartar (TAR tuhr) Hardened plaque that hurts gum health.

Tendon A type of connecting tissue that joins muscles to bones and muscles to muscles.

Tendonitis Painful swelling of a tendon caused by overuse.

Testes The pair of glands that produce sperm.

Therapy Professional counseling.

Time management Strategies for using time efficiently.

Tinnitus A constant ringing in the ears.

Tissue A group of similar cells that do a particular job.

Toddler A child between the ages of one and three.

Tolerance (TAHL er ence) 1. The ability to accept other people as they are. **2.** The body's need for larger and larger amounts of a drug to produce the same effect.

Tornado A whirling, funnel-shaped windstorm that drops from storm clouds to the ground.

Trachea (TRAY kee uh) A passageway in your throat that takes air into and out of your lungs.

Trichomoniasis (TREE koh moh NI ah sis) An STD caused by the protozoan Trichomonas vaginalis.

Tuberculosis (TB) (too ber kyuh LOH sis) A bacterial disease that usually affects the lungs.

Tumor (TOO mer) A group of abnormal cells that forms a mass.

Type 1 diabetes A condition in which the immune system attacks insulin-producing cells in the pancreas.

Type 2 diabetes A condition in which the body cannot effectively use the insulin it produces.

Español

meta ritmo cardiaco Número de latidos del corazón, por minuto, que una persona debe tratar de alcanzar durante una actividad de intensidad moderada a vigorosa, para obtener lo máximo posible de beneficio para el aparato circulatorio.

sarro Placa endurecida que daña la salud de las encías.

tendón Tipo de tejido conjuntivo que une un músculo a otro o un músculo a un hueso.

tendonitis Hinchazón dolorosa de un tendon causada por el uso excesivo.

testículos Las dos glándulas que producen los espermatozoides.

terapia Consejo profesional.

organización del tiempo Estrategias para usar el tiempo eficazmente.

acúifeno Sonido constante en el oído.

tejido Grupo de células similares que tienen una función en particular.

niño que empieza a andar Niño entre uno y tres años.

tolerancia 1. Habilidad de aceptar a otras personas de la forma que son. **2.** Necesidad del cuerpo de mayores cantidades de una droga para obtener el mismo efecto.

tornado Tormenta en forma de torbellino, que gira en grandes círculos y que cae del cielo a la tierra.

tráquea Pasaje en la garganta que lleva el aire de y hacia los pulmones.

trichomoniasis Enfermedad de transmisión sexual causada por los protozoarios Trichomonas vaginalis.

tuberculosis Enfermedad causada por bacterias que generalmente afecta a los pulmones.

tumor Grupo de células anormales que forma una masa.

diabetes tipo 1 Afección por la cual el sistema inmunológico ataca las células productoras de insulina en el páncreas.

diabetes tipo 2 Afección que se caracteriza por la inhabilidad del cuerpo para usar de manera eficaz la insulina que produce.

U

Ulcer (UHL ser) An open sore in the stomach lining.

Ultraviolet (UV) rays Invisible form of radiation that can enter skin cells and change their structure.

Úlcera Llaga abierta en el forro del estómago

rayos ultravioletas Forma invisible de radiación que puede penetrar células de la piel y cambia su estructura.

English

Underweight Less than the appropriate weight for gender, height, age, body frame, and growth pattern.

Universal precautions Actions taken to prevent the spread of disease by treating all blood as if it were contaminated.

Uterus (YOO tuh ruhs) A pear-shaped organ, located within the pelvis, in which the developing baby is nourished and protected.

Vaccine (vak SEEN) A preparation of dead or weakened pathogens that is introduced into the body to cause an immune response.

Values The beliefs that guide the way a person lives.

Vector An organism, such as an insect, that transmits a pathogen.

Veins Blood vessels that carry blood from all parts of the body back to the heart.

Victim Any individual who suffers injury, loss, or death due to violence.

Violence An act of physical force resulting in injury or abuse.

Viruses (VY ruh suhz) The smallest and simplest pathogens.

Vitamins (VY tuh muhns) Compounds that help to regulate body processes.

Warm-up Gentle exercises that get heart muscles ready for moderate-to-vigorous activity.

Warranty A company's or a store's written agreement to repair a product or refund your money if the product does not function properly.

Weather emergency A dangerous situation brought on by changes in the atmosphere.

Wellness A state of well-being or balanced health over a long period of time.

Win-win solution An agreement that gives each party something they want.

Withdrawal A series of painful physical and mental symptoms that a person experiences when he or she stops using an addictive substance.

Español

de peso insuficiente Por debajo del peso apropiado de acuerdo con el sexo, estatura, edad, estructura corporal y ritmo de crecimiento.

precauciones universales Medidas para prevenir la propagación de enfermedades al tratar toda la sangre como si estuviera contaminada.

útero Órgano en forma de pera en el cual se nutre un bebé en desarrollo.

vacuna Fórmula compuesta por gérmenes patógenos muertos o debilitados que es introducida en el cuerpo para causar una reacción inmune.

valores Creencias que guían la forma en la cual vive una persona.

vector Organismo, por ejemplo un insecto, que transmite un agente patógeno.

vena Tipo de vaso sanguíneo que lleva la sangre de todas partes del cuerpo de regreso al corazón.

víctima Cualquier individuo que sufre algún daño, pérdida o muerte debido a la violencia.

violencia Acto de fuerza física que resulta en un daño o abuso.

viruses Patógenos más simples y pequeños.

vitaminas Compuestos que ayudan a regular los procesos del cuerpo.

precalentamiento Ejercicio suave que prepara a los músculos cardíacos para estar listos para actividad física moderada o fuerte.

garantía Promesa escrita de un fabricante o una tienda de reparar un producto o devolver el dinero al comprador, si el producto no funciona debidamente.

emergencia meteorológica Situación peligrosa debido a cambios en la atmósfera.

bienestar Mantener una salud balanceada por un largo período de tiempo.

solución de ganancia doble Acuerdo o resultado que da algo de lo que quieren a cada lado.

síndrome de abstinencia Series de síntomas físicos y mentales dolorosos por los cuales una persona pasa cuando para el uso de una sustancia adictiva.

Glossary/Glosario

English

Español

Y

Youth court A special school program where teens decide punishments for other teens for bullying and other problem behaviors.

corte juvenil Programa escolar especial en el cual adolescentes deciden el castigo para otros adolescentes por intimidar y otros problemas de comportamiento.

Z

Zero tolerance policy A policy that makes no exceptions for anybody for any reason.

normativa de tolerancia Normativa en que no hay excepciones para nadie por ninguna razón.

Index

Index

Index

Index

Index

to quit using tobacco, 341
    for reducing/reusing/recycling, 511
Goal-setting plan, HSH-17
Goals
    encouraging friends toward, 15
    for establishing limits, 29
    fitness, 210–211
    realistic, HSH-17
    taking action to reach, HSH-2
    types of, HSH-17
Gonorrhea, 425
Good citizenship, 54
Government public health care plans, HSH-21
Grains
    as food group, 183
    for good nutrition, 180
    in MyPlate diagram, 182
Greenhouse effect, 503
Green schools, 512–513
Grief, 98–99
    counseling for, 119
    defined, 98
Grief reaction, 98
Groundwater, 510
Group counseling, 117
Group dating, 27
Group pressure, leading to conflict, 136
Group therapy, 120
Growth, during puberty, 229
Growth patterns
    and body image, 191
    and weight, 188
Guardians, character shaped by, 47
Guardians, communicating with, 7
Guardians, conflicts with, 231
    abuse by, 158
    conflicts with, 132
Guilt, about sexual activity, 33
Gum care, 266–269, 302
Guns, 155
Gun safety, 471, 473
Gynecologists, 239

H

Habits, healthy, 432
Hair, 264–265
    care of, 265
    facial and body hair, 257
    scalp problems, 264
Hallucinations, with schizophrenia, 112
Hallucinogens, 378
Hands-Only Cardiopulmonary Resuscitation, 488, 490
Hand washing, 256–257, 302, 412, 432
Hangnails, 257

Harassment, 67
    defined, 67, 152
    responding to, 67
    sexual, 162
    warning signs of, 75
Hazardous wastes, 510
Hazards, 468. *See also* Safety
Head injuries, 476
Head lice, 264
Health care, personal. *See* Personal health care
Health care costs, tobacco use and, 335
Health care plans, HSH-21
    government public, HSH-21
    private, HSH-21
Health care providers, HSH-19, 120
Health care services, HSH-18–HSH-21
    health care providers, HSH-19, HSH-19
    health care settings, HSH-20–HSH-21
    mental health resources, 118–121
    paying for health care, HSH-21
Health care settings, HSH-20–HSH-21
Health care system, HSH-18
Health checkups, 235
Health educators, HSH-19
Healthful food choices, 181–186
    balance of nutrients for, 180
    learning about nutrients for, 176
    meal planning, 184–186
    using MyPlate, 182–183
    when eating out, 185
Health insurance, HSH-21
Health
    behavioral factors in, HSH-7–HSH-9
    defined, HSH-2
    of female reproductive system, 238
    influences on, HSH-5–HSH-6
    making decisions that respect, 49
    of male reproductive system, 234
    mental/emotional, HSH-2. *See also* Mental/emotional
       health
    and mind-body connection, HSH-4
    physical, HSH-2
    social, HSH-3
    taking responsibility for, 51
    total, HSH-12–HSH-3
    wellness vs., HSH-3
Health skills, HSH-10. *See also* individual skills
    accessing information, HSH-12–HSH-13
    communication skills, HSH-14
    decision making, HSH-12, HSH-15–HSH-16
    defined, HSH-10
    goal setting, HSH-12, HSH-17
    practicing healthful behaviors, HSH-11
    self-management skills, HSH-11–HSH-12
    stress management, HSH-11

Index

Immune response
  nonspecific, 415
  specific, 416–417
Immune system, 309–313, 415–417
  antigens and antibodies, 312
  caring for, 312
  function of, 309
  immunity, 313
  inflammation, 310
  lymphatic system, 311
Immune system, tobacco use and, 330
Immunity, 309, 313, 418
Indigestion, 301
Indirect peer pressure, 16
Individual dating, 27
Individual therapy, 120
Infancy, 242–243
Infants, choking emergencies with, 489
Infections, 297, 310
  defense against, 414–418
  defined, 414
  opportunistic, 428
Infertility, 238, 307
Inflammation, 310, 415
Influences
  analyzing, HSH-13, 450
  on body image, 191
  on character, 47
  on food choices, 181
  on health, HSH-5–HSH-6
  peer pressure, 16–17
  peer, 231
  on self-concept, 88
  on self-esteem, 87
  on tobacco use, 338–339
  toward violence, 153
Influenza, 420
Information
  accessing, HSH-12–HSH-13
  accessing. *See* Accessing information
  reliable sources of, HSH-12
  sending and receiving, 5, 6. *See also* Communication
Ingested medications, 396
Ingrown toenails, 257
Inhalants, 380
Inhaled medications, 396
Inherited traits, HSH-4
Inhibitions, 348
Injections, 396
Injuries, emotional, 154
Injuries, sports, 234
  accidental, 468
  head, 476
In-line skating, 476

Insect bites, first aid for, 492
Insects, pathogens spread by, 411
Insulin, 453, 454
Insurance, health, HSH-21
Integrity, 46, 49
Intensity, 211
Internal conflicts, 131
Internet predators, 482
Internet safety, 481–482
Interpersonal conflicts, 131
Interval training, F4F-9
Interventions, 360
Intimidation, 66, 152
Intonation pattern (speech), 122
Iron, 178

J

Jealousy, 135
Jogging, F4F-6
Joint-custody family, 9
Joint replacements, 455
Joints, 207, 314. *See also* Skeletal system
  defined, 282
  functions of, 282–283
Juvenile rheumatoid arthritis (JRA), 454

K

Keratin, 264
Ketamine, 378
Kidneys, 300
  and alcohol use, 353
Kidney stones, 301
KidsHealth, F4F-2
Killer cells, 311
Kindness, 53
Kissing disease, 421
Kitchen safety, 471, 472
Knee guards, 217
Kwanzaa, HSH-6
  accessing information, HSH-12–HSH-13
  communication skills, HSH-14
  decision making, HSH-12, HSH-15–HSH-16
  defined, HSH-10
  goal setting, HSH-12, HSH-17
  practicing healthful behaviors, HSH-11
  self-management skills, HSH-11–HSH-12
  stress management, HSH-11

L

Labeling, 133
Labeling laws, 336
Labia, 236, 306
Landfills, 510, 511

Index

Index

Index

Index

Index